AF290913

Graduate Texts in Physics

For further volumes:
www.springer.com/series/8431

Graduate Texts in Physics

Graduate Texts in Physics publishes core learning/teaching material for graduate- and advanced-level undergraduate courses on topics of current and emerging fields within physics, both pure and applied. These textbooks serve students at the MS- or PhD-level and their instructors as comprehensive sources of principles, definitions, derivations, experiments and applications (as relevant) for their mastery and teaching, respectively. International in scope and relevance, the textbooks correspond to course syllabi sufficiently to serve as required reading. Their didactic style, comprehensiveness and coverage of fundamental material also make them suitable as introductions or references for scientists entering, or requiring timely knowledge of, a research field.

Series Editors

Professor Richard Needs

Cavendish Laboratory
JJ Thomson Avenue
Cambridge CB3 0HE, UK
rn11@cam.ac.uk

Professor William T. Rhodes

Department of Computer and Electrical Engineering and Computer Science
Imaging Science and Technology Center
Florida Atlantic University
777 Glades Road SE, Room 456
Boca Raton, FL 33431, USA
wrhodes@fau.edu

Professor Susan Scott

Department of Quantum Science
Australian National University
Science Road
Acton 0200, Australia
susan.scott@anu.edu.au

Professor H. Eugene Stanley

Center for Polymer Studies Department of Physics
Boston University
590 Commonwealth Avenue, Room 204B
Boston, MA 02215, USA
hes@bu.edu

Professor Martin Stutzmann

Walter Schottky Institut
TU München
85748 Garching, Germany
stutz@wsi.tu-muenchen.de

Philipp O.J. Scherer

Computational Physics

Simulation of Classical and Quantum Systems

Second Edition

 Springer

Philipp O.J. Scherer
Physikdepartment T38
Technische Universität München
Garching, Germany

Additional material to this book can be downloaded from http://extras.springer.com.

ISSN 1868-4513   ISSN 1868-4521 (electronic)
Graduate Texts in Physics
ISBN 978-3-319-00400-6   ISBN 978-3-319-00401-3 (eBook)
DOI 10.1007/978-3-319-00401-3
Springer Cham Heidelberg New York Dordrecht London

Library of Congress Control Number: 2013944508

Printed on acid-free paper

Springer is part of Springer Science+Business Media (www.springer.com)

To Christine

Preface to the Second Edition

This textbook introduces the main principles of computational physics, which include numerical methods and their application to the simulation of physical systems. The first edition was based on a one-year course in computational physics where I presented a selection of only the most important methods and applications. Approximately one-third of this edition is new. I tried to give a larger overview of the numerical methods, traditional ones as well as more recent developments. In many cases it is not possible to pin down the "best" algorithm, since this may depend on subtle features of a certain application, the general opinion changes from time to time with new methods appearing and computer architectures evolving, and each author is convinced that his method is the best one. Therefore I concentrated on a discussion of the prevalent methods and a comparison for selected examples. For a comprehensive description I would like to refer the reader to specialized textbooks like "Numerical Recipes" or elementary books in the field of the engineering sciences.

The major changes are as follows.

A new chapter is dedicated to the discretization of differential equations and the general treatment of boundary value problems. While finite differences are a natural way to discretize differential operators, finite volume methods are more flexible if material properties like the dielectric constant are discontinuous. Both can be seen as special cases of the finite element methods which are omnipresent in the engineering sciences. The method of weighted residuals is a very general way to find the "best" approximation to the solution within a limited space of trial functions. It is relevant for finite element and finite volume methods but also for spectral methods which use global trial functions like polynomials or Fourier series.

Traditionally, polynomials and splines are very often used for interpolation. I included a section on rational interpolation which is useful to interpolate functions with poles but can also be an alternative to spline interpolation due to the recent development of barycentric rational interpolants without poles.

The chapter on numerical integration now discusses Clenshaw-Curtis and Gaussian methods in much more detail, which are important for practical applications due to their high accuracy.

Besides the elementary root finding methods like bisection and Newton-Raphson, also the combined methods by Dekker and Brent and a recent extension by Chandrupatla are discussed in detail. These methods are recommended in most text books. Function minimization is now discussed also with derivative free methods, including Brent's golden section search method. Quasi-Newton methods for root finding and function minimizing are thoroughly explained.

Eigenvalue problems are ubiquitous in physics. The QL-method, which is very popular for not too large matrices is included as well as analytic expressions for several differentiation matrices.

The discussion of the singular value decomposition was extended and its application to low rank matrix approximation and linear fitting is discussed.

For the integration of equations of motion (i.e. of initial value problems) many methods are available, often specialized for certain applications. For completeness, I included the predictor-corrector methods by Nordsieck and Gear which have been often used for molecular dynamics and the backward differentiation methods for stiff problems.

A new chapter is devoted to molecular mechanics, since this is a very important branch of current computational physics. Typical force field terms are discussed as well as the calculation of gradients which are necessary for molecular dynamics simulations.

The simulation of waves now includes three additional two-variable methods which are often used in the literature and are based on generally applicable schemes (leapfrog, Lax-Wendroff, Crank-Nicolson).

The chapter on simple quantum systems was rewritten. Wave packet simulation has become very important in theoretical physics and theoretical chemistry. Several methods are compared for spatial discretization and time integration of the one-dimensional Schrödinger equation. The dissipative two-level system is used to discuss elementary operations on a qubit.

The book is accompanied by many computer experiments. For those readers who are unable to try them out, the essential results are shown by numerous figures.

This book is intended to give the reader a good overview over the fundamental numerical methods and their application to a wide range of physical phenomena. Each chapter now starts with a small abstract, sometimes followed by necessary physical background information. Many references, original work as well as specialized text books, are helpful for more deepened studies.

Garching, Germany   Philipp O.J. Scherer
February 2013

Preface to the First Edition

Computers have become an integral part of modern physics. They help to acquire, store and process enormous amounts of experimental data. Algebra programs have become very powerful and give the physician the knowledge of many mathematicians at hand. Traditionally physics has been divided into experimental physics which observes phenomena occurring in the real world and theoretical physics which uses mathematical methods and simplified models to explain the experimental findings and to make predictions for future experiments. But there is also a new part of physics which has an ever growing importance. Computational physics combines the methods of the experimentalist and the theoretician. Computer simulation of physical systems helps to develop models and to investigate their properties.

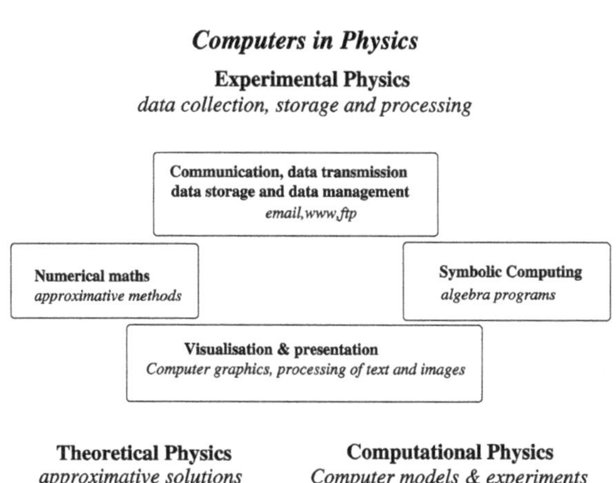

This book is a compilation of the contents of a two-part course on computational physics which I have given at the TUM (Technische Universität München) for several years on a regular basis. It attempts to give the undergraduate physics students a profound background in numerical methods and in computer simulation methods but is also very welcome by students of mathematics and computational science who want to learn about applications of numerical methods in physics. This book may also support lecturers of computational physics and bio-computing. It tries to bridge between simple examples which can be solved analytically and more complicated but instructive applications which provide insight into the underlying physics by doing computer experiments.

The first part gives an introduction into the essential methods of numerical mathematics which are needed for applications in physics. Basic algorithms are explained in detail together with limitations due to numerical inaccuracies. Mathematical explanations are supplemented by numerous numerical experiments.

The second part of the book shows the application of computer simulation methods for a variety of physical systems with a certain focus on molecular biophysics. The main object is the time evolution of a physical system. Starting from a simple rigid rotor or a mass point in a central field, important concepts of classical molecular dynamics are discussed. Further chapters deal with partial differential equations, especially the Poisson-Boltzmann equation, the diffusion equation, nonlinear dynamic systems and the simulation of waves on a 1-dimensional string. In the last chapters simple quantum systems are studied to understand e.g. exponential decay processes or electronic transitions during an atomic collision. A two-state quantum system is studied in large detail, including relaxation processes and excitation by an external field. Elementary operations on a quantum bit (qubit) are simulated.

Basic equations are derived in detail and efficient implications are discussed together with numerical accuracy and stability of the algorithms. Analytical results are given for simple test cases which serve as a benchmark for the numerical methods. Many computer experiments are provided realized as Java applets which can be run in the web browser. For a deeper insight the source code can be studied and modified with the free "netbeans"[1] environment.

Garching, Germany   Philipp O.J. Scherer
April 2010

---

[1] www.netbeans.org.

Contents

Part I
Numerical Methods

Chapter 1
Error Analysis

Several sources of errors are important for numerical data processing:

Experimental uncertainty: Input data from an experiment have a limited precision. Instead of the vector of exact values $\mathbf{x}$ the calculation uses $\mathbf{x} + \Delta\mathbf{x}$, with an uncertainty $\Delta\mathbf{x}$. This can lead to large uncertainties of the calculated results if an unstable algorithm is used or if the unavoidable error inherent to the problem is large.

Rounding errors: The arithmetic unit of a computer uses only a subset of the real numbers, the so called machine numbers $A \subset \mathfrak{R}$. The input data as well as the results of elementary operations have to be represented by machine numbers whereby rounding errors can be generated. This kind of numerical error can be avoided in principle by using arbitrary precision arithmetics[1] or symbolic algebra programs. But this is unpractical in many cases due to the increase in computing time and memory requirements.

Truncation errors: Results from more complex operations like square roots or trigonometric functions can have even larger errors since series expansions have to be truncated and iterations can accumulate the errors of the individual steps.

1.1 Machine Numbers and Rounding Errors

Floating point numbers are internally stored as the product of sign, mantissa and a power of 2. According to the IEEE754 standard [130] single, double and quadruple precision numbers are stored as 32, 64 or 128 bits (Table 1.1):

---

[1]For instance the open source GNU MP bignum library.

P.O.J. Scherer, *Computational Physics*, Graduate Texts in Physics,
DOI 10.1007/978-3-319-00401-3_1,
© Springer International Publishing Switzerland 2013

Table 1.1 Binary floating point formats

Format	Sign	Exponent	Hidden bit	Fraction	Precision ε_M
Float	s	$b_0 \cdots b_7$	1	$a_0 \cdots a_{22}$	$2^{-24} = 5.96\mathrm{E}^{-8}$
Double	s	$b_0 \cdots b_{10}$	1	$a_0 \cdots a_{51}$	$2^{-53} = 1.11\mathrm{E}^{-16}$
Quadruple	s	$b_0 \cdots b_{14}$	1	$a_0 \cdots a_{111}$	$2^{-113} = 9.63\mathrm{E}^{-35}$

Table 1.2 Exponent bias E

Decimal value	Binary value	Hexadecimal value	Data type
127_{10}	1111111_2	\$3F	Single
1023_{10}	1111111111_2	\$3FF	Double
16383_{10}	11111111111111_2	\$3FFF	Quadruple

The sign bit s is 0 for positive and 1 for negative numbers. The exponent b is biased by adding E which is half of its maximum possible value (Table 1.2).[2] The value of a number is given by

$$x = (-)^s \times a \times 2^{b-\mathrm{E}}. \tag{1.1}$$

The mantissa a is normalized such that its first bit is 1 and its value is between 1 and 2

$$1.000_2 \cdots 0 \le a \le 1.111 \cdots 1_2 < 10.0_2 = 2_{10}. \tag{1.2}$$

Since the first bit of a normalized floating point number always is 1, it is not necessary to store it explicitly (hidden bit or J-bit). However, since not all numbers can be normalized, only the range of exponents from \$001 $\cdots$ \$7FE is used for normalized numbers. An exponent of \$000 signals that the number is not normalized (zero is an important example, there exist even two zero numbers with different sign) whereas the exponent \$7FF is reserved for infinite or undefined results (Table 1.3).

The range of normalized double precision numbers is between

$$\mathrm{Min\_Normal} = 2.2250738585072014 \times 10^{-308}$$

and

$$\mathrm{Max\_Normal} = 1.7976931348623157\mathrm{E} \times 10^{308}.$$

Example Consider the following bit pattern which represents a double precision number:

$$\$4059000000000000.$$

---

[2] In the following the usual hexadecimal notation is used which represents a group of 4 bits by one of the digits 0, 1, 2, 3, 4, 5, 6, 7, 8, 9, A, B, C, D, E, F.

Table 1.3 Special double precision numbers

Hexadecimal value	Symbolic value
\$000 0000000000000	$+0$
\$080 00000000000000	-0
\$7FF 0000000000000	$+\mathrm{inf}$
\$FFF 0000000000000	$-\mathrm{inf}$
\$7FF 0000000000001 $\cdots$ \$7FF FFFFFFFFFFFFF	NAN
\$001 0000000000000	Min_Normal
\$7FE FFFFFFFFFFFFF	Max_Normal
\$000 0000000000001	Min_Subnormal
\$000 FFFFFFFFFFFFF	Max_Subnormal

The exponent is $100\,0000\,0101_2 - 011\,1111\,1111_2 = 110_2$ and the mantissa includ-ing the J-bit is $1\,1001\,0000\,0000\cdots_2$. Hence the decimal value is

$$1.5625 \times 2^6 = 100_{10}.$$

Input numbers which are not machine numbers have to be rounded to the nearest machine number. This is formally described by a mapping $\Re \to A$

$$x \to \mathrm{rd}(x)$$

with the property[3]

$$\left| x - \mathrm{rd}(x) \right| \leq |x - g| \quad \text{for all } g \in A. \tag{1.3}$$

For the special case that x is exactly in the middle between two successive ma-chine numbers, a tie-breaking rule is necessary. The simplest rules are to round up always (*round-half-up*) or always down (*round-half-down*). However, these are not symmetric and produce a bias in the average round-off error. The IEEE754 standard [130] recommends the *round-to-nearest-even* method, i.e. the least significant bit of the rounded number should always be zero. Alternatives are *round-to-nearest-odd*, stochastic rounding and alternating rounding.

The cases of *exponent overflow* and *exponent underflow* need special attention:

Whenever the exponent b has the maximum possible value $b = b_{\max}$ and $a = 1.11\cdots 11$ has to be rounded to $a' = 10.00\cdots 0$, the rounded number is not a ma-chine number and the result is $\pm\mathrm{inf}$.

Numbers in the range $2^{b_{\min}} > |x| \geq 2^{b_{\min}-t}$ have to be represented with loss of accuracy by denormalized machine numbers. Their mantissa cannot be normalized since it is $a < 1$ and the exponent has the smallest possible value $b = b_{\min}$. Even smaller numbers with $|x| < 2^{-t+b_{\min}}$ have to be rounded to ± 0.

---

[3]Sometimes rounding is replaced by a simpler truncation operation which, however leads to sig-nificantly larger rounding errors.

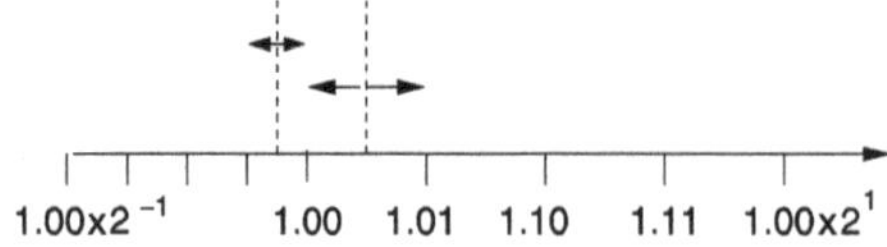

Fig. 1.1 (Round to nearest) Normalized machine numbers with $t = 3$ binary digits are shown. Rounding to the nearest machine number produces a round-off error which is bounded by half the spacing of the machine numbers

The maximum rounding error for normalized numbers with t binary digits

$$a' = s \times 2^{b-\mathrm{E}} \times 1.a_1 a_2 \cdots a_{t-1} \tag{1.4}$$

is given by (Fig. 1.1)

$$\left| a - a' \right| \leq 2^{b-\mathrm{E}} \times 2^{-t} \tag{1.5}$$

and the relative error is bounded by

$$\left| \frac{\mathrm{rd}(x) - x}{x} \right| \leq \frac{2^{-t} \times 2^b}{|a| \times 2^b} \leq 2^{-t}. \tag{1.6}$$

The error bound determines the relative machine precision[4]

$$\varepsilon_M = 2^{-t} \tag{1.7}$$

and the rounding operation can be described by

$$\mathrm{rd}(x) = x(1 + \varepsilon) \quad \text{with } |\varepsilon| \leq \varepsilon_M. \tag{1.8}$$

The round-off error takes its maximum value if the mantissa is close to 1. Consider a number

$$x = 1 + \varepsilon.$$

If $\varepsilon < \varepsilon_M$ then $\mathrm{rd}(x) = 1$ whereas for $\varepsilon > \varepsilon_M$ rounding gives $\mathrm{rd}(x) = 1 + 2^{1-t}$ (Fig. 1.2). Hence ε_M is given by the largest number ε for which $\mathrm{rd}(1.0 + \varepsilon) = 1.0$ and is therefore also called *unit round off*.

1.2 Numerical Errors of Elementary Floating Point Operations

Even for two machine numbers $x, y \in A$ the results of addition, subtraction, multiplication or division are not necessarily machine numbers. We have to expect some additional round-off errors from all these elementary operations [244]. We assume that the results of elementary operations are approximated by machine numbers as precisely as possible. The IEEE754 standard [130] requires that the exact operations

---

[4] Also known as machine epsilon.

Fig. 1.2 (Unit round off)

$$\varepsilon < \varepsilon_M$$

$$\boxed{1.000000 \cdots 0}$$
$$+ \; \boxed{0.000000 \cdots 0}\;\boxed{01111111} \quad\longrightarrow\quad \boxed{1.000000 \cdots 0}$$

$$\varepsilon = \varepsilon_M$$

$$\boxed{1.000000 \cdots 0}$$
$$+ \; \boxed{0.000000 \cdots 0}\;\boxed{10000000} \quad\longrightarrow\quad \boxed{1.000000 \cdots 0}$$

$$\varepsilon > \varepsilon_M$$

$$\boxed{1.000000 \cdots 0}$$
$$+ \; \boxed{0.000000 \cdots 0}\;\boxed{10000001} \quad\longrightarrow\quad \boxed{1.000000 \cdots 1}$$

$x + y, x - y, x \times y, x \div y$ are approximated by floating point operations $A \rightarrow A$ with the property:

$$
\begin{aligned}
fl_+(x, y) &= \mathrm{rd}(x + y) \\
fl_-(x, y) &= \mathrm{rd}(x - y) \\
fl_*(x, y) &= \mathrm{rd}(x \times y) \\
fl_\div(x, y) &= \mathrm{rd}(x \div y).
\end{aligned}
\tag{1.9}
$$

1.2.1 Numerical Extinction

For an addition or subtraction one summand has to be denormalized to line up the exponents (for simplicity we consider only the case $x > 0, y > 0$)

$$
x + y = a_x 2^{b_x - \mathrm{E}} + a_y 2^{b_y - \mathrm{E}} = \left(a_x + a_y 2^{b_y - b_x}\right) 2^{b_x - \mathrm{E}}.
\tag{1.10}
$$

If the two numbers differ much in their magnitude, numerical extinction can happen. Consider the following case:

$$
\begin{aligned}
y &< 2^{b_x - \mathrm{E}} \times 2^{-t} \\
a_y 2^{b_y - b_x} &< 2^{-t}.
\end{aligned}
\tag{1.11}
$$

The mantissa of the exact sum is

$$
a_x + a_y 2^{b_y - b_x} = 1.\alpha_2 \cdots \alpha_{t-1} 01 \beta_2 \cdots \beta_{t-1}.
\tag{1.12}
$$

Rounding to the nearest machine number gives

$$
\mathrm{rd}(x + y) = 2^{b_x} \times (1.\alpha_2 \cdots \alpha_{t-1}) = x
\tag{1.13}
$$

since

$$
\begin{aligned}
|0.01\beta_2 \cdots \beta_{t-1} - 0| &\le |0.011 \cdots 1| = 0.1 - 0.00 \cdots 01 \\
|0.01\beta_2 \cdots \beta_{t-1} - 1| &\ge |0.01 - 1| = 0.11.
\end{aligned}
\tag{1.14}
$$

Consider now the case

$$
y < x \times 2^{-t-1} = a_x \times 2^{b_x - \mathrm{E} - t - 1} < 2^{b_x - \mathrm{E} - t}.
\tag{1.15}
$$

For normalized numbers the mantissa is in the interval

$$1 \leq |a_x| < 2 \tag{1.16}$$

hence we have

$$\mathrm{rd}(x+y) = x \quad \text{if } \frac{y}{x} < 2^{-t-1} = \frac{\varepsilon_M}{2}. \tag{1.17}$$

Especially for $x = 1$ we have

$$\mathrm{rd}(1+y) = 1 \quad \text{if } y < 2^{-t} = 0.00\cdots 0_{t-1}1_t000\cdots \tag{1.18}$$

2^{-t} could be rounded to 0 or to 2^{1-t} since the distance is the same $|2^{-t} - 0| = |2^{-t} - 2^{1-t}| = 2^{-t}$.

The smallest machine number with $fl_+(1, \varepsilon) > 1$ is either $\varepsilon = 0.00\cdots 1_t 0\cdots = 2^{-t}$ or $\varepsilon = 0.00\cdots 1_t 0\cdots 01_{2t-1} = 2^{-t}(1 + 2^{1-t})$. Hence the machine precision ε_M can be determined by looking for the smallest (positive) machine number ε for which $fl_+(1, \varepsilon) > 1$.

1.2.2 Addition

Consider the sum of two floating point numbers

$$y = x_1 + x_2. \tag{1.19}$$

First the input data have to be approximated by machine numbers:

$$\begin{aligned} x_1 &\to \mathrm{rd}(x_1) = x_1(1 + \varepsilon_1) \\ x_2 &\to \mathrm{rd}(x_2) = x_2(1 + \varepsilon_2). \end{aligned} \tag{1.20}$$

The addition of the two summands may produce another error α since the result has to be rounded. The numerical result is

$$\tilde{y} = fl_+\big(\mathrm{rd}(x_1), \mathrm{rd}(x_2)\big) = \big(x_1(1 + \varepsilon_1) + x_2(1 + \varepsilon_2)\big)(1 + \alpha). \tag{1.21}$$

Neglecting higher orders of the error terms we have in first order

$$\tilde{y} = x_1 + x_2 + x_1\varepsilon_1 + x_2\varepsilon_2 + (x_1 + x_2)\alpha \tag{1.22}$$

and the relative error of the numerical sum is

$$\frac{\tilde{y} - y}{y} = \frac{x_1}{x_1 + x_2}\varepsilon_1 + \frac{x_2}{x_1 + x_2}\varepsilon_2 + \alpha. \tag{1.23}$$

If $x_1 \approx -x_2$ then numerical extinction can produce large relative errors and uncertainties of the input data can be strongly enhanced.

1.2.3 Multiplication

Consider the multiplication of two floating point numbers

$$y = x_1 \times x_2. \tag{1.24}$$

The numerical result is

$$\tilde{y} = fl_*\big(\mathrm{rd}(x_1), \mathrm{rd}(x_2)\big) = x_1(1+\varepsilon_1)x_2(1+\varepsilon_2)(1+\mu) \approx x_1x_2(1+\varepsilon_1+\varepsilon_2+\mu) \tag{1.25}$$

with the relative error

$$\frac{\tilde{y}-y}{y} = 1+\varepsilon_1+\varepsilon_2+\mu. \tag{1.26}$$

The relative errors of the input data and of the multiplication just add up to the total relative error. There is no enhancement. Similarly for a division

$$y = \frac{x_1}{x_2} \tag{1.27}$$

the relative error is

$$\frac{\tilde{y}-y}{y} = 1+\varepsilon_1-\varepsilon_2+\mu. \tag{1.28}$$

1.3 Error Propagation

Consider an algorithm consisting of a sequence of elementary operations. From the set of input data which is denoted by the vector

$$\mathbf{x} = (x_1 \cdots x_n) \tag{1.29}$$

a set of output data is calculated

$$\mathbf{y} = (y_1 \cdots y_m). \tag{1.30}$$

Formally this can be denoted by a vector function

$$\mathbf{y} = \varphi(\mathbf{x}) \tag{1.31}$$

which can be written as a product of r simpler functions representing the elementary operations

$$\varphi = \varphi^{(r)} \times \varphi^{(r-1)} \cdots \varphi^{(1)}. \tag{1.32}$$

Starting with $\mathbf{x}$ intermediate results $\mathbf{x}_i = (x_{i1}, \ldots x_{in_i})$ are calculated until the output data $\mathbf{y}$ result from the last step:

$$\mathbf{x}_1 = \varphi^{(1)}(\mathbf{x})$$
$$\mathbf{x}_2 = \varphi^{(2)}(\mathbf{x}_1)$$
$$\vdots$$
$$\mathbf{x}_{r-1} = \varphi^{(r-1)}(\mathbf{x}_{r-2})$$
$$\mathbf{y} = \varphi^{(r)}(\mathbf{x}_{r-1}).$$

(1.33)

In the following we analyze the influence of numerical errors onto the final results. We treat all errors as small quantities and neglect higher orders. Due to round-off errors and possible experimental uncertainties the input data are not exactly given by $\mathbf{x}$ but by

$$\mathbf{x} + \Delta\mathbf{x}.$$

(1.34)

The first step of the algorithm produces the result

$$\tilde{\mathbf{x}}_1 = \mathrm{rd}\big(\varphi^{(1)}(\mathbf{x} + \Delta\mathbf{x})\big).$$

(1.35)

Taylor series expansion gives in first order

$$\tilde{\mathbf{x}}_1 = \big(\varphi^{(1)}(\mathbf{x}) + D\varphi^{(1)}\Delta\mathbf{x}\big)(1 + E_1) + \cdots$$

(1.36)

with the partial derivatives

$$D\varphi^{(1)} = \left(\frac{\partial x_{1i}}{\partial x_j}\right) = \begin{pmatrix} \frac{\partial x_{11}}{\partial x_1} & \cdots & \frac{\partial x_{11}}{\partial x_n} \\ \vdots & \ddots & \vdots \\ \frac{\partial x_{1n_1}}{\partial x_1} & \cdots & \frac{\partial x_{1n_1}}{\partial x_n} \end{pmatrix}$$

(1.37)

and the round-off errors of the first step

$$E_1 = \begin{pmatrix} \varepsilon_1^{(1)} & & \\ & \ddots & \\ & & \varepsilon_{n_1}^{(1)} \end{pmatrix}.$$

(1.38)

The error of the first intermediate result is

$$\Delta\mathbf{x}_1 = \tilde{\mathbf{x}}_1 - \mathbf{x}_1 = D\varphi^{(1)}\Delta\mathbf{x} + \varphi^{(1)}(\mathbf{x})E_1.$$

(1.39)

The second intermediate result is

$$\tilde{\mathbf{x}}_2 = \big(\varphi^{(2)}(\tilde{\mathbf{x}}_1)\big)(1 + E_2) = \varphi^{(2)}(\mathbf{x}_1 + \Delta\mathbf{x}_1)(1 + E_2)$$
$$= \mathbf{x}_2(1 + E_2) + D\varphi^{(2)}D\varphi^{(1)}\Delta\mathbf{x} + D\varphi^{(2)}\mathbf{x}_1 E_1$$

(1.40)

with the error

$$\Delta\mathbf{x}_2 = \mathbf{x}_2 E_2 + D\varphi^{(2)}D\varphi^{(1)}\Delta\mathbf{x} + D\varphi^{(2)}\mathbf{x}_1 E_1.$$

(1.41)

Finally the error of the result is

$$\Delta\mathbf{y} = \mathbf{y}E_r + D\varphi^{(r)}\cdots D\varphi^{(1)}\Delta\mathbf{x} + D\varphi^{(r)}\cdots D\varphi^{(2)}\mathbf{x}_1 E_1 + \cdots + D\varphi^{(r)}\mathbf{x}_{r-1} E_{r-1}.$$

(1.42)

The product of the matrices $D\varphi^{(r)} \cdots D\varphi^{(1)}$ is the matrix which contains the derivatives of the output data with respect to the input data (chain rule)

$$D\varphi = D\varphi^{(r)} \cdots D\varphi^{(1)} = \begin{pmatrix} \frac{\partial y_1}{\partial x_1} & \cdots & \frac{\partial y_1}{\partial x_n} \\ \vdots & \ddots & \vdots \\ \frac{\partial y_m}{\partial x_1} & \cdots & \frac{\partial y_m}{\partial x_n} \end{pmatrix}. \tag{1.43}$$

The first two contributions to the total error do not depend on the way in which the algorithm is divided into elementary steps in contrary to the remaining summands. Hence the inevitable error which is inherent to the problem can be estimated as [244]

$$\Delta^{(in)} y_i = \varepsilon_M |y_i| + \sum_{j=1}^{n} \left| \frac{\partial y_i}{\partial x_j} \right| |\Delta x_j| \tag{1.44}$$

or in case the error of the input data is dominated by the round-off errors $|\Delta x_j| \le \varepsilon_M |x_j|$

$$\Delta^{(in)} y_i = \varepsilon_M |y_i| + \varepsilon_M \sum_{j=1}^{n} \left| \frac{\partial y_i}{\partial x_j} \right| |x_j|. \tag{1.45}$$

Additional errors which are smaller than this inevitable error can be regarded as harmless. If all errors are harmless, the algorithm can be considered well behaved.

1.4 Stability of Iterative Algorithms

Often iterative algorithms are used which generate successive values starting from an initial value $\mathbf{x}_0$ according to an iteration method

$$\mathbf{x}_{j+1} = f(\mathbf{x}_j), \tag{1.46}$$

for instance to solve a large system of equations or to approximate a time evolution $\mathbf{x}_j \approx \mathbf{x}(j \Delta t)$. Consider first a linear iteration equation which can be written in matrix form as

$$\mathbf{x}_{j+1} = A\mathbf{x}_j. \tag{1.47}$$

If the matrix A is the same for all steps we have simply

$$\mathbf{x}_j = A^j \mathbf{x}_0. \tag{1.48}$$

Consider the unavoidable error originating from errors $\Delta \mathbf{x}$ of the start values:

$$\mathbf{x}_j = A^j (\mathbf{x}_0 + \Delta \mathbf{x}) = A^j \mathbf{x}_0 + A^j \Delta \mathbf{x}. \tag{1.49}$$

The initial errors $\Delta \mathbf{x}$ can be enhanced exponentially if A has at least one eigenvalue[5] λ with $|\lambda| > 1$. On the other hand the algorithm is conditionally stable if for all eigenvalues $|\lambda| \le 1$ holds. For a more general nonlinear iteration

$$\mathbf{x}_{j+1} = \varphi(\mathbf{x}_j) \tag{1.50}$$

the error propagates according to

$$\begin{aligned}
\mathbf{x}_1 &= \varphi(\mathbf{x}_0) + D\varphi\,\Delta\mathbf{x} \\
\mathbf{x}_2 &= \varphi(\mathbf{x}_1) = \varphi\big(\varphi(\mathbf{x}_0)\big) + (D\varphi)^2 \Delta\mathbf{x} \\
&\;\;\vdots \\
\mathbf{x}_j &= \varphi\big(\varphi\cdots\varphi(\mathbf{x}_0)\big) + (D\varphi)^j \Delta\mathbf{x}.
\end{aligned} \tag{1.51}$$

The algorithm is conditionally stable if all eigenvalues of the derivative matrix $D\varphi$ have absolute values $|\lambda| \le 1$.

1.5 Example: Rotation

Consider a simple rotation in the complex plane. The equation of motion

$$\dot{z} = i\omega z \tag{1.52}$$

obviously has the exact solution

$$z(t) = z_0 e^{i\omega t}. \tag{1.53}$$

As a simple algorithm for numerical integration we use a time grid

$$t_j = j\Delta t \quad j = 0, 1, 2\ldots \tag{1.54}$$

$$z_j = z(t_j) \tag{1.55}$$

and iterate the function values

$$z_{j+1} = z_j + \dot{z}(t_j) = (1 + i\omega\Delta t)z_j. \tag{1.56}$$

Since

$$|1 + i\omega\Delta t| = \sqrt{1 + \omega^2 \Delta t^2} > 1 \tag{1.57}$$

uncertainties in the initial condition will grow exponentially and the algorithm is not stable. A stable method is obtained by taking the derivative in the middle of the time interval (page 213)

$$\dot{z}\left(t + \frac{\Delta t}{2}\right) = i\omega z\left(t + \frac{\Delta t}{2}\right)$$

---

[5]The eigenvalues of A are solutions of the eigenvalue equation $A\mathbf{x} = \lambda\mathbf{x}$ (Chap. 9).

and making the approximation (page 214)

$$z\left(t + \frac{\Delta t}{2}\right) \approx \frac{z(t) + z(t + \Delta t)}{2}.$$

This gives the implicit equation

$$z_{j+1} = z_j + i\omega \Delta t \frac{z_{j+1} + z_j}{2} \tag{1.58}$$

which can be solved by

$$z_{j+1} = \frac{1 + \frac{i\omega \Delta t}{2}}{1 - \frac{i\omega \Delta t}{2}} z_j. \tag{1.59}$$

Now we have

$$\left| \frac{1 + \frac{i\omega \Delta t}{2}}{1 - \frac{i\omega \Delta t}{2}} \right| = \frac{\sqrt{1 + \frac{\omega^2 \Delta t^2}{4}}}{\sqrt{1 + \frac{\omega^2 \Delta t^2}{4}}} = 1 \tag{1.60}$$

and the calculated orbit is stable.

1.6 Truncation Error

The algorithm in the last example is stable but of course not perfect. Each step produces an error due to the finite time step. The exact solution

$$z(t + \Delta t) = z(t)e^{i\omega \Delta t} = z(t)\left(1 + i\omega \Delta t - \frac{\omega^2 \Delta t^2}{2} + \frac{-i\omega^3 \Delta t^3}{6} \cdots\right) \tag{1.61}$$

is approximated by

$$z(t + \Delta t) \approx z(t)\frac{1 + \frac{i\omega \Delta t}{2}}{1 - \frac{i\omega \Delta t}{2}}$$

$$= z(t)\left(1 + \frac{i\omega \Delta t}{2}\right)\left(1 + \frac{i\omega \Delta t}{2} - \frac{\omega^2 \Delta t^2}{4} - \frac{i\omega^3 \Delta t^3}{8} + \cdots\right) \tag{1.62}$$

$$= z(t)\left(1 + i\omega \Delta t - \frac{\omega^2 \Delta t^2}{2} + \frac{-i\omega^3 \Delta t^3}{4} \cdots\right) \tag{1.63}$$

which deviates from the exact solution by a term of the order $O(\Delta t^3)$, hence the *local error* order of this algorithm is $O(\Delta t^3)$ which is indicated by writing

$$z(t + \Delta t) = z(t)\frac{1 + \frac{i\omega \Delta t}{2}}{1 - \frac{i\omega \Delta t}{2}} + O(\Delta t^3). \tag{1.64}$$

Integration up to a total time $T = N\Delta t$ accumulates a *global error* of the order $N\Delta t^3 = T\Delta t^2$.

Table 1.4 Maximum and minimum integers

Java format	Bit length	Minimum	Maximum
Byte	8	−128	127
Short	16	−32768	32767
Integer	32	−2147483647	2147483648
Long	64	−9223372036854775808	9223372036854775807
Char	16	0	65535

1.7 Problems

Problem 1.1 (Machine precision)   In this computer experiment we determine the machine precision ε_M. Starting with a value of 1.0, x is divided repeatedly by 2 until numerical addition of 1 and $x = 2^{-M}$ gives 1. Compare single and double precision calculations.

Problem 1.2 (Maximum and minimum integers)   Integers are used as counters or to encode elements of a finite set like characters or colors. There are different integer formats available which store signed or unsigned integers of different length (Table 1.4). There is no infinite integer and addition of 1 to the maximum integer gives the minimum integer.

In this computer experiment we determine the smallest and largest integer numbers. Beginning with $I = 1$ we add repeatedly 1 until the condition $I + 1 > I$ becomes invalid or subtract repeatedly 1 until $I - 1 < I$ becomes invalid. For the 64 bit long integer format this takes to long. Here we multiply alternatively I by 2 until $I - 1 < I$ becomes invalid. For the character format the corresponding ordinal number is shown which is obtained by casting the character to an integer.

Problem 1.3 (Truncation error)   This computer experiment approximates the cosine function by a truncated Taylor series

$$\cos(x) \approx \mathrm{mycos}(x, n_{\max}) = \sum_{n=0}^{n_{\max}} (-)^n \frac{x^{2n}}{(2n)!} = 1 - \frac{x^2}{2} + \frac{x^4}{24} - \frac{x^6}{720} + \cdots \quad (1.65)$$

in the interval $-\pi/2 < x < \pi/2$. The function $\mathrm{mycos}(x, n_{\max})$ is numerically compared to the intrinsic cosine function.

Chapter 2
Interpolation

Experiments usually produce a discrete set of data points $(\mathbf{x}_i, f_i)$ which represent the value of a function $f(\mathbf{x})$ for a finite set of arguments $\{\mathbf{x}_0 \cdots \mathbf{x}_n\}$. If additional data points are needed, for instance to draw a continuous curve, interpolation is necessary. Interpolation also can be helpful to represent a complicated function by a simpler one or to develop more sophisticated numerical methods for the calculation of numerical derivatives and integrals. In the following we concentrate on the most important interpolating functions which are polynomials, splines and rational functions. Trigonometric interpolation is discussed in Chap. 7. An interpolating function reproduces the given function values at the interpolation points exactly (Fig. 2.1). The more general procedure of curve fitting, where this requirement is relaxed, is discussed in Chap. 10.

The interpolating polynomial can be explicitly constructed with the Lagrange method. Newton's method is numerically efficient if the polynomial has to be evaluated at many interpolating points and Neville's method has advantages if the polynomial is not needed explicitly and has to be evaluated only at one interpolation point.

Polynomials are not well suited for interpolation over a larger range. Spline functions can be superior which are piecewise defined polynomials. Especially cubic splines are often used to draw smooth curves. Curves with poles can be represented by rational interpolating functions whereas a special class of rational interpolants without poles provides a rather new alternative to spline interpolation.

2.1 Interpolating Functions

Consider the following problem: Given are $n + 1$ sample points (x_i, f_i), $i = 0 \cdots n$ and a function of x which depends on $n + 1$ parameters a_i:

$$\Phi(x; a_0 \cdots a_n). \tag{2.1}$$

P.O.J. Scherer, *Computational Physics*, Graduate Texts in Physics,
DOI 10.1007/978-3-319-00401-3_2,
© Springer International Publishing Switzerland 2013

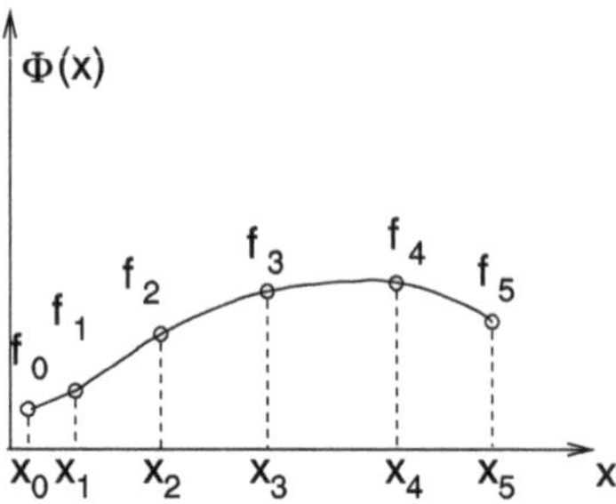

Fig. 2.1 (Interpolating function) The interpolating function $\Phi(x)$ reproduces a given data set $\Phi(x_i) = f_i$ and provides an estimate of the function $f(x)$ between the data points

The parameters are to be determined such that the interpolating function has the proper values at all sample points (Fig. 2.1)

$$\Phi(x_i; a_0 \cdots a_n) = f_i \quad i = 0 \cdots n. \tag{2.2}$$

An interpolation problem is called linear if the interpolating function is a linear combination of functions

$$\Phi(x; a_0 \cdots a_n) = a_0 \Phi_0(x) + a_1 \Phi_1(x) + \cdots + a_n \Phi_n(x). \tag{2.3}$$

Important examples are

- polynomials

$$a_0 + a_1 x + \cdots + a_n x^n \tag{2.4}$$

- trigonometric functions

$$a_0 + a_1 e^{ix} + a_2 e^{2ix} + \cdots + a_n e^{nix} \tag{2.5}$$

- spline functions which are piecewise polynomials, for instance the cubic spline

$$s(x) = \alpha_i + \beta_i(x - x_i) + \gamma_i(x - x_i)^2 + \delta_i(x - x_i)^3 \quad x_i \le x \le x_{i+1}. \tag{2.6}$$

Important examples for nonlinear interpolating functions are

- rational functions

$$\frac{p_0 + p_1 x + \cdots + p_M x^M}{q_0 + q_1 x + \cdots + q_N x^N} \tag{2.7}$$

- exponential functions

$$a_0 e^{\lambda_0 x} + a_1 e^{\lambda_1 x} + \cdots . \tag{2.8}$$

where amplitudes a_i as well as exponents λ_i have to be optimized.

2.2 Polynomial Interpolation

For $n + 1$ sample points (x_i, f_i), $i = 0 \cdots n$, $x_i \ne x_j$ there exists exactly one interpolating polynomial of degree n with

$$p(x_i) = f_i, \quad i = 0 \cdots n. \tag{2.9}$$

2.2.1 Lagrange Polynomials

Lagrange polynomials [137] are defined as

$$L_i(x) = \frac{(x - x_0) \cdots (x - x_{i-1})(x - x_{i+1}) \cdots (x - x_n)}{(x_i - x_0) \cdots (x_i - x_{i-1})(x_i - x_{i+1}) \cdots (x_i - x_n)}. \tag{2.10}$$

They are of degree n and have the property

$$L_i(x_k) = \delta_{i,k}. \tag{2.11}$$

The interpolating polynomial is given in terms of Lagrange polynomials by

$$p(x) = \sum_{i=0}^{n} f_i L_i(x) = \sum_{i=0}^{n} f_i \prod_{k=0,k \neq i}^{n} \frac{x - x_k}{x_i - x_k}. \tag{2.12}$$

2.2.2 Barycentric Lagrange Interpolation

With the polynomial

$$\omega(x) = \prod_{i=0}^{n} (x - x_i) \tag{2.13}$$

the Lagrange polynomial can be written as

$$L_i(x) = \frac{\omega(x)}{x - x_i} \frac{1}{\prod_{k=0,k \neq i}^{n} (x_i - x_k)} \tag{2.14}$$

which, introducing the Barycentric weights [24]

$$u_i = \frac{1}{\prod_{k=0,k \neq i}^{n} (x_i - x_k)} \tag{2.15}$$

becomes the first form of the barycentric interpolation formula

$$L_i(x) = \omega(x) \frac{u_i}{x - x_i}. \tag{2.16}$$

The interpolating polynomial can now be evaluated according to

$$p(x) = \sum_{i=0}^{n} f_i L_i(x) = \omega(x) \sum_{i=0}^{n} f_i \frac{u_i}{x - x_i}. \tag{2.17}$$

Having computed the weights u_i, evaluation of the polynomial only requires $O(n)$ operations whereas calculation of all the Lagrange polynomials requires $O(n^2)$ operations. Calculation of $\omega(x)$ can be avoided considering that

$$p_1(x) = \sum_{i=0}^{n} L_i(x) = \omega(x) \sum_{i=0}^{n} \frac{u_i}{x - x_i} \tag{2.18}$$

is a polynomial of degree n with

$$p_1(x_i) = 1 \quad i = 0 \cdots n. \tag{2.19}$$

But this is only possible if

$$p_1(x) = 1. \tag{2.20}$$

Therefore

$$p(x) = \frac{p(x)}{p_1(x)} = \frac{\sum_{i=0}^{n} f_i \frac{u_i}{x - x_i}}{\sum_{i=0}^{n} \frac{u_i}{x - x_i}} \tag{2.21}$$

which is known as the second form of the barycentric interpolation formula.

2.2.3 Newton's Divided Differences

Newton's method of divided differences [138] is an alternative for efficient numerical calculations [271]. Rewrite

$$f(x) = f(x_0) + \frac{f(x) - f(x_0)}{x - x_0}(x - x_0). \tag{2.22}$$

With the first order divided difference

$$f[x, x_0] = \frac{f(x) - f(x_0)}{x - x_0} \tag{2.23}$$

this becomes

$$f[x, x_0] = f[x_1, x_0] + \frac{f[x, x_0] - f[x_1, x_0]}{x - x_1}(x - x_1) \tag{2.24}$$

and with the second order divided difference

$$f[x, x_0, x_1] = \frac{f[x, x_0] - f[x_1, x_0]}{x - x_1} = \frac{f(x) - f(x_0)}{(x - x_0)(x - x_1)} - \frac{f(x_1) - f(x_0)}{(x_1 - x_0)(x - x_1)}$$

$$= \frac{f(x)}{(x - x_0)(x - x_1)} + \frac{f(x_1)}{(x_1 - x_0)(x_1 - x)} + \frac{f(x_0)}{(x_0 - x_1)(x_0 - x)} \tag{2.25}$$

we have

$$f(x) = f(x_0) + (x - x_0)f[x_1, x_0] + (x - x_0)(x - x_1)f[x, x_0, x_1]. \qquad (2.26)$$

Higher order divided differences are defined recursively by

$$f[x_1 x_2 \cdots x_{r-1} x_r] = \frac{f[x_1 x_2 \cdots x_{r-1}] - f[x_2 \cdots x_{r-1} x_r]}{x_1 - x_r}. \qquad (2.27)$$

They are invariant against permutation of the arguments which can be seen from the explicit formula

$$f[x_1 x_2 \cdots x_r] = \sum_{k=1}^{r} \frac{f(x_k)}{\prod_{i \neq k}(x_k - x_i)}. \qquad (2.28)$$

Finally we have

$$f(x) = p(x) + q(x) \qquad (2.29)$$

with a polynomial of degree n

$$p(x) = f(x_0) + f[x_1, x_0](x - x_0) + f[x_2 x_1 x_0](x - x_0)(x - x_1) + \cdots$$
$$+ f[x_n x_{n-1} \cdots x_0](x - x_0)(x - x_1) \cdots (x - x_{n-1}) \qquad (2.30)$$

and the function

$$q(x) = f[x x_n \cdots x_0](x - x_0) \cdots (x - x_n). \qquad (2.31)$$

Obviously $q(x_i) = 0$, $i = 0 \cdots n$, hence $p(x)$ is the interpolating polynomial.

Algorithm The divided differences are arranged in the following way:

$$
\begin{array}{llll}
f_0 & & & \\
f_1 & f[x_0 x_1] & & \\
\vdots & \vdots & \ddots & \\
f_{n-1} & f[x_{n-2} x_{n-1}] & f[x_{n-3} x_{n-2} x_{n-1}] \cdots f[x_0 \cdots x_{n-1}] & \\
f_n & f[x_{n-1} x_n] & f[x_{n-2} x_{n-1} x_n] \cdots f[x_1 \cdots x_n] \ f[x_0 \cdots x_n] &
\end{array} \qquad (2.32)
$$

Since only the diagonal elements are needed, a one-dimensional data array $t[0] \cdots t[n]$ is sufficient for the calculation of the polynomial coefficients:

```
for i := 0 to n do begin
   t[i] := f[i];
   for k := i − 1 downto 0 do
       t[k] := (t[k + 1] − t[k])/(x[i] − x[k]);
   a[i] := t[0];
end;
```

The value of the polynomial is then evaluated by

```
p := a[n];
for i := n − 1 downto 0 do
   p := p*(x − x[i]) + a[i];
```


2.2.4 Neville Method

The Neville method [180] is advantageous if the polynomial is not needed explicitly and has to be evaluated only at one point. Consider the interpolating polynomial for the points $x_0 \cdots x_k$, which will be denoted as $P_{0,1\ldots k}(x)$. Obviously

$$P_{0,1\ldots k}(x) = \frac{(x - x_0)P_{1\ldots k}(x) - (x - x_k)P_{0\ldots k-1}(x)}{x_k - x_0} \tag{2.33}$$

since for $x = x_1 \cdots x_{k-1}$ the right hand side is

$$\frac{(x - x_0)f(x) - (x - x_k)f(x)}{x_k - x_0} = f(x). \tag{2.34}$$

For $x = x_0$ we have

$$\frac{-(x_0 - x_k)f(x)}{x_k - x_0} = f(x) \tag{2.35}$$

and finally for $x = x_k$

$$\frac{(x_k - x_0)f(x)}{x_k - x_0} = f(x). \tag{2.36}$$

Algorithm We use the following scheme to calculate $P_{0,1\ldots n}(x)$ recursively:

$$
\begin{array}{lllll}
P_0 & & & & \\
P_1 & P_{01} & & & \\
P_2 & P_{12} & P_{012} & & \\
\vdots & \vdots & \vdots & \ddots & \\
P_n & P_{n-1,n} & P_{n-2,n-1,n} & \cdots & P_{01\ldots n}
\end{array}
\tag{2.37}
$$

The first column contains the function values $P_i(x) = f_i$. The value $P_{01\ldots n}$ can be calculated using a 1-dimensional data array $p[0] \cdots p[n]$:

```
for i := 0 to n do begin
   p[i] := f[i];
   for k := i − 1 downto 0 do
      p[k] := (p[k + 1]*(x − x[k]) − p[k]*(x − x[i]))/(x[k] − x[i]);
end;
f := p[0];
```

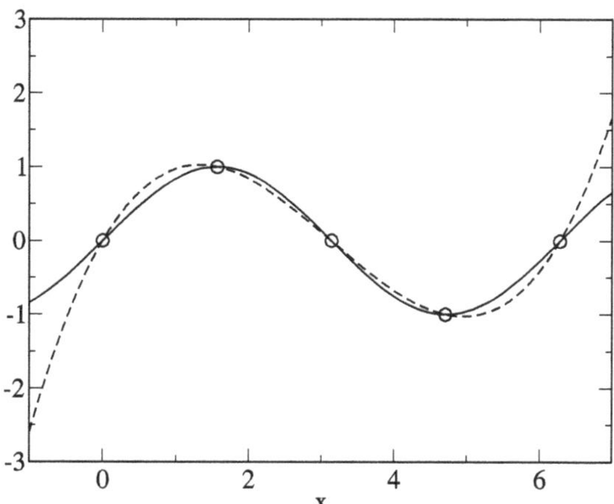

Fig. 2.2 (Interpolating polynomial) The interpolated function (*solid curve*) and the interpolating polynomial (*broken curve*) for the example (2.40) are compared

2.2.5 *Error of Polynomial Interpolation*

The error of polynomial interpolation [12] can be estimated with the help of the following theorem:

If $f(x)$ is $n + 1$ times differentiable then for each $\overline{x}$ there exists ξ within the smallest interval containing $\overline{x}$ as well as all the x_i with

$$q(\overline{x}) = \prod_{i=0}^{n} (\overline{x} - x_i) \frac{f^{(n+1)}(\xi)}{(n+1)!}. \tag{2.38}$$

From a discussion of the function

$$\omega(x) = \prod_{i=0}^{n} (x - x_i) \tag{2.39}$$

it can be seen that the error increases rapidly outside the region of the sample points (extrapolation is dangerous!). As an example consider the sample points (Fig. 2.2)

$$f(x) = \sin(x) \quad x_i = 0, \frac{\pi}{2}, \pi, \frac{3\pi}{2}, 2\pi. \tag{2.40}$$

The maximum interpolation error is estimated by ($|f^{(n+1)}| \leq 1$)

$$\left| f(x) - p(x) \right| \leq \left| \omega(x) \right| \frac{1}{120} \leq \frac{35}{120} \approx 0.3 \tag{2.41}$$

whereas the error increases rapidly outside the interval $0 < x < 2\pi$ (Fig. 2.3).

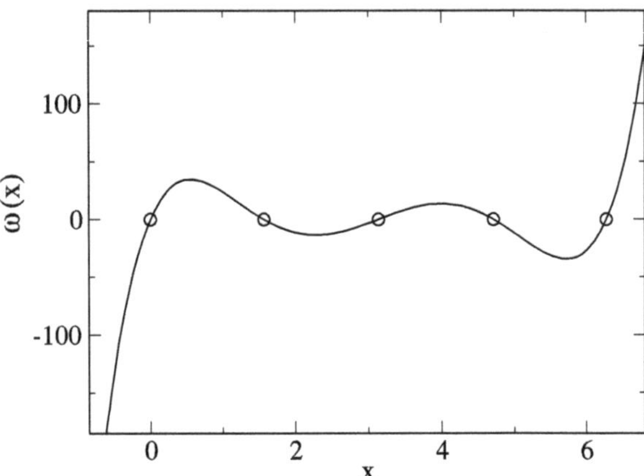

Fig. 2.3 (Interpolation error) The polynomial $\omega(x)$ is shown for the example (2.40). Its roots x_i are given by the x values of the sample points (*circles*). Inside the interval $x_0 \cdots x_4$ the absolute value of ω is bounded by $|\omega(x)| \leq 35$ whereas outside the interval it increases very rapidly

2.3 Spline Interpolation

Polynomials are not well suited for interpolation over a larger range. Often spline functions are superior which are piecewise defined polynomials [186, 228]. The simplest case is a linear spline which just connects the sampling points by straight lines:

$$p_i(x) = y_i + \frac{y_{i+1} - y_i}{x_{i+1} - x_i}(x - x_i) \tag{2.42}$$

$$s(x) = p_i(x) \quad \text{where } x_i \leq x < x_{i+1}. \tag{2.43}$$

The most important case is the cubic spline which is given in the interval $x_i \leq x < x_{i+1}$ by

$$p_i(x) = \alpha_i + \beta_i(x - x_i) + \gamma_i(x - x_i)^2 + \delta_i(x - x_i)^3. \tag{2.44}$$

We want to have a smooth interpolation and assume that the interpolating function and their first two derivatives are continuous. Hence we have for the inner boundaries:

$$i = 0 \cdots n - 1$$

$$p_i(x_{i+1}) = p_{i+1}(x_{i+1}) \tag{2.45}$$

$$p'_i(x_{i+1}) = p'_{i+1}(x_{i+1}) \tag{2.46}$$

$$p''_i(x_{i+1}) = p''_{i+1}(x_{i+1}). \tag{2.47}$$

We have to specify boundary conditions at x_0 and x_n. The most common choice are natural boundary conditions $s''(x_0) = s''(x_n) = 0$, but also periodic boundary conditions $s''(x_0) = s''(x_n)$, $s'(x_0) = s'(x_n)$, $s(x_0) = s(x_n)$ or given derivative values $s'(x_0)$ and $s'(x_n)$ are often used. The second derivative is a linear function [244]

$$p''_i(x) = 2\gamma_i + 6\delta_i(x - x_i) \tag{2.48}$$

which can be written using $h_{i+1} = x_{i+1} - x_i$ and $M_i = s''(x_i)$ as

$$p_i''(x) = M_{i+1}\frac{(x - x_i)}{h_{i+1}} + M_i\frac{(x_{i+1} - x)}{h_{i+1}} \quad i = 0 \cdots n - 1 \tag{2.49}$$

since

$$p_i''(x_i) = M_i\frac{x_{i+1} - x_i}{h_{i+1}} = s''(x_i) \tag{2.50}$$

$$p_i''(x_{i+1}) = M_{i+1}\frac{(x_{i+1} - x_i)}{h_{i+1}} = s''(x_{i+1}). \tag{2.51}$$

Integration gives with the two constants A_i and B_i

$$p_i'(x) = M_{i+1}\frac{(x - x_i)^2}{2h_{i+1}} - M_i\frac{(x_{i+1} - x)^2}{2h_{i+1}} + A_i \tag{2.52}$$

$$p_i(x) = M_{i+1}\frac{(x - x_i)^3}{6h_{i+1}} + M_i\frac{(x_{i+1} - x)^3}{6h_{i+1}} + A_i(x - x_i) + B_i. \tag{2.53}$$

From $s(x_i) = y_i$ and $s(x_{i+1}) = y_{i+1}$ we have

$$M_i\frac{h_{i+1}^2}{6} + B_i = y_i \tag{2.54}$$

$$M_{i+1}\frac{h_{i+1}^2}{6} + A_i h_{i+1} + B_i = y_{i+1} \tag{2.55}$$

and hence

$$B_i = y_i - M_i\frac{h_{i+1}^2}{6} \tag{2.56}$$

$$A_i = \frac{y_{i+1} - y_i}{h_{i+1}} - \frac{h_{i+1}}{6}(M_{i+1} - M_i). \tag{2.57}$$

Now the polynomial is

$$
\begin{aligned}
p_i(x) &= \frac{M_{i+1}}{6h_{i+1}}(x - x_i)^3 - \frac{M_i}{6h_{i+1}}(x - x_i - h_{i+1})^3 + A_i(x - x_i) + B_i \\
&= (x - x_i)^3\left(\frac{M_{i+1}}{6h_{i+1}} - \frac{M_i}{6h_{i+1}}\right) + \frac{M_i}{6h_{i+1}}3h_{i+1}(x - x_i)^2 \\
&\quad + (x - x_i)\left(A_i - \frac{M_i}{6h_{i+1}}3h_{i+1}^2\right) + B_i + \frac{M_i}{6h_{i+1}}h_{i+1}^3.
\end{aligned}
\tag{2.58}
$$

Comparison with

$$p_i(x) = \alpha_i + \beta_i(x - x_i) + \gamma_i(x - x_i)^2 + \delta_i(x - x_i)^3 \tag{2.59}$$

gives

$$\alpha_i = B_i + \frac{M_i}{6}h_{i+1}^2 = y_i \tag{2.60}$$

$$\beta_i = A_i - \frac{h_{i+1}M_i}{2} = \frac{y_{i+1} - y_i}{h_{i+1}} - h_{i+1}\frac{M_{i+1} + 2M_i}{6} \tag{2.61}$$

$$\gamma_i = \frac{M_i}{2} \tag{2.62}$$

$$\delta_i = \frac{M_{i+1} - M_i}{6h_{i+1}}. \tag{2.63}$$

Finally we calculate M_i from the continuity of $s'(x)$. Substituting for A_i in $p_i'(x)$ we have

$$p_i'(x) = M_{i+1}\frac{(x - x_i)^2}{2h_{i+1}} - M_i\frac{(x_{i+1} - x)^2}{2h_{i+1}} + \frac{y_{i+1} - y_i}{h_{i+1}} - \frac{h_{i+1}}{6}(M_{i+1} - M_i) \tag{2.64}$$

and from $p_{i-1}'(x_i) = p_i'(x_i)$ it follows

$$M_i\frac{h_i}{2} + \frac{y_i - y_{i-1}}{h_i} - \frac{h_i}{6}(M_i - M_{i-1})$$

$$= -M_i\frac{h_{i+1}}{2} + \frac{y_{i+1} - y_i}{h_{i+1}} - \frac{h_{i+1}}{6}(M_{i+1} - M_i) \tag{2.65}$$

$$M_i\frac{h_i}{3} + M_{i-1}\frac{h_i}{6} + M_i\frac{h_{i+1}}{3} + M_{i+1}\frac{h_{i+1}}{6} = \frac{y_{i+1} - y_i}{h_{i+1}} - \frac{y_i - y_{i-1}}{h_i} \tag{2.66}$$

which is a system of linear equations for the M_i. Using the abbreviations

$$\lambda_i = \frac{h_{i+1}}{h_i + h_{i+1}} \tag{2.67}$$

$$\mu_i = 1 - \lambda_i = \frac{h_i}{h_i + h_{i+1}} \tag{2.68}$$

$$d_i = \frac{6}{h_i + h_{i+1}}\left(\frac{y_{i+1} - y_i}{h_{i+1}} - \frac{y_i - y_{i-1}}{h_i}\right) \tag{2.69}$$

we have

$$\mu_i M_{i-1} + 2M_i + \lambda_i M_{i+1} = d_i \quad i = 1 \cdots n - 1. \tag{2.70}$$

We define for natural boundary conditions

$$\lambda_0 = 0 \quad \mu_n = 0 \quad d_0 = 0 \quad d_n = 0 \tag{2.71}$$

and in case of given derivative values

$$\lambda_0 = 1 \quad \mu_n = 1 \quad d_0 = \frac{6}{h_1}\left(\frac{y_1 - y_0}{h_1} - y_0'\right) \quad d_n = \frac{6}{h_n}\left(y_n' - \frac{y_n - y_{n-1}}{h_n}\right). \tag{2.72}$$

The system of equation has the form

$$
\begin{bmatrix}
2 & \lambda_0 & & & & \\
\mu_1 & 2 & \lambda_1 & & & \\
& \mu_2 & 2 & \lambda_2 & & \\
& & \ddots & \ddots & \ddots & \\
& & & \mu_{n-1} & 2 & \lambda_{n-1} \\
& & & & \mu_n & 2
\end{bmatrix}
\begin{bmatrix}
M_0 \\ M_1 \\ M_2 \\ \vdots \\ M_{n-1} \\ M_n
\end{bmatrix}
=
\begin{bmatrix}
d_0 \\ d_1 \\ d_2 \\ \vdots \\ d_{n-1} \\ d_n
\end{bmatrix}. \tag{2.73}
$$

For periodic boundary conditions we define

$$
\lambda_n = \frac{h_1}{h_1 + h_n} \quad \mu_n = 1 - \lambda_n \quad d_n = \frac{6}{h_1 + h_n}\left(\frac{y_1 - y_n}{h_1} - \frac{y_n - y_{n-1}}{h_n}\right) \tag{2.74}
$$

and the system of equations is (with $M_n = M_0$)

$$
\begin{bmatrix}
2 & \lambda_1 & & & & \mu_1 \\
\mu_2 & 2 & \lambda_2 & & & \\
& \mu_3 & 2 & \lambda_3 & & \\
& & \ddots & \ddots & \ddots & \\
& & & \mu_{n-1} & 2 & \lambda_{n-1} \\
\lambda_n & & & & \mu_n & 2
\end{bmatrix}
\begin{bmatrix}
M_1 \\ M_2 \\ M_3 \\ \vdots \\ M_{n-1} \\ M_n
\end{bmatrix}
=
\begin{bmatrix}
d_1 \\ d_2 \\ d_3 \\ \vdots \\ d_{n-1} \\ d_n
\end{bmatrix}. \tag{2.75}
$$

All this tridiagonal systems can be easily solved with a special Gaussian elimination method (Sects. 5.3, 5.4).

2.4 Rational Interpolation

The use of rational approximants allows to interpolate functions with poles, where polynomial interpolation can give poor results [244]. Rational approximants without poles [90] are also well suited for the case of equidistant x_i, where higher order polynomials tend to become unstable. The main disadvantages are additional poles which are difficult to control and the appearance of unattainable points. Recent developments using the barycentric form of the interpolating function [25, 90, 227] helped to overcome these difficulties.

2.4.1 Padé Approximant

The Padé approximant [13] of order $[M/N]$ to a function $f(x)$ is the rational function

$$
R_{M/N}(x) = \frac{P_M(x)}{Q_N(x)} = \frac{p_0 + p_1 x + \cdots + p_M x^M}{q_0 + q_1 x + \cdots + q_N x^N} \tag{2.76}
$$

which reproduces the McLaurin series (the Taylor series at $x = 0$) of

$$f(x) = a_0 + a_1 x + a_2 x^2 + \cdots \tag{2.77}$$

up to order $M + N$, i.e.

$$
\begin{aligned}
f(0) &= R(0) \\[4pt]
\frac{\mathrm{d}}{\mathrm{d}x} f(0) &= \frac{\mathrm{d}}{\mathrm{d}x} R(0) \\[4pt]
&\;\;\vdots \\[4pt]
\frac{\mathrm{d}^{(M+N)}}{\mathrm{d}x^{(M+N)}} f(0) &= \frac{\mathrm{d}^{(M+N)}}{\mathrm{d}x^{(M+N)}} R(0).
\end{aligned}
\tag{2.78}
$$

Multiplication gives

$$p_0 + p_1 x + \cdots + p_M x^M = \left(q_0 + q_1 x + \cdots + q_N x^N \right)(a_0 + a_1 x + \cdots) \tag{2.79}$$

and collecting powers of x we find the system of equations

$$
\begin{aligned}
p_0 &= q_0 a_0 \\[4pt]
p_1 &= q_0 a_1 + q_1 a_0 \\[4pt]
p_2 &= q_0 a_2 + a_1 q_1 + a_0 q_2 \\[4pt]
&\;\;\vdots \\[4pt]
p_M &= q_0 a_M + a_{M-1} q_1 + \cdots + a_0 q_M \\[4pt]
0 &= q_0 a_{M+1} + q_1 a_M + \cdots + q_N a_{M-N+1} \\[4pt]
&\;\;\vdots \\[4pt]
0 &= q_0 a_{M+N} + q_1 a_{M+N-1} + \cdots + q_N a_M
\end{aligned}
\tag{2.80}
$$

where

$$a_n = 0 \quad \text{for } n < 0 \tag{2.81}$$

$$q_j = 0 \quad \text{for } j > N. \tag{2.82}$$

Example (Calculate the $[3, 3]$ approximant to $\tan(x)$) The Laurent series of the tangent is

$$\tan(x) = x + \frac{1}{3} x^3 + \frac{2}{15} x^5 + \cdots . \tag{2.83}$$

We set $q_0 = 1$. Comparison of the coefficients of the polynomial

$$p_0 + p_1 x + p_2 x^2 + p_3 x^3 = \left(1 + q_1 x + q_2 x^2 + q_3 x^3\right)\left(x + \frac{1}{3}x^3 + \frac{2}{15}x^5\right) \quad (2.84)$$

gives the equations

$$
\begin{aligned}
x^0: &\quad p_0 = 0 \\
x^1: &\quad p_1 = 1 \\
x^2: &\quad p_2 = q_1 \\
x^3: &\quad p_3 = q_2 + \frac{1}{3} \\
x^4: &\quad 0 = q_3 + \frac{1}{3}q_1 \\
x^5: &\quad 0 = \frac{2}{15} + \frac{1}{3}q_2 \\
x^6: &\quad 0 = \frac{2}{15}q_1 + \frac{1}{3}q_3.
\end{aligned}
\quad (2.85)
$$

We easily find

$$p_2 = q_1 = q_3 = 0 \quad q_2 = -\frac{2}{5} \quad p_3 = -\frac{1}{15} \quad (2.86)$$

and the approximant of order $[3, 3]$ is

$$R_{3,3} = \frac{x - \frac{1}{15}x^3}{1 - \frac{2}{5}x^2}. \quad (2.87)$$

This expression reproduces the tangent quite well (Fig. 2.4). Its pole at $\sqrt{10}/2 \approx 1.581$ is close to the pole of the tangent function at $\pi/2 \approx 1.571$.

2.4.2 Barycentric Rational Interpolation

If the weights of the barycentric form of the interpolating polynomial (2.21) are taken as general parameters $u_i \neq 0$ it becomes a rational function

$$R(x) = \frac{\sum_{i=0}^{n} f_i \frac{u_i}{x - x_i}}{\sum_{i=0}^{n} \frac{u_i}{x - x_i}} \quad (2.88)$$

which obviously interpolates the data points since

$$\lim_{x \to x_i} R(x) = f_i. \quad (2.89)$$

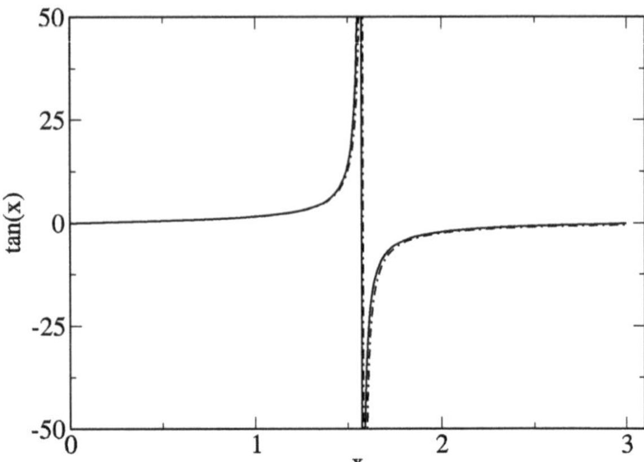

Fig. 2.4 (Padé approximation to $\tan(x)$) The Padé approximant ((2.87), *dash-dotted curve*) reproduces the tangent (*full curve*) quite well

With the polynomials[1]

$$P(x) = \sum_{i=0}^{n} u_i f_i \prod_{j=0;\, j \neq i}^{n} (x - x_j) = \sum_{i=0}^{n} u_i f_i \frac{\omega(x)}{x - x_i}$$

$$Q(x) = \sum_{i=0}^{n} u_i \prod_{j=0;\, j \neq i}^{n} (x - x_j) = \sum_{i=0}^{n} u_i \frac{\omega(x)}{x - x_i}$$

a rational interpolating function is given by[2]

$$R(x) = \frac{P(x)}{Q(x)}.$$

Obviously there are infinitely different rational interpolating functions which differ by the weights $\mathbf{u} = (u_0, u_1 \cdots u_n)$ (an example is shown in Fig. 2.5). To fix the parameters u_i, additional conditions have to be imposed.

2.4.2.1 Rational Interpolation of Order $[M, N]$

One possibility is to assume that $P(x)$ and $Q(x)$ are of order $\leq M$ and $\leq N$, respectively with $M + N = n$. This gives n additional equations for the $2(n + 1)$ polynomial coefficients. The number of unknown equals $n + 1$ and the rational interpolant is uniquely determined up to a common factor in numerator and denominator.

---

[1] $\omega(x) = \prod_{i=0}^{n}(x - x_i)$ as in (2.39).

[2] It can be shown that any rational interpolant can be written in this form.

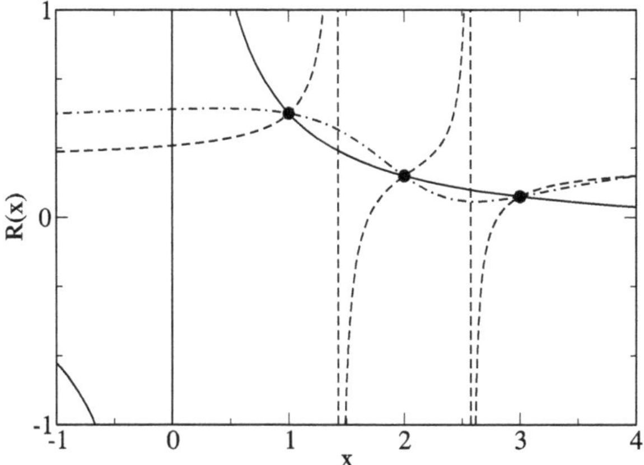

Fig. 2.5 (Rational interpolation) The data points $(1, \frac{1}{2})$, $(2, \frac{1}{5})$, $(3, \frac{1}{10})$ are interpolated by several rational functions. The $[1, 1]$ approximant (2.95) corresponding to $\mathbf{u} = (5, -20, 15)$ is shown by the *solid curve*, the *dashed curve* shows the function $R(x) = \frac{8x^2 - 36x + 38}{10(3x^2 - 12x + 11)}$ which is obtained for $\mathbf{u} = (1, 1, 1)$ and the *dash-dotted curve* shows the function $R(x) = \frac{4x^2 - 20x + 26}{10(5 - 4x + x^2)}$ which follows for $\mathbf{u} = (1, -1, 1)$ and has no real poles

Example (Consider the data points $f(1) = \frac{1}{2}$, $f(2) = \frac{1}{5}$, $f(3) = \frac{1}{10}$) The polynomials are

$$P(x) = \frac{1}{2}u_0(x - 2)(x - 3) + \frac{1}{5}u_1(x - 1)(x - 3) + \frac{1}{10}u_2(x - 1)(x - 2)$$

$$= 3u_0 + \frac{3}{5}u_1 + \frac{1}{5}u_2 + \left[-\frac{5}{2}u_0 - \frac{4}{5}u_1 - \frac{3}{10}u_2 \right] x$$

$$+ \left[\frac{1}{2}u_0 + \frac{1}{5}u_1 + \frac{1}{10}u_2 \right] x^2 \tag{2.90}$$

$$Q(x) = u_0(x - 2)(x - 3) + u_1(x - 1)(x - 3) + u_2(x - 1)(x - 2)$$

$$= 6u_0 + 3u_1 + 2u_2 + [-5u_0 - 4u_1 - 3u_2]x + [u_0 + u_1 + u_2]x^2. \tag{2.91}$$

To obtain a $[1, 1]$ approximant we have to solve the equations

$$\frac{1}{2}u_0 + \frac{1}{5}u_1 + \frac{1}{10}u_2 = 0 \tag{2.92}$$

$$u_0 + u_1 + u_2 = 0 \tag{2.93}$$

which gives

$$u_2 = 3u_0 \quad u_1 = -4u_0 \tag{2.94}$$

and thus

$$R(x) = \frac{\frac{6}{5}u_0 - \frac{1}{5}u_0 x}{2u_0 x} = \frac{6 - x}{10x}. \tag{2.95}$$

General methods to obtain the coefficients u_i for a given data set are described in [25, 227]. They also allow to determine unattainable points corresponding to $u_i = 0$ and to locate the poles. Without loss of generality it can be assumed [227] that $M \geq N$.[3]

Let $P(x)$ be the unique polynomial which interpolates the product $f(x)Q(x)$

$$P(x_i) = f(x_i)Q(x_i) \quad i = 0 \cdots M. \tag{2.96}$$

Then from (2.31) we have

$$f(x)Q(x) - P(x) = (fQ)[x_0 \cdots x_M, x](x - x_0) \cdots (x - x_M). \tag{2.97}$$

Setting

$$x = x_i \quad i = M + 1 \cdots n \tag{2.98}$$

we have

$$f(x_i)Q(x_i) - P(x_i) = (fQ)[x_0 \cdots x_M, x_i](x_i - x_0) \cdots (x - x_M) \tag{2.99}$$

which is zero if $P(x_i)/Q(x_i) = f_i$ for $i = 0 \cdots n$. But then

$$(fQ)[x_0 \cdots x_M, x_i] = 0 \quad i = M + 1 \cdots n. \tag{2.100}$$

The polynomial $Q(x)$ can be written in Newtonian form (2.30)

$$Q(x) = \sum_{i=0}^{N} v_i \prod_{j=0}^{i-1} (x - x_j) = v_0 + v_1(x - x_0) + \cdots + v_N(x - x_0) \cdots (x - x_{N-1}). \tag{2.101}$$

With the abbreviation

$$g_j(x) = x - x_j \quad j = 0 \cdots N \tag{2.102}$$

we find

$$(fg_j)[x_0 \cdots x_M, x_i] = \sum_{k=0 \cdots M, i} \frac{f(x_k)g(x_k)}{\prod_{r \neq k}(x_k - x_r)} = \sum_{k=0 \cdots M, i, k \neq j} \frac{f(x_k)}{\prod_{r \neq k, r \neq j}(x_k - x_r)}$$

$$= f[x_0 \cdots x_{j-1}, x_{j+1} \cdots x_M, x_i] \tag{2.103}$$

---

[3]The opposite case can be treated by considering the reciprocal function values $1/f(x_i)$.

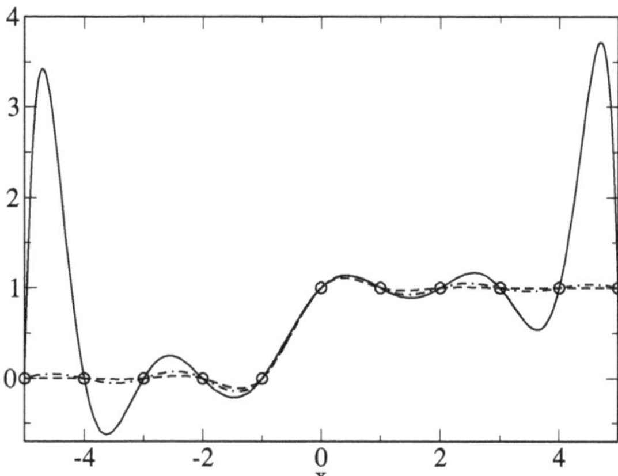

Fig. 2.6 (Interpolation of a step function) A step function with uniform x-values (*circles*) is interpolated by a polynomial (*full curve*), a cubic spline (*dashed curve*) and with the rational Floater-Hormann $d = 1$ function ((2.105), *dash-dotted curve*). The rational function behaves similar to the spline function but provides in addition an analytical function with continuous derivatives

which we apply repeatedly to (2.100) to get the system of $n - M = N$ equations for $N + 1$ unknowns

$$\sum_{j=0}^{N} v_j f[x_j, x_{j+1} \cdots x_M, x_i] = 0 \quad i = M + 1 \cdots n \tag{2.104}$$

from which the coefficients v_j can be found by Gaussian elimination up to a scaling factor. The Newtonian form of $Q(x)$ can then be converted to the barycentric form as described in [271].

2.4.2.2   Rational Interpolation Without Poles

Polynomial interpolation of larger data sets can be ill behaved, especially for the case of equidistant x-values. Rational interpolation without poles can be a much better choice here (Fig. 2.6).

Berrut [23] suggested to choose the following weights

$$u_k = (-1)^k.$$

With this choice $Q(x)$ has no real roots. Floater and Hormann [90] used the different choice

$$u_k = (-1)^{k-1} \left(\frac{1}{x_{k+1} - x_k} + \frac{1}{x_k - x_{k-1}} \right) \quad k = 1 \cdots n - 1$$

$$u_0 = -\frac{1}{x_1 - x_0} \quad u_n = (-1)^{n-1} \frac{1}{x_n - x_{n-1}} \tag{2.105}$$

which becomes very similar for equidistant x-values.

Table 2.1 Floater-Hormann weights for uniform data

$|u_k|$	d
$1, 1, 1, \ldots, 1, 1, 1$	0
$1, 2, 2, 2, \ldots, 2, 2, 2, 1$	1
$1, 3, 4, 4, 4, \ldots, 4, 4, 4, 3, 1$	2
$1, 4, 7, 8, 8, 8, \ldots, 8, 8, 8, 7, 4, 1$	3
$1, 5, 11, 15, 16, 16, 16, \ldots, 16, 16, 16, 15, 11, 5, 1$	4

Floater and Hormann generalized this expression and found a class of rational interpolants without poles given by the weights

$$u_k = (-1)^{k-d} \sum_{i=\max(k-d,0)}^{\min(k,n-d)} \prod_{j=i, j \neq k}^{i+d} \frac{1}{|x_k - x_j|} \tag{2.106}$$

where $0 \leq d \leq n$ and the approximation order increases with d. In the uniform case this simplifies to (Table 2.1)

$$u_k = (-1)^{k-d} \sum_{i=\min(k-d,0)}^{\max(k,n-d)} \binom{d}{k-i}. \tag{2.107}$$

2.5 Multivariate Interpolation

The simplest 2-dimensional interpolation method is bilinear interpolation.[4] It uses linear interpolation for both coordinates within the rectangle $x_i \leq x \leq x_{i+1}$, $y_i \leq y_i \leq y_{i+1}$:

$$p(x_i + h_x, y_i + h_y)$$

$$= p(x_i + h_x, y_i) + h_y \frac{p(x_i + h_x, y_{i+1}) - p(x_i + h_x, y_i)}{y_{i+1} - y_i}$$

$$= f(x_i, y_i) + h_x \frac{f(x_{i+1}, y_i) - f(x_i, y_i)}{x_{i+1} - x_i}$$

$$+ h_y \frac{f(x_i, y_{i+1}) + h_x \frac{f(x_{i+1}, y_{i+1}) - f(x_i, y_{i+1})}{x_{i+1} - x_i} - f(x_i, y_i) - h_x \frac{f(x_{i+1}, y_i) - f(x_i, y_i)}{x_{i+1} - x_i}}{y_{i+1} - y_i}$$

$$\tag{2.108}$$

---

[4]Bilinear means linear interpolation in two dimensions. Accordingly linear interpolation in three dimensions is called trilinear.

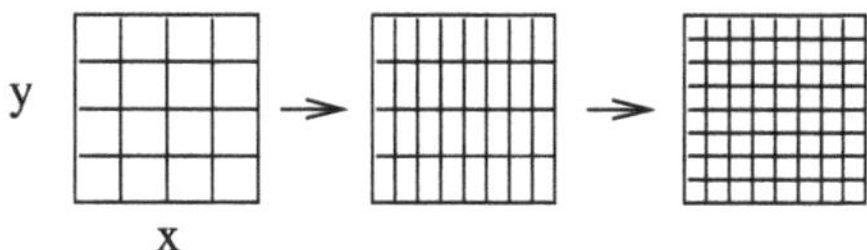

Fig. 2.7 Bispline interpolation

which can be written as a two dimensional polynomial

$$p(x_i + h_x, y_i + h_y) = a_{00} + a_{10}h_x + a_{01}h_y + a_{11}h_x h_y \tag{2.109}$$

with

$$a_{00} = f(x_i, y_i)$$

$$a_{10} = \frac{f(x_{i+1}, y_i) - f(x_i, y_i)}{x_{i+1} - x_i}$$

$$a_{01} = \frac{f(x_i, y_{i+1}) - f(x_i, y_i)}{y_{i+1} - y_i} \tag{2.110}$$

$$a_{11} = \frac{f(x_{i+1}, y_{i+1}) - f(x_i, y_{i+1}) - f(x_{i+1}, y_i) + f(x_i, y_i)}{(x_{i+1} - x_i)(y_{i+1} - y_i)}.$$

Application of higher order polynomials is straightforward. For image processing purposes bicubic interpolation is often used.

If high quality is needed more sophisticated interpolation methods can be applied. Consider for instance two-dimensional spline interpolation on a rectangular mesh of data to create a new data set with finer resolution[5]

$$f_{i,j} = f(ih_x, jh_y) \quad \text{with } 0 \le i < N_x \quad 0 \le j < N_y. \tag{2.111}$$

First perform spline interpolation in x-direction for each data row j to calculate new data sets

$$f_{i',j} = s(x_{i'}, f_{ij}, 0 \le i < N_x) \quad 0 \le j \le N_y \quad 0 \le i' < N_x' \tag{2.112}$$

and then interpolate in y direction to obtain the final high resolution data (Fig. 2.7)

$$f_{i',j'} = s(y_{j'}, f_{i'j}, 0 \le j < N_y) \quad 0 \le i' < N_x' \quad 0 \le j' < N_y'. \tag{2.113}$$

2.6 Problems

Problem 2.1 (Polynomial interpolation) This computer experiment interpolates a given set of n data points by

---

[5]A typical task of image processing.

Table 2.2 Zener diode voltage/current data

Voltage	−1.5	−1.0	−0.5	0.0
Current	−3.375	−1.0	−0.125	0.0

Table 2.3 Additional voltage/current data

Voltage	1.0	2.0	3.0	4.0	4.1	4.2	4.5
Current	0.0	0.0	0.0	0.0	1.0	3.0	10.0

- a polynomial

$$p(x) = \sum_{i=0}^{n} f_i \prod_{k=0, k \neq i}^{n} \frac{x - x_k}{x_i - x_k}, \tag{2.114}$$

- a linear spline which connects successive points by straight lines

$$s_i(x) = a_i + b_i(x - x_i) \quad \text{for } x_i \leq x \leq x_{i+1} \tag{2.115}$$

- a cubic spline with natural boundary conditions

$$s(x) = p_i(x) = \alpha_i + \beta_i(x - x_i) + \gamma_i(x - x_i)^2 + \delta_i(x - x_i)^3 \quad x_i \leq x \leq x_{i+1} \tag{2.116}$$

$$s''(x_n) = s''(x_0) = 0 \tag{2.117}$$

- a rational function without poles

$$R(x) = \frac{\sum_{i=0}^{n} f_i \frac{u_i}{x - x_i}}{\sum_{i=0}^{n} \frac{u_i}{x - x_i}} \tag{2.118}$$

with weights according to Berrut

$$u_k = (-1)^k \tag{2.119}$$

or Floater-Hormann

$$u_k = (-1)^{k-1} \left(\frac{1}{x_{k+1} - x_k} + \frac{1}{x_k - x_{k-1}} \right) \quad k = 1 \cdots n - 1 \tag{2.120}$$

$$u_0 = -\frac{1}{x_1 - x_0} \qquad u_n = (-1)^{n-1} \frac{1}{x_n - x_{n-1}}. \tag{2.121}$$

- Interpolate the data (Table 2.2) in the range $-1.5 < x < 0$.
- Now add some more sample points (Table 2.3) for $-1.5 < x < 4.5$.
- Interpolate the function $f(x) = \sin(x)$ at the points $x = 0, \frac{\pi}{2}, \pi, \frac{3\pi}{2}, 2\pi$. Take more sample points and check if the quality of the fit is improved.

Table 2.4 Pulse and step function data

x	-3	-2	-1	0	1	2	3
y_{pulse}	0	0	0	1	0	0	0
y_{step}	0	0	0	1	1	1	1

Table 2.5 Data set for two-dimensional interpolation

x	0	1	2	0	1	2	0	1	2
y	0	0	0	1	1	1	2	2	2
f	1	0	-1	0	0	0	-1	0	1

- Investigate the oscillatory behavior for a discontinuous pulse or step function as given by the data (Table 2.4).

Problem 2.2 (Two-dimensional interpolation) This computer experiment uses bilinear interpolation or bicubic spline interpolation to interpolate the data (Table 2.5) on a finer grid $\Delta x = \Delta y = 0.1$.

Chapter 3
Numerical Differentiation

For more complex problems analytical derivatives are not always available and have to be approximated by numerical methods. Numerical differentiation is also very important for the discretization of differential equations (Sect. 11.2). The simplest approximation uses a forward difference quotient (Fig. 3.1) and is not very accurate. A symmetric difference quotient improves the quality. Even higher precision is obtained with the extrapolation method. Approximations to higher order derivatives can be obtained systematically with the help of polynomial interpolation.

3.1 One-Sided Difference Quotient

The simplest approximation of a derivative is the ordinary difference quotient which can be taken forward

$$\frac{\mathrm{d}f}{\mathrm{d}x}(x) \approx \frac{\Delta f}{\Delta x} = \frac{f(x+h) - f(x)}{h} \tag{3.1}$$

or backward

$$\frac{\mathrm{d}f}{\mathrm{d}x}(x) \approx \frac{\Delta f}{\Delta x} = \frac{f(x) - f(x-h)}{h}. \tag{3.2}$$

Its truncation error can be estimated from the Taylor series expansion

$$\frac{f(x+h) - f(x)}{h} = \frac{f(x) + hf'(x) + \frac{h^2}{2}f''(x) + \cdots - f(x)}{h}$$

$$= f'(x) + \frac{h}{2}f''(x) + \cdots. \tag{3.3}$$

The error order is $O(h)$. The step width should not be too small to avoid rounding errors. Error analysis gives

$$\widetilde{\Delta f} = fl_-\big(f(x+h)(1 + \varepsilon_1), f(x)(1 + \varepsilon_2)\big)$$

$$= \big(\Delta f + f(x+h)\varepsilon_1 - f(x)\varepsilon_2\big)(1 + \varepsilon_3)$$

$$= \Delta f + \Delta f \varepsilon_3 + f(x+h)\varepsilon_1 - f(x)\varepsilon_2 + \cdots \tag{3.4}$$

P.O.J. Scherer, *Computational Physics*, Graduate Texts in Physics,
DOI 10.1007/978-3-319-00401-3_3,
© Springer International Publishing Switzerland 2013

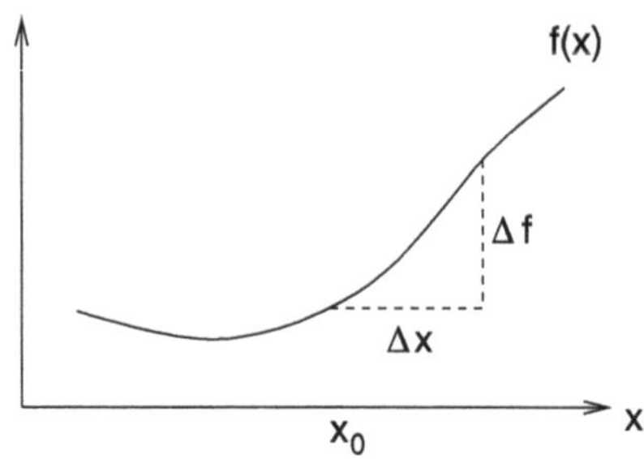

Fig. 3.1 (Numerical differentiation) Numerical differentiation approximates the differential quotient by a difference quotient $\frac{df}{dx} \approx \frac{\Delta f}{\Delta x}$. However, approximation by a simple forward difference $\frac{df}{dx}(x_0) \approx \frac{f(x_0+h)-f(x_0)}{h}$, is not very accurate

$$fl_{\div}\left(\widetilde{\Delta f}, h(1+\varepsilon_4)\right) = \frac{\Delta f + \Delta f \varepsilon_3 + f(x+h)\varepsilon_1 - f(x)\varepsilon_2}{h(1+\varepsilon_4)}(1+\varepsilon_5)$$

$$= \frac{\Delta f}{h}(1 + \varepsilon_5 - \varepsilon_4 + \varepsilon_3) + \frac{f(x+h)}{h}\varepsilon_1$$

$$- \frac{f(x)}{h}\varepsilon_2. \tag{3.5}$$

The errors are uncorrelated and the relative error of the result can be estimated by

$$\frac{\left|\frac{\widetilde{\Delta f}}{\Delta x} - \frac{\Delta f}{\Delta x}\right|}{\frac{\Delta f}{\Delta x}} \leq 3\varepsilon_M + \left|\frac{f(x)}{\frac{\Delta f}{\Delta x}}\right| 2\frac{\varepsilon_M}{h}. \tag{3.6}$$

Numerical extinction produces large relative errors for small step width h. The optimal value of h gives comparable errors from rounding and truncation. It can be found from

$$\frac{h}{2}\left|f''(x)\right| = \left|f(x)\right|\frac{2\varepsilon_M}{h}. \tag{3.7}$$

Assuming that the magnitude of the function and the derivative are comparable, we have the rule of thumb

$$h_o = \sqrt{\varepsilon_M} \approx 10^{-8}$$

(double precision). The corresponding relative error is of the same order.

3.2 Central Difference Quotient

Accuracy is much higher if a symmetric central difference quotient is used (Fig. 3.2):

$$\frac{\Delta f}{\Delta x} = \frac{f\left(x + \frac{h}{2}\right) - f\left(x - \frac{h}{2}\right)}{h}$$

$$= \frac{f(x) + \frac{h}{2}f'(x) + \frac{h^2}{8}f''(x) + \cdots - \left(f(x) - \frac{h}{2}f'(x) + \frac{h^2}{8}f''(x) + \cdots\right)}{h}$$

$$= f'(x) + \frac{h^2}{24}f'''(x) + \cdots. \tag{3.8}$$

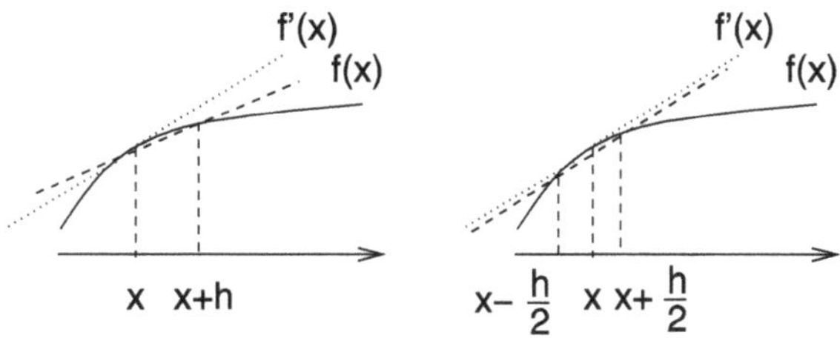

Fig. 3.2 (Difference quotient) The central difference quotient (*right side*) approximates the derivative (*dotted*) much more accurately than the one-sided difference quotient (*left side*)

The error order is $O(h^2)$. The optimal step width is estimated from

$$\frac{h^2}{24}\left|f'''(x)\right| = \left|f(x)\right|\frac{2\varepsilon_M}{h} \tag{3.9}$$

again with the assumption that function and derivatives are of similar magnitude as

$$h_0 = \sqrt[3]{48\varepsilon_M} \approx 10^{-5}. \tag{3.10}$$

The relative error has to be expected in the order of $\frac{h_0^2}{24} \approx 10^{-11}$.

3.3 Extrapolation Methods

The Taylor series of the symmetric difference quotient contains only even powers of h:

$$D(h) = \frac{f(x+h) - f(x-h)}{2h} = f'(x) + \frac{h^2}{3!}f'''(x) + \frac{h^4}{5!}f^{(5)}(x) + \cdots. \tag{3.11}$$

The Extrapolation method [217] uses a series of step widths, e.g.

$$h_{i+1} = \frac{h_i}{2} \tag{3.12}$$

and calculates an estimate of $D(0)$ by polynomial interpolation (Fig. 3.3). Consider $D_0 = D(h_0)$ and $D_1 = D(\frac{h_0}{2})$. The polynomial of degree 1 (with respect to h^2) $p(h) = a + bh^2$ can be found by the Lagrange method

$$p(h) = D_0 \frac{h^2 - \frac{h_0^2}{4}}{h_0^2 - \frac{h_0^2}{4}} + D_1 \frac{h^2 - h_0^2}{\frac{h_0^2}{4} - h_0^2}. \tag{3.13}$$

Extrapolation for $h = 0$ gives

$$p(0) = -\frac{1}{3}D_0 + \frac{4}{3}D_1. \tag{3.14}$$

Taylor series expansion shows

$$p(0) = -\frac{1}{3}\left(f'(x) + \frac{h_0^2}{3!}f'''(x) + \frac{h_0^4}{5!}f^{(5)}(x) + \cdots\right)$$

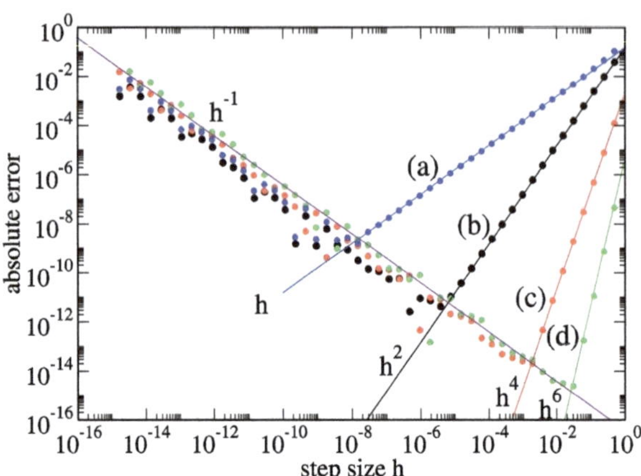

Fig. 3.3 (Numerical differentiation) The derivative $\frac{d}{dx}\sin(x)$ is calculated numerically using algorithms with increasing error order (3.1, 3.8, 3.14, 3.18). For very small step sizes the error increases as h^{-1} due to rounding errors

$$+ \frac{4}{3}\left(f'(x) + \frac{h_0^2}{4\cdot 3!}f'''(x) + \frac{h_0^4}{16\cdot 5!}f^{(5)}(x) + \cdots \right) \qquad (3.15)$$

$$= f'(x) - \frac{1}{4}\frac{h_0^4}{5!}f^{(5)}(x) + \cdots \qquad (3.16)$$

that the error order is $O(h_0^4)$. For 3 step widths $h_0 = 2h_1 = 4h_2$ we obtain the polynomial of second order (in h^2)

$$p(h) = D_0 \frac{(h^2 - \frac{h_0^2}{4})(h^2 - \frac{h_0^2}{16})}{(h_0^2 - \frac{h_0^2}{4})(h_0^2 - \frac{h_0^2}{16})} + D_1 \frac{(h^2 - h_0^2)(h^2 - \frac{h_0^2}{16})}{(\frac{h_0^2}{4} - h_0^2)(\frac{h_0^2}{4} - \frac{h_0^2}{16})}$$

$$+ D_2 \frac{(h^2 - h_0^2)(h^2 - \frac{h_0^2}{4})}{(\frac{h_0^2}{16} - h_0^2)(\frac{h_0^2}{16} - \frac{h_0^2}{4})} \qquad (3.17)$$

and the improved expression

$$p(0) = D_0 \frac{\frac{1}{64}}{\frac{3}{4}\cdot\frac{15}{16}} + D_1 \frac{\frac{1}{16}}{\frac{-3}{4}\cdot\frac{3}{16}} + D_2 \frac{\frac{1}{4}}{\frac{-15}{16}\cdot\frac{-3}{16}}$$

$$= \frac{1}{45}D_0 - \frac{4}{9}D_1 + \frac{64}{45}D_2 = f'(x) + O(h_0^6). \qquad (3.18)$$

Often used is the following series of step widths:

$$h_i^2 = \frac{h_0^2}{2^i}. \qquad (3.19)$$

The Neville method

$$P_{i\ldots k}(h^2) = \frac{(h^2 - \frac{h_0^2}{2^i})P_{i+1\ldots k}(h^2) - (h^2 - \frac{h_0^2}{2^k})P_{i\ldots k-1}(h^2)}{\frac{h_0^2}{2^k} - \frac{h_0^2}{2^i}} \qquad (3.20)$$

gives for $h = 0$

$$P_{i\cdots k} = \frac{P_{i\cdots k-1} - 2^{k-i} P_{i+1\cdots k}}{1 - 2^{k-i}} \tag{3.21}$$

which can be written as

$$P_{i\cdots k} = P_{i+1\cdots k} + \frac{P_{i\cdots k-1} - P_{i+1\cdots k}}{1 - 2^{k-i}} \tag{3.22}$$

and can be calculated according to the following scheme:

$$
\begin{aligned}
P_0 &= D\left(h^2\right) \quad P_{01} \;\; P_{012} \;\; P_{0123} \\[2mm]
P_1 &= D\left(\frac{h^2}{2}\right) P_{12} \;\; P_{123} \\[2mm]
P_2 &= D\left(\frac{h^2}{4}\right) P_{23} \\[2mm]
&\;\; \vdots \qquad\qquad \vdots \;\; \vdots \;\; \ddots
\end{aligned}
\tag{3.23}
$$

Here the values of the polynomials are arranged in matrix form

$$P_{i\cdots k} = T_{i,k-i} = T_{i,j} \tag{3.24}$$

with the recursion formula

$$T_{i,j} = T_{i+1,j-1} + \frac{T_{i,j-1} - T_{i+1,j}}{1 - 2^j}. \tag{3.25}$$

3.4 Higher Derivatives

Difference quotients for higher derivatives can be obtained systematically using polynomial interpolation. Consider equidistant points

$$x_n = x_0 + nh = \cdots = x_0 - 2h, x_0 - h, x_0, x_0 + h, x_0 + 2h, \ldots . \tag{3.26}$$

From the second order polynomial

$$
\begin{aligned}
p(x) &= y_{-1}\frac{(x - x_0)(x - x_1)}{(x_{-1} - x_0)(x_{-1} - x_1)} + y_0\frac{(x - x_{-1})(x - x_1)}{(x_0 - x_{-1})(x_0 - x_1)} \\[2mm]
&\quad + y_1\frac{(x - x_{-1})(x - x_0)}{(x_1 - x_{-1})(x_1 - x_0)} \\[2mm]
&= y_{-1}\frac{(x - x_0)(x - x_1)}{2h^2} + y_0\frac{(x - x_{-1})(x - x_1)}{-h^2} \\[2mm]
&\quad + y_1\frac{(x - x_{-1})(x - x_0)}{2h^2}
\end{aligned}
\tag{3.27}
$$

we calculate the derivatives

$$p'(x) = y_{-1}\frac{2x - x_0 - x_1}{2h^2} + y_0\frac{2x - x_{-1} - x_1}{-h^2} + y_1\frac{2x - x_{-1} - x_0}{2h^2} \tag{3.28}$$

$$p''(x) = \frac{y_{-1}}{h^2} - 2\frac{y_0}{h^2} + \frac{y_1}{h^2} \tag{3.29}$$

which are evaluated at x_0:

$$f'(x_0) \approx p'(x_0) = -\frac{1}{2h}y_{-1} + \frac{1}{2h}y_1 = \frac{f(x_0+h) - f(x_0-h)}{2h} \tag{3.30}$$

$$f''(x_0) \approx p''(x_0) = \frac{f(x_0-h) - 2f(x_0) + f(x_0+h)}{h^2}. \tag{3.31}$$

Higher order polynomials can be evaluated with an algebra program. For five sample points

$$x_0 - 2h, x_0 - h, x_0, x_0 + h, x_0 + 2h$$

we find

$$f'(x_0) \approx \frac{f(x_0-2h) - 8f(x_0-h) + 8f(x_0+h) - f(x_0+2h)}{12h} \tag{3.32}$$

$$f''(x_0) \approx \frac{-f(x_0-2h) + 16f(x_0-h) - 30f(x_0) + 16f(x_0+h) - f(x_0+2h)}{12h^2} \tag{3.33}$$

$$f'''(x_0) \approx \frac{-f(x_0-2h) + 2f(x_0-h) - 2f(x_0+h) + f(x_0+2h)}{2h^3} \tag{3.34}$$

$$f^{(4)}(x_0) \approx \frac{f(x_0-2h) - 4f(x_0-h) + 6f(x_0+h) - 4f(x_0+h) + f(x_0+2h)}{h^4}.$$

3.5 Partial Derivatives of Multivariate Functions

Consider polynomials of more than one variable. In two dimensions we use the Lagrange polynomials

$$L_{i,j}(x, y) = \prod_{k \neq i} \frac{(x - x_k)}{(x_i - x_k)} \prod_{j \neq l} \frac{(y - y_l)}{(y_j - y_l)}. \tag{3.35}$$

The interpolating polynomial is

$$p(x, y) = \sum_{i,j} f_{i,j} L_{i,j}(x, y). \tag{3.36}$$

For the nine sample points

$$
\begin{array}{ccc}
(x_{-1}, y_1) & (x_0, y_1) & (x_1, y_1) \\
(x_{-1}, y_0) & (x_0, y_0) & (x_1, y_0) \\
(x_{-1}, y_{-1}) & (x_0, y_{-1}) & (x_1, y_{-1})
\end{array}
\tag{3.37}
$$

we obtain the polynomial

$$p(x, y) = f_{-1,-1}\frac{(x - x_0)(x - x_1)(y - y_0)(y - y_1)}{(x_{-1} - x_0)(x_{-1} - x_1)(y_{-1} - y_0)(y_{-1} - y_1)} + \cdots \tag{3.38}$$

which gives an approximation to the gradient

$$\operatorname{grad} f(x_0 y_0) \approx \operatorname{grad} p(x_0 y_0) = \begin{pmatrix} \frac{f(x_0+h,y_0)-f(x_0-h,y_0)}{2h} \\ \frac{f(x_0,y_0+h)-f(x_0,y_0-h)}{2h} \end{pmatrix}, \qquad (3.39)$$

the Laplace operator

$$\left(\frac{\partial^2}{\partial x^2} + \frac{\partial^2}{\partial y^2} \right) f(x_0, y_0) \approx \left(\frac{\partial^2}{\partial x^2} + \frac{\partial^2}{\partial y^2} \right) p(x_0, y_0)$$

$$= \frac{1}{h^2} \big(f(x_0, y_0 + h) + f(x_0, y_0 - h)$$

$$+ f(x_0, y_0 + h) + f(x_0, y_0 - h) - 4f(x_0, y_0) \big) \tag{3.40}$$

and the mixed second derivative

$$\frac{\partial^2}{\partial x \partial y} f(x_0, y_0) \approx \frac{\partial^2}{\partial x \partial y} p(x_0, y_0)$$

$$= \frac{1}{4h^2} \big(f(x_0 + h, y_0 + h) + f(x_0 - h, y_0 - h)$$

$$- f(x_0 - h, y_0 + h) - f(x_0 + h, y_0 - h) \big). \tag{3.41}$$

3.6 Problems

Problem 3.1 (Numerical differentiation) In this computer experiment we calculate the derivative of $f(x) = \sin(x)$ numerically with

- the single sided difference quotient

$$\frac{\mathrm{d}f}{\mathrm{d}x} \approx \frac{f(x+h)-f(x)}{h}, \tag{3.42}$$

- the symmetrical difference quotient

$$\frac{\mathrm{d}f}{\mathrm{d}x} \approx D_h f(x) = \frac{f(x+h)-f(x-h)}{2h}, \tag{3.43}$$

- higher order approximations which can be derived using the extrapolation method

$$-\frac{1}{3} D_h f(x) + \frac{4}{3} D_{h/2} f(x) \tag{3.44}$$

$$\frac{1}{45} D_h f(x) - \frac{4}{9} D_{h/2} f(x) + \frac{64}{45} D_{h/4} f(x). \tag{3.45}$$

The error of the numerical approximation is shown on a log-log plot as a function of the step width h.

Chapter 4
Numerical Integration

Physical simulations often involve the calculation of definite integrals over complicated functions, for instance the Coulomb interaction between two electrons. Integration is also the elementary step in solving equations of motion.

An integral over a finite interval $[a, b]$ can always be transformed into an integral over $[0, 1]$ or $[-1, 1]$

$$\int_a^b f(x)dx = \int_0^1 f\big(a + (b - a)t\big)(b - a)dt$$

$$= \int_{-1}^1 f\left(\frac{a+b}{2} + \frac{b-a}{2}t\right)\frac{b-a}{2}dt. \tag{4.1}$$

An Integral over an infinite interval may have to be transformed into an integral over a finite interval by substitution of the integration variable, for example

$$\int_0^\infty f(x)dx = \int_0^1 f\left(\frac{t}{1-t}\right)\frac{dt}{(1-t)^2} \tag{4.2}$$

$$\int_{-\infty}^\infty f(x)dx = \int_{-1}^1 f\left(\frac{t}{1-t^2}\right)\frac{t^2+1}{(t^2-1)^2}dt. \tag{4.3}$$

In general a definite integral can be approximated numerically as the weighted average over a finite number of function values

$$\int_a^b f(x)dx \approx \sum_{x_i} w_i f(x_i). \tag{4.4}$$

Specific sets of quadrature points x_i and quadrature weights w_i are known as "integral rules". Newton-Cotes rules like the trapezoidal rule, the midpoint rule or Simpson's rule, use equidistant points x_i and are easy to apply. Accuracy can be improved by dividing the integration range into sub-intervals and applying composite Newton-Cotes rules. Extrapolation methods reduce the error almost to machine precision but need many function evaluations. Equidistant sample points are convenient but not the best choice. Clenshaw-Curtis expressions use non uniform sample points and

P.O.J. Scherer, *Computational Physics*, Graduate Texts in Physics,
DOI 10.1007/978-3-319-00401-3_4,

a rapidly converging Chebyshev expansion. Gaussian integration fully optimizes the sample points with the help of orthogonal polynomials.

4.1 Equidistant Sample Points

For equidistant points

$$x_i = a + ih \quad i = 0 \cdots N \quad h = \frac{b-a}{N} \tag{4.5}$$

the interpolating polynomial of order N with $p(x_i) = f(x_i)$ is given by the Lagrange method

$$p(x) = \sum_{i=0}^{N} f_i \prod_{k=0,k\neq i}^{N} \frac{x - x_k}{x_i - x_k}. \tag{4.6}$$

Integration of the polynomial gives

$$\int_a^b p(x)dx = \sum_{i=0}^{N} f_i \int_a^b \prod_{k=0,k\neq i}^{N} \frac{x - x_k}{x_i - x_k}dx. \tag{4.7}$$

After substituting

$$x = a + hs$$
$$x - x_k = h(s - k) \tag{4.8}$$
$$x_i - x_k = (i - k)h$$

we have

$$\int_a^b \prod_{k=0,k\neq i}^{N} \frac{x - x_k}{x_i - x_k}dx = \int_0^N \prod_{k=0,k\neq i}^{N} \frac{s - k}{i - k}hds = h\alpha_i \tag{4.9}$$

and hence

$$\int_a^b p(x)dx = (b - a) \sum_{i=0}^{N} f_i\alpha_i. \tag{4.10}$$

The weight factors are given by

$$w_i = (b - a)\alpha_i = Nh\alpha_i. \tag{4.11}$$

4.1.1 Closed Newton-Cotes Formulae

For $N = 1$ the polynomial is

$$p(x) = f_0 \frac{x - x_1}{x_0 - x_1} + f_1 \frac{x - x_0}{x_1 - x_0} \tag{4.12}$$

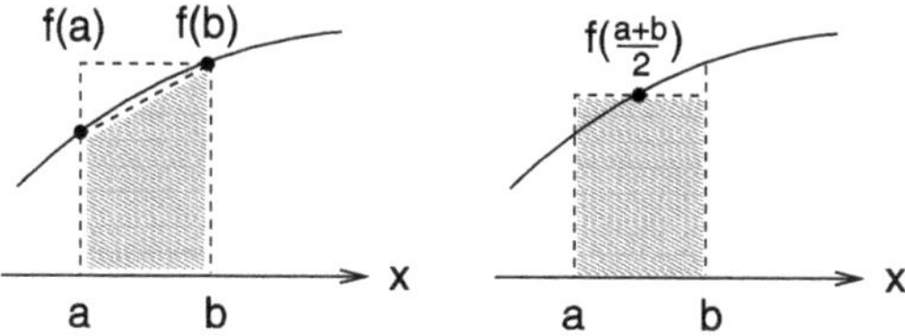

Fig. 4.1 (Trapezoidal rule and midpoint rule) The trapezoidal rule (*left*) approximates the integral by the average of the function values at the boundaries. The midpoint rule (*right*) evaluates the function in the center of the interval and has the same error order

and the integral is

$$\int_a^b p(x)dx = f_0 \int_0^1 \frac{s-1}{0-1} h\, ds + f_1 \int_0^1 \frac{s-0}{1-0} h\, ds$$

$$= -f_0 h \left(\frac{(1-1)^2}{2} - \frac{(0-1)^2}{2} \right) + f_1 h \left(\frac{1^2}{2} - \frac{0^2}{2} \right)$$

$$= h \frac{f_0 + f_1}{2} \tag{4.13}$$

which is known as the trapezoidal rule (Fig. 4.1). $N = 2$ gives Simpson's rule

$$2h \frac{f_0 + 4f_1 + f_2}{6}. \tag{4.14}$$

Larger N give further integration rules

$$3h \frac{f_0 + 3f_1 + 3f_2 + f_3}{8} \qquad\qquad \text{3/8-rule}$$

$$4h \frac{7f_0 + 32f_1 + 12f_2 + 32f_3 + 7f_4}{90} \qquad\qquad \text{Milne-rule}$$

$$5h \frac{19f_0 + 75f_1 + 50f_2 + 50f_3 + 75f_4 + 19f_5}{288}$$

$$6h \frac{41f_0 + 216f_1 + 27f_2 + 272f_3 + 27f_4 + 216f_5 + 41f_6}{840} \qquad \text{Weddle-rule.} \tag{4.15}$$

For even larger N negative weight factors appear and the formulas are not numerically stable.

4.1.2 Open Newton-Cotes Formulae

Alternatively, the integral can be computed from only interior points

$$x_i = a + ih \quad i = 1, 2 \cdots N \quad h = \frac{b-a}{N+1}. \tag{4.16}$$

The simplest case is the midpoint rule (Fig. 4.1)

$$\int_a^b f(x)dx \approx 2hf_1 = (b-a)f\left(\frac{a+b}{2}\right). \tag{4.17}$$

The next two are

$$\frac{3h}{2}(f_1 + f_2) \tag{4.18}$$

$$\frac{4h}{3}(2f_1 - f_2 + 2f_3). \tag{4.19}$$

4.1.3 Composite Newton-Cotes Rules

Newton-Cotes formulae are only accurate, if the step width is small. Usually the integration range is divided into small sub-intervals

$$[x_i, x_{i+1}] \quad x_i = a + ih \quad i = 0 \cdots N \tag{4.20}$$

for which a simple quadrature formula can be used. Application of the trapezoidal rule for each interval

$$I_i = \frac{h}{2}\big(f(x_i) + f(x_{i+1})\big) \tag{4.21}$$

gives the composite trapezoidal rule

$$T = h\left(\frac{f(a)}{2} + f(a+h) + \cdots + f(b-h) + \frac{f(b)}{2}\right) \tag{4.22}$$

with error order $O(h^2)$. Repeated application of Simpson's rule for $[a, a+2h]$, $[a+2h, a+4h]$, $\cdots$ gives the composite Simpson's rule

$$S = \frac{h}{3}\big(f(a) + 4f(a+h) + 2f(a+2h) + 4f(a+3h) + \cdots$$
$$+ 2f(b-2h) + 4f(b-h) + f(b)\big) \tag{4.23}$$

with error order $O(h^4)$.[1]

Repeated application of the midpoint rule gives the composite midpoint rule

$$M = 2h\big(f(a+h) + f(a+3h) + \cdots + f(b-h)\big) \tag{4.24}$$

with error order $O(h^2)$.

---

[1] The number of sample points must be even.

4.1.4 Extrapolation Method (Romberg Integration)

For the trapezoidal rule the Euler-McLaurin expansion exists which for a $2m$ times differentiable function has the form

$$\int_{x_0}^{x_N} f(x)dx - T = \alpha_2 h^2 + \alpha_4 h^4 + \cdots + \alpha_{2m-2} h^{2m-2} + O(h^{2m}). \qquad (4.25)$$

Therefore extrapolation methods are applicable. From the composite trapezoidal rule for h and $h/2$ an approximation of error order $O(h^4)$ results:

$$\int_{x_0}^{x_N} f(x)dx - T(h) = \alpha_2 h^2 + \alpha_4 h^4 + \cdots \qquad (4.26)$$

$$\int_{x_0}^{x_N} f(x)dx - T(h/2) = \alpha_2 \frac{h^2}{4} + \alpha_4 \frac{h^4}{16} + \cdots \qquad (4.27)$$

$$\int_{x_0}^{x_N} f(x)dx - \frac{4T(h/2) - T(h)}{3} = -\alpha_4 \frac{h^4}{4} + \cdots . \qquad (4.28)$$

More generally, for the series of step widths

$$h_k = \frac{h_0}{2^k} \qquad (4.29)$$

the Neville method gives the recursion for the interpolating polynomial

$$P_{i\cdots k}\left(h^2\right) = \frac{(h^2 - \frac{h_0^2}{2^{2i}}) P_{i+1\cdots k}(h^2) - (h^2 - \frac{h_0^2}{2^{2k}}) P_{i\cdots k-1}(h^2)}{\frac{h_0^2}{2^{2k}} - \frac{h_0^2}{2^{2i}}} . \qquad (4.30)$$

which for $h = 0$ becomes the higher order approximation to the integral (Fig. 4.2)

$$P_{i\cdots k} = \frac{2^{-2k} P_{i\cdots k-1} - 2^{-2i} P_{i+1\cdots k}}{2^{-2k} - 2^{-2i}} = \frac{P_{i\cdots k-1} - 2^{2k-2i} P_{i+1\cdots k}}{1 - 2^{2k-2i}}$$

$$= P_{i+1\cdots k} + \frac{P_{i\cdots k-1} - P_{i+1\cdots k}}{1 - 2^{2k-2i}} . \qquad (4.31)$$

The polynomial values can again be arranged in matrix form

$$\begin{matrix} P_0 & P_{01} & P_{012} & \cdots \\ P_1 & P_{12} & & \\ P_2 & & & \\ \vdots & & & \end{matrix} \qquad (4.32)$$

with

$$T_{i,j} = P_{i\cdots i+j} \qquad (4.33)$$

and the recursion formula

$$T_{i,0} = P_i = T_s\left(\frac{h_0}{2^i}\right) \qquad (4.34)$$

$$T_{i,j} = T_{i+1,j-1} + \frac{T_{i,j-1} - T_{i+1,j-1}}{1 - 2^{2j}} . \qquad (4.35)$$

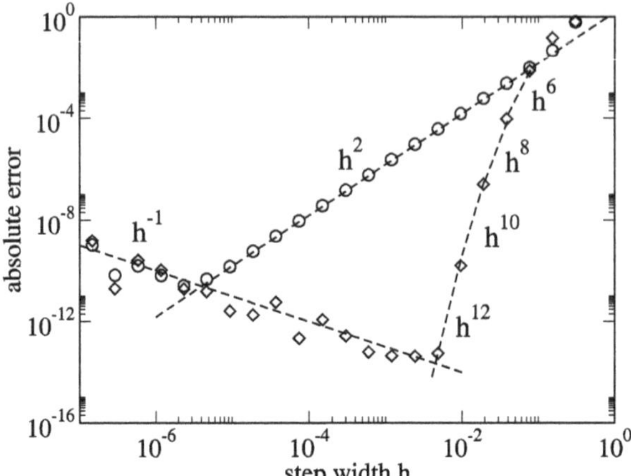

Fig. 4.2 (Romberg integration) The integral $\int_0^{\pi^2} \sin(x^2)dx$ is calculated numerically. *Circles* show the absolute error of the composite trapezoidal rule (4.22) for the step size sequence $h_{i+1} = h_i/2$. *Diamonds* show the absolute error of the extrapolated value (4.31). The error order of the trapezoidal rule is $O(h^2)$ whereas the error order of the Romberg method increases by factors of h^2. For very small step sizes the rounding errors dominate which increase as h^{-1}

4.2 Optimized Sample Points

The Newton-Cotes method integrates polynomials of order up to $N-1$ exactly, using N equidistant sample points. Unfortunately the polynomial approximation converges slowly, at least for not so well behaved integrands. The accuracy of the integration can be improved by optimizing the sample point positions. Gaussian quadrature determines the N positions and N weights such, that a polynomial of order $2N-1$ is integrated exactly. The Clenshaw-Curtis and the related Fejer methods use the roots or the extrema of the Chebyshev polynomials as nodes and determine the weights to integrate polynomials of order N. However, since the approximation by Chebyshev polynomials usually converges very fast, the accuracy is in many cases comparable to the Gaussian method [185, 253]. In the following we restrict the integration interval to $[-1, 1]$. The general case $[a, b]$ is then given by a simple change of variables.

4.2.1 Clenshaw-Curtis Expressions

Clenshaw and Curtis [60] make the variable substitution

$$x = \cos\theta \quad dx = -\sin\theta d\theta \tag{4.36}$$

for the integral

$$\int_{-1}^{1} f(x)dx = \int_0^{\pi} f(\cos t) \sin t\, dt \tag{4.37}$$

and approximate the function by the trigonometric polynomial ((7.19) with $N = 2M, T = 2\pi$)

$$f(\cos t) = \frac{1}{2M}c_0 + \frac{1}{M}\sum_{j=1}^{M-1} c_j \cos(jt) + \frac{1}{2M}c_M \cos(Mt) \tag{4.38}$$

which interpolates (Sect. 7.2.1) $f(\cos t)$ at the sample points

$$t_n = n\Delta t = n\frac{\pi}{M} \quad \text{with } n = 0, 1 \cdots M \tag{4.39}$$

$$x_n = \cos t_n = \cos\left(n\frac{\pi}{M}\right) \tag{4.40}$$

and where the Fourier coefficients are given by (7.17)

$$c_j = f_0 + 2\sum_{n=1}^{M-1} f\big(\cos(t_n)\big)\cos\left(\frac{\pi}{M}jn\right) + f_M \cos(j\pi). \tag{4.41}$$

The function $\cos(jt)$ is related to the Chebyshev polynomials of the first kind which for $-1 \le x \le 1$ are given by the trigonometric definition

$$T_j(x) = \cos\big(j\arccos(x)\big) \tag{4.42}$$

and can be calculated recursively

$$T_0(x) = 1 \tag{4.43}$$

$$T_1(x) = x \tag{4.44}$$

$$T_{j+1}(x) = 2xT_j(x) - T_{j-1}(x). \tag{4.45}$$

Substituting $x = \cos t$ we find

$$T_j(\cos t) = \cos(jt). \tag{4.46}$$

Hence the Fourier series (4.38) corresponds to a Chebyshev approximation

$$f(x) = \sum_{j=0}^{M} a_j T_j(x) = \frac{c_0}{2M}T_0(x) + \sum_{j=1}^{M-1}\frac{c_j}{M}T_j(x) + \frac{c_M}{2M}T_M(x) \tag{4.47}$$

and can be used to approximate the integral

$$\int_{-1}^{1} f(x)dx \approx \int_{0}^{\pi}\left\{\frac{1}{2M}c_0 + \frac{1}{M}\sum_{j=1}^{M-1} c_j \cos(jt) + \frac{1}{2M}c_M \cos(Mt)\right\}\sin\theta\, d\theta \tag{4.48}$$

$$= \frac{1}{M}c_0 + \frac{1}{M}\sum_{j=1}^{M-1} c_j\frac{\cos(j\pi)+1}{1-j^2} + \frac{1}{2M}c_M\frac{\cos(M\pi)+1}{1-M^2} \tag{4.49}$$

where, in fact, only the even j contribute.

Example (Clenshaw Curtis quadrature for $M = 5$) The function has to be evaluated at the sample points $x_k = \cos(\frac{\pi}{5}k) = (1, 0.80902, 0.30902, -0.30902, -0.80902, -1)$. The Fourier coefficients are given by

$$
\begin{pmatrix} c_0 \\ c_1 \\ c_2 \\ c_3 \\ c_4 \\ c_5 \end{pmatrix} = \begin{pmatrix} 1 & 2 & 2 & 2 & 2 & 1 \\ 1 & 1.618 & 0.618 & -0.618 & -1.618 & -1 \\ 1 & 0.618 & -1.618 & -1.618 & 0.618 & 1 \\ 1 & -0.618 & -1.618 & 1.618 & 0.618 & -1 \\ 1 & -1.618 & 0.618 & 0.618 & -1.618 & 1 \\ 1 & -2 & 2 & -2 & 2 & -1 \end{pmatrix} \begin{pmatrix} f_0 \\ f_1 \\ f_2 \\ f_3 \\ f_4 \\ f_5 \end{pmatrix} \tag{4.50}
$$

and the integral is approximately

$$
\int_{-1}^{1} f(x)dx \approx \begin{pmatrix} \dfrac{1}{5} & 0 & -\dfrac{2}{15} & 0 & -\dfrac{2}{75} & 0 \end{pmatrix} \begin{pmatrix} c_0 \\ c_1 \\ c_2 \\ c_3 \\ c_4 \\ c_5 \end{pmatrix}
$$

$$
= 0.0400 f_0 + 0.3607 f_1 + 0.5993 f_2 + 0.5993 f_3
$$

$$
+ 0.3607 f_4 + 0.0400 f_5. \tag{4.51}
$$

Clenshaw Curtis weights of very high order can be calculated efficiently [240, 267] using the FFT algorithm (fast Fourier transformation, Sect. 7.3.2)

4.2.2 Gaussian Integration

Now we will optimize the positions of the N quadrature points x_i to obtain the maximum possible accuracy. We approximate the integral by a sum

$$
\int_a^b f(x)dx \approx \sum_{i=1}^{N} f(x_i)w_i \tag{4.52}
$$

and determine the $2N$ parameters x_i and w_i such that a polynomial of order $2N - 1$ is integrated exactly. This can be achieved with the help of a set of polynomials which are orthogonal with respect to the scalar product

$$
\langle fg \rangle = \int_a^b f(x)g(x)w(x)dx \tag{4.53}
$$

where the weight function $w(x)$ and the interval $[a, b]$ determine a particular set of orthogonal polynomials.

4.2.2.1 Gauss-Legendre Integration

Again we restrict the integration interval to $[-1, 1]$ in the following. For integrals with one or two infinite boundaries see Sect. 4.2.2.2. The simplest choice for the weight function is

$$w(x) = 1. \tag{4.54}$$

An orthogonal system of polynomials on the interval $[-1, 1]$ can be found using the Gram-Schmidt method:

$$P_0 = 1 \tag{4.55}$$

$$P_1 = x - \frac{P_0}{\langle P_0 P_0 \rangle} \int_{-1}^{1} x P_0(x) dx = x \tag{4.56}$$

$$P_2 = x^2 - \frac{P_1}{\langle P_1 P_1 \rangle} \int_{-1}^{1} x^2 P_1(x) dx - \frac{P_0}{\langle P_0 P_0 \rangle} \int_{-1}^{1} x^2 P_0(x) dx$$

$$= x^2 - \frac{1}{3} \tag{4.57}$$

$$P_n = x^n - \frac{P_{n-1}}{\langle P_{n-1} P_{n-1} \rangle} \int_{-1}^{1} x^n P_{n-1}(x) dx$$

$$- \frac{P_{n-2}}{\langle P_{n-2} P_{n-2} \rangle} \int_{-1}^{1} x^n P_{n-2}(x) dx - \cdots . \tag{4.58}$$

The P_n are known as Legendre-polynomials. Consider now a polynomial $p(x)$ of order $2N - 1$. It can be interpolated at the N quadrature points x_i using the Lagrange method by a polynomial $\widetilde{p}(x)$ of order $N - 1$:

$$\widetilde{p}(x) = \sum_{j=1}^{N} L_j(x) p(x_j). \tag{4.59}$$

Then $p(x)$ can be written as

$$p(x) = \widetilde{p}(x) + (x - x_1)(x - x_2) \cdots (x - x_N) q(x). \tag{4.60}$$

Obviously $q(x)$ is a polynomial of order $(2N - 1) - N = N - 1$. Now choose the positions x_i as the roots of the Legendre polynomial of order N

$$(x - x_1)(x - x_2) \cdots (x - x_N) = P_N(x). \tag{4.61}$$

Then we have

$$\int_{-1}^{1} (x - x_1)(x - x_2) \cdots (x - x_N) q(x) dx = 0 \tag{4.62}$$

since P_N is orthogonal to the polynomial of lower order. But now

$$\int_{-1}^{1} p(x) dx = \int_{-1}^{1} \widetilde{p}(x) dx = \int_{-1}^{1} \sum_{j=1}^{N} p(x_j) L_j(x) dx = \sum_{j=1}^{N} w_j p(x_j) \tag{4.63}$$

with the weight factors

$$w_j = \int_{-1}^{1} L_j(x)dx. \tag{4.64}$$

Example (Gauss-Legendre integration with 2 quadrature points) The second order Legendre polynomial

$$P_2(x) = x^2 - \frac{1}{3} \tag{4.65}$$

has two roots

$$x_{1,2} = \pm\sqrt{\frac{1}{3}}. \tag{4.66}$$

The Lagrange polynomials are

$$L_1 = \frac{x - \sqrt{\frac{1}{3}}}{-\sqrt{\frac{1}{3}} - \sqrt{\frac{1}{3}}} \qquad L_2 = \frac{x + \sqrt{\frac{1}{3}}}{\sqrt{\frac{1}{3}} + \sqrt{\frac{1}{3}}} \tag{4.67}$$

and the weights

$$w_1 = \int_{-1}^{1} L_1 dx = -\frac{\sqrt{3}}{2}\left(\frac{x^2}{2} - \sqrt{\frac{1}{3}}x\right)_{-1}^{1} = 1 \tag{4.68}$$

$$w_2 = \int_{-1}^{1} L_2 dx = \frac{\sqrt{3}}{2}\left(\frac{x^2}{2} + \sqrt{\frac{1}{3}}x\right)_{-1}^{1} = 1. \tag{4.69}$$

This gives the integral rule

$$\int_{-1}^{1} f(x)dx \approx f\left(-\sqrt{\frac{1}{3}}\right) + f\left(\sqrt{\frac{1}{3}}\right). \tag{4.70}$$

For a general integration interval we substitute

$$x = \frac{a+b}{2} + \frac{b-a}{2}u \tag{4.71}$$

and find the approximation

$$\begin{aligned}
\int_{a}^{b} f(x)dx &= \int_{-1}^{1} f\left(\frac{a+b}{2} + \frac{b-a}{2}u\right)\frac{b-a}{2}du \\
&\approx \frac{b-a}{2}\left(f\left(\frac{a+b}{2} - \frac{b-a}{2}\sqrt{\frac{1}{3}}\right)\right. \\
&\quad \left. + f\left(\frac{a+b}{2} + \frac{b-a}{2}\sqrt{\frac{1}{3}}\right)\right).
\end{aligned} \tag{4.72}$$

The next higher order Gaussian rule is given by

$$n = 3: \quad w_1 = w_3 = 5/9, \quad w_2 = 8/9, \quad x_3 = -x_1 = 0.77459\cdots, \qquad x_2 = 0. \tag{4.73}$$

4.2.2.2 Other Types of Gaussian Integration

Further integral rules can be obtained by using other sets of orthogonal polynomials, for instance

Chebyshev polynomials

$$w(x) = \frac{1}{\sqrt{1-x^2}} \tag{4.74}$$

$$\int_{-1}^{1} f(x)dx = \int_{-1}^{1} f(x)\sqrt{1-x^2}w(x)dx \tag{4.75}$$

$$T_{n+1}(x) = 2xT_n(x) - T_{n-1}(x) \tag{4.76}$$

$$x_k = \cos\left(\frac{2k-1}{2N}\pi\right) \quad w_k = \frac{\pi}{N}. \tag{4.77}$$

Hermite polynomials

$$w(x) = e^{-x^2} \tag{4.78}$$

$$\int_{-\infty}^{\infty} f(x)dx = \int_{-\infty}^{\infty} f(x)e^{x^2}w(x)dx \tag{4.79}$$

$$H_0(x) = 1, \quad H_1(x) = 2x, \quad H_{n+1}(x) = 2xH_n(x) - 2nH_{n-1}(x).$$

Laguerre polynomials

$$w(x) = e^{-x} \tag{4.80}$$

$$\int_{0}^{\infty} f(x)dx = \int_{0}^{\infty} f(x)e^{x}w(x)dx \tag{4.81}$$

$$L_0(x) = 1, \quad L_1(x) = 1-x,$$

$$L_{n+1}(x) = \frac{1}{n+1}\big((2n+1-x)L_n(x) - nL_{n-1}(x)\big). \tag{4.82}$$

4.2.2.3 Connection with an Eigenvalue Problem

The determination of quadrature points and weights can be formulated as an eigenvalue problem [109, 133]. Any set of orthogonal polynomials satisfies a three term recurrence relation

$$P_{n+1}(x) = (a_{n+1}x + b_{n+1})P_n(x) - c_{n+1}P_{n-1}(x) \tag{4.83}$$

with $a_n > 0, c_n > 0$ which can be written in matrix form [272]

$$x\begin{pmatrix} P_0(x) \\ P_1(x) \\ \\ \\ P_{N-1}(x) \end{pmatrix} = \begin{pmatrix} -\frac{b_1}{a_1} & \frac{1}{a_1} \\ \frac{c_2}{a_2} & -\frac{b_2}{a_2} & \frac{1}{a_2} \\ & \ddots & \ddots & \ddots \\ & & \ddots & \ddots & \ddots \\ & & & \frac{c_{N-1}}{a_{N-1}} & -\frac{b_{N-1}}{a_{N-1}} & \frac{1}{a_{N-1}} \\ & & & & \frac{c_N}{a_N} & -\frac{b_N}{a_N} \end{pmatrix} \begin{pmatrix} P_0(x) \\ P_1(x) \\ \\ \\ P_{N-1}(x) \end{pmatrix}$$

$$+ \begin{pmatrix} 0 \\ 0 \\ \vdots \\ \vdots \\ 0 \\ \frac{1}{a_N} P_N(x) \end{pmatrix} \tag{4.84}$$

or shorter

$$x\mathbf{P}(x) = T\mathbf{P}(x) + \frac{1}{a_N} P_N(x)\mathbf{e}_{N-1} \tag{4.85}$$

with a tridiagonal matrix T. Obviously $P_N(x) = 0$ if and only if

$$x_j\mathbf{P}(x_j) = T\mathbf{P}(x_j), \tag{4.86}$$

hence the roots of $P_N(x)$ are given by the eigenvalues of T. The matrix T is symmetric if the polynomials are orthonormal, otherwise it can be transformed into a symmetric tridiagonal matrix by an orthogonal transformation [272]. Finally the quadrature weight corresponding to the eigenvalue x_j can be calculated from the first component of the corresponding eigenvector [109].

4.3 Problems

Problem 4.1 (Romberg integration) Use the trapezoidal rule

$$T(h) = h\left(\frac{1}{2}f(a) + f(a+h) + \cdots + f(b-h) + \frac{1}{2}f(b)\right) = \int_a^b f(x)dx + \cdots \tag{4.87}$$

with the step sequence

$$h_i = \frac{h_0}{2^i} \tag{4.88}$$

and calculate the elements of the triangular matrix

$$T(i,0) = T(h_i) \tag{4.89}$$

$$T(i,k) = T(i+1,k-1) + \frac{T(i,k-1) - T(i+1,k-1)}{1 - \frac{h_i^2}{h_{i+k}^2}} \tag{4.90}$$

to obtain the approximations

$$T_{01} = P_{01}, \qquad T_{02} = P_{012}, \qquad T_{03} = P_{0123}, \qquad \cdots \qquad (4.91)$$

- calculate

$$\int_0^{\pi^2} \sin(x^2)\,dx = 0.6773089370468890331\cdots \qquad (4.92)$$

and compare the absolute error of the trapezoidal sums $T(h_i) = T_{i,0}$ and the extrapolated values $T_{0,i}$.

- calculate

$$\int_\varepsilon^1 \frac{dx}{\sqrt{x}} \qquad (4.93)$$

for $\varepsilon = 10^{-3}$. Compare with the composite midpoint rule

$$T(h) = h\left(f\left(a + \frac{h}{2}\right) + f\left(a + \frac{3h}{2}\right) + \cdots + f\left(b - \frac{3h}{2}\right) + f\left(b - \frac{h}{2}\right) \right).$$
$$(4.94)$$

Chapter 5
Systems of Inhomogeneous Linear Equations

Many problems in physics and especially computational physics involve systems of linear equations

$$a_{11}x_1 + \cdots + a_{1n}x_n = b_1$$

$$\vdots \qquad \vdots \qquad \qquad \vdots \tag{5.1}$$

$$a_{n1}x_1 + \cdots + a_{nn}x_n = b_n$$

or shortly in matrix form

$$A\mathbf{x} = \mathbf{b} \tag{5.2}$$

which arise e.g. from linearization of a general nonlinear problem like (Sect. 20.2)

$$0 = \begin{pmatrix} F_1(x_1 \cdots x_n) \\ \vdots \\ F_n(x_1 \cdots x_n) \end{pmatrix} = \begin{pmatrix} F_1(x_1^{(0)} \cdots x_n^{(0)}) \\ \vdots \\ F_n(x_1^{(0)} \cdots x_n^{(0)}) \end{pmatrix} + \begin{pmatrix} \frac{\partial F_1}{\partial x_1} & \cdots & \frac{\partial F_1}{\partial x_n} \\ \vdots & \ddots & \vdots \\ \frac{\partial F_n}{\partial x_1} & \cdots & \frac{\partial F_n}{\partial x_n} \end{pmatrix} \begin{pmatrix} x_1 - x_1^{(0)} \\ \vdots \\ x_n - x_n^{(0)} \end{pmatrix} + \cdots \tag{5.3}$$

or from discretization of differential equations like

$$0 = \frac{\partial f}{\partial x} - g(x) \rightarrow \begin{pmatrix} \vdots \\ \frac{f((j+1)\Delta x) - f(j\Delta x)}{\Delta x} - g(j\Delta x) \\ \vdots \end{pmatrix}$$

$$= \begin{pmatrix} \ddots & & & \\ & -\frac{1}{\Delta x} & \frac{1}{\Delta x} & \\ & & -\frac{1}{\Delta x} & \frac{1}{\Delta x} \\ & & & \ddots \end{pmatrix} \begin{pmatrix} \vdots \\ f_j \\ f_{j+1} \\ \vdots \end{pmatrix} - \begin{pmatrix} \vdots \\ g_j \\ g_{j+1} \\ \vdots \end{pmatrix}. \tag{5.4}$$

If the dimension of the system is not too large standard methods like Gaussian elimination or QR decomposition are sufficient. Systems with a tridiagonal matrix

P.O.J. Scherer, *Computational Physics*, Graduate Texts in Physics,
DOI 10.1007/978-3-319-00401-3_5,

are important for cubic spline interpolation and numerical second derivatives. They can be solved very efficiently with a specialized Gaussian elimination method. Practical applications often involve very large dimensions and require iterative methods. Convergence of Jacobi and Gauss-Seidel methods is slow and can be improved by relaxation or over-relaxation. An alternative for large systems is the method of conjugate gradients.

5.1 Gaussian Elimination Method

A series of linear combinations of the equations transforms the matrix A into an upper triangular matrix. Start with subtracting a_{i1}/a_{11} times the first row from rows $2 \cdots n$

$$\begin{pmatrix} \mathbf{a}_1^T \\ \mathbf{a}_2^T \\ \vdots \\ \mathbf{a}_n^T \end{pmatrix} \rightarrow \begin{pmatrix} \mathbf{a}_1^T \\ \mathbf{a}_2^T - l_{21}\mathbf{a}_1^T \\ \vdots \\ \mathbf{a}_n^T - l_{n1}\mathbf{a}_1^T \end{pmatrix} \tag{5.5}$$

which can be written as a multiplication

$$A^{(1)} = L_1 A \tag{5.6}$$

with the Frobenius matrix

$$L_1 = \begin{pmatrix} 1 & & & & \\ -l_{21} & 1 & & & \\ -l_{31} & & 1 & & \\ \vdots & & & \ddots & \\ -l_{n1} & & & & 1 \end{pmatrix} \qquad l_{i1} = \frac{a_{i1}}{a_{11}}. \tag{5.7}$$

The result has the form

$$A^{(1)} = \begin{pmatrix} a_{11} & a_{12} & \cdots & a_{1n-1} & a_{1n} \\ 0 & a_{22}^{(1)} & \cdots & a_{2n-1}^{(1)} & a_{2n}^{(1)} \\ 0 & a_{32}^{(1)} & \cdots & \cdots & a_{3n}^{(1)} \\ 0 & \vdots & & & \vdots \\ 0 & a_{n2}^{(1)} & \cdots & \cdots & a_{nn}^{(1)} \end{pmatrix}. \tag{5.8}$$

Now subtract $\frac{a_{i2}}{a_{22}}$ times the second row from rows $3 \cdots n$. This can be formulated as

$$A^{(2)} = L_2 A^{(1)} = L_2 L_1 A \tag{5.9}$$

with

$$L_2 = \begin{pmatrix} 1 & & & & \\ 0 & 1 & & & \\ 0 & -l_{32} & 1 & & \\ \vdots & \vdots & & \ddots & \\ 0 & -l_{n2} & & & 1 \end{pmatrix} \qquad l_{i2} = \frac{a_{i2}^{(1)}}{a_{22}^{(1)}}. \tag{5.10}$$

The result is

$$A^{(2)} = \begin{pmatrix} a_{11}^{(2)} & a_{12}^{(2)} & a_{13}^{(2)} & \cdots & a_{1n}^{(2)} \\ 0 & a_{22}^{(2)} & a_{23}^{(2)} & \cdots & a_{2n}^{(2)} \\ 0 & 0 & a_{33}^{(2)} & \cdots & a_{3n}^{(2)} \\ \vdots & \vdots & \vdots & & \vdots \\ 0 & 0 & a_{n3}^{(2)} & \cdots & a_{nn}^{(2)} \end{pmatrix}. \tag{5.11}$$

Continue until an upper triangular matrix results after $n-1$ steps:

$$A^{(n-1)} = L_{n-1} A^{(n-2)} \tag{5.12}$$

$$L_{n-1} = \begin{pmatrix} 1 & & & & \\ & 1 & & & \\ & & \ddots & & \\ & & & 1 & \\ & & & -l_{n,n-1} & 1 \end{pmatrix} \qquad l_{n,n-1} = \frac{a_{n,n-1}^{(n-2)}}{a_{n-1,n-1}^{(n-2)}} \tag{5.13}$$

$$A^{(n-1)} = L_{n-1} L_{n-2} \cdots L_2 L_1 A = U \tag{5.14}$$

$$U = \begin{pmatrix} u_{11} & u_{12} & u_{13} & \cdots & u_{1n} \\ & u_{22} & u_{23} & \cdots & u_{2n} \\ & & u_{33} & \cdots & u_{3n} \\ & & & \ddots & \vdots \\ & & & & u_{nn} \end{pmatrix}. \tag{5.15}$$

The transformed system of equations

$$U\mathbf{x} = \mathbf{y} \qquad \mathbf{y} = L_{n-1} L_{n-1} \cdots L_2 L_1 \mathbf{b} \tag{5.16}$$

can be solved easily by backward substitution:

$$x_n = \frac{1}{u_{nn}} y_n \tag{5.17}$$

$$x_{n-1} = \frac{y_{n-1} - x_n u_{n-1,n}}{u_{n-1,n-1}} \tag{5.18}$$

$$\vdots \tag{5.19}$$

Alternatively the matrices L_i can be inverted:

$$L_1^{-1} = \begin{pmatrix} 1 & & & & \\ l_{21} & 1 & & & \\ l_{31} & & 1 & & \\ \vdots & & & \ddots & \\ l_{n1} & & & & 1 \end{pmatrix} \quad \cdots \quad L_{n-1}^{-1} = \begin{pmatrix} 1 & & & & \\ & 1 & & & \\ & & \ddots & & \\ & & & 1 & \\ & & & l_{n,n-1} & 1 \end{pmatrix}. \tag{5.20}$$

This gives

$$A = L_1^{-1} L_2^{-1} \cdots L_{n-1}^{-1} U. \tag{5.21}$$

The product of the inverted matrices is a lower triangular matrix:

$$L_1^{-1} L_2^{-1} = \begin{pmatrix} 1 & & & \\ l_{21} & 1 & & \\ l_{31} & l_{32} & 1 & \\ \vdots & \vdots & & \ddots \\ l_{n1} & l_{n2} & & & 1 \end{pmatrix}$$

$$\vdots \tag{5.22}$$

$$L = L_1^{-1} L_2^{-1} \cdots L_{n-1}^{-1} = \begin{pmatrix} 1 & & & & \\ l_{21} & 1 & & & \\ \vdots & \vdots & \ddots & & \\ l_{n-1,1} & l_{n-1,2} & \cdots & 1 & \\ l_{n1} & l_{n2} & \cdots & l_{n,n-1} & 1 \end{pmatrix}.$$

Hence the matrix A becomes decomposed into a product of a lower and an upper triangular matrix

$$A = LU \tag{5.23}$$

which can be used to solve the system of equations (5.2).

$$A\mathbf{x} = LU\mathbf{x} = \mathbf{b} \tag{5.24}$$

in two steps:

$$L\mathbf{y} = \mathbf{b} \tag{5.25}$$

which can be solved from the top

$$y_1 = b_1 \tag{5.26}$$

$$y_2 = b_2 - l_{21} y_1 \tag{5.27}$$

$$\vdots \tag{5.28}$$

and

$$U\mathbf{x} = \mathbf{y} \tag{5.29}$$

which can be solved from the bottom

$$x_n = \frac{1}{u_{nn}} y_n \tag{5.30}$$

$$x_{n-1} = \frac{y_{n-1} - x_n u_{n-1,n}}{u_{n-1,n-1}}. \tag{5.31}$$

$$\vdots \tag{5.32}$$

5.1.1 Pivoting

To improve numerical stability and to avoid division by zero pivoting is used. Most common is partial pivoting. In every step the order of the equations is changed in order to maximize the pivoting element $a_{k,k}$ in the denominator. This gives LU decomposition of the matrix PA where P is a permutation matrix. P is not needed explicitly. Instead an index vector is used which stores the new order of the equations

$$P \begin{pmatrix} 1 \\ \vdots \\ N \end{pmatrix} = \begin{pmatrix} i_1 \\ \vdots \\ i_N \end{pmatrix}. \tag{5.33}$$

Total pivoting exchanges rows and columns of A. This can be time consuming for larger matrices.

If the elements of the matrix are of different orders of magnitude it can be necessary to balance the matrix, for instance by normalizing all rows of A. This can be also achieved by selecting the maximum of

$$\frac{a_{ik}}{\sum_j |a_{ij}|} \tag{5.34}$$

as the pivoting element.

5.1.2 Direct LU Decomposition

LU decomposition can be also performed in a different order [203]. For symmetric positive definite matrices there exists the simpler and more efficient Cholesky method decomposes the matrix into the product LL^T of a lower triangular matrix and its transpose [204].

5.2 QR Decomposition

The Gaussian elimination method can become numerically unstable [254]. An alternative method to solve a system of linear equations uses the decomposition [108]

$$A = QR \tag{5.35}$$

with a unitary matrix $Q^\dagger Q = 1$ (an orthogonal matrix $Q^T Q = 1$ if A is real) and an upper right triangular matrix R. The system of linear equations (5.2) is simplified by multiplication with $Q^\dagger = Q^{-1}$

$$QR\mathbf{x} = A\mathbf{x} = \mathbf{b} \tag{5.36}$$

$$R\mathbf{x} = Q^\dagger \mathbf{b}. \tag{5.37}$$

Such a system with upper triangular matrix is easily solved (see (5.29)).

5.2.1 QR Decomposition by Orthogonalization

Gram-Schmidt orthogonalization [108, 244] provides a simple way to perform a QR decomposition. It is used for symbolic calculations and also for least square fitting (10.1.2) but can become numerically unstable.

From the decomposition $A = QR$ we have

$$a_{ik} = \sum_{j=1}^{k} q_{ij} r_{jk} \tag{5.38}$$

$$\mathbf{a}_k = \sum_{j=1}^{k} r_{jk} \mathbf{q}_j \tag{5.39}$$

which gives the k-th column vector $\mathbf{a}_k$ of A as a linear combination of the orthonormal vectors $\mathbf{q}_1 \cdots \mathbf{q}_k$. Similarly $\mathbf{q}_k$ is a linear combination of the first k columns of A. With the help of the Gram-Schmidt method r_{jk} and $\mathbf{q}_j$ are calculated as follows:

$$r_{11} := |a_1| \tag{5.40}$$

$$\mathbf{q}_1 := \frac{\mathbf{a}_1}{r_{11}}. \tag{5.41}$$

For $k = 2 \cdots n$:

$$r_{ik} := \mathbf{q}_i \mathbf{a}_k \qquad i = 1 \cdots k - 1 \tag{5.42}$$

$$\mathbf{b}_k := \mathbf{a}_k - r_{1k}\mathbf{q}_1 - \cdots - r_{k-1,k}\mathbf{q}_{k-1} \tag{5.43}$$

$$r_{kk} := |\mathbf{b}_k| \tag{5.44}$$

$$\mathbf{q}_k := \frac{\mathbf{b}_k}{r_{kk}}. \tag{5.45}$$

Obviously now

$$\mathbf{a}_k = r_{kk}\mathbf{q}_k + r_{k-1,k}\mathbf{q}_{k-1} + \cdots + r_{1k}\mathbf{q}_1 \tag{5.46}$$

since per definition

$$\mathbf{q}_i \mathbf{a}_k = r_{ik} \qquad i = 1 \cdots k \tag{5.47}$$

and

$$r_{kk}^2 = |\mathbf{b}_k|^2 = |\mathbf{a}_k|^2 + r_{1k}^2 + \cdots + r_{k-1,k}^2 - 2r_{1k}^2 - \cdots - 2r_{k-1,k}^2. \tag{5.48}$$

Hence

$$\mathbf{q}_k \mathbf{a}_k = \frac{1}{r_{kk}}(\mathbf{a}_k - r_{1k}\mathbf{q}_1 \cdots r_{k-1,k}\mathbf{q}_{k-1})\mathbf{a}_k = \frac{1}{r_{kk}}\left(|a_k|^2 - r_{1k}^2 - \cdots - r_{k-1,k}^2\right) = r_{kk}. \tag{5.49}$$

Orthogonality gives

$$\mathbf{q}_i \mathbf{a}_k = 0 \quad i = k + 1 \cdots n. \tag{5.50}$$

In matrix notation we have finally

$$A = (\mathbf{a}_1 \cdots \mathbf{a}_n) = (\mathbf{q}_1 \cdots \mathbf{q}_n) \begin{pmatrix} r_{11} & r_{12} & \cdots & r_{1n} \\ & r_{22} & \cdots & r_{2n} \\ & & \ddots & \vdots \\ & & & r_{nn} \end{pmatrix}. \tag{5.51}$$

If the columns of A are almost linearly dependent, numerical stability can be improved by an additional orthogonalization step

$$\mathbf{b}_k \to \mathbf{b}_k - (\mathbf{q}_1\mathbf{b}_k)\mathbf{q}_1 - \cdots - (\mathbf{q}_{k-1}\mathbf{b}_k)\mathbf{q}_{k-1} \tag{5.52}$$

after (5.43) which can be iterated several times to improve the results [66, 244].

Fig. 5.1 (Householder
transformation)
Geometrically the
Householder transformation
(5.53) is a mirror operation
with respect to a plane with
normal vector **u**

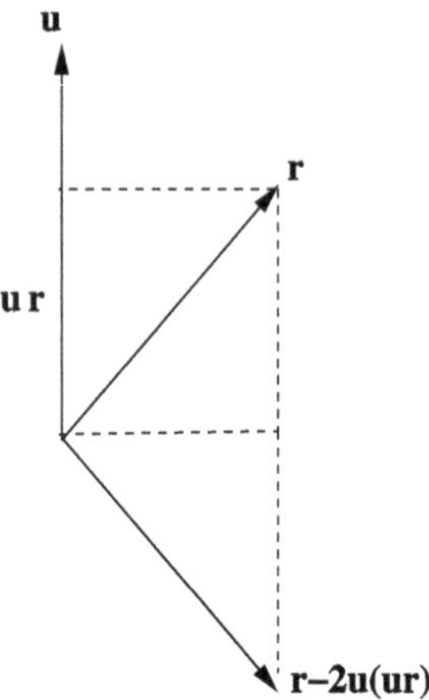

5.2.2 QR Decomposition by Householder Reflections

Numerically stable algorithms use a series of transformations with unitary matrices,
mostly Householder reflections (Fig. 5.1) [244][1] which have the form

$$P = P^T = 1 - 2\mathbf{u}\mathbf{u}^T \tag{5.53}$$

with a unit vector

$$|\mathbf{u}| = 1. \tag{5.54}$$

Obviously P is an orthogonal matrix since

$$P^T P = \left(1 - 2\mathbf{u}\mathbf{u}^T\right)\left(1 - 2\mathbf{u}\mathbf{u}^T\right) = 1 - 4\mathbf{u}\mathbf{u}^T + 4\mathbf{u}\mathbf{u}^T\mathbf{u}\mathbf{u}^T = 1. \tag{5.55}$$

In the first step we try to find a vector **u** such that the first column vector of A

$$\mathbf{a}_1 = \begin{pmatrix} a_{11} \\ \vdots \\ a_{n1} \end{pmatrix} \tag{5.56}$$

is transformed into a vector along the 1-axis

$$P\mathbf{a}_1 = \left(1 - 2\mathbf{u}\mathbf{u}^T\right)\mathbf{a}_1 = k\mathbf{e}_1 = \begin{pmatrix} k \\ 0 \\ \vdots \\ 0 \end{pmatrix}. \tag{5.57}$$

---

[1] Alternatively Givens rotations [108] can be employed which need slightly more floating point
operations.

Multiplication with the transpose vector gives

$$k^2 = (P\mathbf{a}_1)^T P\mathbf{a}_1 = \mathbf{a}_1^T P^T P\mathbf{a}_1 = |\mathbf{a}_1|^2 \tag{5.58}$$

and

$$k = \pm|\mathbf{a}_1| \tag{5.59}$$

can have both signs. From (5.57) we have

$$\mathbf{a}_1 - 2\mathbf{u}(\mathbf{u}\mathbf{a}_1) = k\mathbf{e}_1. \tag{5.60}$$

Multiplication with $\mathbf{a}_1^T$ gives

$$2(\mathbf{u}\mathbf{a}_1)^2 = |\mathbf{a}_1|^2 - k(\mathbf{a}_1\mathbf{e}_1) \tag{5.61}$$

and since

$$|\mathbf{a}_1 - k\mathbf{e}_1|^2 = |\mathbf{a}_1|^2 + k^2 - 2k(\mathbf{a}_1\mathbf{e}_1) = 2|\mathbf{a}_1|^2 - 2k(\mathbf{a}_1\mathbf{e}_1) \tag{5.62}$$

we have

$$2(\mathbf{u}\mathbf{a}_1)^2 = \frac{1}{2}|\mathbf{a}_1 - k\mathbf{e}_1|^2 \tag{5.63}$$

and from (5.60) we find

$$\mathbf{u} = \frac{\mathbf{a}_1 - k\mathbf{e}_1}{2\mathbf{u}\mathbf{a}_1} = \frac{\mathbf{a}_1 - k\mathbf{e}_1}{|\mathbf{a}_1 - k\mathbf{e}_1|}. \tag{5.64}$$

To avoid numerical extinction the sign of k is chosen such that

$$\sigma = \operatorname{sign}(k) = -\operatorname{sign}(a_{11}). \tag{5.65}$$

Then,

$$\mathbf{u} = \frac{1}{\sqrt{2(a_{11}^2 + \cdots + a_{n1}^2) + 2|a_{11}|\sqrt{a_{11}^2 + \cdots + a_{n1}^2}}}$$
$$\times \begin{pmatrix} \operatorname{sign}(a_{11})(|a_{11}| + \sqrt{a_{11}^2 + a_{21}^2 + \cdots + a_{n1}^2}) \\ a_{21} \\ \vdots \\ a_{n1} \end{pmatrix} \tag{5.66}$$

$$2\mathbf{u}\mathbf{u}^T\mathbf{a}_1 = \begin{pmatrix} \operatorname{sign}(a_{11})(|a_{11}| + \sqrt{a_{11}^2 + a_{21}^2 + \cdots + a_{n1}^2}) \\ a_{21} \\ \vdots \\ a_{n1} \end{pmatrix}$$

$$\times \frac{1}{(a_{11}^2 + \cdots + a_{n1}^2) + |a_{11}|\sqrt{a_{11}^2 + \cdots + a_{n1}^2}}$$

$$\times \left(a_{11}^2 + |a_{11}|\sqrt{a_{11}^2 + \cdots + a_{n1}^2} + a_{21}^2 + \cdots + a_{n1}^2\right) \tag{5.67}$$

and the Householder transformation of the first column vector of A gives

$$(1 - 2\mathbf{u}\mathbf{u}^T)\mathbf{a}_1 = \begin{pmatrix} -\mathrm{sign}(a_{11})\sqrt{a_{11}^2 \cdots a_{n1}^2} \\ 0 \\ \vdots \\ 0 \end{pmatrix}. \tag{5.68}$$

Thus after the first step a matrix results of the form

$$A^{(1)} = P_1 A = \begin{pmatrix} a_{11}^{(1)} & a_{12}^{(1)} & \cdots & a_{1n}^{(1)} \\ 0 & a_{22}^{(1)} & & a_{2n}^{(1)} \\ \vdots & \vdots & & \vdots \\ 0 & a_{n2}^{(1)} & \cdots & a_{nn}^{(1)} \end{pmatrix}.$$

In the following $(n - 2)$ steps further Householder reflections are applied in the subspace $k \le i, j \le n$ to eliminate the elements

$$a_{k+1,k} \cdots a_{n,k}$$

of the k-th row vector below the diagonal of the matrix

$$A^{(k-1)} = P_{k-1} \cdots P_1 A = \begin{pmatrix} a_{11}^{(1)} & \cdots & a_{1,k-1}^{(1)} & a_{1,k}^{(1)} & \cdots & a_{1,n}^{(1)} & \\ 0 & \ddots & \vdots & \vdots & & \vdots \\ \vdots & \ddots & a_{k-1,k-1}^{(k-1)} & a_{k-1,k}^{(k-1)} & & a_{k-1,n}^{(k-1)} \\ \vdots & \vdots & 0 & a_{k,k}^{(k-1)} & & a_{k,n}^{(k-1)} \\ \vdots & \vdots & \vdots & a_{k+1,k}^{(k-1)} & & a_{k+1,n}^{(k-1)} \\ \vdots & \vdots & \vdots & \vdots & & \vdots \\ 0 & \cdots & 0 & a_{n,k}^{(k-1)} & \cdots & a_{n,n}^{(k-1)} \end{pmatrix}$$

$$\tag{5.69}$$

$$P_k = \begin{pmatrix} 1_1 & & & \\ & \ddots & & \\ & & 1_{k-1} & \\ & & & 1 - 2\mathbf{u}\mathbf{u}^T \end{pmatrix}.$$

Finally an upper triangular matrix results

$$
A^{(n-1)} = (P_{n-1} \cdots P_1)A = R =
\begin{pmatrix}
a_{11}^{(1)} & a_{12}^{(1)} & \cdots & a_{1,n-1}^{(1)} & a_{1,n}^{(1)} \\
0 & a_{22}^{(2)} & \cdots & a_{2,n-1}^{(2)} & a_{2,n}^{(2)} \\
\vdots & 0 & \ddots & \vdots & \vdots \\
\vdots & \vdots & \vdots & a_{n-1,n-1}^{(n-1)} & a_{n-1,n}^{(n-1)} \\
0 & 0 & \cdots & 0 & a_{n,n}^{(n-1)}
\end{pmatrix} .
\tag{5.70}
$$

If the orthogonal matrix Q is needed explicitly additional numerical operations are necessary to form the product

$$
Q = (P_{n-1} \cdots P_1)^T .
\tag{5.71}
$$

5.3 Linear Equations with Tridiagonal Matrix

Linear equations with the form

$$
b_1 x_1 + c_1 x_2 = r_1
\tag{5.72}
$$

$$
a_i x_{i-1} + b_i x_i + c_i x_{i+1} = r_i \quad i = 2 \cdots (n-1)
\tag{5.73}
$$

$$
a_n x_{n-1} + b_n x_n = r_n
\tag{5.74}
$$

can be solved very efficiently with a specialized Gaussian elimination method.[2] They are important for cubic spline interpolation or second derivatives. We begin by eliminating a_2. To that end we multiply the first line with a_2/b_1 and subtract it from the first line. The result is the equation

$$
\beta_2 x_2 + c_2 x_3 = \rho_2
\tag{5.75}
$$

with the abbreviations

$$
\beta_2 = b_2 - \frac{c_1 a_2}{b_1} \quad \rho_2 = r_2 - \frac{r_1 a_2}{b_1} .
\tag{5.76}
$$

We iterate this procedure

$$
\beta_i x_i + c_i x_{i+1} = \rho_i
\tag{5.77}
$$

$$
\beta_i = b_i - \frac{c_{i-1} a_i}{\beta_{i-1}} \quad \rho_i = r_i - \frac{\rho_{i-1} a_i}{\beta_{i-1}}
\tag{5.78}
$$

---

[2]This algorithm is only well behaved if the matrix is diagonal dominant $|b_i| > |a_i| + |c_i|$.

until we reach the n-th equation, which becomes simply

$$\beta_n x_n = \rho_n \tag{5.79}$$

$$\beta_n = b_n - \frac{c_{n-1} a_n}{\beta_{n-1}} \qquad \rho_n = r_n - \frac{\rho_{n-1} a_n}{\beta_{n-1}}. \tag{5.80}$$

Now we immediately have

$$x_n = \frac{\rho_n}{\beta_n} \tag{5.81}$$

and backward substitution gives

$$x_{i-1} = \frac{\rho_{i-1} - c_{i-1} x_i}{\beta_{i-1}} \tag{5.82}$$

and finally

$$x_1 = \frac{r_1 - c_1 x_2}{\beta_2}. \tag{5.83}$$

This algorithm can be formulated as LU decomposition: Multiplication of the matrices

$$L = \begin{pmatrix} 1 & & & & \\ l_2 & 1 & & & \\ & l_3 & 1 & & \\ & & \ddots & \ddots & \\ & & & l_n & 1 \end{pmatrix} \qquad U = \begin{pmatrix} \beta_1 & c_1 & & & \\ & \beta_2 & c_2 & & \\ & & \beta_3 & c_3 & \\ & & & \ddots & \\ & & & & \beta_n \end{pmatrix} \tag{5.84}$$

gives

$$LU = \begin{pmatrix} \beta_1 & c_1 & & & & \\ & \ddots & \ddots & & & \\ & & \ddots & \ddots & & \\ & & l_i \beta_{i-1} & (l_i c_{i-1} + \beta_i) & c_i & \\ & & & \ddots & \ddots & \ddots \\ & & & & l_n \beta_{n-1} & (l_n c_{n-1} + \beta_n) \end{pmatrix} \tag{5.85}$$

which coincides with the matrix

$$
A = \begin{pmatrix}
b_1 & c_1 \\
a_2 & \ddots & \ddots \\
 & \ddots & \ddots & \ddots \\
 & & a_i & b_i & c_i \\
 & & & \ddots & \ddots & \ddots \\
 & & & & a_{n-1} & b_{n-1} & c_{n-1} \\
 & & & & & a_n & b_n
\end{pmatrix}
\tag{5.86}
$$

if we choose

$$
l_i = \frac{a_i}{\beta_{i-1}}
\tag{5.87}
$$

since then from (5.78)

$$
b_i = \beta_i + l_i c_{i-1}
\tag{5.88}
$$

and

$$
l_i \beta_{i-1} = a_i .
\tag{5.89}
$$

5.4 Cyclic Tridiagonal Systems

Periodic boundary conditions lead to a small perturbation of the tridiagonal matrix

$$
A = \begin{pmatrix}
b_1 & c_1 & & & & & & a_1 \\
a_2 & \ddots & \ddots \\
 & \ddots & \ddots & \ddots \\
 & & a_i & b_i & c_i \\
 & & & \ddots & \ddots & \ddots \\
 & & & & a_{n-1} & b_{n-1} & c_{n-1} \\
c_n & & & & & a_n & b_n
\end{pmatrix} .
\tag{5.90}
$$

The system of equations

$$
A\mathbf{x} = \mathbf{r}
\tag{5.91}
$$

can be reduced to a tridiagonal system [205] with the help of the Sherman-Morrison formula [236], which states that if A_0 is an invertible matrix and $\mathbf{u}, \mathbf{v}$ are vectors and

$$
1 + \mathbf{v}^T A_0^{-1} \mathbf{u} \neq 0
\tag{5.92}
$$

then the inverse of the matrix[3]

$$A = A_0 + \mathbf{u}\mathbf{v}^T \tag{5.93}$$

is given by

$$A^{-1} = A_0^{-1} - \frac{A_0^{-1}\mathbf{u}\mathbf{v}^T A_0^{-1}}{1 + \mathbf{v}^T A_0^{-1}\mathbf{u}}. \tag{5.94}$$

We choose

$$\mathbf{u}\mathbf{v}^T = \begin{pmatrix} \alpha \\ 0 \\ \vdots \\ 0 \\ c_n \end{pmatrix} \begin{pmatrix} 1 & 0 & \cdots & 0 & \frac{a_1}{\alpha} \end{pmatrix} = \begin{pmatrix} \alpha & & & & a_1 \\ & & & & \\ & & & & \\ c_n & & & & \frac{a_1 c_n}{\alpha} \end{pmatrix}. \tag{5.95}$$

Then

$$A_0 = A - \mathbf{u}\mathbf{v}^T = \begin{pmatrix} (b_1 - \alpha) & c_1 & & & & & 0 \\ a_2 & \ddots & \ddots & & & & \\ & \ddots & \ddots & \ddots & & & \\ & & a_i & b_i & c_i & & \\ & & & \ddots & \ddots & \ddots & \\ & & & & a_{n-1} & b_{n-1} & c_{n-1} \\ 0 & & & & & a_n & (b_n - \frac{a_1 c_n}{\alpha}) \end{pmatrix} \tag{5.96}$$

is tridiagonal. The free parameter α has to be chosen such that the diagonal elements do not become too small. We solve the system (5.91) by solving the two tridiagonal systems

$$A_0 \mathbf{x}_0 = \mathbf{r}$$
$$A_0 \mathbf{q} = \mathbf{u} \tag{5.97}$$

and compute $\mathbf{x}$ from

$$\mathbf{x} = A^{-1}\mathbf{r} = A_0^{-1}\mathbf{r} - \frac{(A_0^{-1}\mathbf{u})\mathbf{v}^T(A_0^{-1}\mathbf{r})}{1 + \mathbf{v}^T(A_0^{-1}\mathbf{u})} = \mathbf{x}_0 - \mathbf{q}\frac{\mathbf{v}^T\mathbf{x}_0}{1 + \mathbf{v}^T\mathbf{q}}. \tag{5.98}$$

---

[3] Here $\mathbf{u}\mathbf{v}^T$ is the outer or matrix product of the two vectors.

5.5 Iterative Solution of Inhomogeneous Linear Equations

Discretized differential equations often lead to systems of equations with large sparse matrices, which have to be solved by iterative methods.

5.5.1 General Relaxation Method

We want to solve

$$Ax = b \tag{5.99}$$

iteratively. To that end we divide the matrix A into two (non singular) parts [244]

$$A = A_1 + A_2 \tag{5.100}$$

and rewrite (5.99) as

$$A_1 x = b - A_2 x \tag{5.101}$$

which we use to define the iteration

$$\mathbf{x}^{(n+1)} = \Phi\left(\mathbf{x}^{(n)}\right) \tag{5.102}$$

$$\Phi(\mathbf{x}) = -A_1^{-1} A_2 \mathbf{x} + A_1^{-1} \mathbf{b}. \tag{5.103}$$

A fixed point $\mathbf{x}$ of this equation fulfills

$$\mathbf{x}_{fp} = \Phi(\mathbf{x}_{fp}) = -A_1^{-1} A_2 \mathbf{x}_{fp} + A_1^{-1} \mathbf{b} \tag{5.104}$$

and is obviously a solution of (5.99). The iteration can be written as

$$\begin{aligned}
\mathbf{x}^{(n+1)} &= -A_1^{-1}(A - A_1)\mathbf{x}^{(n)} + A_1^{-1}\mathbf{b} \\
&= \left(E - A_1^{-1}A\right)\mathbf{x}^{(n)} + A_1^{-1}\mathbf{b} = \mathbf{x}^{(n)} - A_1^{-1}\left(A\mathbf{x}^{(n)} - \mathbf{b}\right)
\end{aligned} \tag{5.105}$$

or

$$A_1\left(\mathbf{x}^{(n+1)} - \mathbf{x}^{(n)}\right) = -\left(A\mathbf{x}^{(n)} - \mathbf{b}\right). \tag{5.106}$$

5.5.2 Jacobi Method

Jacobi divides the matrix A into its diagonal and two triangular matrices [38]:

$$A = L + U + D. \tag{5.107}$$

For A_1 the diagonal part is chosen

$$A_1 = D \tag{5.108}$$

giving

$$\mathbf{x}^{(n+1)} = -D^{-1}(A - D)\mathbf{x}^{(n)} + D^{-1}\mathbf{b} \tag{5.109}$$

which reads explicitly

$$x_i^{(n+1)} = -\frac{1}{a_{ii}} \sum_{j \neq i} a_{ij} x_j^{(n)} + \frac{1}{a_{ii}} b_i. \tag{5.110}$$

This method is stable but converges rather slowly. Reduction of the error by a factor of 10^{-p} needs about $\frac{pN}{2}$ iterations. N grid points have to be evaluated in each iteration and the method scales with $O(N^2)$ [206].

5.5.3 Gauss-Seidel Method

With

$$A_1 = D + L \tag{5.111}$$

the iteration becomes

$$(D + L)\mathbf{x}^{(n+1)} = -U\mathbf{x}^{(n)} + \mathbf{b} \tag{5.112}$$

which has the form of a system of equations with triangular matrix [139]. It reads explicitly

$$\sum_{j \leq i} a_{ij} x_j^{(n+1)} = -\sum_{j > i} a_{ij} x_j^{(n)} + b_i. \tag{5.113}$$

Forward substitution starting from x_1 gives

$$i = 1: \quad x_1^{(n+1)} = \frac{1}{a_{11}} \left(-\sum_{j \geq 2} a_{ij} x_j^{(n)} + b_1 \right)$$

$$i = 2: \quad x_2^{(n+1)} = \frac{1}{a_{22}} \left(-a_{21} x_1^{(n+1)} - \sum_{j \geq 3} a_{ij} x_j^{(n)} + b_2 \right)$$

$$i = 3: \quad x_3^{(n+1)} = \frac{1}{a_{33}} \left(-a_{31} x_1^{(n+1)} - a_{32} x_2^{(n+1)} - \sum_{j \geq 4} a_{ij} x_j^{(n)} + b_3 \right)$$

$$\vdots$$

$$x_i^{(n+1)} = \frac{1}{a_{ii}}\left(-\sum_{j<i} a_{ij}x_j^{(n+1)} - \sum_{j>i} a_{ij}x_j^{(n)} + b_i\right). \tag{5.114}$$

This looks very similar to the Jacobi method. But here all changes are made immediately. Convergence is slightly better (roughly a factor of 2) and the numerical effort is reduced [206].

5.5.4 Damping and Successive Over-Relaxation

Convergence can be improved [206] by combining old and new values. Starting from the iteration

$$A_1\mathbf{x}^{(n+1)} = (A_1 - A)\mathbf{x}^{(n)} + \mathbf{b} \tag{5.115}$$

we form a linear combination with

$$D\mathbf{x}^{(n+1)} = D\mathbf{x}^{(n)} \tag{5.116}$$

giving the new iteration equation

$$\big((1-\omega)D + \omega A_1\big)\mathbf{x}^{(n+1)} = \big((1-\omega)D + \omega A_1 - \omega A\big)\mathbf{x}^{(n)} + \omega\mathbf{b}. \tag{5.117}$$

In case of the Jacobi method with $D = A_1$ we have

$$D\mathbf{x}^{(n+1)} = (D - \omega A)\mathbf{x}^{(n)} + \omega\mathbf{b} \tag{5.118}$$

or explicitly

$$x_i^{(n+1)} = (1-\omega)x_i^{(n)} + \frac{\omega}{a_{ii}}\left(-\sum_{j\neq i} a_{ij}x_j^{(n)} + b_i\right). \tag{5.119}$$

The changes are damped ($0 < \omega < 1$) or exaggerated[4] ($1 < \omega < 2$).

In case of the Gauss-Seidel method with $A_1 = D + L$ the new iteration (5.117) is

$$(D + \omega L)\mathbf{x}^{(n+1)} = (D + \omega L - \omega A)\mathbf{x}^{(n)} + \omega\mathbf{b} = (1-\omega)D\mathbf{x}^{(n)} - \omega U\mathbf{x}^{(n)} + \omega\mathbf{b} \tag{5.120}$$

or explicitly

$$x_i^{(n+1)} = (1-\omega)x_i^{(n)} + \frac{\omega}{a_{ii}}\left(-\sum_{j<i} a_{ij}x_j^{(n+1)} - \sum_{j>i} a_{ij}x_j^{(n)} + \mathbf{b}\right). \tag{5.121}$$

---

[4]This is also known as the method of successive over-relaxation (SOR).

It can be shown that the successive over-relaxation method converges only for $0 < \omega < 2$.

For optimal choice of ω about $\frac{1}{3}p\sqrt{N}$ iterations are needed to reduce the error by a factor of 10^{-p}. The order of the method is $O(N^{\frac{3}{2}})$ which is comparable to the most efficient matrix inversion methods [206].

5.6 Conjugate Gradients

At the minimum of the quadratic function

$$h(\mathbf{x}) = h_0 + \mathbf{b}^T\mathbf{x} + \frac{1}{2}\mathbf{x}^T A\mathbf{x} \tag{5.122}$$

the gradient

$$\mathbf{g} = \nabla h(\mathbf{x}) = A\mathbf{x} + \mathbf{b} \tag{5.123}$$

is zero and therefore the minimum of h is also a solution of the linear system of equations

$$A\mathbf{x} = -\mathbf{b}. \tag{5.124}$$

The stationary point can be found especially efficient with the method of conjugate gradients (page 107). The function h is minimized along the search direction

$$\mathbf{s}_{r+1} = -\mathbf{g}_{r+1} + \beta_{r+1}\mathbf{s}_r$$

by solving

$$0 = \frac{\partial}{\partial\lambda}\left(\mathbf{b}^T(\mathbf{x}_r + \lambda\mathbf{s}_r) + \frac{1}{2}(\mathbf{x}_r^T + \lambda\mathbf{s}_r^T)A(\mathbf{x}_r + \lambda\mathbf{s}_r)\right)$$

$$= \mathbf{b}^T\mathbf{s}_r + \mathbf{x}_r^T A\mathbf{s}_r + \lambda\mathbf{s}_r^T A\mathbf{s}_r \tag{5.125}$$

$$\lambda_r = -\frac{\mathbf{b}^T\mathbf{s}_r + \mathbf{x}^T A\mathbf{s}_r}{\mathbf{s}_r^T A\mathbf{s}_r} = -\frac{\mathbf{g}_r\mathbf{s}_r}{\mathbf{s}_r^T A\mathbf{s}_r}. \tag{5.126}$$

The parameter β is chosen as

$$\beta_{r+1} = \frac{g_{r+1}^2}{g_r^2}. \tag{5.127}$$

The gradient of h is the residual vector and is iterated according to

$$\mathbf{g}_{r+1} = A(\mathbf{x}_r + \lambda_r\mathbf{s}_r) + \mathbf{b} = \mathbf{g}_r + \lambda_r A\mathbf{s}_r. \tag{5.128}$$

This method [121] solves a linear system without storing the matrix A itself. Only the product $A\mathbf{s}$ is needed. In principle the solution is reached after $N = \dim(A)$ steps, but due to rounding errors more steps can be necessary and the method has to be considered as an iterative one.

5.7 Matrix Inversion

LU and QR decomposition can be also used to calculate the inverse of a nonsingular matrix

$$AA^{-1} = 1. \tag{5.129}$$

The decomposition is performed once and then the column vectors of A^{-1} are calculated similar to (5.24)

$$L\left(UA^{-1}\right) = 1 \tag{5.130}$$

or (5.37)

$$RA^{-1} = Q^{\dagger}. \tag{5.131}$$

Consider now a small variation of the right hand side of (5.2)

$$\mathbf{b} + \Delta\mathbf{b}. \tag{5.132}$$

Instead of

$$A^{-1}\mathbf{b} = \mathbf{x} \tag{5.133}$$

the resulting vector is

$$A^{-1}(\mathbf{b} + \Delta\mathbf{b}) = \mathbf{x} + \Delta\mathbf{x} \tag{5.134}$$

and the deviation can be measured by[5]

$$\|\Delta\mathbf{x}\| = \|A^{-1}\| \|\Delta\mathbf{b}\| \tag{5.135}$$

and since

$$\|A\| \|\mathbf{x}\| = \|\mathbf{b}\| \tag{5.136}$$

the relative error becomes

$$\frac{\|\Delta\mathbf{x}\|}{\|\mathbf{x}\|} = \|A\| \|A^{-1}\| \frac{\|\Delta\mathbf{b}\|}{\|\mathbf{b}\|}. \tag{5.137}$$

The relative error of $\mathbf{b}$ is multiplied by the condition number for inversion

$$\mathrm{cond}(A) = \|A\| \|A^{-1}\|. \tag{5.138}$$

---

[5]The vector norm used here is not necessarily the Euclidian norm.

5.8 Problems

Problem 5.1 (Comparison of different direct Solvers, Fig. 5.2) In this computer experiment we solve the system of equations

$$Ax = b \tag{5.139}$$

with several methods:

- Gaussian elimination without pivoting (Sect. 5.1),
- Gaussian elimination with partial pivoting (Sect. 5.1.1),
- QR decomposition with Householder reflections (Sect. 5.2.2),
- QR decomposition with Gram-Schmidt orthogonalization (Sect. 5.2.1),
- QR decomposition with Gram-Schmidt orthogonalization with extra orthogonalization step (5.52).

The right hand side is chosen as

$$\mathbf{b} = A \begin{pmatrix} 1 \\ 2 \\ \vdots \\ n \end{pmatrix} \tag{5.140}$$

hence the exact solution is

$$\mathbf{x} = \begin{pmatrix} 1 \\ 2 \\ \vdots \\ n \end{pmatrix}. \tag{5.141}$$

Several test matrices can be chosen:

- Gaussian elimination is theoretically unstable but is stable in many practical cases. The instability can be demonstrated with the example [254]

$$A = \begin{pmatrix} 1 & & & & 1 \\ -1 & 1 & & & 1 \\ -1 & -1 & 1 & & 1 \\ \vdots & \vdots & \vdots & \ddots & \vdots \\ -1 & -1 & -1 & -1 & 1 \end{pmatrix}. \tag{5.142}$$

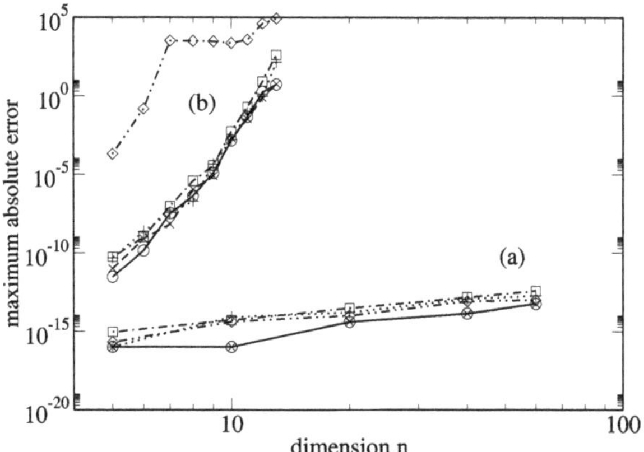

Fig. 5.2 (Comparison of different direct solvers) Gaussian elimination without (*circles*) and with (*x*) pivoting, QR decomposition with Householder reflections (*squares*), with Gram-Schmidt orthogonalization (*diamonds*) and including extra orthogonalization (+) are compared. The maximum difference $\max_{i=1\cdots n}(|x_i - x_i^{exact}|)$ increases only slightly with the dimension n for the well behaved matrix ((5.146), *a*) but quite dramatically for the ill conditioned Hilbert matrix ((5.148), *b*)

No pivoting takes place in the LU decomposition of this matrix and the entries in the last column double in each step:

$$A^{(1)} = \begin{pmatrix} 1 & & & & 1 \\ & 1 & & & 2 \\ & -1 & 1 & & 2 \\ & \vdots & \vdots & \ddots & \vdots \\ & -1 & -1 & -1 & 2 \end{pmatrix} \qquad A^{(2)} = \begin{pmatrix} 1 & & & & 1 \\ & 1 & & & 2 \\ & & 1 & & 4 \\ & & \vdots & \ddots & \vdots \\ & & -1 & -1 & 4 \end{pmatrix} \cdots$$

$$A^{(n-1)} = \begin{pmatrix} 1 & & & & 1 \\ & 1 & & & 2 \\ & & \ddots & & 4 \\ & & & \ddots & \vdots \\ & & & & 2^{n-1} \end{pmatrix} . \tag{5.143}$$

Since the machine precision is $\epsilon_M = 2^{-53}$ for double precision calculations we have to expect numerical inaccuracy for dimension $n > 53$.

- Especially well conditioned are matrices [273] which are symmetric

$$A_{ij} = A_{ji} \tag{5.144}$$

and also diagonal dominant

$$\sum_{j \neq i} |A_{ij}| < |A_{ii}|. \tag{5.145}$$

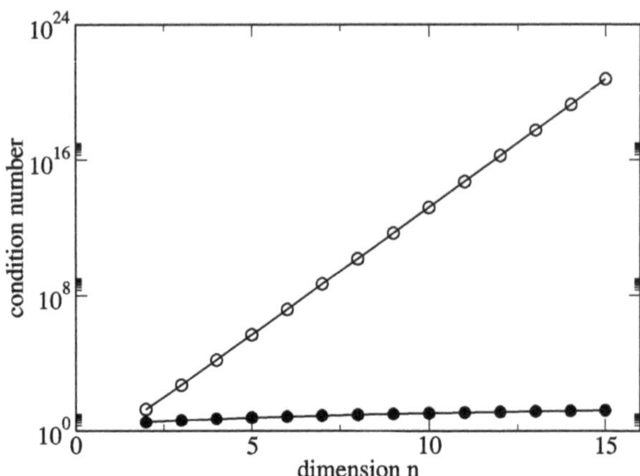

Fig. 5.3 (Condition numbers) The condition number cond(A) increases only linearly with the dimension n for the well behaved matrix ((5.146), *full circles*) but exponentially for the ill conditioned Hilbert matrix ((5.148), *open circles*)

We use the matrix

$$
A = \begin{pmatrix}
n & 1 & \cdots & 1 & 1 \\
1 & n & \cdots & 1 & 1 \\
\vdots & & \ddots & & \vdots \\
1 & 1 & 1 & n & 1 \\
1 & 1 & 1 & 1 & n
\end{pmatrix}
\tag{5.146}
$$

which can be inverted explicitly by

$$
A^{-1} = \begin{pmatrix}
a & b & \cdots & b & b \\
b & a & & b & b \\
\vdots & & \ddots & & \\
b & b & b & a & b \\
b & b & b & b & a
\end{pmatrix}
\qquad a = \frac{1}{n - \frac{1}{2}}, \; b = -\frac{1}{2n^2 - 3n + 1}
\tag{5.147}
$$

and has a condition number[6] which is proportional to the dimension n (Fig. 5.3).

- The Hilbert matrix [22, 122]

$$
A = \begin{pmatrix}
1 & \frac{1}{2} & \frac{1}{3} & \cdots & \frac{1}{n} \\
\frac{1}{2} & \frac{1}{3} & \frac{1}{4} & & \frac{1}{n+1} \\
\frac{1}{3} & \frac{1}{4} & \frac{1}{5} & & \frac{1}{n+2} \\
\vdots & & & \ddots & \vdots \\
\frac{1}{n} & \frac{1}{n+1} & \frac{1}{n+2} & \cdots & \frac{1}{2n-1}
\end{pmatrix}
\tag{5.148}
$$

is especially ill conditioned [16] even for moderate dimension. It is positive definite and therefore the inverse matrix exists and even can be written down explicitly [222]. Its column vectors are very close to linearly dependent and the

---

[6]Using the Frobenius norm $\|A\| = \sqrt{\sum_{ij} A_{ij}^2}$.

condition number grows exponentially with its dimension (Fig. 5.3). Numerical errors are large for all methods compared (Fig. 5.2).

- Random matrices

$$A_{ij} = \xi \in [-1, 1]. \tag{5.149}$$

Chapter 6
Roots and Extremal Points

In computational physics very often roots of a function, i.e. solutions of an equation
like

$$f(x_1 \cdots x_N) = 0 \tag{6.1}$$

have to be determined. A related problem is the search for local extrema (Fig. 6.1)

$$\max f(x_1 \cdots x_N) \quad \min f(x_1 \cdots x_N) \tag{6.2}$$

which for a smooth function are solutions of the equations

$$\frac{\partial f(x_1 \cdots x_N)}{\partial x_i} = 0, \quad i = 1 \ldots N. \tag{6.3}$$

In one dimension bisection is a very robust but rather inefficient root finding method.
If a good starting point close to the root is available and the function smooth enough,
the Newton-Raphson method converges much faster. Special strategies are necessary to find roots of not so well behaved functions or higher order roots. The combination of bisection and interpolation like in Dekker's and Brent's methods provides
generally applicable algorithms. In multidimensions calculation of the Jacobian matrix is not always possible and Quasi-Newton methods are a good choice. Whereas
local extrema can be found as the roots of the gradient, at least in principle, direct
optimization can be more efficient. In one dimension the ternary search method or
Brent's more efficient golden section search method can be used. In multidimensions the class of direction set search methods is very popular which includes the
methods of steepest descent and conjugate gradients, the Newton-Raphson method
and, if calculation of the full Hessian matrix is too expensive, the Quasi-Newton
methods.

6.1 Root Finding

If there is exactly one root in the interval $a_0 < x < b_0$ then one of the following
methods can be used to locate the position with sufficient accuracy. If there are

P.O.J. Scherer, *Computational Physics*, Graduate Texts in Physics,
DOI 10.1007/978-3-319-00401-3_6,
© Springer International Publishing Switzerland 2013

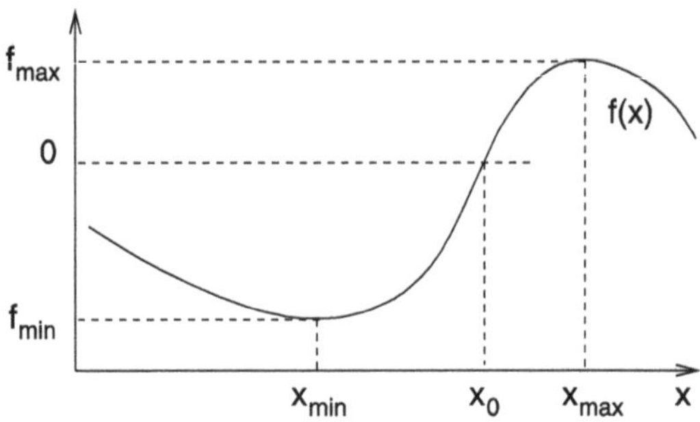

Fig. 6.1 Roots and local extrema of a function

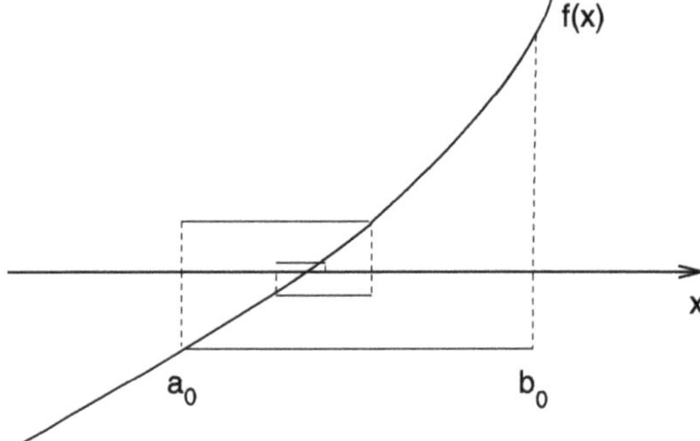

Fig. 6.2 Root finding by bisection

multiple roots, these methods will find one of them and special care has to be taken to locate the other roots.

6.1.1 Bisection

The simplest method [45] to solve

$$f(x) = 0 \qquad (6.4)$$

uses the following algorithm (Fig. 6.2):

(1) Determine an interval $[a_0, b_0]$, which contains a sign change of $f(x)$. If no such interval can be found then $f(x)$ does not have any zero crossings.
(2) Divide the interval into $[a_0, a_0 + \frac{b_0 - a_0}{2}]$ $[a_0 + \frac{b_0 - a_0}{2}, b_0]$ and choose that interval $[a_1, b_1]$, where $f(x)$ changes its sign.
(3) repeat until the width $b_n - a_n < \varepsilon$ is small enough.[1]

The bisection method needs two starting points which bracket a sign change of the function. It converges but only slowly since each step reduces the uncertainty by a factor of 2.

---

[1] Usually a combination like $\varepsilon = 2\varepsilon_M + |b_n|\varepsilon_r$ of an absolute and a relative tolerance is taken.

Fig. 6.3 Regula falsi method

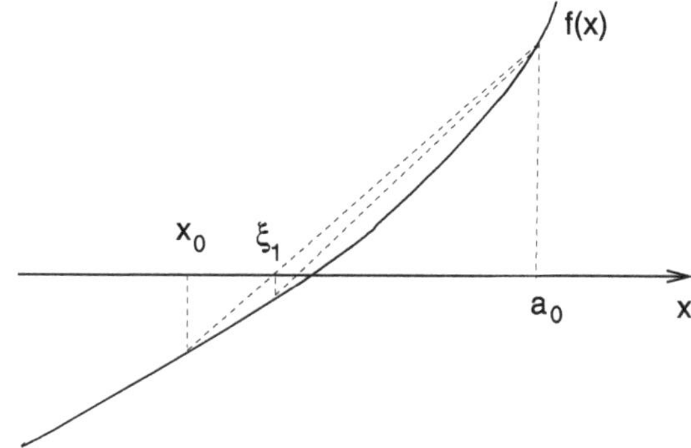

6.1.2 Regula Falsi (False Position) Method

The regula falsi [92] method (Fig. 6.3) is similar to the bisection method [45]. However, polynomial interpolation is used to divide the interval $[x_r, a_r]$ with $f(x_r)f(a_r) < 0$. The root of the linear polynomial

$$p(x) = f(x_r) + (x - x_r)\frac{f(a_r) - f(x_r)}{a_r - x_r} \tag{6.5}$$

is given by

$$\xi_r = x_r - f(x_r)\frac{a_r - x_r}{f(a_r) - f(x_r)} = \frac{a_r f(x_r) - x_r f(a_r)}{f(x_r) - f(a_r)} \tag{6.6}$$

which is inside the interval $[x_r, a_r]$. Choose the sub-interval which contains the sign change:

$$f(x_r)f(\xi_r) < 0 \rightarrow [x_{r+1}, a_{r+1}] = [x_r, \xi_r]$$
$$f(x_r)f(\xi_r) > 0 \rightarrow [x_{r+1}, a_{r+1}] = [\xi_r, a_r]. \tag{6.7}$$

Then ξ_r provides a series of approximations with increasing precision to the root of $f(x) = 0$.

6.1.3 Newton-Raphson Method

Consider a function which is differentiable at least two times around the root ξ. Taylor series expansion around a point x_0 in the vicinity

$$f(x) = f(x_0) + (x - x_0)f'(x_0) + \frac{1}{2}(x - x_0)^2 f''(x_0) + \cdots \tag{6.8}$$

gives for $x = \xi$

$$0 = f(x_0) + (\xi - x_0)f'(x_0) + \frac{1}{2}(\xi - x_0)^2 f''(x_0) + \cdots. \tag{6.9}$$

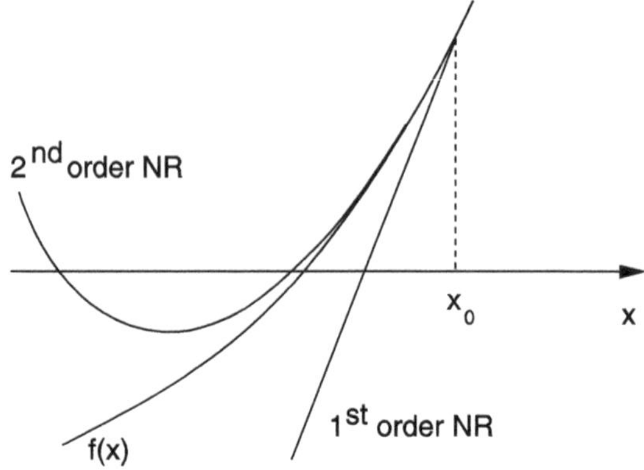

Fig. 6.4 Newton-Raphson method

Truncation of the series and solving for ξ gives the first order Newton-Raphson [45, 252] method (Fig. 6.4)

$$x_{r+1} = x_r - \frac{f(x_r)}{f'(x_r)} \tag{6.10}$$

and the second order Newton-Raphson method (Fig. 6.4)

$$x_{r+1} = x_r - \frac{f'(x_r) \pm \sqrt{f'(x_r)^2 - 2f(x_r)f''(x_r)}}{f''(x_r)}. \tag{6.11}$$

The Newton-Raphson method converges fast if the starting point is close enough to the root. Analytic derivatives are needed. It may fail if two or more roots are close by.

6.1.4 Secant Method

Replacing the derivative in the first order Newton Raphson method by a finite difference quotient gives the secant method [45] (Fig. 6.5) which has been known for thousands of years before [196]

$$x_{r+1} = x_r - f(x_r) \frac{x_r - x_{r-1}}{f(x_r) - f(x_{r-1})}. \tag{6.12}$$

Round-off errors can become important as $|f(x_r) - f(x_{r-1})|$ gets small. At the beginning choose a starting point x_0 and determine

$$x_1 = x_0 - f(x_0) \frac{2h}{f(x_0 + h) - f(x_0 - h)} \tag{6.13}$$

using a symmetrical difference quotient.

Fig. 6.5 Secant method

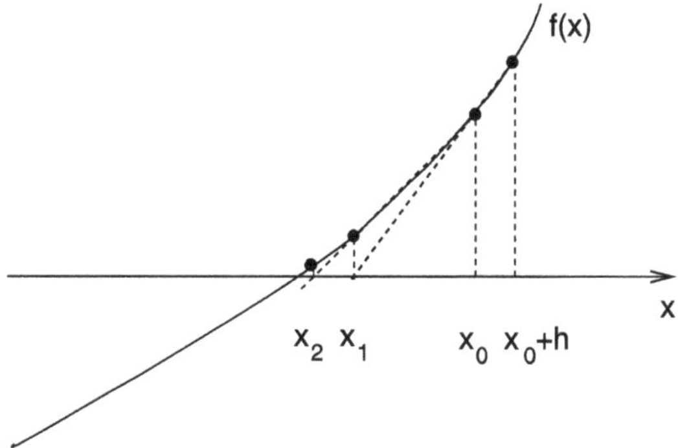

6.1.5 Interpolation

The secant method is also obtained by linear interpolation

$$p(x) = \frac{x - x_r}{x_{r-1} - x_r} f_{r-1} + \frac{x - x_{r-1}}{x_r - x_{r-1}} f_r. \tag{6.14}$$

The root of the polynomial $p(x_{r+1}) = 0$ determines the next iterate x_{r+1}

$$x_{r+1} = \frac{1}{f_{r-1} - f_r}(x_r f_{r-1} - x_{r-1} f_r) = x_r - f_r \frac{x_r - x_{r-1}}{f_r - f_{r-1}}. \tag{6.15}$$

Quadratic interpolation of three function values is known as Muller's method [177]. Newton's form of the interpolating polynomial is

$$p(x) = f_r + (x - x_r)f[x_r, x_{r-1}] + (x - x_r)(x - x_{r-1})f[x_r, x_{r-1}, x_{r-2}] \tag{6.16}$$

which can be rewritten

$$\begin{aligned}
p(x) &= f_r + (x - x_r)f[x_r, x_{r-1}] + (x - x_r)^2 f[x_r, x_{r-1}, x_{r-2}] \\
&\quad + (x_r - x_{r-1})(x - x_r)f[x_r, x_{r-1}, x_{r-2}] \\
&= f_r + (x - x_r)^2 f[x_r, x_{r-1}, x_{r-2}] \\
&\quad + (x - x_r)\big(f[x_r, x_{r-1}] + f[x_r, x_{r-2}] - f[x_{r-1}, x_{r-2}]\big) \\
&= f_r + A(x - x_r) + B(x - x_r)^2 \tag{6.17}
\end{aligned}$$

and has the roots

$$x_{r+1} = x_r - \frac{A}{2B} \pm \sqrt{\frac{A^2}{4B^2} - \frac{f_r}{B}}. \tag{6.18}$$

To avoid numerical cancellation, this is rewritten

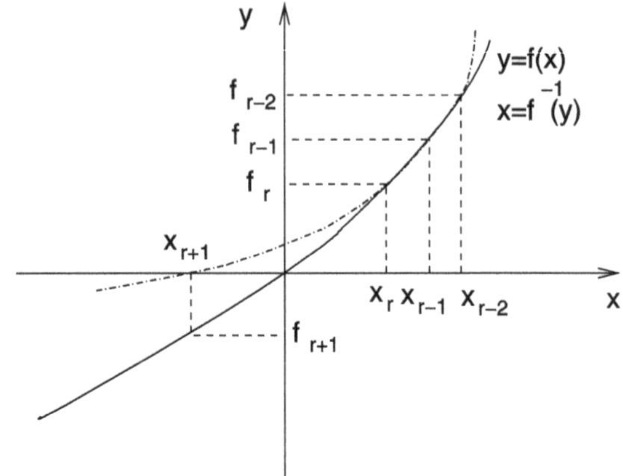

Fig. 6.6 Root finding by interpolation of the inverse function

$$x_{r+1} = x_r + \frac{1}{2B}\left(-A \pm \sqrt{A^2 - 4Bf_r}\right)$$

$$= x_r + \frac{-2f_r}{A^2 - (A^2 - 4Bf_r)}\left(A \mp \sqrt{A^2 - 4Bf_r}\right)$$

$$= x_r + \frac{-2f_r}{A \pm \sqrt{A^2 - 4Bf_r}}. \tag{6.19}$$

The sign in the denominator is chosen such that x_{r+1} is the root closer to x_r. The roots of the polynomial can become complex valued and therefore this method is useful to find complex roots.

6.1.6 Inverse Interpolation

Complex values of x_r can be avoided by interpolation of the inverse function instead

$$x = f^{-1}(y). \tag{6.20}$$

Using the two points x_r, x_{r-1} the Lagrange method gives

$$p(y) = x_{r-1}\frac{y - f_r}{f_{r-1} - f_r} + x_r\frac{y - f_{r-1}}{f_r - f_{r-1}} \tag{6.21}$$

and the next approximation of the root corresponds again to the secant method (6.12)

$$x_{r+1} = p(0) = \frac{x_{r-1}f_r - x_r f_{r-1}}{f_r - f_{r-1}} = x_r + \frac{(x_{r-1} - x_r)}{f_r - f_{r-1}}f_r. \tag{6.22}$$

Inverse quadratic interpolation needs three starting points x_r, x_{r-1}, x_{r-2} together with the function values f_r, f_{r-1}, f_{r-2} (Fig. 6.6). The inverse function $x = f^{-1}(y)$ is interpolated with the Lagrange method

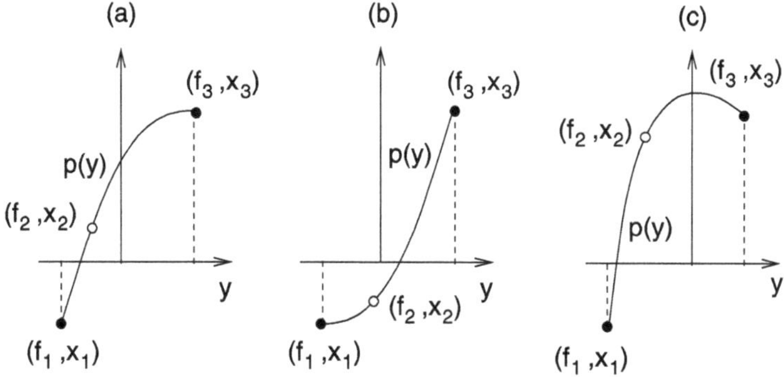

Fig. 6.7 (Validity of inverse quadratic interpolation) Inverse quadratic interpolation is only applicable if the interpolating polynomial $p(y)$ is monotonous in the range of the interpolated function values $f_1 \ldots f_3$. (**a**) and (**b**) show the limiting cases where the polynomial has a horizontal tangent at f_1 or f_3. (**c**) shows the case where the extremum of the parabola is inside the interpolation range and interpolation is not feasible

$$p(y) = \frac{(y - f_{r-1})(y - f_r)}{(f_{r-2} - f_{r-1})(f_{r-2} - f_r)} x_{r-2} + \frac{(y - f_{r-2})(y - f_r)}{(f_{r-1} - f_{r-2})(f_{r-1} - f_r)} x_{r-1}$$
$$+ \frac{(y - f_{r-1})(y - f_{r-2})}{(f_r - f_{r-1})(f_r - f_{r-2})} x_r. \tag{6.23}$$

For $y = 0$ we find the next iterate

$$x_{r+1} = p(0) = \frac{f_{r-1} f_r}{(f_{r-2} - f_{r-1})(f_{r-2} - f_r)} x_{r-2} + \frac{f_{r-2} f_r}{(f_{r-1} - f_{r-2})(f_{r-1} - f_r)} x_{r-1}$$
$$+ \frac{f_{r-1} f_{r-2}}{(f_r - f_{r-1})(f_r - f_{r-2})} x_r. \tag{6.24}$$

Inverse quadratic interpolation is only a good approximation if the interpolating parabola is single valued and hence if it is a monotonous function in the range of f_r, f_{r-1}, f_{r-2}. For the following discussion we assume that the three values of x are renamed such that $x_1 < x_2 < x_3$.

Consider the limiting case (a) in Fig. 6.7 where the polynomial has a horizontal tangent at $y = f_3$ and can be written as

$$p(y) = x_3 + (x_1 - x_3) \frac{(y - f_3)^2}{(f_1 - f_3)^2}. \tag{6.25}$$

Its value at $y = 0$ is

$$p(0) = x_3 + (x_1 - x_3) \frac{f_3^2}{(f_1 - f_3)^2}$$
$$= x_1 + (x_3 - x_1)\left(1 - \frac{f_3^2}{(f_1 - f_3)^2}\right). \tag{6.26}$$

If f_1 and f_3 have different sign and $|f_1| < |f_3|$ (Sect. 6.1.7.2) we find

$$1 - \frac{f_3^2}{(f_1 - f_3)^2} < \frac{3}{4}. \tag{6.27}$$

Brent [36] used this as a criterion for the applicability of the inverse quadratic interpolation. However, this does not include all possible cases where interpolation is applicable. Chandrupatla [54] gave a more general discussion. The limiting condition is that the polynomial $p(y)$ has a horizontal tangent at one of the boundaries $x_{1,3}$. The derivative values are

$$\begin{aligned}
\frac{dp}{dy}(y = f_1) &= \frac{x_2(f_1 - f_3)}{(f_2 - f_1)(f_2 - f_3)} + \frac{x_3(f_1 - f_2)}{(f_3 - f_1)(f_3 - f_2)} + \frac{x_1}{f_1 - f_2} + \frac{x_1}{f_1 - f_3} \\
&= \frac{(f_2 - f_1)}{(f_3 - f_1)(f_3 - f_2)}\left[\frac{x_2(f_3 - f_1)^2}{(f_2 - f_1)^2} - x_3\right.\\
&\quad \left. - \frac{x_1(f_3 - f_1)^2 - x_1(f_2 - f_1)^2}{(f_2 - f_1)^2}\right] \\
&= \frac{(f_2 - f_1)(x_2 - x_1)}{(f_3 - f_1)(f_3 - f_2)}\left[\Phi^{-2} - \xi^{-1}\right]
\end{aligned} \tag{6.28}$$

$$\begin{aligned}
\frac{dp}{dy}(y = f_3) &= \frac{x_2(f_3 - f_1)}{(f_2 - f_1)(f_2 - f_3)} + \frac{x_1(f_3 - f_2)}{(f_1 - f_2)(f_1 - f_3)} + \frac{x_3}{f_3 - f_2} + \frac{x_3}{f_3 - f_1} \\
&= \frac{(f_3 - f_2)}{(f_2 - f_1)(f_3 - f_1)}\left[-\frac{x_2(f_3 - f_1)^2}{(f_3 - f_2)^2} + x_3\frac{(f_3 - f_1)^2}{(f_3 - f_2)^2}\right.\\
&\quad \left. - x_3\frac{(f_3 - f_2)^2}{(f_3 - f_2)^2} + x_1\right] \\
&= \frac{(f_3 - f_2)(x_3 - x_2)}{(f_2 - f_1)(f_3 - f_1)}\left[\left(\frac{1}{\Phi - 1}\right)^2 - \frac{1}{1 - \xi}\right]
\end{aligned} \tag{6.29}$$

with [54]

$$\xi = \frac{x_2 - x_1}{x_3 - x_1} \qquad \Phi = \frac{f_2 - f_1}{f_3 - f_1} \tag{6.30}$$

$$\xi - 1 = \frac{x_2 - x_3}{x_3 - x_1} \qquad \Phi - 1 = \frac{f_2 - f_3}{f_3 - f_1}. \tag{6.31}$$

Since for a parabola either $f_1 < f_2 < f_3$ or $f_1 > f_2 > f_3$ the conditions for applicability of inverse interpolation finally become

$$\Phi^2 < \xi \tag{6.32}$$

$$1 - \xi > (1 - \Phi)^2 \tag{6.33}$$

which can be combined into

$$1 - \sqrt{1 - \xi} < |\Phi| < \sqrt{\xi}. \tag{6.34}$$

This method is usually used in combination with other methods (Sect. 6.1.7.2).

6.1.7 Combined Methods

Bisection converges slowly. The interpolation methods converge faster but are less reliable. The combination of both gives methods which are reliable and converge faster than pure bisection.

6.1.7.1 Dekker's Method

Dekker's method [47, 70] combines bisection and secant method. The root is bracketed by intervals $[c_r, b_r]$ with decreasing width where b_r is the best approximation to the root found and c_r is an earlier guess for which $f(c_r)$ and $f(b_r)$ have different sign. First an attempt is made to use linear interpolation between the points $(b_r, f(b_r))$ and $(a_r, f(a_r))$ where a_r is usually the preceding approximation $a_r = b_{r-1}$ and is set to the second interval boundary $a_r = c_{r-1}$ if the last iteration did not lower the function value (Fig. 6.8).

Starting from an initial interval $[x_0, x_1]$ with $\mathrm{sign}(f(x_0)) \neq \mathrm{sign}(f(x_1))$ the method proceeds as follows [47]:

- *initialization*

$$f_1 = f(x_1) \quad f_0 = f(x_0)$$
$$\text{if } |f_1| < |f_0| \text{ then } \{$$
$$b = x_1 \quad c = a = x_0$$
$$f_b = f_1 \quad f_c = f_a = f_0\}$$
$$\text{else } \{$$
$$b = x_0 \quad c = a = x_1$$
$$f_b = f_0 \quad f_c = f_a = f_1\}$$

- *iteration*

$$x_s = b - f_b \frac{b - a}{f_b - f_a}$$
$$x_m = \frac{c + b}{2}.$$

If x_s is very close to the last b then increase the distance to avoid too small steps else choose x_s if it is between b and x_m, otherwise choose x_m (thus choosing the smaller interval)

$$x_r = \begin{cases} b + \delta\,\mathrm{sign}(c - b) & \text{if } \mathrm{abs}(x_s - b) < \delta \\ x_s & \text{if } b + \delta < x_s < x_m \text{ or } b - \delta > x_s > x_m \\ x_m & \text{else.} \end{cases}$$

Fig. 6.8 Dekker's method

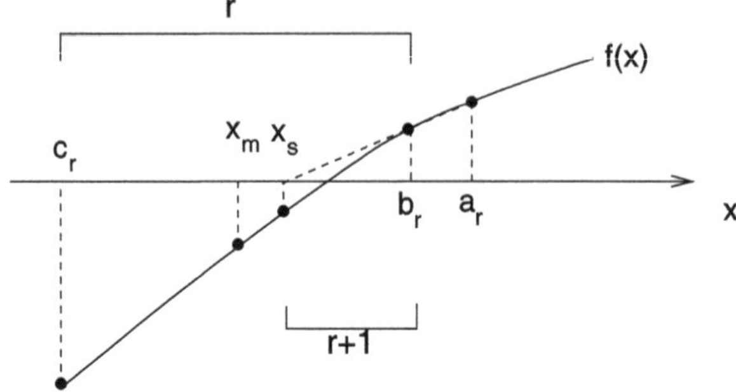

Determine x_k as the latest of the previous iterates $x_0 \ldots x_{r-1}$ for which $\mathrm{sign}(f(x_k)) \neq \mathrm{sign}(f(x_r))$.

If the new function value is lower update the approximation to the root

$$f_r = f(x_r)$$
$$\text{if } |f_r| < |f_k| \text{ then } \{$$
$$a = b \quad b = x_r \quad c = x_k$$
$$f_a = f_b \quad f_b = f_r \quad f_c = f_k\}$$

otherwise keep the old approximation and update the second interval boundary

$$\text{if } |f_r| \geq |f_k| \text{ then } \{$$
$$b = x_k \quad a = c = x_r$$
$$f_b = f_k \quad f_a = f_c = f_r\}$$

repeat until $|c - b| < \varepsilon$ or $f_r = 0$.

6.1.7.2 Brent's Method

In certain cases Dekker's method converges very slowly making many small steps of the order ε. Brent [36, 37, 47] introduced some modifications to reduce such problems and tried to speed up convergence by the use of inverse quadratic interpolation (Sect. 6.1.6). To avoid numerical problems the iterate (6.24) is written with the help of a quotient

$$x_{r+1} = \frac{f_b f_c}{(f_a - f_b)(f_a - f_c)} a + \frac{f_a f_c}{(f_b - f_a)(f_b - f_c)} b$$
$$+ \frac{f_b f_a}{(f_c - f_b)(f_c - f_a)} c$$
$$= b + \frac{p}{q} \tag{6.35}$$

with

$$p = \frac{f_b}{f_a}\left((c-b)\frac{f_a}{f_c}\left(\frac{f_a}{f_c}-\frac{f_b}{f_c}\right)-(b-a)\left(\frac{f_b}{f_c}-1\right)\right)$$

$$= (c-b)\frac{f_b(f_a-f_b)}{f_c^2}-(b-a)\frac{f_b(f_b-f_c)}{f_a f_c}$$

$$= \frac{a f_b f_c(f_b-f_c)+b[f_a f_b(f_b-f_a)+f_b f_c(f_c-f_b)]+c f_a f_b(f_a-f_b)}{f_a f_c^2}$$

$$(6.36)$$

$$q = -\left(\frac{f_a}{f_c}-1\right)\left(\frac{f_b}{f_c}-1\right)\left(\frac{f_b}{f_a}-1\right) = -\frac{(f_a-f_c)(f_b-f_c)(f_b-f_a)}{f_a f_c^2}.$$

$$(6.37)$$

If only two points are available linear interpolation is used. The iterate (6.22)
then is written as

$$x_{r+1} = b + \frac{(a-b)}{f_b-f_a}f_b = b + \frac{p}{q} \tag{6.38}$$

with

$$p = (a-b)\frac{f_b}{f_a} \qquad q = \left(\frac{f_b}{f_a}-1\right). \tag{6.39}$$

The division is only performed if interpolation is appropriate and division by zero
cannot happen. Brent's method is fast and robust at the same time. It is often recommended by text books. The algorithm is summarized in the following [209].

Start with an initial interval $[x_0, x_1]$ with $f(x_0)f(x_1) \leq 0$

- *initialization*

$$a = x_0 \quad b = x_1 \quad c = a$$
$$f_a = f(a) \quad f_b = f(b) \quad f_c = f_a$$
$$e = d = b - a$$

- *iteration*

If c is a better approximation than b exchange values

$$\text{if } |f_c| < |f_b| \text{ then } \{$$
$$a = b \quad b = c \quad c = a$$
$$f_a = f_b \quad f_b = f_c \quad f_c = f_a\}$$

calculate midpoint relative to b

$$x_m = 0.5(c-b)$$

stop if accuracy is sufficient

$$\text{if } |x_m| < \varepsilon \text{ or } f_b = 0 \text{ then exit}$$

use bisection if the previous step width e was too small or the last step did not
improve

$$\text{if } |e| < \varepsilon \text{ or } |f_a| \le |f_b| \text{ then } \{$$

$$e = d = x_m\}$$

otherwise try interpolation

$$\text{else } \{$$

$$\text{if } a = c \text{ then } \left\{$$

$$p = 2x_m \frac{f_b}{f_a} \quad q = \frac{f_b - f_a}{f_a}\right\}$$

$$\text{else } \Bigg\{$$

$$p = 2x_m \frac{f_b(f_a - f_b)}{f_c^2} - (b - a)\frac{f_b(f_b - f_c)}{f_a f_c}$$

$$q = \left(\frac{f_a}{f_c} - 1\right)\left(\frac{f_b}{f_c} - 1\right)\left(\frac{f_b}{f_a} - 1\right)\Bigg\}$$

make p a positive quantity

$$\text{if } p > 0 \text{ then } \{q = -q\} \text{ else } \{p = -p\}$$

update previous step width

$$s = e \quad e = d$$

use interpolation if applicable, otherwise use bisection

$$\text{if } 2p < 3x_m q - |\varepsilon q| \text{ and } p < |0.5\,sq| \text{ then } \left\{$$

$$d = \frac{p}{q}\right\}$$

$$\text{else } \{e = d = x_m\}$$

$$a = b \quad f_a = f_b$$

$$\text{if } |d| > \varepsilon \text{ then } \{$$

$$b = b + d\}$$

$$\text{else } \big\{b = b + \varepsilon\,\text{sign}(x_m)\big\}$$

calculate new function value

$$f_b = f(b)$$

be sure to bracket the root

$$\text{if } \text{sign}(f_b) = \text{sign}(f_c) \text{ then } \{$$

$$c = a \quad f_c = f_a$$

$$e = d = b - a\}$$

Fig. 6.9 Chandrupatla's method

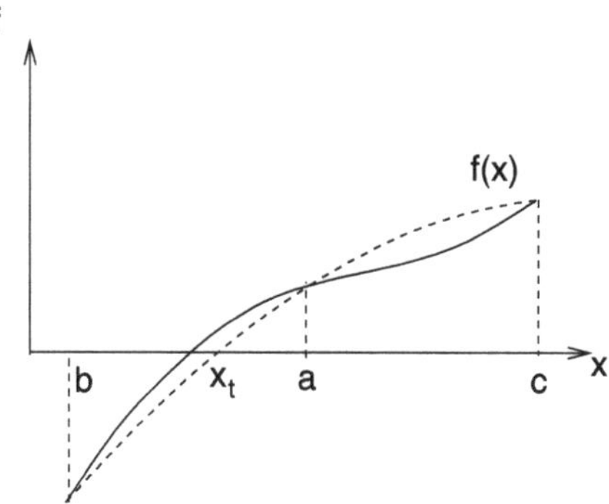

6.1.7.3 Chandrupatla's method

In 1997 Chandrupatla [54] published a method which tries to use inverse quadratic interpolation whenever possible according to (6.34). He calculates the relative position of the new iterate as (Fig. 6.9)

$$
\begin{aligned}
t &= \frac{x-c}{b-c} \\
&= \frac{1}{b-c}\left[\frac{f_c f_b}{(f_a-f_b)(f_a-f_c)}a + \frac{f_a f_c}{(f_b-f_a)(f_b-f_c)}b \right. \\
&\quad \left. + \frac{f_b f_a}{(f_c-f_b)(f_c-f_a)}c - c\right] \\
&= \frac{a-c}{b-c}\frac{f_c}{f_c-f_a}\frac{f_b}{f_b-f_a} + \frac{f_a f_c}{(f_b-f_a)(f_b-f_c)}.
\end{aligned}
\tag{6.40}
$$

The algorithm proceeds as follows:

Start with an initial interval $[x_0, x_1]$ with $f(x_0)f(x_1) \le 0$

- *initialization*

$$
\begin{aligned}
b &= x_0 \quad a = c = x_1 \\
f_b &= f(b) \quad f_a = f_c = f(c) \\
t &= 0.5
\end{aligned}
$$

- *iteration*

$$
\begin{aligned}
x_t &= a + t(b-a) \\
f_t &= f(x_t)
\end{aligned}
$$

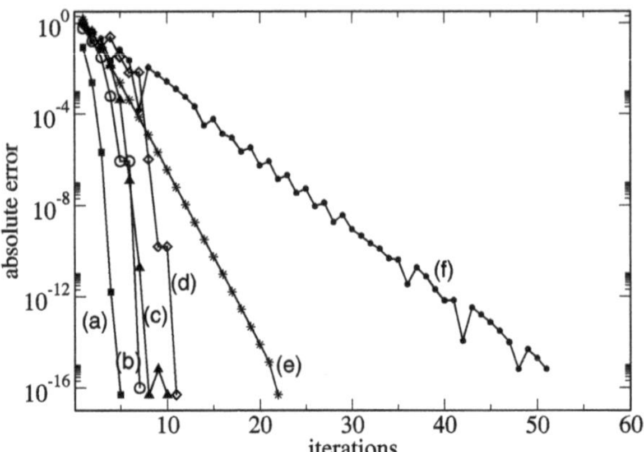

Fig. 6.10 (Comparison of different solvers) The root of the equation $f(x) = x^2 - 2$ is determined with different methods: Newton-Raphson (a, *squares*), Chandrupatla (b, *circles*), Brent (c, *triangle up*), Dekker (d, *diamonds*), regula falsi (e, *stars*), pure bisection (f, *dots*). Starting values are $x_1 = -1, x_2 = 2$. The absolute error is shown as function of the number of iterations. For $x_1 = -1$, the Newton-Raphson method converges against $-\sqrt{2}$

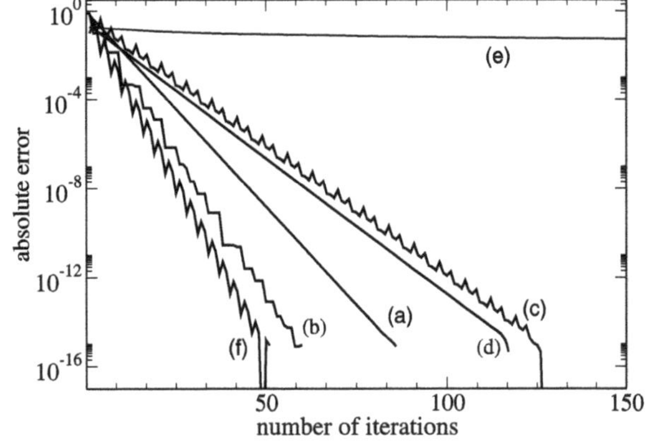

Fig. 6.11 (Comparison of different solvers for a third order root) The root of the equation $f(x) = (x - 1)^3$ is determined with different methods: Newton-Raphson (a), Chandrupatla (b), Brent (c), Dekker (d), regula falsi (e), pure bisection (f). Starting values are $x_1 = 0$, $x_2 = 1.8$. The absolute error is shown as function of the number of iterations

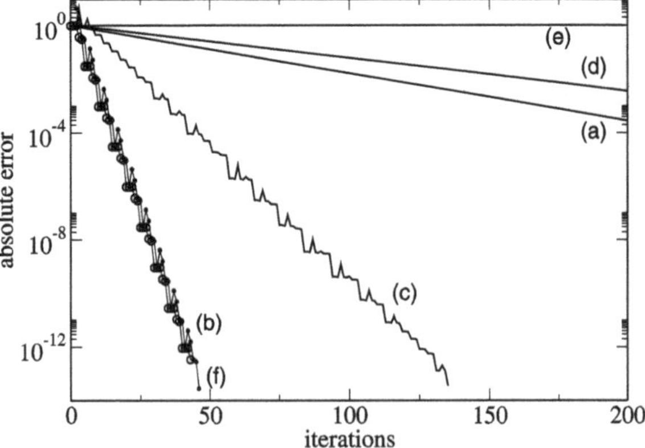

Fig. 6.12 (Comparison of different solvers for a high order root) The root of the equation $f(x) = x^{25}$ is determined with different methods: Newton-Raphson (a), Chandrupatla (b, *circles*), Brent (c), Dekker (d), regula falsi (e), pure bisection (f, *dots*). Starting values are $x_1 = -1, x_2 = 2$. The absolute error is shown as function of the number of iterations

$$\text{if } \text{sign}(f_t) = \text{sign}(f_a) \{$$
$$c = a \quad f_c = f_a$$
$$a = x_t \quad f_a = F_t\}$$
$$\text{else } \{$$
$$c = b \quad b = a \quad a = x_t$$
$$f_c = f_b \quad f_b = f_a \quad f_a = f_t\}$$
$$x_m = a \quad f_m = f_a$$
$$\text{if } \text{abs}(f_b) < \text{abs}(f_a) \{$$
$$x_m = b \quad f_m = f_b\}$$
$$tol = 2\varepsilon_M |x_m| + \varepsilon_a$$
$$t_l = \frac{tol}{|b - c|}$$
$$\text{if } t_l > 0.5 \text{ or } f_m = 0 \text{ exit}$$
$$\xi = \frac{a - b}{c - b} \quad \Phi = \frac{f_a - f_b}{f_c - f_b}$$
$$\text{if } 1 - \sqrt{1 - \xi} < \Phi < \sqrt{\xi} \left\{\right.$$
$$t = \frac{f_a}{f_b - f_a} \frac{f_c}{f_b - f_c} + \frac{c - a}{b - a} \frac{f_a}{f_c - f_a} \frac{f_b}{f_c - f_b}\right\}$$
$$\text{else } \{t = 0.5\}$$
$$\text{if } t < t_l \{t = t_l\}$$
$$\text{if } t > (1 - t_l) \{t = 1 - t_l\}$$

Chandrupatlas's method is more efficient than Dekker's and Brent's, especially for higher order roots (Figs. 6.10, 6.11, 6.12).

6.1.8 Multidimensional Root Finding

The Newton-Raphson method can be easily generalized for functions of more than one variable. We search for the solution of a system of n nonlinear equations in n variables x_i

$$\mathbf{f(x)} = \begin{pmatrix} f_1(x_1 \cdots x_n) \\ \vdots \\ f_n(x_1 \cdots x_n) \end{pmatrix} = 0. \tag{6.41}$$

The first order Newton-Raphson method results from linearization of

$$0 = \mathbf{f(x)} = \mathbf{f}(\mathbf{x}^0) + J(\mathbf{x}^0)(\mathbf{x} - \mathbf{x}^0) + \cdots \tag{6.42}$$

with the Jacobian matrix

$$J = \begin{pmatrix} \frac{\partial f_1}{\partial x_1} & \cdots & \frac{\partial f_1}{\partial x_n} \\ \vdots & \ddots & \vdots \\ \frac{\partial f_n}{\partial x_1} & \cdots & \frac{\partial f_n}{\partial x_n} \end{pmatrix}. \tag{6.43}$$

If the Jacobian matrix is not singular the equation

$$0 = \mathbf{f}(\mathbf{x}^0) + J(\mathbf{x}^0)(\mathbf{x} - \mathbf{x}^0) \tag{6.44}$$

can be solved by

$$\mathbf{x} = \mathbf{x}^0 - \left(J(\mathbf{x}^0)\right)^{-1}\mathbf{f}(\mathbf{x}^0). \tag{6.45}$$

This can be repeated iteratively

$$\mathbf{x}^{(r+1)} = \mathbf{x}^{(r)} - \left(J(\mathbf{x}^{(r)})\right)^{-1}\mathbf{f}(\mathbf{x}^{(r)}). \tag{6.46}$$

6.1.9 Quasi-Newton Methods

Calculation of the Jacobian matrix can be very time consuming. Quasi-Newton methods use instead an approximation to the Jacobian which is updated during each iteration. Defining the differences

$$\mathbf{d}^{(r)} = \mathbf{x}^{(r+1)} - \mathbf{x}^{(r)} \tag{6.47}$$

$$\mathbf{y}^{(r)} = \mathbf{f}(\mathbf{x}^{(r+1)}) - \mathbf{f}(\mathbf{x}^{(r)}) \tag{6.48}$$

we obtain from the truncated Taylor series

$$\mathbf{f}(\mathbf{x}^{(r+1)}) = \mathbf{f}(\mathbf{x}^{(r)}) + J(\mathbf{x}^{(r)})(\mathbf{x}^{(r+1)} - \mathbf{x}^{(r)}) \tag{6.49}$$

the so called Quasi-Newton or secant condition

$$\mathbf{y}^{(r)} = J(\mathbf{x}^{(r)})\mathbf{d}^{(r)}. \tag{6.50}$$

We attempt to construct a family of successive approximation matrices J_r so that, if J were a constant, the procedure would become consistent with the Quasi-Newton condition. Then for the new update J_{r+1} we have

$$J_{r+1}\mathbf{d}^{(r)} = \mathbf{y}^{(r)}. \tag{6.51}$$

Since $\mathbf{d}^{(r)}, \mathbf{y}^{(r)}$ are already known, these are only n equations for the n^2 elements of J_{r+1}. To specify J_{r+1} uniquely, additional conditions are required. For instance, it is reasonable to assume, that

$$J_{r+1}\mathbf{u} = J_r\mathbf{u} \quad \text{for all } \mathbf{u} \perp \mathbf{d}^{(r)}. \tag{6.52}$$

Then J_{r+1} differs from J_r only by a rank one updating matrix

$$J_{r+1} = J_r + \mathbf{u}\mathbf{d}^{(r)T}. \tag{6.53}$$

From the secant condition we obtain

$$J_{r+1}\mathbf{d}^{(r)} = J_r\mathbf{d}^{(r)} + \mathbf{u}\big(\mathbf{d}^{(r)}\mathbf{d}^{(r)T}\big) = \mathbf{y}^{(r)} \tag{6.54}$$

hence

$$\mathbf{u} = \frac{1}{|d^{(r)}|^2}\big(\mathbf{y}^{(r)} - J_r\mathbf{d}^{(r)}\big). \tag{6.55}$$

This gives Broyden's update formula [42]

$$J_{r+1} = J_r + \frac{1}{|d^{(r)}|^2}\big(\mathbf{y}^{(r)} - J_r\mathbf{d}^{(r)}\big)\mathbf{d}^{(r)T}. \tag{6.56}$$

To update the inverse Jacobian matrix, the Sherman-Morrison formula [236]

$$\big(A + \mathbf{u}\mathbf{v}^T\big)^{-1} = A^{-1} - \frac{A^{-1}\mathbf{u}\mathbf{v}^T A^{-1}}{1 + \mathbf{v}^T A^{-1}\mathbf{u}} \tag{6.57}$$

can be applied to have

$$\begin{aligned}
J_{r+1}^{-1} &= J_r^{-1} - \frac{J_r^{-1}\frac{1}{|d^{(r)}|^2}\big(\mathbf{y}^{(r)} - J_r\mathbf{d}^{(r)}\big)\mathbf{d}^{(r)T} J_r^{-1}}{1 + \frac{1}{|d^{(r)}|^2}\mathbf{d}^{(r)T} J_r^{-1}\big(\mathbf{y}^{(r)} - J_r\mathbf{d}^{(r)}\big)} \\
&= J_r^{-1} - \frac{\big(J_r^{-1}\mathbf{y}^{(r)} - \mathbf{d}^{(r)}\big)\mathbf{d}^{(r)T} J_r^{-1}}{\mathbf{d}^{(r)T} J_r^{-1}\mathbf{y}^{(r)}}.
\end{aligned} \tag{6.58}$$

6.2 Function Minimization

Minimization or maximization of a function[2] is a fundamental task in numerical mathematics and closely related to root finding. If the function $f(x)$ is continuously differentiable then at the extremal points the derivative is zero

$$\frac{df}{dx} = 0. \tag{6.59}$$

Hence, in principle root finding methods can be applied to locate local extrema of a function. However, in some cases the derivative cannot be easily calculated or the function even is not differentiable. Then derivative free methods similar to bisection for root finding have to be used.

6.2.1 The Ternary Search Method

Ternary search is a simple method to determine the minimum of a unimodal function $f(x)$. Initially we have to find an interval $[a_0, b_0]$ which is certain to contain the minimum. Then the interval is divided into three equal parts $[a_0, c_0]$,

---

[2] In the following we consider only a minimum since a maximum could be found as the minimum of $-f(x)$.

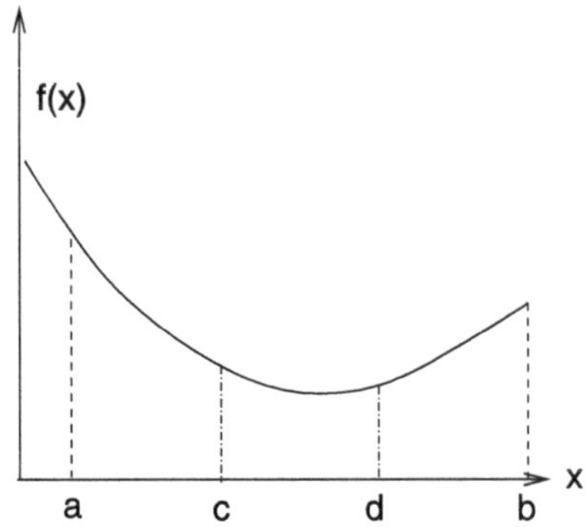

$[c_0, d_0]$, $[d_0, b_0]$ and either the first or the last of the three intervals is excluded (Fig. 6.13). The procedure is repeated with the remaining interval $[a_1, b_1] = [a_0, d_0]$ or $[a_1, b_1] = [c_0, b_0]$.

Each step needs two function evaluations and reduces the interval width by a factor of 2/3 until the maximum possible precision is obtained. It can be determined by considering a differentiable function which can be expanded around the minimum x_0 as

$$f(x) = f(x_0) + \frac{(x - x_0)^2}{2} f''(x_0) + \cdots. \tag{6.60}$$

Numerically calculated function values $f(x)$ and $f(x_0)$ only differ, if

$$\frac{(x - x_0)^2}{2} f''(x_0) > \varepsilon_M f(x_0) \tag{6.61}$$

which limits the possible numerical accuracy to

$$\varepsilon(x_0) = \min|x - x_0| = \sqrt{\frac{2 f(x_0)}{f''(x_0)} \varepsilon_M} \tag{6.62}$$

and for reasonably well behaved functions (Fig. 6.14) we have the rule of thumb [207]

$$\varepsilon(x_0) \approx \sqrt{\varepsilon_M}. \tag{6.63}$$

However, it may be impossible to reach even this precision, if the quadratic term of the Taylor series vanishes (Fig. 6.15).

The algorithm can be formulated as follows:

- *iteration*

$$\text{if } (b - a) < \delta \text{ then exit}$$

$$c = a + \frac{1}{3}(b - a) \quad d = a + \frac{2}{3}(b - a)$$

$$f_c = f(c) \quad f_d = f(d)$$

$$\text{if } f_c < f_d \text{ then } b = d \text{ else } a = c$$

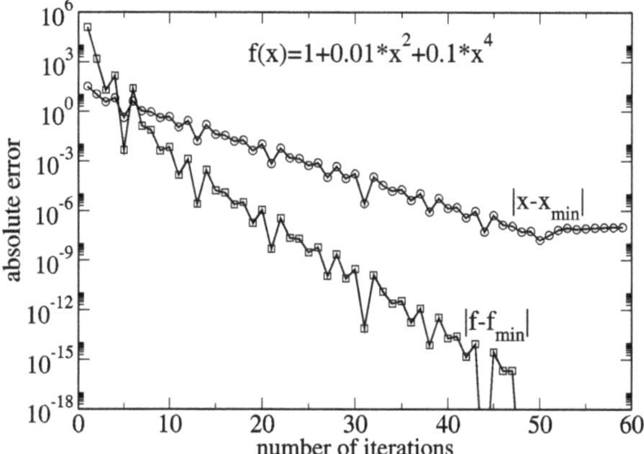

Fig. 6.14 (Ternary search method) The minimum of the function $f(x) = 1 + 0.01x^2 + 0.1x^4$ is determined with the ternary search method. Each iteration needs two function evaluations. After 50 iterations the function minimum $f_{min} = 1$ is reached to machine precision $\varepsilon_M \approx 10^{-16}$. The position of the minimum x_{min} cannot be determined with higher precision than $\sqrt{\varepsilon_M} \approx 10^{-8}$ (6.63)

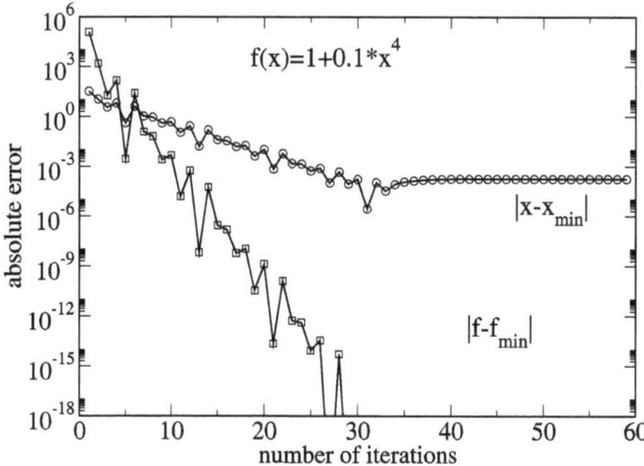

Fig. 6.15 (Ternary search method for a higher order minimum) The minimum of the function $f(x) = 1 + 0.1\,x^4$ is determined with the ternary search method. Each iteration needs two function evaluations. After 30 iterations the function minimum $f_{min} = 1$ is reached to machine precision $\varepsilon_M \approx 10^{-16}$. The position of the minimum x_{min} cannot be determined with higher precision than $\sqrt[4]{\varepsilon_M} \approx 10^{-4}$

6.2.2 The Golden Section Search Method (Brent's Method)

To bracket a local minimum of a unimodal function $f(x)$ three points a, b, c are necessary (Fig. 6.16) with

$$f(a) > f(b) \quad f(c) > f(b). \tag{6.64}$$

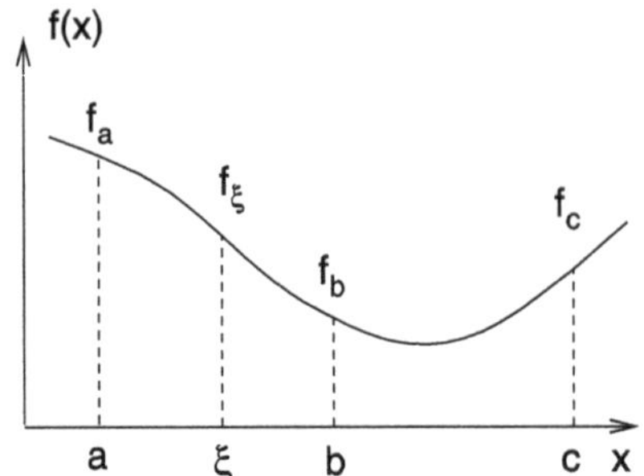

Fig. 6.16 (Golden section search method) A local minimum of the function $f(x)$ is bracketed by three points a, b, c. To reduce the uncertainty of the minimum position a new point ξ is chosen in the interval $a < \xi < c$ and either a or c is dropped according to the relation of the function values. For the example shown a has to be replaced by ξ

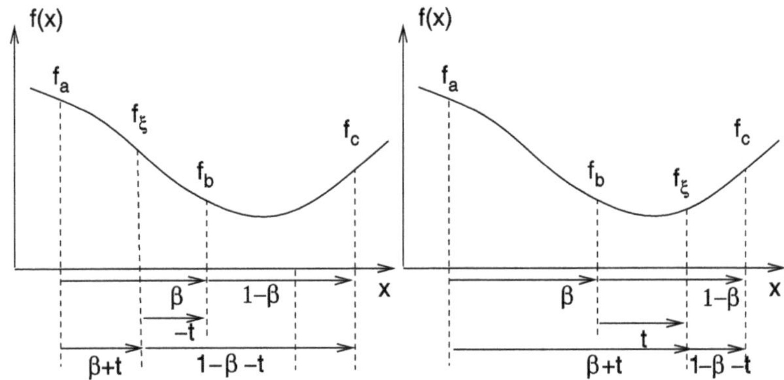

Fig. 6.17 Golden section search method

The position of the minimum can be determined iteratively by choosing a new value ξ in the interval $a < \xi < c$ and dropping either a or c, depending on the ratio of the function values. A reasonable choice for ξ can be found as follows (Fig. 6.17) [147, 207]. Let us denote the relative positions of the middle point and the trial point as

$$\frac{b-a}{c-a} = \beta \quad \frac{c-b}{c-a} = 1 - \beta \quad \frac{b-a}{c-b} = \frac{\beta}{1-\beta} \quad \frac{\xi - b}{c-a} = t$$

$$\frac{\xi - a}{c-a} = \frac{\xi - b + b - a}{c-a} = t + \beta. \tag{6.65}$$

The relative width of the new interval will be

$$\frac{c-\xi}{c-a} = (1 - \beta - t) \quad \text{or} \quad \frac{b-a}{c-a} = \beta \quad \text{if } a < \xi < b \tag{6.66}$$

$$\frac{\xi - a}{c-a} = (t + \beta) \quad \text{or} \quad \frac{c-b}{c-a} = (1 - \beta) \quad \text{if } b < \xi < c. \tag{6.67}$$

The golden search method requires that

$$t = 1 - 2\beta = \frac{c + a - 2b}{c - a} = \frac{(c - b) - (b - a)}{c - a}. \tag{6.68}$$

Otherwise it would be possible that the larger interval width is selected many times slowing down the convergence. The value of t is positive if $c - b > b - a$ and negative if $c - b < b - a$, hence the trial point always is in the larger of the two intervals. In addition the golden search method requires that the ratio of the spacing remains constant. Therefore we set

$$\frac{\beta}{1 - \beta} = -\frac{t + \beta}{t} = -\frac{1 - \beta}{t} \quad \text{if } a < \xi < b \tag{6.69}$$

$$\frac{\beta}{1 - \beta} = \frac{t}{1 - \beta - t} = \frac{t}{\beta} \quad \text{if } b < \xi < c. \tag{6.70}$$

Eliminating t we obtain for $a < \xi < b$ the equation

$$\frac{(\beta - 1)}{\beta}(\beta^2 + \beta - 1) = 0. \tag{6.71}$$

Besides the trivial solution $\beta = 1$ there is only one positive solution

$$\beta = \frac{\sqrt{5} - 1}{2} \approx 0.618. \tag{6.72}$$

For $b < \xi < c$ we end up with

$$\frac{\beta}{\beta - 1}(\beta^2 - 3\beta + 1) = 0 \tag{6.73}$$

which has the nontrivial positive solution

$$\beta = \frac{3 - \sqrt{5}}{2} \approx 0.382. \tag{6.74}$$

Hence the lengths of the two intervals $[a, b]$, $[b, c]$ have to be in the golden ratio $\varphi = \frac{1+\sqrt{5}}{2} \approx 1.618$ which gives the method its name. Using the golden ratio the width of the interval bracketing the minimum reduces by a factor of 0.618 (Figs. 6.18, 6.19).

The algorithm can be formulated as follows:

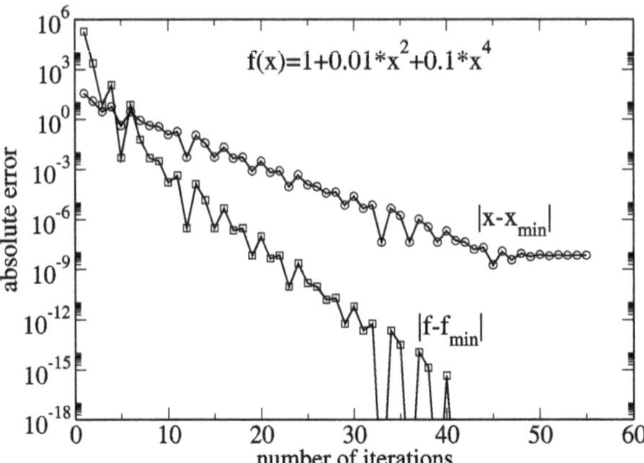

Fig. 6.18 (Golden section search method) The minimum of the function $f(x) = 1 + 0.01x^2 + 0.1x^4$ is determined with the golden section search method. Each iteration needs only one function evaluation. After 40 iterations the function minimum $f_{min} = 1$ is reached to machine precision $\varepsilon_M \approx 10^{-16}$. The position of the minimum x_{min} cannot be determined to higher precision than $\sqrt{\varepsilon_M} \approx 10^{-8}$ (6.63)

$$\text{if } c - a < \delta \text{ then exit}$$
$$\text{if } (b - a) \geq (c - b) \text{ then } \{$$
$$x = 0.618b + 0.382a$$
$$f_x = f(x)$$
$$\text{if } f_x < f_b \text{ then } \{c = b \quad b = x \quad f_c = f_b \quad f_b = f_x\}$$
$$\text{else } a = x \quad f_a = f_x\}$$
$$\text{if } (b - a) < (c - b) \text{ then } \{$$
$$x = 0.618b + 0.382c$$
$$f_x = f(x)$$
$$\text{if } f_x < f_b \text{ then } \{a = b \quad b = x \quad f_a = f_b \quad f_b = f_x\}$$
$$\text{else } c = x \quad f_c = f_x\}$$

To start the method we need three initial points which can be found by Brent's exponential search method (Fig. 6.20). Begin with three points

$$a_0, b_0 = a_0 + h, c_0 + 1.618h \tag{6.75}$$

where h_0 is a suitable initial step width which depends on the problem. If the minimum is not already bracketed then if necessary exchange a_0 and b_0 to have

$$f(a_0) > f(b_0) > f(c_0). \tag{6.76}$$

Then replace the three points by

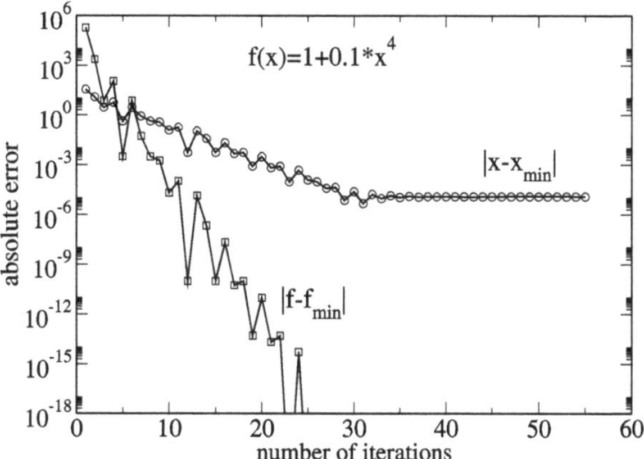

Fig. 6.19 (Golden section search for a higher order minimum) The minimum of the function $f(x) = 1 + 0.1\,x^4$ is determined with the golden section search method. Each iteration needs only one function evaluation. After 28 iterations the function minimum $f_{min} = 1$ is reached to machine precision $\varepsilon_M \approx 10^{-16}$. The position of the minimum x_{min} cannot be determined to higher precision than $\sqrt[4]{\varepsilon_M} \approx 10^{-4}$

Fig. 6.20 Brent's exponential search

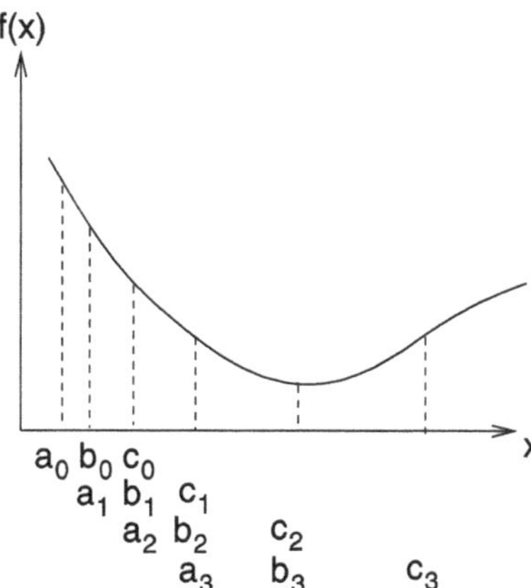

$$a_1 = b_0 \quad b_1 = c_0 \quad c_1 = c_0 + 1.618(c_0 - b_0) \tag{6.77}$$

and repeat this step until

$$f(b_n) < f(c_n) \tag{6.78}$$

or n exceeds a given maximum number. In this case no minimum can be found and we should check if the initial step width was too large.

Brent's method can be improved by making use of derivatives and by combining the golden section search with parabolic interpolation [207].

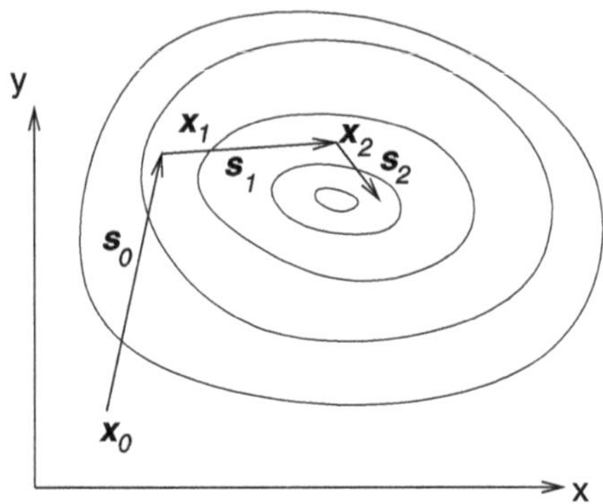

Fig. 6.21 (Direction set minimization) Starting from an initial guess $\mathbf{x}_0$ a local minimum is approached by making steps along a set of direction vectors $\mathbf{s}_r$

6.2.3 Minimization in Multidimensions

We search for local minima (or maxima) of a function

$$h(\mathbf{x})$$

which is at least two times differentiable. In the following we denote the gradient vector by

$$\mathbf{g}^T(\mathbf{x}) = \left(\frac{\partial h}{\partial x_1}, \dots, \frac{\partial h}{\partial x_n} \right) \tag{6.79}$$

and the matrix of second derivatives (Hessian) by

$$H = \left(\frac{\partial^2}{\partial x_i \partial x_j} h \right). \tag{6.80}$$

The very popular class of direction set methods proceeds as follows (Fig. 6.21). Starting from an initial guess $\mathbf{x}_0$ a set of direction vectors $\mathbf{s}_r$ and step lengths λ_r is determined such that the series of vectors

$$\mathbf{x}_{r+1} = \mathbf{x}_r + \lambda_r \mathbf{s}_r \tag{6.81}$$

approaches the minimum of $h(\mathbf{x})$. The method stops if the norm of the gradient becomes sufficiently small or if no lower function value can be found.

6.2.4 Steepest Descent Method

The simplest choice, which is known as the method of gradient descent or steepest descent[3] is to go in the direction of the negative gradient

---

[3] Which should not be confused with the method of steepest descent for the approximate calculation of integrals.

$$\mathbf{s}_r = -\mathbf{g}_r \tag{6.82}$$

and to determine the step length by minimizing h along this direction

$$h(\lambda) = h(\mathbf{x}_r - \lambda \mathbf{g}_r) = \min. \tag{6.83}$$

Obviously two consecutive steps are orthogonal to each other since

$$0 = \frac{\partial}{\partial \lambda} h(\mathbf{x}_{r+1} - \lambda \mathbf{g}_r)_{|\lambda=0} = -\mathbf{g}_{r+1}^T \mathbf{g}_r. \tag{6.84}$$

This can lead to a zig-zagging behavior and a very slow convergence of this method (Figs. 6.22, 6.23).

6.2.5 Conjugate Gradient Method

This method is similar to the steepest descent method but the search direction is iterated according to

$$\mathbf{s}_0 = -\mathbf{g}_0 \tag{6.85}$$

$$\mathbf{x}_{r+1} = \mathbf{x}_r + \lambda_r \mathbf{s}_r \tag{6.86}$$

$$\mathbf{s}_{r+1} = -\mathbf{g}_{r+1} + \beta_{r+1} \mathbf{s}_r \tag{6.87}$$

where λ_r is chosen to minimize $h(\mathbf{x}_{r+1})$ and the simplest choice for β is made by Fletcher and Rieves [89]

$$\beta_{r+1} = \frac{g_{r+1}^2}{g_r^2}. \tag{6.88}$$

6.2.6 Newton-Raphson Method

The first order Newton-Raphson method uses the iteration

$$\mathbf{x}_{r+1} = \mathbf{x}_r - H(\mathbf{x}_r)^{-1} \mathbf{g}(\mathbf{x}_r). \tag{6.89}$$

The search direction is

$$\mathbf{s} = H^{-1} \mathbf{g} \tag{6.90}$$

and the step length is $\lambda = 1$. This method converges fast if the starting point is close to the minimum. However, calculation of the Hessian can be very time consuming (Fig. 6.22).

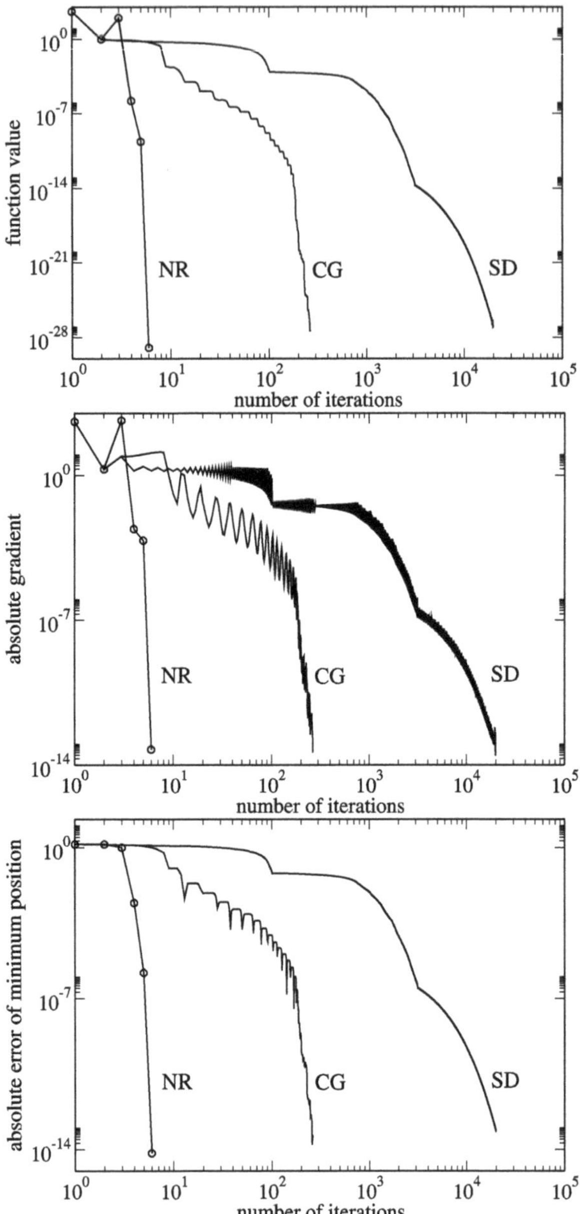

Fig. 6.22 (Function minimization) The minimum of the Rosenbrock function $h(x, y) = 100(y - x^2)^2 + (1 - x)^2$ is determined with different methods. Conjugate gradients converge much faster than steepest descent. Starting at $(x, y) = (0, 2)$, Newton-Raphson reaches the minimum at $x = y = 1$ within only 5 iterations to machine precision

6.2.7 *Quasi-Newton Methods*

Calculation of the full Hessian matrix as needed for the Newton-Raphson method can be very time consuming. Quasi-Newton methods use instead an approximation to the Hessian which is updated during each iteration. From the Taylor series

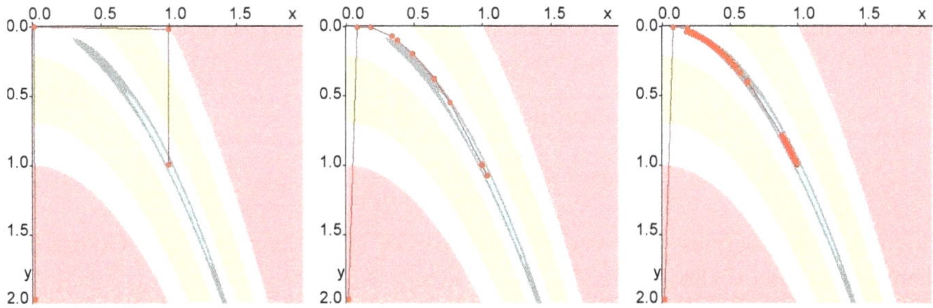

Fig. 6.23 (Minimization of the Rosenbrock function) *Left*: Newton-Raphson finds the minimum after 5 steps within machine precision. *Middle*: conjugate gradients reduce the gradient to 4×10^{-14} after 265 steps. *Right*: The steepest descent method needs 20000 steps to reduce the gradient to 5×10^{-14}. *Red lines* show the minimization pathway. *Colored areas* indicate the function value (*light blue*: < 0.1, *grey*: $0.1\ldots0.5$, *green*: $5\ldots50$, *pink*: > 100). Screenshots taken from Problem 6.2

$$h(\mathbf{x}) = h_0 + \mathbf{b}^T \mathbf{x} + \frac{1}{2}\mathbf{x}^T H \mathbf{x} + \cdots \tag{6.91}$$

we obtain the gradient

$$\mathbf{g}(\mathbf{x}_r) = \mathbf{b} + H\mathbf{x}_r + \cdots = \mathbf{g}(\mathbf{x}_{r-1}) + H(\mathbf{x}_r - \mathbf{x}_{r-1}) + \cdots. \tag{6.92}$$

Defining the differences

$$\mathbf{d}_r = \mathbf{x}_{r+1} - \mathbf{x}_r \tag{6.93}$$

$$\mathbf{y}_r = \mathbf{g}_{r+1} - \mathbf{g}_r \tag{6.94}$$

and neglecting higher order terms we obtain the Quasi-Newton or secant condition

$$H\mathbf{d}_r = \mathbf{y}_r. \tag{6.95}$$

We want to approximate the true Hessian by a series of matrices H_r which are updated during each iteration to sum up all the information gathered so far. Since the Hessian is symmetric and positive definite, this also has to be demanded for the H_r.[4] This cannot be achieved by a rank one update matrix. Popular methods use a symmetric rank two update of the form

$$H_{r+1} = H_r + \alpha \mathbf{u}\mathbf{u}^T + \beta \mathbf{v}\mathbf{v}^T. \tag{6.96}$$

The Quasi-Newton condition then gives

$$H_{r+1}\mathbf{d}_r = H_r\mathbf{d}_r + \alpha\left(\mathbf{u}^T\mathbf{d}_r\right)\mathbf{u} + \beta\left(\mathbf{v}^T\mathbf{d}_r\right)\mathbf{v} = \mathbf{y}_r \tag{6.97}$$

hence $H_r\mathbf{d}_r - \mathbf{y}_r$ must be a linear combination of $\mathbf{u}$ and $\mathbf{v}$. Making the simple choice

$$\mathbf{u} = \mathbf{y}_r \quad \mathbf{v} = H_r\mathbf{d}_r \tag{6.98}$$

---

[4]This is a major difference to the Quasi-Newton methods for root finding (6.1.9).

and assuming that these two vectors are linearly independent, we find

$$\beta = -\frac{1}{(\mathbf{v}^T \mathbf{d}_r)} = -\frac{1}{(\mathbf{d}_r^T H_r \mathbf{d}_r)} \tag{6.99}$$

$$\alpha = \frac{1}{(\mathbf{u}^T \mathbf{d}_r)} = \frac{1}{(\mathbf{y}_r^T \mathbf{d}_r)} \tag{6.100}$$

which together defines the very popular BFGS (Broyden, Fletcher, Goldfarb, Shanno) method [43, 86, 105, 235]

$$H_{r+1} = H_r + \frac{\mathbf{y}_r \mathbf{y}_r^T}{\mathbf{y}_r^T \mathbf{d}_r} - \frac{(H_r \mathbf{d}_r)(H_r \mathbf{d}_r)^T}{\mathbf{d}_r^T H_r \mathbf{d}_r}. \tag{6.101}$$

Alternatively the DFP method by Davidon, Fletcher and Powell, directly updates the inverse Hessian matrix $B = H^{-1}$ according to

$$B_{r+1} = B_r + \frac{\mathbf{d}_r \mathbf{d}_r^T}{\mathbf{y}_r^T \mathbf{d}_r} - \frac{(B_r \mathbf{y}_r)(B_r \mathbf{y}_r)^T}{\mathbf{y}_r^T B_r \mathbf{y}_r}. \tag{6.102}$$

Both of these methods can be inverted with the help of the Sherman-Morrison formula to give

$$B_{r+1} = B_r + \frac{(\mathbf{d}_r - B_r \mathbf{y}_r)\mathbf{d}_r^T + \mathbf{d}_r (\mathbf{d}_r - B \mathbf{y}_r)^T}{\mathbf{y}_r^T \mathbf{d}_r} - \frac{(\mathbf{d}_r - B_r \mathbf{y}_r)^T \mathbf{y}}{(\mathbf{y}_r^T \mathbf{d})^2} \mathbf{d} \mathbf{d}^T \tag{6.103}$$

$$H_{r+1} = H_r + \frac{(\mathbf{y}_r - H_r \mathbf{d}_r)\mathbf{y}_r^T + \mathbf{y}_r (\mathbf{y}_r - H_r \mathbf{d}_r)^T}{\mathbf{y}_r^T \mathbf{d}_r} - \frac{(\mathbf{y}_r - H_r \mathbf{d}_r)\mathbf{d}_r}{(\mathbf{y}_r^T \mathbf{d}_r)^2} \mathbf{y}_r \mathbf{y}_r^T. \tag{6.104}$$

6.3 Problems

Problem 6.1 (Root finding methods) This computer experiment searches roots of several test functions:

$$f(x) = x^n - 2 \quad n = 1, 2, 3, 4 \quad \text{(Fig. 6.10)}$$
$$f(x) = 5\sin(5x)$$
$$f(x) = \left(\cos(2x)\right)^2 - x^2$$
$$f(x) = 5\left(\sqrt{|x+2|} - 1\right)$$
$$f(x) = e^{-x} \ln x$$
$$f(x) = (x-1)^3 \quad \text{(Fig. 6.11)}$$
$$f(x) = x^{25} \quad \text{(Fig. 6.12)}$$

You can vary the initial interval or starting value and compare the behavior of different methods:

- bisection
- regula falsi
- Dekker's method
- Brent's method
- Chandrupatla's method
- Newton-Raphson method

Problem 6.2 (Stationary points) This computer experiment searches a local minimum of the Rosenbrock function[5]

$$h(x, y) = 100(y - x^2)^2 + (1 - x)^2. \tag{6.105}$$

- The method of steepest descent minimizes $h(x, y)$ along the search direction

$$s_x^{(n)} = -g_x^{(n)} = -400x(x_n^2 - y_n) - 2(x_n - 1) \tag{6.106}$$

$$s_y^{(n)} = -g_y^{(n)} = -200(y_n - x_n^2). \tag{6.107}$$

- Conjugate gradients make use of the search direction

$$s_x^{(n)} = -g_x^{(n)} + \beta_n s_x^{(n-1)} \tag{6.108}$$

$$s_y^{(n)} = -g_y^{(n)} + \beta_n s_y^{(n-1)}. \tag{6.109}$$

- The Newton-Raphson method needs the inverse Hessian

$$H^{-1} = \frac{1}{\det(H)} \begin{pmatrix} h_{yy} & -h_{xy} \\ -h_{xy} & h_{xx} \end{pmatrix} \tag{6.110}$$

$$\det(H) = h_{xx}h_{yy} - h_{xy}^2 \tag{6.111}$$

$$h_{xx} = 1200x^2 - 400y + 2 \quad h_{yy} = 200 \quad h_{xy} = -400x \tag{6.112}$$

and iterates according to

$$\begin{pmatrix} x_{n+1} \\ y_{n+1} \end{pmatrix} = \begin{pmatrix} x_n \\ y_n \end{pmatrix} - H^{-1} \begin{pmatrix} g_x^n \\ q_y^n \end{pmatrix}. \tag{6.113}$$

You can choose an initial point (x_0, y_0). The iteration stops if the gradient norm falls below 10^{-14} or if the line search fails to find a lower function value.

---

[5]A well known test function for minimization algorithms.

Chapter 7
Fourier Transformation

Fourier transformation is a very important tool for signal analysis but also helpful to simplify the solution of differential equations or the calculation of convolution integrals. In this chapter we discuss the discrete Fourier transformation as a numerical approximation to the continuous Fourier integral. It can be realized efficiently by Goertzel's algorithm or the family of fast Fourier transformation methods. Computer experiments demonstrate trigonometric interpolation and nonlinear filtering as applications.

7.1 Fourier Integral and Fourier Series

We use the symmetric definition of the Fourier transformation:

$$\widetilde{f}(\omega) = \mathcal{F}[f](\omega) = \frac{1}{\sqrt{2\pi}} \int_{-\infty}^{\infty} f(t)\mathrm{e}^{-\mathrm{i}\omega t}\,dt. \tag{7.1}$$

The inverse Fourier transformation

$$f(t) = \mathcal{F}^{-1}[\widetilde{f}](t) = \frac{1}{\sqrt{2\pi}} \int_{-\infty}^{\infty} \widetilde{f}(\omega)\mathrm{e}^{\mathrm{i}\omega t}\,d\omega \tag{7.2}$$

decomposes $f(t)$ into a superposition of oscillations. The Fourier transform of a convolution integral

$$g(t) = f(t) \otimes h(t) = \int_{-\infty}^{\infty} f(t')h(t - t')\,dt' \tag{7.3}$$

becomes a product of Fourier transforms:

$$\widetilde{g}(\omega) = \frac{1}{\sqrt{2\pi}} \int_{-\infty}^{\infty} dt'\, f(t')\mathrm{e}^{-\mathrm{i}\omega t'} \int_{-\infty}^{\infty} h(t - t')\mathrm{e}^{-\mathrm{i}\omega(t-t')}\,d(t - t')$$

$$= \sqrt{2\pi}\, \widetilde{f}(\omega)\widetilde{h}(\omega). \tag{7.4}$$

P.O.J. Scherer, *Computational Physics*, Graduate Texts in Physics,
DOI 10.1007/978-3-319-00401-3_7,

A periodic function with $f(t+T) = f(t)$[1] is transformed into a Fourier series

$$f(t) = \sum_{n=-\infty}^{\infty} e^{i\omega_n t}\, \hat{f}(\omega_n) \text{ with } \omega_n = n\frac{2\pi}{T}, \quad \hat{f}(\omega_n) = \frac{1}{T}\int_0^T f(t)e^{-i\omega_n t}\,dt. \quad (7.5)$$

For a periodic function which in addition is real valued $f(t) = f(t)^*$ and even $f(t) = f(-t)$, the Fourier series becomes a cosine series

$$f(t) = \hat{f}(\omega_0) + 2\sum_{n=1}^{\infty} \hat{f}(\omega_n)\cos\omega_n t \qquad (7.6)$$

with real valued coefficients

$$\hat{f}(\omega_n) = \frac{1}{T}\int_0^T f(t)\cos\omega_n t\,dt. \qquad (7.7)$$

7.2 Discrete Fourier Transformation

We divide the time interval $0 \le t < T$ by introducing a grid of N equidistant points

$$t_n = n\Delta t = n\frac{T}{N} \quad \text{with } n = 0, 1 \cdots N - 1. \qquad (7.8)$$

The function values (samples)

$$f_n = f(t_n) \qquad (7.9)$$

are arranged as components of a vector

$$\mathbf{f} = \begin{pmatrix} f_0 \\ \vdots \\ f_{N-1} \end{pmatrix}.$$

With respect to the orthonormal basis

$$\mathbf{e}_n = \begin{pmatrix} \delta_{0,n} \\ \vdots \\ \delta_{N-1,n} \end{pmatrix}, \quad n = 0, 1 \cdots N - 1 \qquad (7.10)$$

$\mathbf{f}$ is expressed as a linear combination

$$\mathbf{f} = \sum_{n=0}^{N-1} f_n \mathbf{e}_n. \qquad (7.11)$$

---

[1]This could also be the periodic continuation of a function which is only defined for $0 < t < T$.

The discrete Fourier transformation is the transformation to an orthogonal base in frequency space

$$\mathbf{e}_{\omega_j} = \sum_{n=0}^{N-1} e^{i\omega_j t_n} \mathbf{e}_n = \begin{pmatrix} 1 \\ e^{i\frac{2\pi}{N}j} \\ \vdots \\ e^{i\frac{2\pi}{N}j(N-1)} \end{pmatrix} \tag{7.12}$$

with

$$\omega_j = \frac{2\pi}{T}j. \tag{7.13}$$

These vectors are orthogonal

$$\mathbf{e}_{\omega_j}\mathbf{e}^*_{\omega_{j'}} = \sum_{n=0}^{N-1} e^{i(j-j')\frac{2\pi}{N}n} = \begin{cases} \frac{1-e^{i(j-j')2\pi}}{1-e^{i(j-j')2\pi/N}} = 0 & \text{for } j-j' \neq 0 \\ N & \text{for } j-j' = 0 \end{cases} \tag{7.14}$$

$$\mathbf{e}_{\omega_j}\mathbf{e}^*_{\omega_{j'}} = N\delta_{j,j'}. \tag{7.15}$$

Alternatively a real valued basis can be defined:

$$\begin{aligned} &\cos\left(\frac{2\pi}{N}jn\right) && j = 0, 1 \cdots j_{\max} \\ &\sin\left(\frac{2\pi}{N}jn\right) && j = 1, 2 \cdots j_{\max} \\ &j_{\max} = \frac{N}{2} \text{ (even } N) \quad j_{\max} = \frac{N-1}{2} \text{ (odd } N). \end{aligned} \tag{7.16}$$

The components of $\mathbf{f}$ in frequency space are given by the scalar product

$$\tilde{f}_{\omega_j} = \mathbf{f}\mathbf{e}_{\omega_j} = \sum_{n=0}^{N-1} f_n e^{-i\omega_j t_n} = \sum_{n=0}^{N-1} f_n e^{-ij\frac{2\pi}{T}n\frac{T}{N}} = \sum_{n=0}^{N-1} f_n e^{-i\frac{2\pi}{N}jn}. \tag{7.17}$$

From

$$\sum_{j=0}^{N-1} \tilde{f}_{\omega_j} e^{i\omega_j t_n} = \sum_{n'}\sum_{\omega_j} f_{n'} e^{-i\omega_j t_{n'}} e^{i\omega_j t_n} = N f_n \tag{7.18}$$

we find the inverse transformation

$$f_n = \frac{1}{N}\sum_{j=0}^{N-1} \tilde{f}_{\omega_j} e^{i\omega_j t_n} = \frac{1}{N}\sum_{j=0}^{N-1} \tilde{f}_{\omega_j} e^{i\frac{2\pi}{N}nj}. \tag{7.19}$$

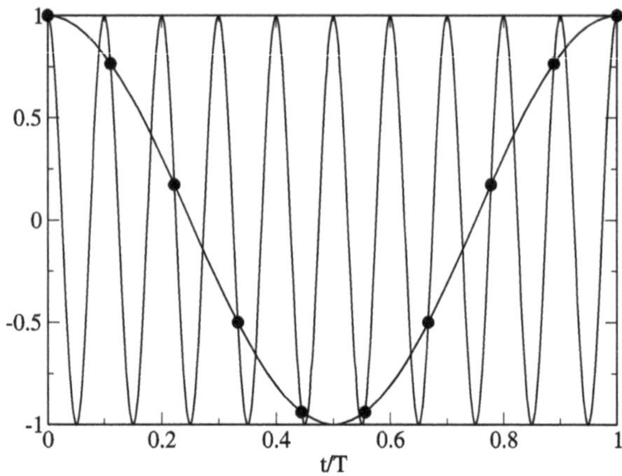

Fig. 7.1 (Equivalence of ω_1 and ω_{N-1}) The two functions $\cos \omega t$ and $\cos(N-1)\omega t$ have the same values at the sample points t_n but are very different in between

7.2.1 Trigonometric Interpolation

The last equation can be interpreted as an interpolation of the function $f(t)$ at the sampling points t_n by a linear combination of trigonometric functions

$$f(t) = \frac{1}{N} \sum_{j=0}^{N-1} \tilde{f}_{\omega_j} \left(e^{i\frac{2\pi}{T}t} \right)^j \tag{7.20}$$

which is a polynomial of

$$q = e^{i\frac{2\pi}{T}t}. \tag{7.21}$$

Since

$$e^{-i\omega_j t_n} = e^{-i\frac{2\pi}{N}jn} = e^{i\frac{2\pi}{N}(N-j)n} = e^{i\omega_{N-j}t_n} \tag{7.22}$$

the frequencies ω_j and ω_{N-j} are equivalent (Fig. 7.1)

$$\tilde{f}_{\omega_{N-j}} = \sum_{n=0}^{N-1} f_n e^{-i\frac{2\pi}{N}(N-j)n} = \sum_{n=0}^{N-1} f_n e^{i\frac{2\pi}{N}jn} = \tilde{f}_{\omega_{-j}}. \tag{7.23}$$

If we use trigonometric interpolation to approximate $f(t)$ between the grid points, the two frequencies are no longer equivalent and we have to restrict the frequency range to avoid unphysical high frequency components (Fig. 7.2):

$$\begin{aligned} -\frac{2\pi}{T}\frac{N-1}{2} &\leq \omega_j \leq \frac{2\pi}{T}\frac{N-1}{2} \quad N \text{ odd} \\ -\frac{2\pi}{T}\frac{N}{2} &\leq \omega_j \leq \frac{2\pi}{T}\left(\frac{N}{2}-1\right) \quad N \text{ even.} \end{aligned} \tag{7.24}$$

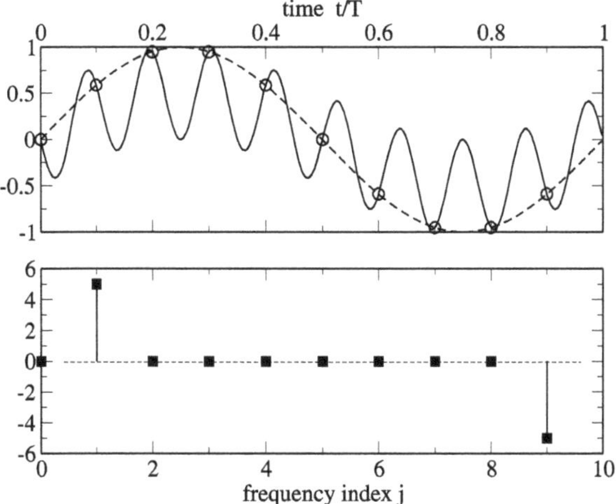

Fig. 7.2 (Trigonometric interpolation) For trigonometric interpolation the high frequencies have to be replaced by the corresponding negative frequencies to provide meaningful results between the sampling points. The *circles* show sampling points which are fitted using only positive frequencies (*full curve*) or replacing the unphysical high frequency by its negative counterpart (*broken curve*). The *squares* show the calculated Fourier spectrum

The interpolating function (N even) is

$$f(t) = \frac{1}{N} \sum_{j=-\frac{N}{2}}^{\frac{N}{2}-1} \widetilde{f}_{\omega_j} e^{i\omega_j t} \quad \text{even } N \tag{7.25}$$

$$f(t) = \frac{1}{N} \sum_{j=-\frac{N-1}{2}}^{\frac{N-1}{2}} \widetilde{f}_{\omega_j} e^{i\omega_j t} \quad \text{odd } N. \tag{7.26}$$

The maximum frequency is

$$\omega_{\max} = \frac{2\pi}{T} \frac{N}{2} \tag{7.27}$$

and hence

$$f_{\max} = \frac{1}{2\pi} \omega_{\max} = \frac{N}{2T} = \frac{f_s}{2}. \tag{7.28}$$

This is known as the sampling theorem which states that the sampling frequency f_s must be larger than twice the maximum frequency present in the signal.

7.2.2 Real Valued Functions

For a real valued function

$$f_n = f_n^*$$ (7.29)

and hence

$$\tilde{f}_{\omega_j}^* = \left(\sum_{n=0}^{N-1} f_n e^{-i\omega_j t_n}\right)^* = \sum_{n=0}^{N-1} f_n e^{i\omega_j t_n} = \tilde{f}_{\omega_{-j}}.$$ (7.30)

Here it is sufficient to calculate the sums for $j = 0 \cdots N/2$. If the function is real valued and also even

$$f_{-n} = f_n$$ (7.31)

$$\tilde{f}_{\omega_j} = \sum_{n=0}^{N-1} f_{-n} e^{-i\omega_j t_n} = \sum_{n=0}^{N-1} f_n e^{-i(-\omega_j)t_n} = \tilde{f}_{\omega_{-j}}$$ (7.32)

and the Fourier sum (7.19) turns into a cosine sum

$$f_n = \frac{1}{2M-1}\tilde{f}_{\omega_0} + \frac{2}{2M-1}\sum_{j=1}^{M-1} \tilde{f}_{\omega_j} \cos\left(\frac{2\pi}{2M-1}jn\right) \quad \text{odd } N = 2M-1 \quad (7.33)$$

$$f_n = \frac{1}{2M}\tilde{f}_{\omega_0} + \frac{1}{M}\sum_{j=1}^{M-1} \tilde{f}_{\omega_j} \cos\left(\frac{\pi}{M}jn\right) + \frac{1}{2M}\tilde{f}_{\omega_M}\cos(n\pi) \quad \text{even } N = 2M$$

(7.34)

which correspond to two out of eight different versions [268] of the discrete cosine transformation [2, 211].

Equation (7.34) can be used to define the interpolating function

$$f(t) = \frac{1}{2M}\tilde{f}_{\omega_0} + \frac{1}{M}\sum_{j=1}^{M-1} \tilde{f}_{\omega_j} \cos(\omega_j t) + \frac{1}{2M}\tilde{f}_{\omega_M}\cos\left(\frac{2\pi M}{T}t\right).$$ (7.35)

The real valued Fourier coefficients are given by

$$\tilde{f}_{\omega_j} = f_0 + 2\sum_{n=1}^{M-1} f_n \cos(\omega_j t_n) \quad \text{odd } N = 2M-1$$ (7.36)

$$\tilde{f}_{\omega_j} = f_0 + 2\sum_{n=1}^{M-1} f_n \cos(\omega_j t_n) + f_M \cos(j\pi) \quad \text{even } N = 2M.$$ (7.37)

7.2.3 Approximate Continuous Fourier Transformation

We continue the function $f(t)$ periodically by setting

$$f_N = f_0 \tag{7.38}$$

and write

$$\tilde{f}_{\omega_j} = \sum_{n=0}^{N-1} f_n e^{-i\omega_j n} = \frac{1}{2} f_0 + e^{-i\omega_j} f_1 + \cdots + e^{-i\omega_j (N-1)} f_{N-1} + \frac{1}{2} f_N. \tag{7.39}$$

Comparing with the trapezoidal rule (4.13) for the integral

$$\int_0^T e^{-i\omega_j t} f(t)dt \approx \frac{T}{N}\left(\frac{1}{2}e^{-i\omega_j 0} f(0) + e^{-i\omega_j \frac{T}{N}} f\left(\frac{T}{N}\right) + \cdots \right.$$

$$\left. + e^{-i\omega_j \frac{T}{N}(N-1)} f\left(\frac{T}{N}(N-1)\right) + \frac{1}{2}f(T)\right) \tag{7.40}$$

we find

$$\hat{f}(\omega_j) = \frac{1}{T}\int_0^T e^{-i\omega_j t} f(t)dt \approx \frac{1}{N}\tilde{f}_{\omega_j} \tag{7.41}$$

which shows that the discrete Fourier transformation is an approximation to the Fourier series of a periodic function with period T which coincides with $f(t)$ in the interval $0 < t < T$. The range of the integral can be formally extended to $\pm\infty$ by introducing a windowing function

$$W(t) = \begin{cases} 1 & \text{for } 0 < t < T \\ 0 & \text{else.} \end{cases} \tag{7.42}$$

The discrete Fourier transformation approximates the continuous Fourier transformation but windowing leads to a broadening of the spectrum. For practical purposes smoother windowing functions are used like a triangular window or one of the following [119]:

$$W(t_n) = e^{-\frac{1}{2}\left(\frac{n-(N-1)/2}{\sigma(N-1)/2}\right)^2} \quad \sigma \leq 0.5 \quad \text{Gaussian window}$$

$$W(t_n) = 0.53836 - 0.46164\cos\left(\frac{2\pi n}{N-1}\right) \quad \text{Hamming window}$$

$$W(t_n) = 0.5\left(1 - \cos\left(\frac{2\pi n}{N-1}\right)\right) \quad \text{Hann(ing) window.}$$

7.3 Fourier Transform Algorithms

Straightforward evaluation of the sum

$$\widetilde{f}_{\omega_j} = \sum_{n=0}^{N-1} \cos\left(\frac{2\pi}{N} jn\right) f_n + \mathrm{i}\sin\left(\frac{2\pi}{N} jn\right) f_n \tag{7.43}$$

needs $O(N^2)$ additions, multiplications and trigonometric functions.

7.3.1 Goertzel's Algorithm

Goertzel's method [103] is very useful if not the whole Fourier spectrum is needed but only some of the Fourier components, for instance to demodulate a frequency shift key signal or the dial tones which are used in telephony.

The Fourier transform can be written as

$$\sum_{n=0}^{N-1} f_n e^{-\mathrm{i}\frac{2\pi}{N} jn} = f_0 + e^{-\frac{2\pi\mathrm{i}}{N} j}\left(f_1 + e^{-\frac{2\pi\mathrm{i}}{N} j} f_2 \cdots \left(f_{N-2} + e^{-\frac{2\pi\mathrm{i}}{N} j} f_{N-1}\right) \cdots \right) \tag{7.44}$$

which can be evaluated recursively

$$\begin{aligned} y_{N-1} &= f_{N-1} \\ y_n &= f_n + e^{-\frac{2\pi\mathrm{i}}{N} j} y_{n+1} \quad n = N-2 \cdots 0 \end{aligned} \tag{7.45}$$

to give the result

$$\hat{f}_{\omega_j} = y_0. \tag{7.46}$$

Equation (7.45) is a simple discrete filter function. Its transmission function is obtained by application of the z-transform [144]

$$u(z) = \sum_{n=0}^{\infty} u_n z^{-n} \tag{7.47}$$

(the discrete version of the Laplace transform) which yields

$$y(z) = \frac{f(z)}{1 - z e^{-\frac{2\pi\mathrm{i}}{N} j}}. \tag{7.48}$$

One disadvantage of this method is that it uses complex numbers. This can be avoided by the following more complicated recursion

$$\begin{aligned} u_{N+1} &= u_N = 0 \\ u_n &= f_n + 2u_{n+1}\cos\frac{2\pi}{N}k - u_{n+2} \quad \text{for } n = N-1 \cdots 0 \end{aligned} \tag{7.49}$$

with the transmission function

$$\frac{u(z)}{f(z)} = \frac{1}{1 - z(e^{\frac{2\pi i}{N}j} + e^{-\frac{2\pi i}{N}j}) + z^2}$$

$$= \frac{1}{(1 - ze^{-\frac{2\pi i}{N}j})(1 - ze^{\frac{2\pi i}{N}j})}. \tag{7.50}$$

A second filter removes one factor in the denominator

$$\frac{y(z)}{u(z)} = \left(1 - ze^{\frac{2\pi i}{N}j}\right) \tag{7.51}$$

which in the time domain corresponds to the simple expression

$$y_n = u_n - e^{\frac{2\pi i}{N}j} u_{n+1}.$$

The overall filter function finally again is (7.48).

$$\frac{y(z)}{f(z)} = \frac{1}{1 - ze^{-\frac{2\pi i}{N}j}}. \tag{7.52}$$

Hence the Fourier component of $\mathbf{f}$ is given by

$$\hat{f}_{\omega_j} = y_0 = u_0 - e^{\frac{2\pi i}{N}j} u_1. \tag{7.53}$$

The order of the iteration (7.44) can be reversed by writing

$$\hat{f}_{\omega_j} = f_0 \cdots e^{\frac{2\pi i}{N}(N-1)} f_{N-1} = e^{-\frac{2\pi i}{N}j(N-1)}\left(f_0 e^{\frac{2\pi i}{N}j(N-1)} \cdots f_{N-1}\right) \tag{7.54}$$

which is very useful for real time filter applications.

7.3.2 Fast Fourier Transformation

If the number of samples is $N = 2^p$, the Fourier transformation can be performed very efficiently by this method.[2] The phase factor

$$e^{-i\frac{2\pi}{N}jm} = W_N^{jm} \tag{7.55}$$

can take only N different values. The number of trigonometric functions can be reduced by reordering the sum. Starting from a sum with N samples

$$F_N(f_0 \cdots f_{N-1}) = \sum_{n=0}^{N-1} f_n W_N^{jn} \tag{7.56}$$

---

[2]There exist several Fast Fourier Transformation algorithms [74, 187]. We consider only the simplest one here [62].

we separate even and odd powers of the unit root

$$F_N(f_0 \cdots f_{N-1}) = \sum_{m=0}^{\frac{N}{2}-1} f_{2m} W_N^{j2m} + \sum_{m=0}^{\frac{N}{2}-1} f_{2m+1} W_N^{j(2m+1)}$$

$$= \sum_{m=0}^{\frac{N}{2}-1} f_{2m} e^{-i\frac{2\pi}{N/2} jm} + W_N^j \sum_{m=0}^{\frac{N}{2}-1} f_{2m+1} e^{-i\frac{2\pi}{N/2} jm}$$

$$= F_{N/2}(f_0, f_2 \cdots f_{N-2}) + W_N^j F_{N/2}(f_1, f_3 \cdots f_{N-1}).$$

$$(7.57)$$

This division is repeated until only sums with one summand remain

$$F_1(f_n) = f_n. \tag{7.58}$$

For example, consider the case $N = 8$:

$$F_8(f_0 \cdots f_7) = F_4(f_0 f_2 f_4 f_6) + W_8^j F_4(f_1 f_3 f_5 f_7)$$
$$- - -$$
$$F_4(f_0 f_2 f_4 f_6) = F_2(f_0 f_4) + W_4^j F_2(f_2 f_6)$$
$$F_4(f_1 f_3 f_5 f_7) = F_2(f_1 f_5) + W_4^j F_2(f_3 f_7)$$
$$- - -$$
$$F_2(f_0 f_4) = f_0 + W_2^j f_4$$
$$F_2(f_2 f_6) = f_2 + W_2^j f_6$$
$$F_2(f_1 f_5) = f_1 + W_2^j f_5$$
$$F_2(f_3 f_7) = f_3 + W_2^j f_7.$$

Expansion gives

$$F_8 = f_0 + W_2^j f_4 + W_4^j f_2 + W_4^j W_2^j f_6$$
$$+ W_8^j f_1 + W_8^j W_2^j f_5 + W_8^j W_4^j f_3 + W_8^j W_4^j W_2^j f_7. \tag{7.59}$$

Generally a summand of the Fourier sum can be written using the binary representation of n

$$n = \sum l_i \quad l_i = 1, 2, 4, 8 \cdots \tag{7.60}$$

in the following way:

$$f_n e^{-i\frac{2\pi}{N} jn} = f_n e^{-i\frac{2\pi}{N}(l_1 + l_2 + \cdots)j} = f_n W_{N/l_1}^j W_{N/l_2}^j \cdots. \tag{7.61}$$

The function values are reordered according to the following algorithm

(i)   count from 0 to $N - 1$ using binary numbers $m = 000, 001, 010, \cdots$
(ii) bit reversal gives the binary numbers $n = 000, 100, 010, \cdots$
(iii) store f_n at the position m. This will be denoted as $s_m = f_n$

As an example for $N = 8$ the function values are in the order

$$
\begin{pmatrix} s_0 \\ s_1 \\ s_2 \\ s_3 \\ s_4 \\ s_5 \\ s_6 \\ s_7 \end{pmatrix} = \begin{pmatrix} f_0 \\ f_4 \\ f_2 \\ f_6 \\ f_1 \\ f_5 \\ f_3 \\ f_7 \end{pmatrix}.
\tag{7.62}
$$

Now calculate sums with two summands. Since W_2^j can take only two different values

$$
W_2^j = \begin{cases} 1 & \text{for } j = 0, 2, 4, 6 \\ -1 & \text{for } j = 1, 3, 5, 7 \end{cases}
\tag{7.63}
$$

a total of 8 sums have to be calculated which can be stored again in the same workspace:

$$
\begin{pmatrix} f_0 + f_4 \\ f_0 - f_4 \\ f_2 + f_6 \\ f_2 - f_6 \\ f_1 + f_5 \\ f_1 - f_5 \\ f_3 + f_7 \\ f_3 - f_7 \end{pmatrix} = \begin{pmatrix} s_0 + W_2^0 s_1 \\ s_0 + W_2^1 s_1 \\ s_2 + W_2^2 s_3 \\ s_2 + W_2^3 s_3 \\ s_4 + W_2^4 s_5 \\ s_4 + W_2^5 s_5 \\ s_6 + W_2^6 s_7 \\ s_6 + W_2^7 s_7 \end{pmatrix}.
\tag{7.64}
$$

Next calculate sums with four summands. W_4^j can take one of four values

$$
W_4^j = \begin{cases} 1 & \text{for } j = 0, 4 \\ -1 & \text{for } j = 2, 6 \\ W_4 & \text{for } j = 1, 5 \\ -W_4 & \text{for } j = 3, 7. \end{cases}
\tag{7.65}
$$

The following combinations are needed:

$$
\begin{pmatrix}
f_0 + f_4 + (f_2 + f_6) \\
f_0 + f_4 - (f_2 + f_6) \\
(f_0 - f_4) + W_4(f_2 - f_6) \\
(f_0 - f_4) - W_4(f_2 - f_6) \\
f_1 + f_5 + (f_3 + f_7) \\
f_1 + f_5 - (f_3 + f_7) \\
(f_1 - f_5) \pm W_4(f_3 - f_7) \\
(f_1 - f_5) \pm W_4(f_3 - f_7)
\end{pmatrix}
=
\begin{pmatrix}
s_0 + W_4^0 s_2 \\
s_1 + W_4^1 s_3 \\
s_0 + W_4^2 s_2 \\
s_1 + W_4^3 s_3 \\
s_4 + W_4^4 s_6 \\
s_5 + W_4^5 s_7 \\
s_4 + W_4^6 s_6 \\
s_5 + W_4^7 s_7
\end{pmatrix} .
\tag{7.66}
$$

The next step gives the sums with eight summands. With

$$
W_8^j =
\begin{cases}
1 & j = 0 \\
W_8 & j = 1 \\
W_8^2 & j = 2 \\
W_8^3 & j = 3 \\
-1 & j = 4 \\
-W_8 & j = 5 \\
-W_8^2 & j = 6 \\
-W_8^3 & j = 7
\end{cases}
\tag{7.67}
$$

we calculate

$$
\begin{pmatrix}
f_0 + f_4 + (f_2 + f_6) + (f_1 + f_5 + (f_3 + f_7)) \\
f_0 + f_4 - (f_2 + f_6) + W_8(f_1 + f_5 - (f_3 + f_7)) \\
(f_0 - f_4) + W_4(f_2 - f_6) + W_8^2(f_1 - f_5) \pm W_4(f_3 - f_7) \\
(f_0 - f_4) - W_4(f_2 - f_6) + W_8^3((f_1 - f_5) \pm W_4(f_3 - f_7)) \\
f_0 + f_4 + (f_2 + f_6) - (f_1 + f_5 + (f_3 + f_7)) \\
f_0 + f_4 - (f_2 + f_6) - W_8(f_1 + f_5 - (f_3 + f_7)) \\
(f_0 - f_4) + W_4(f_2 - f_6) - W_8^2((f_1 - f_5) \pm W_4(f_3 - f_7)) \\
(f_0 - f_4) - W_4(f_2 - f_6) - W_8^3((f_1 - f_5) \pm W_4(f_3 - f_7))
\end{pmatrix}
=
\begin{pmatrix}
s_0 + W_8^0 s_4 \\
s_1 + W_8^1 s_5 \\
s_2 + W_8^2 s_6 \\
s_3 + W_8^3 s_7 \\
s_0 + W_8^4 s_4 \\
s_1 + W_8^5 s_5 \\
s_2 + W_8^6 s_6 \\
s_3 + W_8^7 s_7
\end{pmatrix}
$$
$$
\tag{7.68}
$$

which is the final result.

The following shows a simple Fast Fourier Transformation algorithm. The number of trigonometric function evaluations can be reduced but this reduces the readability. At the beginning $Data[k]$ are the input data in bit reversed order.

```
size := 2
first := 0
While first < Number_of_Samples do begin
```

```
   for n := 0 to size/2 − 1 do begin
      j := first + n
      k := j + size/2 − 1
      T := exp(−2∗Pi∗i∗n/Number_of_Samples)∗Data[k]
      Data[j] := Data[j] + T
      Data[k] := Data[k] − T
   end;
   first := first∗2
   size := size∗2
end;
```

7.4 Problems

Problem 7.1 (Discrete Fourier transformation) In this computer experiment for a given set of input samples

$$f_n = f\left(n\frac{T}{N}\right) \quad n = 0 \cdots N - 1 \tag{7.69}$$

- the Fourier coefficients

$$\widetilde{f}_{\omega_j} = \sum_{n=0}^{N-1} f_n e^{-i\omega_j t_n} \quad \omega_j = \frac{2\pi}{T} j, \quad j = 0 \cdots N - 1 \tag{7.70}$$

  are calculated with Goertzel's method (7.3.1).
- The results from the inverse transformation

$$f_n = \frac{1}{N} \sum_{j=0}^{N-1} \widetilde{f}_{\omega_j} e^{i\frac{2\pi}{N}nj} \tag{7.71}$$

  are compared with the original function values $f(t_n)$.
- The Fourier sum is used for trigonometric interpolation with only positive frequencies

$$f(t) = \frac{1}{N} \sum_{j=0}^{N-1} \widetilde{f}_{\omega_j} \left(e^{i\frac{2\pi}{T}t}\right)^j . \tag{7.72}$$

- Finally the unphysical high frequencies are replaced by negative frequencies (7.24). The results can be studied for several kinds of input data.

Problem 7.2 (Noise filter) This computer experiment demonstrates a nonlinear filter.

First a noisy input signal is generated.

The signal can be chosen as

- monochromatic $\sin(\omega t)$
- the sum of two monochromatic signals $a_1 \sin \omega_1 t + a_2 \sin \omega_2 t$
- a rectangular signal with many harmonic frequencies $\mathrm{sign}(\sin \omega t)$

Different kinds of white noise can be added

- dichotomous ± 1
- constant probability density in the range $[-1, 1]$
- Gaussian probability density

The amplitudes of signal and noise can be varied. All Fourier components are removed which are below a threshold value and the filtered signal is calculated by inverse Fourier transformation.

Chapter 8
Random Numbers and Monte Carlo Methods

Many-body problems often involve the calculation of integrals of very high dimension which cannot be treated by standard methods. For the calculation of thermodynamic averages Monte Carlo methods [49, 85, 174, 220] are very useful which sample the integration volume at randomly chosen points. In this chapter we discuss algorithms for the generation of pseudo-random numbers with given probability distribution which are essential for all Monte Carlo methods. We show how the efficiency of Monte Carlo integration can be improved by sampling preferentially the important configurations. Finally the famous Metropolis algorithm is applied to classical many-particle systems and nonlinear optimization problems.

8.1 Some Basic Statistics

In the following we discuss some important concepts which are used to analyze experimental data sets [218]. Repeated measurements of some observable usually give slightly different results due to fluctuations of the observable in time and errors of the measurement process. The distribution of the measured data is described by a probability distribution, which in many cases approximates a simple mathematical form like the Gaussian normal distribution. The moments of the probability density give important information about the statistical properties, especially the mean and the standard deviation of the distribution. If the errors of different measurements are uncorrelated, the average value of a larger number of measurements is a good approximation to the "exact" value.

8.1.1 Probability Density and Cumulative Probability Distribution

Consider an observable ξ, which is measured in a real or a computer experiment. Repeated measurements give a statistical distribution of values.

P.O.J. Scherer, *Computational Physics*, Graduate Texts in Physics,
DOI 10.1007/978-3-319-00401-3_8,
© Springer International Publishing Switzerland 2013

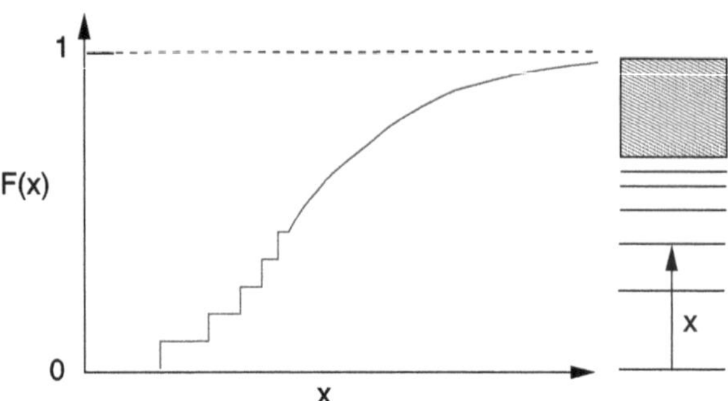

Fig. 8.1 (Cumulative probability distribution of transition energies) The figure shows schematically the distribution of transition energies for an atom which has a discrete and a continuous part

The cumulative probability distribution (Fig. 8.1) is given by the function

$$F(x) = P\{\xi \le x\} \tag{8.1}$$

and has the following properties:

- $F(x)$ is monotonously increasing
- $F(-\infty) = 0$, $F(\infty) = 1$
- $F(x)$ can be discontinuous (if there are discrete values of ξ)

The probability to measure a value in the interval $x_1 < \xi \le x_2$ is

$$P(x_1 < \xi \le x_2) = F(x_2) - F(x_1). \tag{8.2}$$

The height of a jump gives the probability of a discrete value

$$P(\xi = x_0) = F(x_0 + 0) - F(x_0 - 0). \tag{8.3}$$

In regions where $F(x)$ is continuous, the probability density can be defined as

$$f(x_0) = F'(x_0) = \lim_{\Delta x \to 0} \frac{1}{\Delta x} P(x_0 < \xi \le x_0 + \Delta x). \tag{8.4}$$

8.1.2 Histogram

From an experiment $F(x)$ cannot be determined directly. Instead a finite number N of values x_i are measured. By

$$Z_N(x)$$

we denote the number of measurements with $x_i \le x$. The cumulative probability distribution is the limit

$$F(x) = \lim_{N \to \infty} \frac{1}{N} Z_N(x). \tag{8.5}$$

A histogram (Fig. 8.2) counts the number of measured values which are in the interval $x_i < x \le x_{i+1}$:

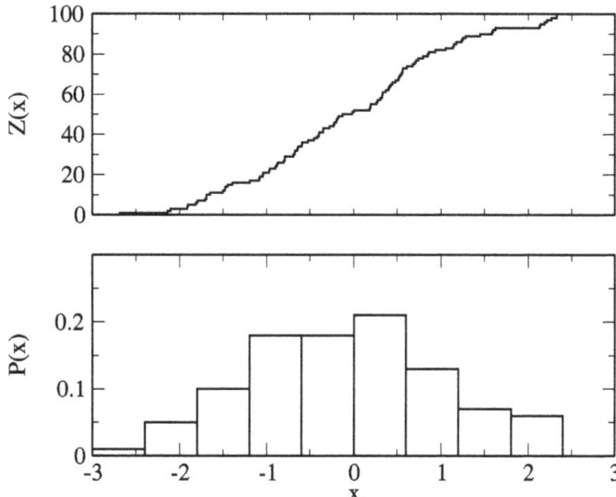

Fig. 8.2 (Histogram) The cumulative distribution of 100 Gaussian random numbers is shown together with a histogram with bin width $\Delta x = 0.6$

$$\frac{1}{N}\big(Z_N(x_{i+1}) - Z_N(x_i)\big) \approx F(x_{i+1}) - F(x_i) = P(x_i < \xi \le x_{i+1}). \qquad (8.6)$$

Contrary to $Z_N(x)$ itself, the histogram depends on the choice of the intervals.

8.1.3 Expectation Values and Moments

The expectation value of the random variable ξ is defined by

$$E[\xi] = \int_{-\infty}^{\infty} x \, dF(x) = \lim_{a \to -\infty, b \to \infty} \int_{a}^{b} x \, dF(x) \qquad (8.7)$$

with the Riemann-Stieltjes integral [202]

$$\int_{a}^{b} x \, dF(x) = \lim_{N \to \infty} \sum_{i=1}^{N} x_i \big(F(x_i) - F(x_{i-1})\big)\Big|_{x_i = a + \frac{b-a}{N} i}. \qquad (8.8)$$

Higher moments are defined as

$$E\big[\xi^k\big] = \int_{-\infty}^{\infty} x^k \, dF(x) \qquad (8.9)$$

if these integrals exist. Most important are the expectation value

$$\bar{x} = E[\xi] \qquad (8.10)$$

and the variance, which results from the first two moments

$$\sigma^2 = \int_{-\infty}^{\infty} (x - \bar{x})^2 \, dF = \int x^2 \, dF + \int \bar{x}^2 \, dF - 2\bar{x} \int x \, dF$$

$$= E\big[\xi^2\big] - \big(E[\xi]\big)^2. \qquad (8.11)$$

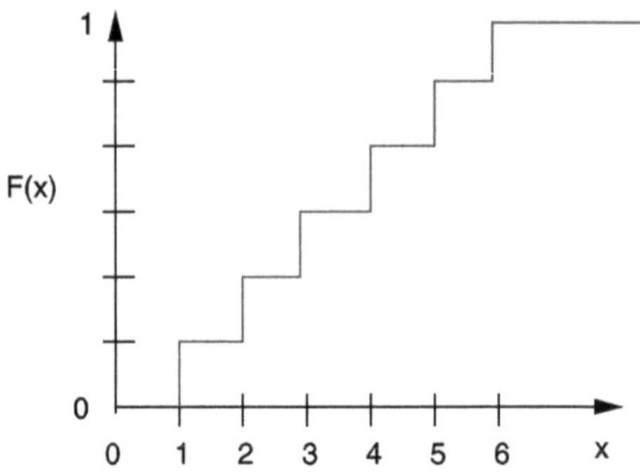

Fig. 8.3 Cumulative probability distribution of a fair die

The standard deviation σ is a measure of the width of the distribution. The expectation value of a function $\varphi(x)$ is defined by

$$E\big[\varphi(x)\big] = \int_{-\infty}^{\infty} \varphi(x)\, dF(x). \tag{8.12}$$

For continuous $F(x)$ we have with $dF(x) = f(x)\, dx$ the ordinary integral

$$E\big[\xi^k\big] = \int_{-\infty}^{\infty} x^k f(x)\, dx \tag{8.13}$$

$$E\big[\varphi(x)\big] = \int_{-\infty}^{\infty} \varphi(x) f(x)\, dx \tag{8.14}$$

whereas for a pure step function $F(x)$ (only discrete values x_i are observed with probabilities $p(x_i) = F(x_i + 0) - F(x_i - 0)$)

$$E\big[\xi^k\big] = \sum x_i^k p(x_i) \tag{8.15}$$

$$E\big[\varphi(x)\big] = \sum \varphi(x_i) p(x_i). \tag{8.16}$$

8.1.4 Example: Fair Die

When a six-sided fair die is rolled, each of its sides shows up with the same probability of $1/6$. The cumulative probability distribution $F(x)$ is a pure step function (Fig. 8.3) and

$$\bar{x} = \int_{-\infty}^{\infty} x\, dF = \sum_{i=1}^{6} x_i \big(F(x_i + 0) - F(x_i - 0)\big) = \frac{1}{6}\sum_{i=1}^{6} x_i = \frac{21}{6} = 3.5$$
$$\tag{8.17}$$

$$\overline{x^2} = \sum_{i=1}^{6} x_i^2 \big(F(x_i + 0) - F(x_i - 0)\big) = \frac{1}{6}\sum_{i=1}^{6} x_i^2 = \frac{91}{6} = 15.1666\cdots \tag{8.18}$$

$$\sigma = \sqrt{\overline{x^2} - \bar{x}^2} = 2.9. \tag{8.19}$$

8.1.5 Normal Distribution

The Gaussian normal distribution is defined by the cumulative probability distribution

$$\Phi(x) = \frac{1}{\sqrt{2\pi}} \int_{-\infty}^{x} e^{-t^2/2}\, dt \tag{8.20}$$

and the probability density

$$\varphi(x) = \frac{1}{\sqrt{2\pi}} e^{-x^2/2} \tag{8.21}$$

with the properties

$$\int_{-\infty}^{\infty} \varphi(x)\, dx = \Phi(\infty) = 1 \tag{8.22}$$

$$\overline{x} = \int_{-\infty}^{\infty} x\varphi(x)\, dx = 0 \tag{8.23}$$

$$\sigma^2 = \overline{x^2} = \int_{-\infty}^{\infty} x^2\varphi(x)\, dx = 1. \tag{8.24}$$

Since $\Phi(0) = \frac{1}{2}$ and with the definition

$$\Phi_0(x) = \frac{1}{\sqrt{2\pi}} \int_{0}^{x} e^{-t^2/2}\, dt \tag{8.25}$$

we have

$$\Phi(x) = \frac{1}{2} + \Phi_0(x) \tag{8.26}$$

which can be expressed in terms of the error function[1]

$$\mathrm{erf}(x) = \frac{2}{\sqrt{\pi}} \int_{0}^{x} e^{-t^2 dt}\, dt = 2\Phi_0(\sqrt{2}x) \tag{8.27}$$

as

$$\Phi_0(x) = \frac{1}{2}\mathrm{erf}\left(\frac{x}{\sqrt{2}}\right). \tag{8.28}$$

A general Gaussian distribution with mean value $\overline{x}$ and standard deviation σ has the probability distribution

$$\varphi_{\overline{x},\sigma} = \frac{1}{\sigma\sqrt{2\pi}} \exp\left(-\frac{(x'-\overline{x})^2}{2\sigma^2}\right) \tag{8.29}$$

and the cumulative distribution

---

[1] $\mathrm{erf}(x)$ is an intrinsic function in FORTRAN or C.

$$\Phi_{\bar{x},\sigma}(x) = \Phi\left(\frac{x-\bar{x}}{\sigma}\right) = \int_{-\infty}^{x} dx' \frac{1}{\sigma\sqrt{2\pi}} \exp\left(-\frac{(x'-\bar{x})^2}{2\sigma^2}\right) \tag{8.30}$$

$$= \frac{1}{2}\left(1 + \mathrm{erf}\left(\frac{x-\bar{x}}{\sigma\sqrt{2}}\right)\right). \tag{8.31}$$

8.1.6 Multivariate Distributions

Consider now two quantities which are measured simultaneously. ξ and η are the corresponding random variables. The cumulative distribution function is

$$F(x, y) = P(\xi \le x \text{ and } \eta \le y). \tag{8.32}$$

Expectation values are defined as

$$E\big[\varphi(x, y)\big] = \int_{-\infty}^{\infty} \int_{-\infty}^{\infty} \varphi(x, y)\, d^2 F(x, y). \tag{8.33}$$

For continuous $F(x, y)$ the probability density is

$$f(x, y) = \frac{\partial^2 F}{\partial x \partial y} \tag{8.34}$$

and the expectation value is simply

$$E\big[\varphi(x, y)\big] \int_{-\infty}^{\infty} dx \int_{-\infty}^{\infty} dy\, \varphi(x, y) f(x, y). \tag{8.35}$$

The moments of the distribution are the expectation values

$$M_{k,l} = E\big[\xi^k \eta^l\big]. \tag{8.36}$$

Most important are the averages

$$\bar{x} = E[\xi] \quad \bar{y} = E[\eta] \tag{8.37}$$

and the covariance matrix

$$\begin{pmatrix} E[(\xi-\bar{x})^2] & E[(\xi-\bar{x})(\eta-\bar{y})] \\ E[(\xi-\bar{x})(\eta-\bar{y})] & E[(\eta-\bar{y})^2] \end{pmatrix} = \begin{pmatrix} \overline{x^2}-\bar{x}^2 & \overline{xy}-\bar{x}\,\bar{y} \\ \overline{xy}-\bar{x}\,\bar{y} & \overline{y^2}-\bar{y}^2 \end{pmatrix}. \tag{8.38}$$

The correlation coefficient is defined as

$$\rho = \frac{\overline{xy}-\bar{x}\,\bar{y}}{\sqrt{(\overline{x^2}-\bar{x}^2)(\overline{y^2}-\bar{y}^2)}}. \tag{8.39}$$

If there is no correlation then $\rho = 0$ and $F(x, y) = F_1(x)F_2(y)$.

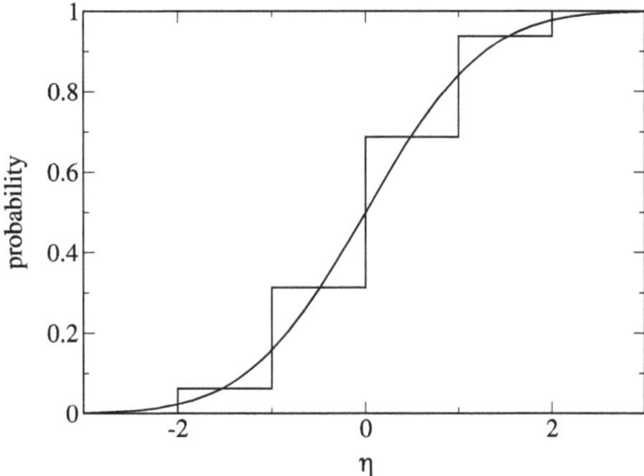

Fig. 8.4 (Central limit theorem) The cumulative distribution function of η (8.42) is shown for $N = 4$ and compared to the normal distribution (8.20)

8.1.7 Central Limit Theorem

Consider N independent random variables ξ_i with the same cumulative distribution function $F(x)$, for which $E[\xi] = 0$ and $E[\xi^2] = 1$. Define a new random variable

$$\eta_N = \frac{\xi_1 + \xi_2 + \cdots + \xi_N}{\sqrt{N}} \tag{8.40}$$

with the cumulative distribution function $F_N(x)$. In the limit $N \to \infty$ this distribution approaches (Fig. 8.4) a cumulative normal distribution [215]

$$\lim_{N \to \infty} F_N(x) = \Phi(x) = \frac{1}{\sqrt{2\pi}} \int_{-\infty}^{x} e^{-t^2/2}\, dt. \tag{8.41}$$

8.1.8 Example: Binomial Distribution

Toss a coin N times giving $\xi_i = 1$ (heads) or $\xi_i = -1$ (tails) with equal probability $P = \frac{1}{2}$. Then $E[\xi_i] = 0$ and $E[\xi_i^2] = 1$. The distribution of

$$\eta = \frac{1}{\sqrt{N}} \sum_{i=1}^{N} \xi_i \tag{8.42}$$

can be derived from the binomial distribution

$$1 = \left[\frac{1}{2} + \left(-\frac{1}{2}\right)\right]^N = 2^{-N} \sum_{p=0}^{N} (-1)^{N-p} \binom{N}{N-p} \tag{8.43}$$

where p counts the number of tosses with $\xi = +1$. Since

$$n = p \cdot 1 + (N - p) \cdot (-1) = 2p - N \in [-N, N] \tag{8.44}$$

the probability of finding $\eta = \frac{n}{\sqrt{N}}$ is given by the binomial coefficient

$$P\left(\eta = \frac{2p - N}{\sqrt{N}}\right) = 2^{-N}\binom{N}{N - p}$$ (8.45)

or

$$P\left(\eta = \frac{n}{\sqrt{N}}\right) = 2^{-N}\binom{N}{\frac{N-n}{2}}.$$ (8.46)

8.1.9 Average of Repeated Measurements

A quantity X is measured N times. The results $X_1 \cdots X_N$ are independent random numbers with the same distribution function $f(X_i)$. Their expectation value is the exact value $E[X_i] = \int dX_i\, X_i f(X_i) = X$ and the standard deviation due to measurement uncertainties is $\sigma_X = \sqrt{E[X_i^2] - X^2}$. The new random variables

$$\xi_i = \frac{X_i - X}{\sigma_X}$$ (8.47)

have zero mean

$$E[\xi_i] = \frac{E[X_i] - X}{\sigma_X} = 0$$ (8.48)

and unit standard deviation

$$\sigma_\xi^2 = E\left[\xi_i^2\right] - E[\xi_i]^2 = E\left[\frac{X_i^2 + X^2 - 2XX_i}{\sigma_X^2}\right] = \frac{E[X_i^2] - X^2}{\sigma_X^2} = 1.$$ (8.49)

Hence the quantity

$$\eta = \frac{\sum_1^N \xi_i}{\sqrt{N}} = \frac{\sum_1^N X_i - NX}{\sqrt{N}\sigma_X} = \frac{\sqrt{N}}{\sigma_X}(\overline{X} - X)$$ (8.50)

obeys a normal distribution

$$f(\eta) = \frac{1}{\sqrt{2\pi}}e^{-\eta^2/2}.$$ (8.51)

From

$$f(\overline{X})d\overline{X} = f(\eta)\,d\eta = f\left(\eta(\overline{X})\right)\frac{\sqrt{N}}{\sigma_X}\,d\overline{X}$$ (8.52)

we obtain

$$f(\overline{X}) = \frac{\sqrt{N}}{\sqrt{2\pi}\sigma_X}\exp\left\{-\frac{N}{2\sigma_X^2}(\overline{X} - X)^2\right\}.$$ (8.53)

The average of N measurements obeys a Gaussian distribution around the exact value X with a reduced standard deviation of

$$\sigma_{\overline{X}} = \frac{\sigma_X}{\sqrt{N}}.$$ (8.54)

8.2 Random Numbers

True random numbers of high quality can be generated using physical effects like thermal noise in a diode or from a light source. Special algorithms are available to generate pseudo random numbers which have comparable statistical properties but are not unpredictable since they depend on some initial seed values. Often an iteration

$$r_{i+1} = f(r_i) \tag{8.55}$$

is used to calculate a series of pseudo random numbers. Using 32 Bit integers there are 2^{32} different numbers, hence the period cannot exceed 2^{32}.

8.2.1 Linear Congruent Mapping

A simple algorithm is the linear congruent mapping

$$r_{i+1} = (a r_i + c) \bmod m \tag{8.56}$$

with maximum period m. Larger periods can be achieved if the random number depends on several predecessors. A function of the type

$$r_i = f(r_{i-1}, r_{i-2} \cdots r_{i-t}) \tag{8.57}$$

using 32 Bit integers has a maximum period of 2^{32t}.

Example For $t = 2$ and generating 10^6 numbers per second the period is 584942 years.

8.2.2 Marsaglia-Zamann Method

A high quality random number generator can be obtained from the combination of two generators [168]. The first one

$$r_i = (r_{i-2} - r_{i-3} - c) \bmod \left(2^{32} - 18\right) \tag{8.58}$$

with

$$c = \begin{cases} 1 & \text{for } r_{n-2} - r_{n-3} < 0 \\ 0 & \text{else} \end{cases} \tag{8.59}$$

has a period of 2^{95}. The second one

$$r_i = (69069 r_{i-1} + 1013904243) \bmod 2^{32} \tag{8.60}$$

has a period of 2^{32}. The period of the combination is 2^{127}. Here is a short subroutine in C:

```c
#define N 100000
typedef unsigned long int unlong /* 32 Bit */
unlong x=521288629, y=362436069, z=16163801, c=1, n=1131199209;
unlong mzran()
{ unlong s;
   if (y>x+c) {s=y-(x+c)-18; c=0;}
   else {s=y-(x+c)-18;c=1;}
   x=y; y=z; z=s; n=69069*n+1013904243;
   return(z+n);
}
```

8.2.3 Random Numbers with Given Distribution

Assume we have a program that generates random numbers in the interval $[0, 1]$ like in C:

$$rand()/(double)RAND\_MAX.$$

The corresponding cumulative distribution function is

$$F_0(x) = \begin{cases} 0 & \text{for } x < 0 \\ x & \text{for } 0 \leq x \leq 1 \\ 1 & \text{for } x > 1. \end{cases} \tag{8.61}$$

Random numbers with cumulative distribution $F(x)$ can be obtained as follows:

$$\text{choose an } RN \; r \in [0, 1] \text{ with } P(r \leq x) = F_0(x)$$
$$\text{let } \xi = F^{-1}(r)$$

$F(x)$ increases monotonously and therefore

$$P(\xi \leq x) = P\big(F(\xi) \leq F(x)\big) = P\big(r \leq F(x)\big) = F_0\big(F(x)\big) \tag{8.62}$$

but since $0 \leq F(x) \leq 1$ we have

$$P(\xi \leq x) = F(x). \tag{8.63}$$

This method of course is applicable only if F^{-1} can be expressed analytically.

8.2.4 Examples

8.2.4.1 Fair Die

A six-sided fair die can be simulated as follows:

choose a random number $r \in [0, 1]$

$$\text{Let } \xi = F^{-1}(r) = \begin{cases} 1 & \text{for } 0 \le r < \frac{1}{6} \\ 2 & \text{for } \frac{1}{6} \le r < \frac{2}{6} \\ 3 & \text{for } \frac{2}{6} \le r < \frac{3}{6} \\ 4 & \text{for } \frac{3}{6} \le r < \frac{4}{6} \\ 5 & \text{for } \frac{4}{6} \le r < \frac{5}{6} \\ 6 & \text{for } \frac{5}{6} \le r < 1. \end{cases}$$

8.2.4.2 Exponential Distribution

The cumulative distribution function

$$F(x) = 1 - e^{-x/\lambda} \tag{8.64}$$

which corresponds to the exponential probability density

$$f(x) = \frac{1}{\lambda} e^{-x/\lambda} \tag{8.65}$$

can be inverted by solving

$$r = 1 - e^{-x/\lambda} \tag{8.66}$$

for x:

choose a random number $r \in [0, 1]$

let $x = F^{-1}(r) = -\lambda \ln(1 - r)$.

8.2.4.3 Random Points on the Unit Sphere

We consider the surface element

$$\frac{1}{4\pi} R^2 \, d\varphi \sin\theta \, d\theta. \tag{8.67}$$

Our aim is to generate points on the unit sphere (θ, φ) with the probability density

$$f(\theta, \varphi) \, d\varphi \, d\theta = \frac{1}{4\pi} d\varphi \sin\theta \, d\theta = -\frac{1}{4\pi} d\varphi d\cos\theta. \tag{8.68}$$

The corresponding cumulative distribution is

$$F(\theta, \varphi) = -\frac{1}{4\pi} \int_1^{\cos\theta} d\cos\theta \int_0^{\varphi} d\varphi = \frac{\varphi}{2\pi} \frac{1 - \cos\theta}{2} = F_\varphi F_\theta. \tag{8.69}$$

Since this factorizes, the two angles can be determined independently:

choose a first random number $r_1 \in [0, 1]$

let $\varphi = F_\varphi^{-1}(r_1) = 2\pi r_1$

choose a second random number $r_2 \in [0, 1]$

let $\theta = F_\theta^{-1}(r_2) = \arccos(1 - 2r_2)$.

8.2.4.4 Gaussian Distribution (Box Muller)

For a Gaussian distribution the inverse F^{-1} has no simple analytical form. The famous Box Muller method [34] is based on a 2-dimensional normal distribution with probability density

$$f(x, y) = \frac{1}{2\pi} \exp\left\{ -\frac{x^2 + y^2}{2} \right\} \tag{8.70}$$

which reads in polar coordinates

$$f(x, y)dxdy = f_p(\rho, \varphi)d\rho\, d\varphi \frac{1}{2\pi} e^{-\rho^2/2} \rho\, d\rho\, d\varphi. \tag{8.71}$$

Hence

$$f_p(\rho, \varphi) = \frac{1}{2\pi} \rho e^{-\rho^2/2} \tag{8.72}$$

and the cumulative distribution factorizes:

$$F_p(\rho, \varphi) = \frac{1}{2\pi} \varphi \cdot \int_0^\rho \rho' e^{-\rho'^2/2} d\rho' = \frac{\varphi}{2\pi} \left(1 - e^{-\rho^2}\right) = F_\varphi(\varphi) F_\rho(\rho). \tag{8.73}$$

The inverse of F_ρ is

$$\rho = \sqrt{-\ln(1 - r)} \tag{8.74}$$

and the following algorithm generates Gaussian random numbers:

$$r_1 = RN \in [0, 1]$$
$$r_2 = RN \in [0, 1]$$
$$\rho = \sqrt{-\ln(1 - r_1)}$$
$$\varphi = 2\pi r_2$$
$$x = \rho \cos\varphi.$$

8.3 Monte Carlo Integration

Physical problems often involve high dimensional integrals (for instance path integrals, thermodynamic averages) which cannot be evaluated by standard methods. Here Monte Carlo methods can be very useful. Let us start with a very basic example.

8.3.1 Numerical Calculation of π

The area of a unit circle ($r = 1$) is given by $r^2\pi = \pi$. Hence π can be calculated by numerical integration. We use the following algorithm:

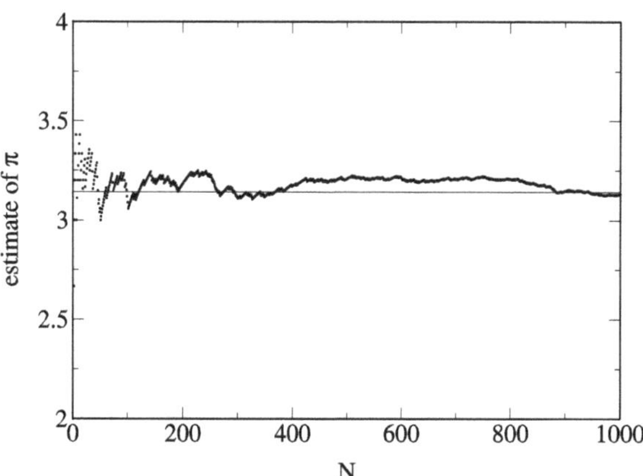

Fig. 8.5 Convergence of the numerical integration

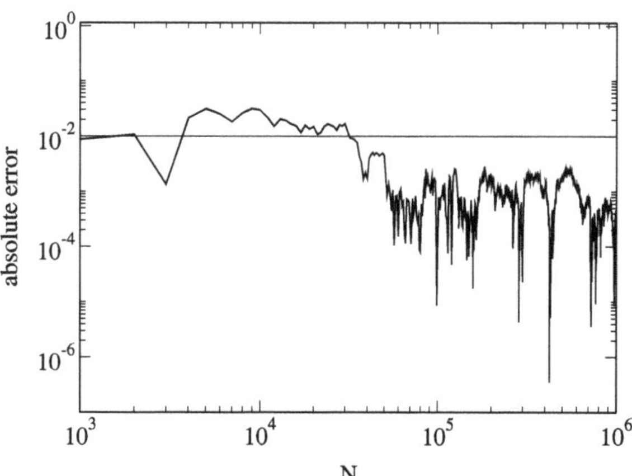

Fig. 8.6 Error of the numerical integration

choose N points randomly in the first quadrant, for instance N independent pairs $x, y \in [0, 1]$
calculate $r^2 = x^2 + y^2$
count the number of points within the circle, i.e. the number of points $Z(r^2 \leq 1)$.
$\frac{\pi}{4}$ is approximately given by $\frac{Z(r^2 \leq 1)}{N}$.

The result converges rather slowly (Figs. 8.5, 8.6)

8.3.2 Calculation of an Integral

Let ξ be a random variable in the interval $[a, b]$ with the distribution

$$P(x < \xi \leq x + dx) = f(x)\,dx = \begin{cases} \frac{1}{b-a} & \text{for } x \in [a, b] \\ 0 & \text{else.} \end{cases} \qquad (8.75)$$

The expectation value of a function $g(x)$ is

$$E[g(x)] = \int_{-\infty}^{\infty} g(x) f(x)\, dx = \int_{a}^{b} g(x)\, dx \qquad (8.76)$$

hence the average of N randomly taken function values approximates the integral

$$\int_{a}^{b} g(x)\, dx \approx \frac{1}{N} \sum_{i=1}^{N} g(\xi_i) = \overline{g(\xi)}. \qquad (8.77)$$

To estimate the error we consider the new random variable

$$\gamma = \frac{1}{N} \sum_{i=1}^{N} g(\xi). \qquad (8.78)$$

Its average is

$$\overline{\gamma} = E[\gamma] = \frac{1}{N} \sum_{i=1}^{N} E[g(x)] = E[g(x)] = \int_{a}^{b} g(x)\, dx \qquad (8.79)$$

and the variance follows from

$$\sigma_{\gamma}^2 = E\big[(\gamma - \overline{\gamma})^2\big] = E\left[\left(\frac{1}{N}\sum g(\xi_i) - \overline{\gamma}\right)^2\right] = E\left[\left(\frac{1}{N}\sum (g(\xi_i) - \overline{\gamma})\right)^2\right] \qquad (8.80)$$

$$= \frac{1}{N^2} E\left[\sum (g(\xi_i) - \overline{\gamma})^2\right] = \frac{1}{N}\left(\overline{g(\xi)^2} - \overline{g(\xi)}^2\right) = \frac{1}{N} \sigma_{g(\xi)}^2. \qquad (8.81)$$

The width of the distribution and hence the uncertainty falls off as $1/\sqrt{N}$.

8.3.3 More General Random Numbers

Consider now random numbers $\xi \in [a, b]$ with arbitrary (but within $[a, b]$ not vanishing) probability density $f(x)$. The integral is approximated by

$$\frac{1}{N} \sum_{i=1}^{N} \frac{g(\xi_i)}{f(\xi_i)} = E\left[\frac{g(x)}{f(x)}\right] = \int_{a}^{b} \frac{g(x)}{f(x)} f(x)\, dx = \int_{a}^{b} g(x)\, dx. \qquad (8.82)$$

The new random variable

$$\tau = \frac{1}{N} \sum_{i=1}^{N} \frac{g(\xi_i)}{f(\xi_i)} \qquad (8.83)$$

according to (8.81) has a standard deviation given by

$$\sigma_{\tau} = \frac{1}{\sqrt{N}} \sigma\left(\frac{g(\xi)}{f(\xi)}\right) \qquad (8.84)$$

which can be reduced by choosing f similar to g. Then preferentially ξ are generated in regions where the integrand is large (importance sampling).

8.4 Monte Carlo Method for Thermodynamic Averages

Consider the partition function of a classical N particle system

$$Z_{NVT} = \frac{1}{N!}\frac{1}{h^{3N}} \int dp^{3N} \int dq^{3N} e^{-\beta H(p_1\cdots p_N, q_1\cdots q_N)} \tag{8.85}$$

with an energy function

$$H = \sum_{i=1}^{N} \frac{p_i^2}{2m_i} + V(q_1\cdots q_N). \tag{8.86}$$

If the potential energy does not depend on the momenta and for equal masses the partition function simplifies to

$$\begin{aligned} Z_{NVT} &= \frac{1}{N!}\frac{1}{h^{3N}} \int dp^{3N} e^{-\beta \frac{p_i^2}{2m}} \int dq^{3N} e^{-\beta V(q)} \\ &= \frac{1}{N!}\left(\frac{2\pi m k T}{h^2}\right)^{3N/2} \int dq^{3N} e^{-\beta V(q)} \end{aligned} \tag{8.87}$$

and it remains the calculation of the configuration integral

$$Z_{NVT}^{conf} = \int dq^{3N} e^{-\beta V(q)}. \tag{8.88}$$

In the following we do not need the partition function itself but only averages of some quantity $A(q)$ given by

$$\langle A \rangle = \frac{\int dq^{3N} A(q) e^{-\beta V(q)}}{\int dq^{3N} e^{-\beta V(q)}}. \tag{8.89}$$

8.4.1 Simple Sampling

Let ξ be a random variable with probability distribution

$$P\big(\xi \in [q, q+dq]\big) = f(q)\,dq \tag{8.90}$$

$$\int f(q)\,dq = 1. \tag{8.91}$$

We choose M random numbers ξ and calculate the expectation value of $A(\xi)$ from

$$E\big[A(\xi)\big] = \lim_{M\to\infty} \frac{1}{M} \sum_{m=1}^{M} A(\xi^{(m)}) = \int A(q) f(q)\,dq. \tag{8.92}$$

Consider now the case of random numbers ξ equally distributed over the range $q_{min}\cdots q_{max}$:

$$f(q) = \begin{cases} \frac{1}{q_{\max}-q_{\min}} & q \in [q_{\min}, q_{\max}] \\ 0 & \text{else.} \end{cases} \tag{8.93}$$

Define a sample by choosing one random value ξ for each of the $3N$ coordinates. The average over a large number M of samples gives the expectation value

$$E\left(A(\xi_1 \cdots \xi_{3N})e^{-\beta V(\xi_1 \cdots \xi_{3N})}\right) = \lim_{M \to \infty} \frac{1}{M} \sum_{m=1}^{M} A\left(\xi_i^{(m)}\right)e^{-\beta V(\xi_i^{(m)})} \tag{8.94}$$

as

$$\int A(q_i)e^{-\beta V(q_i)} f(q_1) \cdots f(q_{3N})dq_1 \cdots dq_{3N}$$

$$= \frac{1}{(q_{\max} - q_{\min})^{3N}} \int_{q_{\min}}^{q_{\max}} \cdots \int_{q_{\min}}^{q_{\max}} A(q_i)e^{-\beta V(q_i)}\, dq^{3N}. \tag{8.95}$$

Hence

$$\frac{E(A(\xi_i)e^{-\beta V(\xi_i)})}{E(e^{-\beta V(\xi_i)})} = \frac{\int_{q_{\min}}^{q_{\max}} A(q_i)e^{-\beta V(q_i)}\, dq^{3N}}{\int_{q_{\min}}^{q_{\max}} e^{-\beta V(q_i)}\, dq^{3N}} \approx \langle A \rangle \tag{8.96}$$

is an approximation to the average of $A(q_i)$, if the range of the q_i is sufficiently large. However, many of the samples will have small weight and contribute only little.

8.4.2 Importance Sampling

Let us try to sample preferentially the most important configurations. Choose the distribution function as

$$f(q_1 \cdots q_{3N}) = \frac{e^{-\beta V(q_1 \cdots q_{3N})}}{\int e^{-\beta V(q_1 \cdots q_{3N})}}. \tag{8.97}$$

The expectation value of $A(q)$ approximates the thermal average

$$E\left(A(\xi_1 \cdots \xi_{3N})\right) = \lim_{M \to \infty} \frac{1}{M} \sum_{m=1}^{M} A\left(\xi_i^{(m)}\right) = \frac{\int A(q_i)e^{-\beta V(q_i)}\, dq^{3N}}{\int e^{-\beta V(q_i)}\, dq^{3N}} = \langle A \rangle \tag{8.98}$$

and the partition function is not needed explicitly for the calculation.

8.4.3 Metropolis Algorithm

The algorithm by Metropolis [175] can be used to select the necessary configurations. Starting from an initial configuration $\mathbf{q}_0 = (q_1^{(0)} \cdots q_{3N}^{(0)})$ a chain of configurations is generated. Each configuration depends only on its predecessor, hence the configurations form a Markov chain.

Fig. 8.7 Principle of detailed
balance

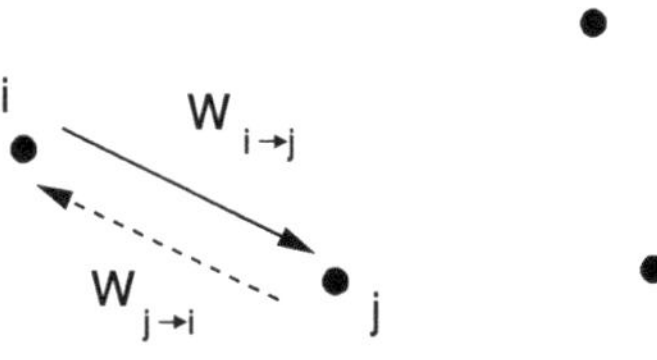

The transition probabilities

$$W_{i \to j} = P(\mathbf{q}_i \to \mathbf{q}_j) \qquad (8.99)$$

are chosen to fulfill the condition of detailed balance (Fig. 8.7)

$$\frac{W_{i \to j}}{W_{j \to i}} = e^{-\beta(V(\mathbf{q}_j) - V(\mathbf{q}_i))}. \qquad (8.100)$$

This is a sufficient condition that the configurations are generated with probabilities
given by their Boltzmann factors. This can be seen from consideration of an ensemble of such Markov chains: Let $N_n(\mathbf{q}_i)$ denote the number of chains which are in
the configuration $\mathbf{q}_i$ after n steps. The changes during the following step are

$$\Delta N(\mathbf{q}_i) = N_{n+1}(\mathbf{q}_i) - N_n(\mathbf{q}_i) = \sum_{\mathbf{q}_j \in conf.} N_n(\mathbf{q}_j) W_{j \to i} - N_n(\mathbf{q}_i) W_{i \to j}.$$

$$(8.101)$$

In thermal equilibrium

$$N_{eq}(\mathbf{q}_i) = N_0 e^{-\beta V(\mathbf{q}_i)}$$

and the changes (8.101) vanish:

$$\Delta N(\mathbf{q}_i) = N_0 \sum_{\mathbf{q}_j} e^{-\beta V(\mathbf{q}_j)} W_{j \to i} - e^{-\beta V(\mathbf{q}_i)} W_{i \to j}$$

$$= N_0 \sum_{\mathbf{q}_j} e^{-\beta V(\mathbf{q}_j)} W_{j \to i} - e^{-\beta V(\mathbf{q}_i)} W_{j \to i} e^{-\beta(V(\mathbf{q}_j) - V(\mathbf{q}_i))}$$

$$= 0. \qquad (8.102)$$

A solution of

$$\Delta N(\mathbf{q}_i) = \sum_{\mathbf{q}_j \in conf.} N_n(\mathbf{q}_j) W_{j \to i} - N_n(\mathbf{q}_i) W_{i \to j} = 0 \qquad (8.103)$$

corresponds to a zero eigenvalue of the system of equations

$$\sum_{\mathbf{q}_j} N(\mathbf{q}_j) W_{j\to i} - N(\mathbf{q}_i) \sum_{\mathbf{q}_j} W_{i\to j} = \lambda N(\mathbf{q}_i). \qquad (8.104)$$

One solution of this eigenvalue equation is given by

$$\frac{N_{eq}(\mathbf{q}_j)}{N_{eq}(\mathbf{q}_i)} = e^{-\beta(V(\mathbf{q}_j) - V(\mathbf{q}_i))}. \qquad (8.105)$$

However, there may be other solutions. For instance if not all configurations are connected by possible transitions and some isolated configurations are occupied initially.

Metropolis Algorithm This famous algorithm consists of the following steps:

(a) choose a new configuration randomly (trial step) with probability

$$T(\mathbf{q}_i \to \mathbf{q}_{trial}) = T(\mathbf{q}_{trial} \to \mathbf{q}_i)$$

(b) calculate

$$R = \frac{e^{-\beta V(\mathbf{q}_{trial})}}{e^{-\beta V(\mathbf{q}_i)}}$$

(c) if $R \geq 1$ the trial step is accepted $\mathbf{q}_{i+1} = \mathbf{q}_{trial}$
(d) if $R < 1$ the trial step is accepted only with probability R. Choose a random number $\xi \in [0, 1]$ and the next configuration according to

$$\mathbf{q}_{i+1} = \begin{cases} \mathbf{q}_{trial} & \text{if } \xi < R \\ \mathbf{q}_i & \text{if } \xi \geq R. \end{cases}$$

The transition probability is the product

$$W_{i\to j} = T_{i\to j} A_{i\to j} \qquad (8.106)$$

of the probability $T_{i\to j}$ to select $i \to j$ as a trial step and the probability $A_{i\to j}$ to accept the trial step. Now we have

$$\begin{aligned} &\text{for } R \geq 1 \to A_{i\to j} = 1, \ A_{j\to i} = R^{-1} \\ &\text{for } R < 1 \to A_{i\to j} = R, \ A_{j\to i} = 1. \end{aligned} \qquad (8.107)$$

Since $T_{i\to j} = T_{j\to i}$, in both cases

$$\frac{N_{eq}(\mathbf{q}_j)}{N_{eq}(\mathbf{q}_i)} = \frac{W_{i\to j}}{W_{j\to i}} = \frac{A_{i\to j}}{A_{j\to i}} = R = e^{-\beta(V(\mathbf{q}_j) - V(\mathbf{q}_i))}. \qquad (8.108)$$

8.5 Problems

Problem 8.1 (Central limit theorem)　This computer experiment draws a histogram for the random variable τ, which is calculated from N random numbers $\xi_1 \cdots \xi_N$:

$$\tau = \frac{\sum_{i=1}^{N} \xi_i}{\sqrt{N}}.$$ (8.109)

The ξ_i are random numbers with zero mean and unit variance and can be chosen as

- $\xi_i = \pm 1$ (coin tossing)
- Gaussian random numbers

Investigate how a Gaussian distribution is approached for large N.

Problem 8.2 (Nonlinear optimization)   MC methods can be used for nonlinear optimization (Traveling salesman problem, structure optimization etc.) [31]. Consider an energy function depending on many coordinates

$$E(q_1, q_2 \cdots q_N).$$ (8.110)

Introduce a fictitious temperature T and generate configurations with probabilities

$$P(q_1 \cdots q_N) = \frac{1}{Z} e^{-E(q_1 \cdots q_N)/T}.$$ (8.111)

Slow cooling drives the system into a local minimum. By repeated heating and cooling other local minima can be reached (simulated annealing).

In this computer experiment we try to find the shortest path which visits each of up to $N = 50$ given points. The fictitious Boltzmann factor for a path with total length L is

$$P(L) = e^{-L/T}$$ (8.112)

Starting from an initial path $S = (i_1, i_2 \cdots i_N)$ $n < 5$ and p are chosen randomly and a new path $S' = (i_1 \cdots i_{p-1}, i_{p+n} \cdots i_p, i_{p+n+1} \cdots i_N)$ is generated by reverting the sub-path

$$i_p \cdots i_{p+n} \to i_{p+n} \cdots i_p.$$

Start at high temperature $T > L$ and cool down slowly.

Chapter 9
Eigenvalue Problems

Eigenvalue problems are omnipresent in physics. Important examples are the time independent Schrödinger equation in a finite orthogonal basis (Chap. 9)

$$\sum_{j=1}^{M} \langle \phi_{j'} | H | \phi_j \rangle C_j = E C_{j'} \tag{9.1}$$

or the harmonic motion of a molecule around its equilibrium structure (Sect. 14.4.1)

$$\omega^2 m_i \left(\xi_i - \xi_i^{eq} \right) = \sum_j \frac{\partial^2 U}{\partial \xi_i \partial \xi_j} \left(\xi_j - \xi_j^{eq} \right). \tag{9.2}$$

Most important are ordinary eigenvalue problems,[1] which involve the solution of a homogeneous system of linear equations

$$\sum_{j=1}^{N} a_{ij} x_j = \lambda x_i \tag{9.3}$$

with a Hermitian (or symmetric, if real) matrix [198]

$$a_{ji} = a_{ij}^*. \tag{9.4}$$

Matrices of small dimension can be diagonalized directly by determining the roots of the characteristic polynomial and solving a homogeneous system of linear equations. The Jacobi method uses successive rotations to diagonalize a matrix with a unitary transformation. A very popular method for not too large symmetric matrices reduces the matrix to tridiagonal form which can be diagonalized efficiently with the QL algorithm. Some special tridiagonal matrices can be diagonalized analytically. Special algorithms are available for matrices of very large dimension, for instance the famous Lanczos method.

---

[1] We do not consider general eigenvalue problems here.

P.O.J. Scherer, *Computational Physics*, Graduate Texts in Physics,
DOI 10.1007/978-3-319-00401-3_9,
© Springer International Publishing Switzerland 2013

9.1 Direct Solution

For matrices of very small dimension (2, 3) the determinant

$$\det |a_{ij} - \lambda\delta_{ij}| = 0 \tag{9.5}$$

can be written explicitly as a polynomial of λ. The roots of this polynomial are the eigenvalues. The eigenvectors are given by the system of equations

$$\sum_j (a_{ij} - \lambda\delta_{ij})u_j = 0. \tag{9.6}$$

9.2 Jacobi Method

Any symmetric 2×2 matrix

$$A = \begin{pmatrix} a_{11} & a_{12} \\ a_{12} & a_{22} \end{pmatrix} \tag{9.7}$$

can be diagonalized by a rotation of the coordinate system. Rotation by the angle φ corresponds to an orthogonal transformation with the rotation matrix

$$R_\varphi = \begin{pmatrix} \cos\varphi & -\sin\varphi \\ \sin\varphi & \cos\varphi \end{pmatrix}. \tag{9.8}$$

In the following we use the abbreviations

$$c = \cos\varphi, \quad s = \sin\varphi, \quad t = \tan\varphi. \tag{9.9}$$

The transformed matrix is

$$\begin{aligned}
RAR^{-1} &= \begin{pmatrix} c & -s \\ s & c \end{pmatrix} \begin{pmatrix} a_{11} & a_{12} \\ a_{12} & a_{22} \end{pmatrix} \begin{pmatrix} c & s \\ -s & c \end{pmatrix} \\
&= \begin{pmatrix} c^2 a_{11} + s^2 a_{22} - 2csa_{12} & cs(a_{11} - a_{22}) + (c^2 - s^2)a_{12} \\ cs(a_{11} - a_{22}) + (c^2 - s^2)a_{12} & s^2 a_{11} + c^2 a_{22} + 2csa_{12} \end{pmatrix}.
\end{aligned} \tag{9.10}$$

It is diagonal if

$$0 = cs(a_{11} - a_{22}) + (c^2 - s^2)a_{12} = \frac{a_{11} - a_{22}}{2}\sin(2\varphi) + a_{12}\cos(2\varphi) \tag{9.11}$$

or

$$\tan(2\varphi) = \frac{2a_{12}}{a_{22} - a_{11}}. \tag{9.12}$$

Calculation of φ is not necessary since only its cosine and sine appear in (9.10). From [198]

$$\frac{1-t^2}{t} = \frac{c^2-s^2}{2cs} = \cot(2\varphi) = \frac{a_{22}-a_{11}}{2a_{12}} \tag{9.13}$$

we see that t is a root of

$$t^2 + \frac{a_{22}-a_{11}}{a_{12}}t - 1 = 0 \tag{9.14}$$

hence

$$t = -\frac{a_{22}-a_{11}}{2a_{12}} \pm \sqrt{1+\left(\frac{a_{22}-a_{11}}{2a_{12}}\right)^2} = \frac{1}{\frac{a_{22}-a_{11}}{2a_{12}} \pm \sqrt{1+(\frac{a_{22}-a_{11}}{2a_{12}})^2}}. \tag{9.15}$$

For reasons of convergence [198] the solution with smaller magnitude is chosen which can be written as

$$t = \frac{\operatorname{sign}(\frac{a_{22}-a_{11}}{2a_{12}})}{|\frac{a_{22}-a_{11}}{2a_{12}}| + \sqrt{1+(\frac{a_{22}-a_{11}}{2a_{12}})^2}} \tag{9.16}$$

again for reasons of convergence the smaller solution φ is preferred and therefore we take

$$c = \frac{1}{\sqrt{1+t^2}} \quad s = \frac{t}{\sqrt{1+t^2}}. \tag{9.17}$$

The diagonal elements of the transformed matrix are

$$\tilde{a}_{11} = c^2 a_{11} + s^2 a_{22} - 2csa_{12} \tag{9.18}$$

$$\tilde{a}_{22} = s^2 a_{11} + c^2 a_{22} + 2csa_{12}. \tag{9.19}$$

The trace of the matrix is invariant

$$\tilde{a}_{11} + \tilde{a}_{22} = a_{11} + a_{22} \tag{9.20}$$

whereas the difference of the diagonal elements is

$$\tilde{a}_{11} - \tilde{a}_{22} = (c^2-s^2)(a_{11}-a_{22}) - 4csa_{12}$$

$$= \frac{1-t^2}{1+t^2}(a_{11}-a_{22}) - 4\frac{a_{12}t}{1+t^2}$$

$$= (a_{11}-a_{22}) + \left(-a_{12}\frac{1-t^2}{t}\right)\frac{-2t^2}{1+t^2} - 4\frac{a_{12}t}{1+t^2}$$

$$= (a_{11}-a_{22}) - 2ta_{12} \tag{9.21}$$

and the transformed matrix has the simple form

$$
\begin{pmatrix} a_{11} - a_{12}t & \\ & a_{22} + a_{12}t \end{pmatrix}.
\tag{9.22}
$$

For larger dimension $N > 2$ the Jacobi method uses the following algorithm:

(1) look for the dominant non-diagonal element $\max_{i \neq j} |a_{ij}|$
(2) perform a rotation in the (ij)-plane to cancel the element $\tilde{a}_{ij}$ of the transformed matrix $\tilde{A} = R^{(ij)} \cdot A \cdot R^{(ij)-1}$. The corresponding rotation matrix has the form

$$
R^{(ij)} = \begin{pmatrix}
1 & & & & & & \\
& \ddots & & & & & \\
& & c & & s & & \\
& & & \ddots & & & \\
& & -s & & c & & \\
& & & & & \ddots & \\
& & & & & & 1
\end{pmatrix}
\tag{9.23}
$$

(3) repeat (1–2) until convergence (if possible).

The sequence of Jacobi rotations gives the over all transformation

$$
RAR^{-1} = \cdots R_2 R_1 A R_1^{-1} R_2^{-1} \cdots = \begin{pmatrix} \lambda_1 & & \\ & \ddots & \\ & & \lambda_N \end{pmatrix}.
\tag{9.24}
$$

Hence

$$
AR^{-1} = R^{-1} \begin{pmatrix} \lambda_1 & & \\ & \ddots & \\ & & \lambda_N \end{pmatrix}
\tag{9.25}
$$

and the column vectors of $R^{-1} = (\mathbf{v}_1, \mathbf{v}_2 \cdots \mathbf{v}_N)$ are the eigenvectors of A:

$$
A(\mathbf{v}_1, \mathbf{v}_2 \cdots \mathbf{v}_N) = (\lambda_1 \mathbf{v}_1, \lambda_2 \mathbf{v}_2 \cdots \lambda_N \mathbf{v}_N).
\tag{9.26}
$$

9.3 Tridiagonal Matrices

A tridiagonal matrix has nonzero elements only in the main diagonal and the first diagonal above and below. Many algorithms simplify significantly when applied to tridiagonal matrices.

9.3.1 *Characteristic Polynomial of a Tridiagonal Matrix*

The characteristic polynomial of a tridiagonal matrix

$$
P_A(\lambda) = \det \begin{vmatrix} a_{11} - \lambda & a_{12} & & \\ a_{21} & a_{22} - \lambda & & \\ & & \ddots & a_{N-1N} \\ & & a_{NN-1} & a_{NN} - \lambda \end{vmatrix}
\tag{9.27}
$$

can be calculated recursively:

$$
\begin{aligned}
P_0 &= 1 \\
P_1(\lambda) &= a_{11} - \lambda \\
P_2(\lambda) &= (a_{22} - \lambda)P_1(\lambda) - a_{12}a_{21} \\
&\vdots \\
P_N(\lambda) &= (a_{NN} - \lambda)P_{N-1}(\lambda) - a_{N,N-1}a_{N-1,N}P_{N-2}(\lambda).
\end{aligned}
\tag{9.28}
$$

9.3.2 *Special Tridiagonal Matrices*

Certain classes of tridiagonal matrices can be diagonalized exactly [55, 151, 281].

9.3.2.1 Discretized Second Derivatives

Discretization of a second derivative involves, under Dirichlet boundary conditions $f(x_0) = f(x_{N+1}) = 0$, the differentiation matrix (Sect. 18.2)

$$
M = \begin{pmatrix} -2 & 1 & & & \\ 1 & -2 & 1 & & \\ & \ddots & \ddots & \ddots & \\ & & 1 & -2 & 1 \\ & & & 1 & -2 \end{pmatrix}.
\tag{9.29}
$$

Its eigenvectors have the form

$$
\mathbf{f} = \begin{pmatrix} f_1 \\ \vdots \\ f_n \\ \vdots \\ f_N \end{pmatrix} = \begin{pmatrix} \sin k \\ \vdots \\ \sin(nk) \\ \vdots \\ \sin(Nk) \end{pmatrix}.
\tag{9.30}
$$

This can be seen by inserting (9.30) into the n-th line of the eigenvalue equation (9.31)

$$M\mathbf{f} = \lambda \mathbf{f} \tag{9.31}$$

$$(M\mathbf{f})_n = \left(\sin\big((n-1)k\big) + \sin\big((n+1)k\big) - 2\sin(nk)\right)$$
$$= 2\sin(nk)\big(\cos(k) - 1\big) = \lambda(\mathbf{f})_n \tag{9.32}$$

with the eigenvalue

$$\lambda = 2(\cos k - 1) = -4\sin^2\left(\frac{k}{2}\right). \tag{9.33}$$

The first line of the eigenvalue equation (9.31) reads

$$(M\mathbf{f})_1 = \left(-2\sin(k) + \sin(2k)\right)$$
$$= 2\sin(k)\big(\cos(k) - 1\big) = \lambda(\mathbf{f})_n \tag{9.34}$$

and from the last line we have

$$(M\mathbf{f})_N = \left(-2\sin(Nk) + \sin\big([N-1]k\big)\right)$$
$$= \lambda(\mathbf{f})_N = 2\big(\cos(k) - 1\big)\sin(Nk) \tag{9.35}$$

which holds if

$$\sin\big((N-1)k\big) = 2\sin(Nk)\cos(k). \tag{9.36}$$

This simplifies to

$$\sin(Nk)\cos(k) - \cos(Nk)\sin(k) = 2\sin(Nk)\cos(k)$$
$$\sin(Nk)\cos(k) + \cos(Nk)\sin(k) = 0 \tag{9.37}$$
$$\sin\big((N+1)k\big) = 0.$$

Hence the possible values of k are

$$k = \frac{\pi}{(N+1)}l \quad \text{with } l = 1, 2 \cdots N \tag{9.38}$$

and the eigenvectors are explicitly (Fig. 9.1)

$$\mathbf{f} = \begin{pmatrix} \sin(\frac{\pi}{N+1}l) \\ \vdots \\ \sin(\frac{\pi}{N+1}l\,n) \\ \vdots \\ \sin(\frac{\pi}{N+1}l\,N) \end{pmatrix}. \tag{9.39}$$

Fig. 9.1 (Lowest eigenvector for fixed and open boundaries) *Top*: for fixed boundaries $f_n = \sin(nk)$ which is zero at the additional points x_0, x_{N+1}. For open boundaries $f_n = \cos((n-1)k)$ with horizontal tangent at x_1, x_N due to the boundary conditions $f_2 = f_0$, $f_{N-1} = f_{N+1}$

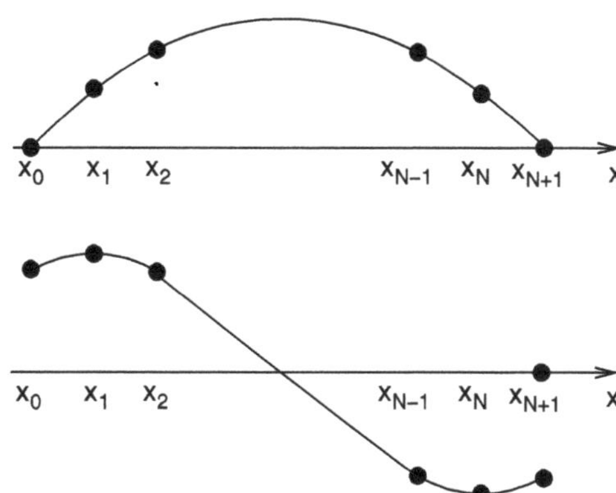

For Neumann boundary conditions $\frac{\partial f}{\partial x}(x_1) = \frac{\partial f}{\partial x}(x_N) = 0$ the matrix is slightly different (Sect. 18.2)

$$
M = \begin{pmatrix}
-2 & 2 & & & \\
1 & -2 & 1 & & \\
 & \ddots & \ddots & \ddots & \\
 & & 1 & -2 & 1 \\
 & & & 2 & -2
\end{pmatrix}. \tag{9.40}
$$

Its eigenvalues are also given by the expression (9.33). To obtain the eigenvectors, we try a more general ansatz with a phase shift

$$
\mathbf{f} = \begin{pmatrix}
\sin \Phi_1 \\
\vdots \\
\sin(\Phi_1 + (n-1)k) \\
\vdots \\
\sin(\Phi_1 + (N-1)k)
\end{pmatrix}. \tag{9.41}
$$

Obviously

$$
\sin\big(\Phi_1 + (n-1)k - k\big) + \sin\big(\Phi_1 + (n-1)k + k\big) - 2\sin\big(\Phi_1 + (n-1)k\big)
$$
$$
= 2(\cos k - 1)\sin\big(\Phi_1 + (n-1)k\big). \tag{9.42}
$$

The first and last lines of the eigenvalue equation give

$$
0 = -2\sin(\Phi_1) + 2\sin(\Phi_1 + k) - 2(\cos k - 1)\sin(\Phi_1)
$$
$$
= 2\cos\Phi_1 \sin k \tag{9.43}
$$

and

$$0 = -2\sin\big(\Phi_1 + (N-1)k\big) + 2\sin\big(\Phi_1 + (N-1)k - k\big)$$
$$- 2(\cos k - 1)\sin\big(\Phi_1 + (N-1)k\big)$$
$$= 2\cos\big(\Phi_1 + (N-1)k\big)\sin k \tag{9.44}$$

which is solved by

$$\Phi_1 = \frac{\pi}{2} \quad k = \frac{\pi}{N-1}l, \quad l = 1, 2 \cdots N \tag{9.45}$$

hence finally the eigenvector is (Fig. 9.1)

$$\mathbf{f} = \begin{pmatrix} 1 \\ \vdots \\ \cos(\frac{n-1}{N-1}\pi l) \\ \vdots \\ (-1)^l \end{pmatrix}. \tag{9.46}$$

Even simpler is the case of the corresponding cyclic tridiagonal matrix

$$M = \begin{pmatrix} -2 & 1 & & & & 1 \\ 1 & -2 & 1 & & & \\ & & \ddots & \ddots & \ddots & \\ & & & 1 & -2 & 1 \\ 1 & & & & 1 & -2 \end{pmatrix} \tag{9.47}$$

which has eigenvectors

$$\mathbf{f} = \begin{pmatrix} e^{ik} \\ \vdots \\ e^{ink} \\ \vdots \\ e^{iNk} \end{pmatrix} \tag{9.48}$$

and eigenvalues

$$\lambda = -2 + e^{-ik} + e^{ik} = 2\big(\cos(k) - 1\big) = -\sin^2\left(\frac{k}{2}\right) \tag{9.49}$$

where the possible k − values again follow from the first and last line

$$-2e^{ik} + e^{i2k} + e^{iNk} = \big(-2 + e^{-ik} + e^{ik}\big)e^{ik} \tag{9.50}$$
$$e^{ik} + e^{i(N-1)k} - 2e^{iNk} = \big(-2 + e^{-ik} + e^{ik}\big)e^{iNk} \tag{9.51}$$

which both lead to

$$e^{iNk} = 1 \tag{9.52}$$

$$k = \frac{2\pi}{N}l, \quad l = 0, 1 \cdots N - 1. \tag{9.53}$$

9.3.2.2 Discretized First Derivatives

Using symmetric differences to discretize a first derivative in one dimension leads
to the matrix[2]

$$D = \begin{pmatrix} & 1 & & & & \\ -1 & & 1 & & & \\ & \ddots & & \ddots & & \\ & & \ddots & & \ddots & \\ & & & -1 & & 1 \\ & & & & -1 & \end{pmatrix}. \tag{9.54}$$

The characteristic polynomial of the Hermitian matrix iD is given by the recursion
(9.3.1)

$$\begin{aligned} P_0 &= 1 \\ P_1 &= -\lambda \\ &\ \vdots \\ P_N &= -\lambda P_{N-1} - P_{N-2} \end{aligned} \tag{9.55}$$

which after the substitution $x = -\lambda/2$ is exactly the recursion for the Chebyshev
polynomial of the second kind $U_N(x)$. Hence the eigenvalues of D are given by the
roots x_k of $U_N(x)$ as

$$\lambda_D = 2ix_k = 2i\cos\left(\frac{k\pi}{N+1}\right) \quad k = 1, 2 \cdots N. \tag{9.56}$$

The eigenvalues of the corresponding cyclic tridiagonal matrix

$$D = \begin{pmatrix} & 1 & & & & -1 \\ -1 & & 1 & & & \\ & \ddots & & \ddots & & \\ & & \ddots & & \ddots & \\ & & & -1 & & 1 \\ 1 & & & & -1 & \end{pmatrix} \tag{9.57}$$

---

[2]This matrix is skew symmetric, hence iT is Hermitian and has real eigenvalues $i\lambda$.

are easy to find. Inserting the ansatz for the eigenvector

$$\begin{pmatrix} \exp ik \\ \vdots \\ \exp iNk \end{pmatrix} \tag{9.58}$$

we find the eigenvalues

$$e^{i(m+1)k} - e^{i(m-1)k} = \lambda e^{imk} \tag{9.59}$$

$$\lambda = 2i \sin k \tag{9.60}$$

and from the first and last equation

$$1 = e^{iNk} \tag{9.61}$$

$$e^{ik} = e^{i(N+1)k} \tag{9.62}$$

the possible k-values

$$k = \frac{2\pi}{N} l, \quad l = 0, 1 \cdots N - 1. \tag{9.63}$$

9.3.3 The QL Algorithm

Any real matrix A can be decomposed into the product of a lower triangular matrix L and an orthogonal matrix $Q^T = Q^{-1}$ (this is quite similar to the QR factorization with an upper triangular matrix which is discussed in Sect. 5.2)

$$A = QL. \tag{9.64}$$

For symmetric tridiagonal matrices this factorization can be efficiently realized by multiplication with a sequence of rotation matrices which eliminate the off-diagonal elements in the lower part

$$Q = R^{(N-1,N)} \cdots R^{(2,3)} R^{(1,2)}. \tag{9.65}$$

An orthogonal transformation of A is given by

$$Q^T A Q = Q^T Q L Q = L Q. \tag{9.66}$$

The QL algorithm is an iterative algorithm. It uses the transformation

$$A_{n+1} = Q_n^T A_n Q_n = Q_n^T (A_n - \sigma_n) Q_n + \sigma_n = Q_n^T Q_n L_n Q_n + \sigma_n$$
$$= L_n Q_n + \sigma_n \tag{9.67}$$

where the shift parameter σ_n was introduced to improve convergence and the QL factorization is applied to $A_n - \sigma_n$. This transformation conserves symmetry and tridiagonal form. Repeated transformation gives a sequence of tridiagonal matrices, which converge to a diagonal matrix if the shifts σ_n are properly chosen. A very popular choice [198] is Wilkinson's shift

$$\sigma = a_{11} - \text{sign}(\delta)\frac{a_{12}^2}{|\delta| + \sqrt{\delta^2 + a_{12}^2}} \qquad \delta = \frac{a_{22} - a_{11}}{2} \tag{9.68}$$

which is that eigenvalue of the matrix $\begin{pmatrix} a_{11} & a_{12} \\ a_{12} & a_{22} \end{pmatrix}$ which is closer to a_{11} (the smaller one if $a_{11} = a_{22}$).

9.4 Reduction to a Tridiagonal Matrix

Any symmetric matrix can be transformed to a tridiagonal matrix by a series of Householder transformations (5.53)

$$A' = PAP \quad \text{with } P = P^T = 1 - 2\frac{\mathbf{u}\mathbf{u}^T}{|\mathbf{u}|^2}. \tag{9.69}$$

The following orthogonal transformation P_1 brings the first row and column to tridiagonal form. We divide the matrix A according to

$$A = \begin{pmatrix} a_{11} & \boldsymbol{\alpha}^T \\ \boldsymbol{\alpha} & A_{rest} \end{pmatrix} \tag{9.70}$$

with the $(N-1)$-dimensional vector

$$\boldsymbol{\alpha} = \begin{pmatrix} a_{12} \\ \vdots \\ a_{1n} \end{pmatrix}.$$

Now let

$$\mathbf{u} = \begin{pmatrix} 0 \\ a_{12} + \lambda \\ \vdots \\ a_{1N} \end{pmatrix} = \begin{pmatrix} 0 \\ \boldsymbol{\alpha} \end{pmatrix} + \lambda \mathbf{e}^{(2)} \quad \text{with } \mathbf{e}^{(2)} = \begin{pmatrix} 0 \\ 1 \\ 0 \\ \vdots \\ 0 \end{pmatrix}. \tag{9.71}$$

Then

$$|\mathbf{u}|^2 = |\boldsymbol{\alpha}|^2 + \lambda^2 + 2\lambda a_{12} \tag{9.72}$$

and

$$\mathbf{u}^T \begin{pmatrix} a_{11} \\ \boldsymbol{\alpha} \end{pmatrix} = |\boldsymbol{\alpha}|^2 + \lambda a_{12}. \tag{9.73}$$

The first row of A is transformed by multiplication with P_1 according to

$$P_1 \begin{pmatrix} a_{11} \\ \boldsymbol{\alpha} \end{pmatrix} = \begin{pmatrix} a_{11} \\ \boldsymbol{\alpha} \end{pmatrix} - 2 \frac{|\boldsymbol{\alpha}|^2 + \lambda a_{12}}{|\boldsymbol{\alpha}|^2 + \lambda^2 + 2\lambda a_{12}} \left[\begin{pmatrix} 0 \\ \boldsymbol{\alpha} \end{pmatrix} + \lambda \mathbf{e}^{(2)} \right]. \tag{9.74}$$

The elements number $3 \cdots N$ are eliminated if we choose[3]

$$\lambda = \pm|\boldsymbol{\alpha}| \tag{9.75}$$

because then

$$2 \frac{|\boldsymbol{\alpha}|^2 + \lambda a_{12}}{|\boldsymbol{\alpha}|^2 + \lambda^2 + 2\lambda a_{12}} = 2 \frac{|\boldsymbol{\alpha}|^2 \pm |\boldsymbol{\alpha}| a_{12}}{|\boldsymbol{\alpha}|^2 + |\boldsymbol{\alpha}|^2 \pm 2|\boldsymbol{\alpha}| a_{12}} = 1 \tag{9.76}$$

and

$$P_1 \begin{pmatrix} a_{11} \\ \boldsymbol{\alpha} \end{pmatrix} = \begin{pmatrix} a_{11} \\ \boldsymbol{\alpha} \end{pmatrix} - \begin{pmatrix} 0 \\ \boldsymbol{\alpha} \end{pmatrix} - \lambda \mathbf{e}^{(2)} = \begin{pmatrix} a_{11} \\ \mp|\boldsymbol{\alpha}| \\ 0 \\ \vdots \\ 0 \end{pmatrix}. \tag{9.77}$$

Finally we have

$$A^{(2)} = P_1 A P_1 = \begin{pmatrix} a_{11} & a_{12}^{(2)} & 0 & \cdots & 0 \\ a_{12}^{(2)} & a_{22}^{(2)} & a_{23}^{(2)} & \cdots & a_{2N}^{(2)} \\ 0 & a_{23}^{(2)} & \ddots & & a_{3N}^{(2)} \\ \vdots & \vdots & & \ddots & \vdots \\ 0 & a_{2N}^{(2)} & a_{3N}^{(2)} & \cdots & a_{NN}^{(2)} \end{pmatrix} \tag{9.78}$$

as desired.

For the next step we choose

$$\boldsymbol{\alpha} = \begin{pmatrix} a_{22}^{(2)} \\ \vdots \\ a_{2N}^{(2)} \end{pmatrix}, \qquad \mathbf{u} = \begin{pmatrix} 0 \\ 0 \\ \boldsymbol{\alpha} \end{pmatrix} \pm |\boldsymbol{\alpha}| \mathbf{e}^{(3)} \tag{9.79}$$

---

[3]To avoid numerical extinction we choose the sign to be that of A_{12}.

to eliminate the elements $a_{24} \cdots a_{2N}$. Note that P_2 does not change the first row and column of $A^{(2)}$ and therefore

$$A^{(3)} = P_2 A^{(2)} P_2 = \begin{pmatrix} a_{11} & a_{12}^{(2)} & 0 & \cdots & \cdots & 0 \\ a_{12}^{(2)} & a_{22}^{(2)} & a_{23}^{(3)} & 0 & \cdots & 0 \\ 0 & a_{23}^{(3)} & a_{33}^{(3)} & \cdots & \cdots & a_{3N}^{(3)} \\ \vdots & 0 & \vdots & & & \vdots \\ \vdots & \vdots & \vdots & & & \vdots \\ 0 & 0 & a_{3N}^{(3)} & \cdots & \cdots & a_{NN}^{(3)} \end{pmatrix}. \tag{9.80}$$

After $N - 1$ transformations finally a tridiagonal matrix is obtained.

9.5 Large Matrices

Special algorithms are available for matrices of very large dimension to calculate only some eigenvalues and eigenvectors. The famous Lanczos method [153] diagonalizes the matrix in a subspace which is constructed from the vectors

$$\mathbf{x}_0, A\mathbf{x}_0, A^2\mathbf{x}_0 \cdots A^N\mathbf{x}_0 \tag{9.81}$$

which, starting from an initial normalized guess vector $\mathbf{x}_0$ are orthonormalized to obtain a tridiagonal matrix,

$$\mathbf{x}_1 = \frac{A\mathbf{x}_0 - (\mathbf{x}_0 A\mathbf{x}_0)\mathbf{x}_0}{|A\mathbf{x}_0 - (\mathbf{x}_0 A\mathbf{x}_0)\mathbf{x}_0|} = \frac{A\mathbf{x}_0 - a_0\mathbf{x}_0}{b_0}$$

$$\mathbf{x}_2 = \frac{A\mathbf{x}_1 - b_0\mathbf{x}_0 - (\mathbf{x}_1 A\mathbf{x}_1)\mathbf{x}_1}{|A\mathbf{x}_1 - b_0\mathbf{x}_0 - (\mathbf{x}_1 A\mathbf{x}_1)\mathbf{x}_1|} = \frac{A\mathbf{x}_1 - b_0\mathbf{x}_0 - a_1\mathbf{x}_1}{b_1}$$

$$\vdots \tag{9.82}$$

$$\mathbf{x}_N = \frac{A\mathbf{x}_{N-1} - b_{N-2}\mathbf{x}_{N-2} - (\mathbf{x}_{N-1} A\mathbf{x}_{N-1})\mathbf{x}_{N-1}}{|A\mathbf{x}_{N-1} - b_{N-2}\mathbf{x}_{N-2} - (\mathbf{x}_{N-1} A\mathbf{x}_{N-1})\mathbf{x}_{N-1}|}$$

$$= \frac{A\mathbf{x}_{N-1} - b_{N-2}\mathbf{x}_{N-2} - a_{N-1}\mathbf{x}_{N-1}}{b_{N-1}} = \frac{\mathbf{r}_{N-1}}{b_{N-1}}.$$

This series is truncated by setting

$$a_N = (\mathbf{x}_N A\mathbf{x}_N) \tag{9.83}$$

and neglecting

$$\mathbf{r}_N = A\mathbf{x}_N - b_{N-1}\mathbf{x}_{N-1} - a_N\mathbf{x}_N. \tag{9.84}$$

Within the subspace of the $\mathbf{x}_1 \cdots \mathbf{x}_N$ the matrix A is represented by the tridiagonal matrix

$$T = \begin{pmatrix} a_0 & b_0 & & & & \\ b_0 & a_1 & b_1 & & & \\ & & \ddots & \ddots & & \\ & & \ddots & a_{N-1} & b_{N-1} \\ & & & b_{N-1} & a_N \end{pmatrix} \tag{9.85}$$

which can be diagonalized with standard methods. The whole method can be iterated using an eigenvector of T as the new starting vector and increasing N until the desired accuracy is achieved. The main advantage of the Lanczos method is that the matrix A will not be stored in memory. It is sufficient to calculate scalar products with A.

9.6 Problems

Problem 9.1 (Computer experiment: disorder in a tight-binding model)   We consider a two-dimensional lattice of interacting particles. Pairs of nearest neighbors have an interaction V and the diagonal energies are chosen from a Gaussian distribution

$$P(E) = \frac{1}{\Delta\sqrt{2\pi}} e^{-E^2/2\Delta^2}. \tag{9.86}$$

The wave function of the system is given by a linear combination

$$\psi = \sum_{ij} C_{ij}\psi_{ij} \tag{9.87}$$

where on each particle (i, j) one basis function ψ_{ij} is located. The nonzero elements of the interaction matrix are given by

$$H(ij|ij) = E_{ij} \tag{9.88}$$

$$H(ij|i \pm 1, j) = H(ij|i, j \pm 1) = V. \tag{9.89}$$

The matrix H is numerically diagonalized and the amplitudes C_{ij} of the lowest state are shown as circles located at the grid points. As a measure of the degree of localization the quantity

$$\sum_{ij} |C_{ij}|^4 \tag{9.90}$$

is evaluated. Explore the influence of coupling V and disorder Δ.

Chapter 10
Data Fitting

Often a set of data points has to be fitted by a continuous function, either to obtain approximate function values in between the data points or to describe a functional relationship between two or more variables by a smooth curve, i.e. to fit a certain model to the data. If uncertainties of the data are negligibly small, an exact fit is possible, for instance with polynomials, spline functions or trigonometric functions (Chap. 2). If the uncertainties are considerable, a curve has to be constructed that fits the data points approximately. Consider a two-dimensional data set

$$(x_i, y_i) \quad i = 1 \cdots m \tag{10.1}$$

and a model function

$$f(x, a_1 \cdots a_n) \quad m \ge n \tag{10.2}$$

which depends on the variable x and $n \le m$ additional parameters a_j. The errors of the fitting procedure are given by the residuals

$$r_i = y_i - f(x_i, a_1 \cdots a_n). \tag{10.3}$$

The parameters a_j have to be determined such, that the overall error is minimized, which in most practical cases is measured by the mean square difference[1]

$$S_{sd}(a_1 \cdots a_n) = \frac{1}{m} \sum_{i=1}^{m} r_i^2. \tag{10.4}$$

The optimal parameters are determined by solving the system of normal equations. If the model function depends linearly on the parameters, orthogonalization offers a numerically more stable method. The dimensionality of a data matrix can be reduced with the help of singular value decomposition, which allows to approximate a matrix by another matrix of lower rank and is also useful for linear regression, especially if the columns of the data matrix are linearly dependent.

---

[1] Minimization of the sum of absolute errors $\sum |r_i|$ is much more complicated.

P.O.J. Scherer, *Computational Physics*, Graduate Texts in Physics,
DOI 10.1007/978-3-319-00401-3_10,
© Springer International Publishing Switzerland 2013

10.1 Least Square Fit

A (local) minimum of (10.4) corresponds to a stationary point with zero gradient. For n model parameters there are n, generally nonlinear, equations which have to be solved [275]. From the general condition

$$\frac{\partial S_{sd}}{\partial a_j} = 0 \quad j = 1 \cdots n \tag{10.5}$$

we find

$$\sum_{i=1}^{m} r_i \frac{\partial f(x_i, a_1 \cdots a_n)}{\partial a_j} = 0 \tag{10.6}$$

which can be solved with the methods discussed in Chap. 6. For instance, the Newton-Raphson method starts from a suitable initial guess of parameters

$$\left(a_1^0 \cdots a_n^0\right) \tag{10.7}$$

and tries to improve the fit iteratively by making small changes to the parameters

$$a_j^{s+1} = a_j^s + \Delta a_j^s. \tag{10.8}$$

The changes Δa_j^s are determined approximately by expanding the model function

$$f\left(x_i, a_1^{s+1} \cdots a_n^{s+1}\right) = f\left(x_i, a_1^s \cdots a_n^s\right) + \sum_{j=1}^{n} \frac{\partial f(x_i, a_1^s \cdots a_n^s)}{\partial a_j} \Delta a_j^s + \cdots \tag{10.9}$$

to approximate the new residuals

$$r_i^{s+1} = r_i^s - \sum_{j=1}^{n} \frac{\partial f(x_i, a_1^s \cdots a_m^s)}{\partial a_j} \Delta a_j^s \tag{10.10}$$

and the derivatives

$$\frac{\partial r_i^s}{\partial a_j} = -\frac{\partial f(x_i, a_1^s \cdots a_m^s)}{\partial a_j}. \tag{10.11}$$

Equation (10.6) now becomes

$$\sum_{i=1}^{m} \left(r_i^s - \sum_{j=1}^{n} \frac{\partial f(x_i)}{\partial a_j} \Delta a_j^s \right) \frac{\partial f(x_i)}{\partial a_k} \tag{10.12}$$

which is a system of n (usually overdetermined) linear equations for the Δa_j, the so-called normal equations:

$$\sum_{i=1}^{m} \sum_{j=1}^{n} \frac{\partial f(x_i)}{\partial a_j} \frac{\partial f(x_i)}{\partial a_k} \Delta a_j^s = \sum_{i=1}^{m} r_i^s \frac{\partial f(x_i)}{\partial a_k}. \tag{10.13}$$

With the definition

$$A_{kj} = \frac{1}{m} \sum_{i=1}^{m} \frac{\partial f(x_i)}{\partial a_k} \frac{\partial f(x_i)}{\partial a_j} \tag{10.14}$$

$$b_k = \frac{1}{m} \sum_{i=1}^{m} y_i \frac{\partial f(x_i)}{\partial a_k} \tag{10.15}$$

the normal equations can be written as

$$\sum_{j=1}^{n} A_{kj} \Delta a_j = b_k. \tag{10.16}$$

10.1.1 Linear Least Square Fit

Especially important are model functions which depend linearly on all parameters (Fig. 10.1 shows an example which is discussed in Problem 10.1)

$$f(x, a_1 \cdots a_n) = \sum_{j=1}^{n} a_j f_j(x). \tag{10.17}$$

The derivatives are

$$\frac{\partial f(x_i)}{\partial a_j} = f_j(x_i) \tag{10.18}$$

and the minimum of (10.4) is given by the solution of the normal equations

$$\frac{1}{m} \sum_{j=1}^{n} \sum_{i=1}^{m} f_k(x_i) f_j(x_i) a_j = \frac{1}{m} \sum_{i=1}^{m} y_i f_k(x_i) \tag{10.19}$$

which for a linear fit problem become

$$\sum_{j=1}^{n} A_{kj} a_j = b_k \tag{10.20}$$

with

$$A_{kj} = \frac{1}{m} \sum_{i=1}^{m} f_k(x_i) f_j(x_i) \tag{10.21}$$

$$b_k = \frac{1}{m} \sum_{i=1}^{m} y_i f_k(x_i). \tag{10.22}$$

Fig. 10.1 (Least square fit)
The polynomial
$C(T) = aT + bT^3$ (*full
curve*) is fitted to a set of data
points which are distributed
randomly around the "exact"
values $C(T) = a_0 T + b_0 T^3$
(*dashed curve*). For more
details see Problem 10.1

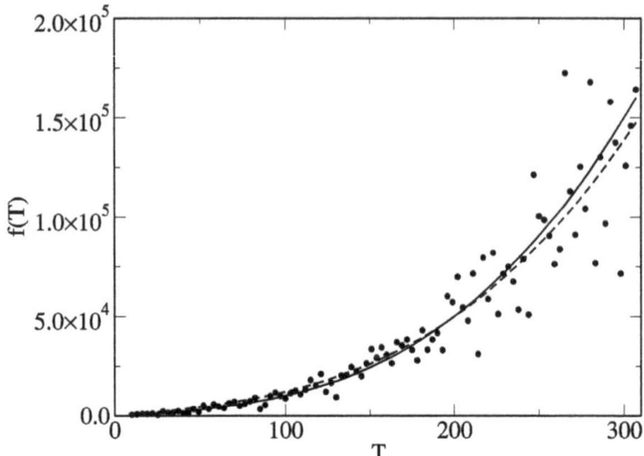

Example (Linear regression) For a linear fit function

$$f(x) = a_0 + a_1 x \tag{10.23}$$

the mean square difference is

$$S_{sd} = \frac{1}{m} \sum_{i=1}^{m} (y_i - a_0 - a_1 x_i)^2 \tag{10.24}$$

and we have to solve the equations

$$0 = \frac{\partial S_{sd}}{\partial a_0} = \frac{1}{m} \sum_{i=1}^{m} (y_i - a_0 - a_1 x_i) = \overline{y} - a_0 - a_1 \overline{x}$$

$$0 = \frac{\partial S_{sd}}{\partial a_1} = \frac{1}{m} \sum_{i=1}^{m} (y_i - a_0 - a_1 x_i) x_i = \overline{xy} - a_0 \overline{x} - a_1 \overline{x^2} \tag{10.25}$$

which can be done here with determinants

$$a_0 = \frac{\begin{vmatrix} \overline{y} & \overline{x} \\ \overline{xy} & \overline{x^2} \end{vmatrix}}{\begin{vmatrix} 1 & \overline{x} \\ \overline{x} & \overline{x^2} \end{vmatrix}} = \frac{\overline{y}\,\overline{x^2} - \overline{x}\,\overline{xy}}{\overline{x^2} - \overline{x}^2} \tag{10.26}$$

$$a_1 = \frac{\begin{vmatrix} 1 & \overline{y} \\ \overline{x} & \overline{xy} \end{vmatrix}}{\begin{vmatrix} 1 & \overline{x} \\ \overline{x} & \overline{x^2} \end{vmatrix}} = \frac{\overline{xy} - \overline{x}\,\overline{y}}{\overline{x^2} - \overline{x}^2}. \tag{10.27}$$

10.1.2 Linear Least Square Fit with Orthogonalization

With the definitions

$$\mathbf{x} = \begin{pmatrix} a_1 \\ \vdots \\ a_n \end{pmatrix} \qquad \mathbf{b} = \begin{pmatrix} y_1 \\ \vdots \\ y_m \end{pmatrix} \tag{10.28}$$

and the $m \times n$ matrix

$$A = \begin{pmatrix} a_{11} & \cdots & a_{1n} \\ \vdots & \ddots & \vdots \\ a_{m1} & \cdots & a_{mn} \end{pmatrix} = \begin{pmatrix} f_1(x_1) & \cdots & f_n(x_1) \\ \vdots & \ddots & \vdots \\ f_1(x_m) & \cdots & f_n(x_m) \end{pmatrix} \tag{10.29}$$

the linear least square fit problem (10.20) can be formulated as a search for the minimum of

$$|A\mathbf{x} - \mathbf{b}| = \sqrt{(A\mathbf{x} - \mathbf{b})^T (A\mathbf{x} - \mathbf{b})}. \tag{10.30}$$

In the last section we calculated the gradient

$$\frac{\partial |A\mathbf{x} - \mathbf{b}|^2}{\partial \mathbf{x}} = A^T (A\mathbf{x} - \mathbf{b}) + (A\mathbf{x} - \mathbf{b})^T A = 2A^T A\mathbf{x} - 2A^T \mathbf{b} \tag{10.31}$$

and solved the normal equations

$$A^T A\mathbf{x} = A^T \mathbf{b}. \tag{10.32}$$

This method can become numerically unstable. Alternatively we use orthogonalization of the n column vectors $\mathbf{a}_k$ of A to have

$$A = (\mathbf{a}_1 \cdots \mathbf{a}_n) = (\mathbf{q}_1 \cdots \mathbf{q}_n) \begin{pmatrix} r_{11} & r_{12} & \cdots & r_{1n} \\ & r_{22} & \cdots & r_{2n} \\ & & \ddots & \vdots \\ & & & r_{nn} \end{pmatrix} \tag{10.33}$$

where $\mathbf{a}_k$ and $\mathbf{q}_k$ are now vectors of dimension m. Since the $\mathbf{q}_k$ are orthonormal $\mathbf{q}_i^T \mathbf{q}_k = \delta_{ik}$ we have

$$\begin{pmatrix} \mathbf{q}_1^T \\ \vdots \\ \mathbf{q}_n^T \end{pmatrix} A = \begin{pmatrix} r_{11} & r_{12} & \cdots & r_{1n} \\ & r_{22} & \cdots & r_{2n} \\ & & \ddots & \vdots \\ & & & r_{nn} \end{pmatrix}. \tag{10.34}$$

The $\mathbf{q}_k$ can be augmented by another $(m - n)$ vectors to provide an orthonormal basis of $\mathbb{R}^m$. These will not be needed explicitly. They are orthogonal to the first

n vectors and hence to the column vectors of A. All vectors $\mathbf{q}_k$ together form an orthogonal matrix

$$Q = \begin{pmatrix} \mathbf{q}_1 & \cdots & \mathbf{q}_n & \mathbf{q}_{n+1} & \cdots & \mathbf{q}_m \end{pmatrix} \tag{10.35}$$

and we can define the transformation of the matrix A:

$$\tilde{A} = \begin{pmatrix} \mathbf{q}_1^T \\ \vdots \\ \mathbf{q}_n^T \\ \mathbf{q}_{n+1}^T \\ \vdots \\ \mathbf{q}_m^T \end{pmatrix} (\mathbf{a}_1 \cdots \mathbf{a}_n) = Q^T A = \begin{pmatrix} R \\ 0 \end{pmatrix} \quad R = \begin{pmatrix} r_{11} & \cdots & r_{1n} \\ & \ddots & \vdots \\ & & r_{nn} \end{pmatrix}. \tag{10.36}$$

The vector $\mathbf{b}$ transforms as

$$\tilde{\mathbf{b}} = Q^T \mathbf{b} = \begin{pmatrix} \mathbf{b}_u \\ \mathbf{b}_l \end{pmatrix} \quad \mathbf{b}_u = \begin{pmatrix} \mathbf{q}_1^T \\ \vdots \\ \mathbf{q}_n^T \end{pmatrix} \mathbf{b} \quad \mathbf{b}_l = \begin{pmatrix} \mathbf{q}_{n+1}^T \\ \vdots \\ \mathbf{q}_m^T \end{pmatrix} \mathbf{b}. \tag{10.37}$$

Since the norm of a vector is not changed by unitary transformations

$$|\mathbf{b} - A\mathbf{x}| = \sqrt{(\mathbf{b}_u - R\mathbf{x})^2 + \mathbf{b}_l^2} \tag{10.38}$$

which is minimized if

$$R\mathbf{x} = \mathbf{b}_u. \tag{10.39}$$

The error of the fit is given by

$$|\mathbf{b} - A\mathbf{x}| = |\mathbf{b}_l|. \tag{10.40}$$

Example (Linear regression) Consider again the fit function

$$f(x) = a_0 + a_1 x \tag{10.41}$$

for the measured data (x_i, y_i). The fit problem is to determine

$$\left| \begin{pmatrix} 1 & x_1 \\ \vdots & \vdots \\ 1 & x_m \end{pmatrix} \begin{pmatrix} a_0 \\ a_1 \end{pmatrix} - \begin{pmatrix} y_1 \\ \vdots \\ y_m \end{pmatrix} \right| = \min. \tag{10.42}$$

Orthogonalization of the column vectors

$$\mathbf{a}_1 = \begin{pmatrix} 1 \\ \vdots \\ 1 \end{pmatrix} \quad \mathbf{a}_2 = \begin{pmatrix} x_1 \\ \vdots \\ x_m \end{pmatrix} \tag{10.43}$$

with the Schmidt method gives:

$$r_{11} = \sqrt{m} \tag{10.44}$$

$$\mathbf{q}_1 = \begin{pmatrix} \frac{1}{\sqrt{m}} \\ \vdots \\ \frac{1}{\sqrt{m}} \end{pmatrix} \tag{10.45}$$

$$r_{12} = \frac{1}{\sqrt{m}} \sum_{i=1}^{m} x_i = \sqrt{m}\,\overline{x} \tag{10.46}$$

$$\mathbf{b}_2 = (x_i - \overline{x}) \tag{10.47}$$

$$r_{22} = \sqrt{\sum (x_i - \overline{x})^2} = \sqrt{m}\,\sigma_x \tag{10.48}$$

$$\mathbf{q}_2 = \left(\frac{x_i - \overline{x}}{\sqrt{m}\,\sigma_x} \right). \tag{10.49}$$

Transformation of the right hand side gives

$$\begin{pmatrix} \mathbf{q}_1^T \\ \mathbf{q}_2^T \end{pmatrix} \begin{pmatrix} y_1 \\ \vdots \\ y_m \end{pmatrix} = \begin{pmatrix} \sqrt{m}\,\overline{y} \\ \sqrt{m}\,\frac{\overline{xy} - \overline{x}\,\overline{y}}{\sigma_x} \end{pmatrix} \tag{10.50}$$

and we have to solve the system of linear equations

$$R\mathbf{x} = \begin{pmatrix} \sqrt{m} & \sqrt{m}\,\overline{x} \\ 0 & \sqrt{m}\sigma \end{pmatrix} \begin{pmatrix} a_0 \\ a_1 \end{pmatrix} = \begin{pmatrix} \sqrt{m}\,\overline{y} \\ \sqrt{m}\,\frac{\overline{xy} - \overline{x}\,\overline{y}}{\sigma_x} \end{pmatrix}. \tag{10.51}$$

The solution

$$a_1 = \frac{\overline{xy} - \overline{x}\,\overline{y}}{\overline{(x - \overline{x})^2}} \tag{10.52}$$

$$a_0 = \overline{y} - \overline{x}a_1 = \frac{\overline{y}\,\overline{x^2} - \overline{x}\,\overline{xy}}{\overline{(x - \overline{x})^2}} \tag{10.53}$$

coincides with the earlier results since

$$\overline{(x - \overline{x})^2} = \overline{x^2} - \overline{x}^2. \tag{10.54}$$

10.2 Singular Value Decomposition

Computational physics often has to deal with large amounts of data. Singular value decomposition is a very useful tool to reduce redundancies and to extract the most important information from data. It has been used for instance for image compression [216], it is very useful to extract the essential dynamics from molecular dynamics simulations [99, 221] and it is an essential tool of bio-informatics [71].

10.2.1 Full Singular Value Decomposition

For $m \geq n$,[2] any real[3] $m \times n$ matrix A of rank $r \leq n$ can be decomposed into a product

$$A = U \Sigma V^T \tag{10.55}$$

$$
\begin{pmatrix} a_{11} & \cdots & a_{1n} \\ \vdots & \ddots & \vdots \\ a_{m1} & \cdots & a_{mn} \end{pmatrix} = \begin{pmatrix} u_{11} & \cdots & u_{1m} \\ \vdots & \ddots & \vdots \\ u_{m1} & \cdots & u_{mm} \end{pmatrix} \begin{pmatrix} s_1 & & \\ & \ddots & \\ & & s_n \\ 0 & \cdots & 0 \end{pmatrix} \begin{pmatrix} v_{11} & \cdots & v_{n1} \\ \vdots & \ddots & \vdots \\ v_{1n} & \cdots & v_{nn} \end{pmatrix} \tag{10.56}
$$

where U is an $m \times m$ orthogonal matrix, Σ is an $m \times n$ matrix, in which the upper part is an $n \times n$ diagonal matrix and V is an orthogonal $n \times n$ matrix.

The diagonal elements s_i are called singular values. Conventionally, they are sorted in descending order and the last $n - r$ of them are zero. For a square $n \times n$ matrix singular value decomposition (10.56) is equivalent to diagonalization

$$A = U S U^T. \tag{10.57}$$

10.2.2 Reduced Singular Value Decomposition

We write

$$U = (U_n, U_{m-n}) \tag{10.58}$$

with the $m \times n$ matrix U_n and the $m \times (m - n)$ matrix U_{m-n} and

$$\Sigma = \begin{pmatrix} S \\ 0 \end{pmatrix} \tag{10.59}$$

with the diagonal $n \times n$ matrix S. The singular value decomposition then becomes

$$A = (U_n, U_{m-n}) \begin{pmatrix} S \\ 0 \end{pmatrix} V^T = U_n S V^T \tag{10.60}$$

---

[2]Otherwise consider the transpose matrix.

[3]Generalization to complex matrices is straightforward.

which is known as reduced singular value decomposition. U_n (usually simply denoted by U) is not unitary but its column vectors, called the left singular vectors, are orthonormal

$$\sum_{i=1}^{m} u_{i,r} u_{i,s} = \delta_{r,s} \tag{10.61}$$

as well as the column vectors of V which are called the right singular vectors

$$\sum_{i=1}^{n} v_{i,r} v_{i,s} = \delta_{r,s}. \tag{10.62}$$

Hence the products

$$U_n^T U_n = V^T V = E_n \tag{10.63}$$

give the $n \times n$ unit matrix.

In principle, U and V can be obtained from diagonalization of $A^T A$ and $A A^T$, since

$$A^T A = \left(V \Sigma^T U^T \right) \left(U \Sigma V^T \right) = V(S,0) \begin{pmatrix} S \\ 0 \end{pmatrix} V^T = V S^2 V^T \tag{10.64}$$

$$A A^T = \left(U \Sigma V^T \right) \left(V \Sigma^T U^T \right) = U \begin{pmatrix} S \\ 0 \end{pmatrix} (S,0) U^T = U_n S^2 U_n^T. \tag{10.65}$$

However, calculation of U by diagonalization is very inefficient, since usually only the first n rows are needed (i.e. U_n). To perform a reduced singular value decomposition, we first diagonalize

$$A^T A = V D V^T \tag{10.66}$$

which has positive eigenvalues $d_i \geq 0$, sorted in descending order and obtain the singular values

$$S = D^{1/2} = \begin{pmatrix} \sqrt{d_1} & & \\ & \ddots & \\ & & \sqrt{d_n} \end{pmatrix}. \tag{10.67}$$

Now we determine a matrix U such, that

$$A = U S V^T \tag{10.68}$$

or, since V is unitary

$$Y = AV = US. \tag{10.69}$$

The last $n - r$ singular values are zero if $r < n$. Therefore we partition the matrices (indices denote the number of rows)

$$\begin{pmatrix} Y_r & 0_{n-r} \end{pmatrix} = \begin{pmatrix} U_r & U_{n-r} \end{pmatrix} \begin{pmatrix} S_r & \\ & 0_{n-r} \end{pmatrix} = \begin{pmatrix} U_r S_r & 0 \end{pmatrix}. \tag{10.70}$$

We retain only the first r columns and obtain a system of equations

$$
\begin{pmatrix} y_{11} & \cdots & y_{1r} \\ \vdots & \ddots & \vdots \\ y_{m1} & \cdots & y_{mr} \end{pmatrix} = \begin{pmatrix} u_{11} & \cdots & u_{1r} \\ \vdots & \ddots & \vdots \\ u_{m1} & \cdots & u_{mr} \end{pmatrix} \begin{pmatrix} s_1 & & \\ & \ddots & \\ & & s_r \end{pmatrix} \tag{10.71}
$$

which can be easily solved to give the first r rows of U

$$
\begin{pmatrix} u_{11} & \cdots & u_{1r} \\ \vdots & \ddots & \vdots \\ u_{m1} & \cdots & u_{mr} \end{pmatrix} = \begin{pmatrix} y_{11} & \cdots & y_{1n} \\ \vdots & \ddots & \vdots \\ y_{m1} & \cdots & y_{mn} \end{pmatrix} \begin{pmatrix} s_1^{-1} & & \\ & \ddots & \\ & & s_r^{-1} \end{pmatrix} . \tag{10.72}
$$

The remaining $n - r$ column vectors of U have to be orthogonal to the first r columns but are otherwise arbitrary. They can be obtained for instance by the Gram Schmidt method.

For larger matrices direct decomposition algorithms are available, for instance [72], which is based on a reduction to bidiagonal form and a variant of the QL algorithm as first introduced by Golub and Kahan [107].

10.2.3 Low Rank Matrix Approximation

Component-wise (10.60) reads

$$
a_{i,j} = \sum_{k=1}^{r} u_{i,k} s_k v_{j,k}. \tag{10.73}
$$

Approximations to A of lower rank are obtained by reducing the sum to only the largest singular values (the smaller singular values are replaced by zero). It can be shown [243] that the matrix of rank $l \leq r$

$$
a_{i,j}^{(l)} = \sum_{k=1}^{l} u_{i,k} s_k v_{j,k} \tag{10.74}
$$

is the rank-l matrix which minimizes

$$
\sum_{i,j} |a_{i,j} - a_{i,j}^{(l)}|^2. \tag{10.75}
$$

If only the largest singular value is taken into account, A is approximated by the rank-1 matrix

$$
a_{i,j}^{(1)} = s_1 u_{i,1} v_{j,1}. \tag{10.76}
$$

As an example, consider an $m \times n$ matrix

$$A = \begin{pmatrix} x_1(t_1) & \cdots & x_n(t_1) \\ \vdots & & \vdots \\ x_1(t_m) & \cdots & x_n(t_m) \end{pmatrix} \tag{10.77}$$

which contains the values of certain quantities $x_1 \cdots x_n$ observed at different times $t_1 \cdots t_m$. For convenience, we assume that the average values have been subtracted, such that $\sum_{j=1}^{m} x_i = 0$. Approximation (10.76) reduces the dimensionality to 1, i.e. a linear relation between the data. The i-th row of A,

$$\begin{pmatrix} x_1(t_i) & \cdots & x_n(t_i) \end{pmatrix} \tag{10.78}$$

is approximated by

$$s_1 u_{i,1} \begin{pmatrix} v_{1,1} & \cdots & v_{n,1} \end{pmatrix} \tag{10.79}$$

which describes a direct proportionality of different observables

$$\frac{1}{v_{j,1}} x_j(t_i) = \frac{1}{v_{k,1}} x_k(t_i). \tag{10.80}$$

According to (10.75) this linear relation minimizes the mean square distance between the data points (10.78) and their approximation (10.79).

Example (Linear approximation [165]) Consider the data matrix

$$A^T = \begin{pmatrix} 1 & 2 & 3 & 4 & 5 \\ 1 & 2.5 & 3.9 & 3.5 & 4.0 \end{pmatrix}. \tag{10.81}$$

First subtract the row averages

$$\overline{x} = 3 \quad \overline{y} = 2.98 \tag{10.82}$$

to obtain

$$A^T = \begin{pmatrix} -2 & -1 & 0 & 1 & 2 \\ -1.98 & -0.48 & 0.92 & 0.52 & 1.02 \end{pmatrix}. \tag{10.83}$$

Diagonalization of

$$A^T A = \begin{pmatrix} 10.00 & 7.00 \\ 7.00 & 6.308 \end{pmatrix} \tag{10.84}$$

gives the eigenvalues

$$d_1 = 15.393 \quad d_2 = 0.915 \tag{10.85}$$

and the eigenvectors

$$V = \begin{pmatrix} 0.792 & -0.610 \\ 0.610 & -0.792 \end{pmatrix}. \tag{10.86}$$

Since there are no zero singular values we find

$$U = AVS^{-1} = \begin{pmatrix} -0.181 & -0.380 \\ -0.070 & 0.252 \\ 0.036 & 0.797 \\ 0.072 & -0.217 \\ 0.143 & -0.451 \end{pmatrix}. \tag{10.87}$$

This gives the decomposition[4]

$$A = \begin{pmatrix} \mathbf{u}_1 & \mathbf{u}_2 \end{pmatrix} \begin{pmatrix} s_1 & \\ & s_2 \end{pmatrix} \begin{pmatrix} \mathbf{v}_1^T \\ \mathbf{v}_2^T \end{pmatrix} = s_1 \mathbf{u}_1 \mathbf{v}_1^T + s_2 \mathbf{u}_2 \mathbf{v}_2^T$$

$$= \begin{pmatrix} -2.212 & -1.704 \\ -0.860 & -0.662 \\ 0.445 & 0.343 \\ 0.879 & 0.677 \\ 1.748 & 1.347 \end{pmatrix} + \begin{pmatrix} 0.212 & -0.276 \\ -0.140 & 0.182 \\ -0.445 & 0.577 \\ 0.121 & -0.157 \\ 0.252 & -0.327 \end{pmatrix}. \tag{10.88}$$

If we neglect the second contribution corresponding to the small singular value s_2 we have an approximation of the data matrix by a rank-1 matrix. The column vectors of the data matrix, denoted as $\mathbf{x}$ and $\mathbf{y}$, are approximated by

$$\mathbf{x} = s_1 v_{11} \mathbf{u}_1 \quad \mathbf{y} = s_1 v_{21} \mathbf{u}_1 \tag{10.89}$$

which describes a proportionality between $\mathbf{x}$ and $\mathbf{y}$ (Fig. 10.2).

10.2.4 Linear Least Square Fit with Singular Value Decomposition

The singular value decomposition can be used for linear regression [165]. Consider a set of data, which have to be fitted to a linear function

$$y = c_0 + c_1 x_1 + \cdots + c_n x_n \tag{10.90}$$

with the residual

$$r_i = c_0 + c_1 x_{i,1} + \cdots + c_n x_{i,n} - y_i. \tag{10.91}$$

Let us subtract the averages

$$r_i - \bar{r} = c_1(x_{i,1} - \bar{x}_1) + \cdots + c_n(x_{i,n} - \bar{x}_n) - (y_i - \bar{y}) \tag{10.92}$$

---

[4]$\mathbf{u}_i \mathbf{v}_i^T$ is the outer or matrix product of two vectors.

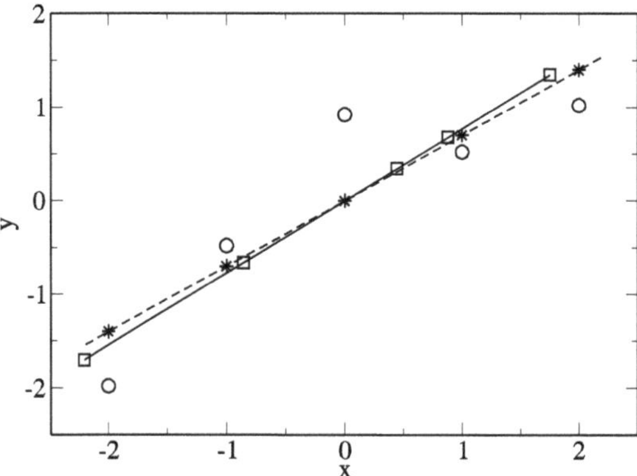

Fig. 10.2 (Linear approximation by singular value decomposition) The data set (10.81) is shown as *circles*. The linear approximation which is obtained by retaining only the dominant singular value is shown by the *squares* and the *full line*. It minimizes the mean square distance to the data points. *Stars* and the *dashed line* show the approximation by linear regression, which minimizes the mean square distance in vertical direction

which we write in matrix notation as

$$\begin{pmatrix} r_1 - \bar{r} \\ \vdots \\ r_m - \bar{r} \end{pmatrix} = \begin{pmatrix} x_{1,1} - \bar{x}_1 & \cdots & x_{1,n} - \bar{x}_n \\ \vdots & & \\ x_{m,1} - \bar{x}_1 & \cdots & x_{m,n} - \bar{x}_n \end{pmatrix} \begin{pmatrix} c_1 \\ \vdots \\ c_n \end{pmatrix} - \begin{pmatrix} y_1 - \bar{y} \\ \vdots \\ y_m - \bar{y} \end{pmatrix} \tag{10.93}$$

or shorter

$$\mathbf{r} = X\mathbf{c} - \mathbf{y}. \tag{10.94}$$

Now let us insert the full decomposition of X

$$\mathbf{r} = U \Sigma V^T \mathbf{c} - \mathbf{y}. \tag{10.95}$$

Since U is orthogonal

$$U^T \mathbf{r} = \Sigma V^T \mathbf{c} - U^T \mathbf{y} = \Sigma \mathbf{a} - \mathbf{b} \tag{10.96}$$

where we introduce the abbreviations

$$\mathbf{a} = V^T \mathbf{c} \quad \mathbf{b} = U^T \mathbf{y}. \tag{10.97}$$

The sum of squared residuals has the form

$$|\mathbf{r}|^2 = |U^T \mathbf{r}|^2 = \left| \begin{pmatrix} S_r & 0 \\ 0 & 0 \end{pmatrix} \begin{pmatrix} \mathbf{a}_r \\ \mathbf{a}_{n-r} \end{pmatrix} - \begin{pmatrix} \mathbf{b}_r \\ \mathbf{b}_{n-r} \end{pmatrix} \right|^2$$

$$= |S_r \mathbf{a}_r - \mathbf{b}_r|^2 + \mathbf{b}_{n-r}^2 \leq |S_r \mathbf{a}_r - \mathbf{b}_r|^2. \tag{10.98}$$

Hence $\mathbf{a}_{n-r}$ is arbitrary and one minimum of S_D is given by

$$\mathbf{a}_r = S_r^{-1} \mathbf{b}_r \quad \mathbf{a}_{n-r} = 0 \tag{10.99}$$

which can be written more compactly as

$$\mathbf{a} = \Sigma^+ \mathbf{b} \tag{10.100}$$

with the Moore-Penrose pseudoinverse [20] of Σ

$$\Sigma^+ = \begin{pmatrix} s_1^{-1} & & & \\ & \ddots & & \\ & & s_r^{-1} & \\ & & & 0 \end{pmatrix}. \tag{10.101}$$

Finally we have

$$\mathbf{c} = V \Sigma^+ U^T \mathbf{y} = X^+ \mathbf{y} \tag{10.102}$$

where

$$X^+ = V \Sigma^+ U^T \tag{10.103}$$

is the Moore-Penrose pseudoinverse of X.

Example The following data matrix has rank 2

$$X = \begin{pmatrix} -3 & -4 & -5 \\ -2 & -3 & -4 \\ 0 & 0 & 0 \\ 2 & 3 & 4 \\ 3 & 4 & 5 \end{pmatrix} \quad \mathbf{y} = \begin{pmatrix} 1.0 \\ 1.1 \\ 0 \\ -1.0 \\ -1.1 \end{pmatrix}. \tag{10.104}$$

A solution to the linear fit problem is given by

$$\mathbf{c} = X^+ \mathbf{y} = \begin{pmatrix} -0.917 & 1.167 & 0 & -1.167 & 0.917 \\ -0.167 & 0.167 & 0 & -0.167 & 0.167 \\ 0.583 & -0.833 & 0 & 0.833 & -0.583 \end{pmatrix} \begin{pmatrix} 1.0 \\ 1.1 \\ 0 \\ -1.0 \\ -1.1 \end{pmatrix}$$

$$= \begin{pmatrix} 0.525 \\ 0.000 \\ -0.525 \end{pmatrix}. \tag{10.105}$$

The fit function is

$$y = 0.525(x_1 - x_3) \tag{10.106}$$

and the residuals are

$$X\mathbf{c} - \mathbf{y} = \begin{pmatrix} 0.05 \\ -0.05 \\ 0 \\ -0.05 \\ 0.05 \end{pmatrix}. \tag{10.107}$$

10.3 Problems

Problem 10.1 (Least square fit)   At temperatures far below Debye and Fermi temperatures the specific heat of a metal contains contributions from electrons and lattice vibrations and can be described by

$$C(T) = aT + bT^3. \tag{10.108}$$

The computer experiment generates data

$$T_j = T_0 + j\Delta t \tag{10.109}$$

$$C_j = \left(a_0 T_j + b_0 T_j^3\right)(1 + \varepsilon_j) \tag{10.110}$$

with relative error

$$\varepsilon_j = \varepsilon \xi_j. \tag{10.111}$$

Random numbers ξ_j are taken from a Gaussian normal distribution function (Sect. 8.2.4.4).

The fit parameters a, b are determined from minimization of the sum of squares

$$S = \frac{1}{n} \sum_{j=1}^{n} \left(C_j - aT_i - bT_i^3\right)^2. \tag{10.112}$$

Compare the "true values" a_0, b_0 with the fitted values a, b.

Chapter 11
Discretization of Differential Equations

Many processes in science and technology can be described by differential equations involving the rate of changes in time or space of a continuous variable, the unknown function. While the simplest differential equations can be solved exactly, a numerical treatment is necessary in most cases and the equations have to be discretized to turn them into a finite system of equations which can be solved by computers [6, 155, 200]. In this chapter we discuss different methods to discretize differential equations. The simplest approach is the method of finite differences, which replaces the differential quotients by difference quotients (Chap. 3). It is often used for the discretization of time. Finite difference methods for the space variables work best on a regular grid. Finite volume methods are very popular in computational fluid dynamics. They take averages over small control volumes and can be easily used with irregular grids. Finite differences and finite volumes belong to the general class of finite element methods which are prominent in the engineering sciences and use an expansion in piecewise polynomials with small support. Spectral methods, on the other hand, expand the solution as a linear combination of global basis functions like polynomials or trigonometric functions. A general concept for the discretization of differential equations is the method of weighted residuals which minimizes the weighted residual of a numerical solution. Most popular is Galerkin's method which uses the expansion functions also as weight functions. Simpler are the point collocation and sub-domain collocation methods which fulfill the differential equation only at certain points or averaged over certain control volumes. More demanding is the least-squares method which has become popular in computational fluid dynamics and computational electrodynamics. The least-square integral provides a measure for the quality of the solution which can be used for adaptive grid size control.

If the Green's function is available for a problem, the method of boundary elements is an interesting alternative. It reduces the dimensionality and is, for instance, very popular in chemical physics to solve the Poisson-Boltzmann equation.

P.O.J. Scherer, *Computational Physics*, Graduate Texts in Physics,
DOI 10.1007/978-3-319-00401-3_11,
© Springer International Publishing Switzerland 2013

11.1 Classification of Differential Equations

An ordinary differential equation (ODE) is a differential equation for a function of
one single variable, like Newton's law for the motion of a body under the influence
of a force field

$$m\frac{\mathrm{d}^2}{\mathrm{d}t^2}\mathbf{x}(t) = \mathbf{F}(\mathbf{x}, t), \tag{11.1}$$

a typical initial value problem where the solution in the domain $t_0 \leq t \leq T$ is determined by position and velocity at the initial time

$$\mathbf{x}(t = t_0) = \mathbf{x}_0 \quad \frac{\mathrm{d}}{\mathrm{d}t}\mathbf{x}(t = t_0) = \mathbf{v}_0. \tag{11.2}$$

Such equations of motion are discussed in Chap. 12. They also appear if the
spatial derivatives of a partial differential equation have been discretized. Usually
this kind of equation is solved by numerical integration over finite time steps $\Delta t =
t_{n+1} - t_n$. Boundary value problems, on the other hand, require certain boundary
conditions[1] to be fulfilled, for instance the linearized Poisson-Boltzmann equation
in one dimension (Chap. 17)

$$\frac{\mathrm{d}^2}{\mathrm{d}x^2}\Phi - \kappa^2\Phi = -\frac{1}{\varepsilon}\rho(x) \tag{11.3}$$

where the value of the potential is prescribed on the boundary of the domain $x_0 \leq
x \leq x_1$

$$\Phi(x_0) = \Phi_0 \quad \Phi(x_1) = \Phi_1. \tag{11.4}$$

Partial differential equations (PDE) finally involve partial derivatives with respect
to at least two different variables, in many cases time and spatial coordinates.

11.1.1 Linear Second Order PDE

A very important class are second order linear partial differential equations of the
general form

$$\left[\sum_{i=1}^{N}\sum_{j=1}^{N}a_{ij}\frac{\partial^2}{\partial x_i \partial x_j} + \sum_{i=1}^{N}b_i\frac{\partial}{\partial x_i} + c\right]f(x_1 \ldots x_N) + d = 0 \tag{11.5}$$

where the coefficients a_{ij}, b_i, c, d are functions of the variables $x_1 \ldots x_N$ but do not
depend on the function f itself. The equation is classified according to the eigenvalues of the coefficient matrix a_{ij} as [141]

---

[1]Dirichlet b.c. concern the function values, Neumann b.c. the derivative, Robin b.c. a linear combination of both, Cauchy b.c. the function value and the normal derivative and mixed b.c. have
different character on different parts of the boundary.

- *elliptical* if all eigenvalues are positive or all eigenvalues are negative, like for the Poisson equation (Chap. 17)

$$\left(\frac{\partial^2}{\partial x^2} + \frac{\partial^2}{\partial y^2} + \frac{\partial^2}{\partial z^2} \right) \Phi(x, y, z) = -\frac{1}{\varepsilon} \rho(x, y, z) \qquad (11.6)$$

- *hyperbolic* if one eigenvalue is negative and all the other eigenvalues are positive or vice versa, for example the wave equation in one spatial dimension (Chap. 18)

$$\frac{\partial^2}{\partial t^2} f - c^2 \frac{\partial^2}{\partial x^2} f = 0 \qquad (11.7)$$

- *parabolic* if at least one eigenvalue is zero, like for the diffusion equation (Chap. 19)

$$\frac{\partial}{\partial t} f(x, y, z, t) - D\left(\frac{\partial^2}{\partial x^2} + \frac{\partial^2}{\partial y^2} + \frac{\partial^2}{\partial z^2} \right) f(x, y, z, t) = S(x, y, z, t)$$

$$(11.8)$$

- *ultra-hyperbolic* if there is no zero eigenvalue and more than one positive as well as more than one negative eigenvalue. Obviously the dimension then must be 4 at least.

11.1.2 Conservation Laws

One of the simplest first order partial differential equations is the advection equation

$$\frac{\partial}{\partial t} f(x, t) + u \frac{\partial}{\partial x} f(x, t) = 0 \qquad (11.9)$$

which describes transport of a conserved quantity with density f (for instance mass, number of particles, charge etc.) in a medium streaming with velocity u. This is a special case of the class of conservation laws (also called continuity equations)

$$\frac{\partial}{\partial t} f(\mathbf{x}, t) + \operatorname{div} \mathbf{J}(\mathbf{x}, t) = g(\mathbf{x}, t) \qquad (11.10)$$

which are very common in physics. Here $\mathbf{J}$ describes the corresponding flux and g is an additional source (or sink) term. For instance the advection-diffusion equation (also known as convection equation) has this form which describes quite general transport processes:

$$\frac{\partial}{\partial t} C = \operatorname{div}(D \operatorname{grad} C - \mathbf{u}C) + S(\mathbf{x}, t) = -\operatorname{div} \mathbf{J} + S(\mathbf{x}, t) \qquad (11.11)$$

where one contribution to the flux

$$\mathbf{J} = -D \operatorname{grad} C + \mathbf{u}C \qquad (11.12)$$

is proportional to the gradient of the concentration C (Fick's first law) and the second part depends on the velocity field $\mathbf{u}$ of a streaming medium. The source term

S represents the effect of chemical reactions. Equation (11.11) is also similar to the drift-diffusion equation in semiconductor physics and closely related to the Navier Stokes equations which are based on the Cauchy momentum equation [1]

$$\rho \frac{d\mathbf{u}}{dt} = \rho\left(\frac{\partial \mathbf{u}}{\partial t} + \mathbf{u}\operatorname{grad}\mathbf{u}\right) = \operatorname{div}\sigma + \mathbf{f} \tag{11.13}$$

where σ denotes the stress tensor. Equation (11.10) is the strong or differential form of the conservation law. The requirements on the smoothness of the solution are reduced by using the integral form which is obtained with the help of Gauss' theorem

$$\int_V \left(\frac{\partial}{\partial t} f(\mathbf{x},t) - g(\mathbf{x},t)\right) dV + \oint_{\partial V} \mathbf{J}(\mathbf{x},t)\, d\mathbf{A} = 0. \tag{11.14}$$

An alternative integral form results from Galerkin's [98] method of weighted residuals which introduces a weight function $w(\mathbf{x})$ and considers the equation

$$\int_V \left(\frac{\partial}{\partial t} f(\mathbf{x},t) + \operatorname{div}\mathbf{J}(\mathbf{x},t) - g(\mathbf{x},t)\right) w(\mathbf{x})\, dV = 0 \tag{11.15}$$

or after applying Gauss' theorem

$$\int_V \left\{\left(\frac{\partial}{\partial t} f(\mathbf{x},t) - g(\mathbf{x},t)\right) w(\mathbf{x}) - \mathbf{J}(\mathbf{x},t)\operatorname{grad} w(\mathbf{x})\right\} dV$$

$$+ \oint_{\partial V} w(\mathbf{x})\mathbf{J}(\mathbf{x},t)\, d\mathbf{A} = 0. \tag{11.16}$$

The so called weak form of the conservation law states that this equation holds for arbitrary weight functions w.

11.2 Finite Differences

The simplest method to discretize a differential equation is to introduce a grid of equidistant points and to discretize the differential operators by finite differences (FDM) as described in Chap. 3. For instance, in one dimension the first and second derivatives can be discretized by

$$x \to x_m = m\Delta x \quad m = 1\ldots M \tag{11.17}$$

$$f(x) \to f_m = f(x_m) \quad m = 1\ldots M \tag{11.18}$$

$$\frac{\partial f}{\partial x} \to \left(\frac{\partial}{\partial x} f\right)_m = \frac{f_{m+1} - f_m}{\Delta x} \quad \text{or} \quad \left(\frac{\partial}{\partial x} f\right)_m = \frac{f_{m+1} - f_{m-1}}{2\Delta x}$$

$$\tag{11.19}$$

$$\frac{\partial^2 f}{\partial x^2} \to \left(\frac{\partial^2}{\partial x^2} f\right)_m = \frac{f_{m+1} + f_{m-1} - 2f_m}{\Delta x^2}. \tag{11.20}$$

These expressions are not well defined at the boundaries of the grid $m = 1, M$ unless the boundary conditions are taken into account. For instance, in case of a Dirichlet problem f_0 and f_{M+1} are given boundary values and

$$\left(\frac{\partial}{\partial x}f\right)_1 = \frac{f_2 - f_0}{2\Delta x} \quad \left(\frac{\partial^2}{\partial x^2}f\right)_1 = \frac{f_2 - 2f_1 + f_0}{\Delta x^2} \tag{11.21}$$

$$\left(\frac{\partial}{\partial x}f\right)_M = \frac{f_{M+1} - f_M}{\Delta x} \quad \text{or} \quad \frac{f_{M+1} - f_{M-1}}{2\Delta x}$$

$$\left(\frac{\partial^2}{\partial x^2}f\right)_M = \frac{f_{M-1} - 2f_M + f_{M+1}}{\Delta x^2}. \tag{11.22}$$

Other kinds of boundary conditions can be treated in a similar way.

11.2.1 Finite Differences in Time

Time derivatives can be treated similarly using an independent time grid

$$t \rightarrow t_n = n\Delta t \quad n = 1 \ldots N \tag{11.23}$$

$$f(t, x) \rightarrow f_m^n = f(t_n, x_m) \tag{11.24}$$

and finite differences like the first order forward difference quotient

$$\frac{\partial f}{\partial t} \rightarrow \frac{f_m^{n+1} - f_m^n}{\Delta t} \tag{11.25}$$

or the symmetric difference quotient

$$\frac{\partial f}{\partial t} \rightarrow \frac{f_m^{n+1} - f_m^{n-1}}{2\Delta t} \tag{11.26}$$

to obtain a system of equations for the function values at the grid points f_m^n. For instance for the diffusion equation in one spatial dimension

$$\frac{\partial f(x,t)}{\partial t} = D\frac{\partial^2}{\partial x^2}f(x,t) + S(x,t) \tag{11.27}$$

the simplest discretization is the FTCS (forward in time, centered in space) scheme

$$\left(f_m^{n+1} - f_m^n\right) = D\frac{\Delta t}{\Delta x^2}\left(f_{m+1}^n + f_{m-1}^n - 2f_m^n\right) + S_m^n \Delta t \tag{11.28}$$

which can be written in matrix notation as

$$\mathbf{f}_{n+1} - \mathbf{f}_n = D\frac{\Delta t}{\Delta x^2}M\mathbf{f}_n + \mathbf{S}_n \Delta t \tag{11.29}$$

with

$$\mathbf{f}_n = \begin{pmatrix} f_1^n \\ f_2^n \\ f_3^n \\ \vdots \\ f_M^n \end{pmatrix} \quad \text{and} \quad M = \begin{pmatrix} -2 & 1 & & & \\ 1 & -2 & 1 & & \\ & 1 & -2 & 1 & \\ & & \ddots & \ddots & \ddots \\ & & & 1 & -2 \end{pmatrix}. \tag{11.30}$$

11.2.2 Stability Analysis

Fully discretized linear differential equations provide an iterative algorithm of the type[2]

$$\mathbf{f}_{n+1} = A\mathbf{f}_n + \mathbf{S}_n \Delta t \tag{11.31}$$

which propagates numerical errors according to

$$\mathbf{f}_{n+1} + \varepsilon_{n+1} = A(\mathbf{f}_n + \varepsilon_n) + \mathbf{S}_n \Delta t \tag{11.32}$$

$$\varepsilon_{j+1} = A\varepsilon_j. \tag{11.33}$$

Errors are amplified exponentially if the absolute value of at least one eigenvalue of A is larger than one. The algorithm is stable if all eigenvalues of A are smaller than one in absolute value (Sect. 1.4). If the eigenvalue problem is difficult to solve, the von Neumann analysis is helpful which decomposes the errors into a Fourier series and considers the Fourier components individually by setting

$$\mathbf{f}_n = g^n(k) \begin{pmatrix} e^{ik} \\ \vdots \\ e^{ikM} \end{pmatrix} \tag{11.34}$$

and calculating the amplification factor

$$\left| \frac{f_m^{n+1}}{f_m^n} \right| = \left| g(k) \right|. \tag{11.35}$$

The algorithm is stable if $|g(k)| \le 1$ for all k.

Example For the discretized diffusion equation (11.28) we find

$$g^{n+1}(k) = g^n(k) + 2D\frac{\Delta t}{\Delta x^2} g^n(k)(\cos k - 1) \tag{11.36}$$

$$g(k) = 1 + 2D\frac{\Delta t}{\Delta x^2}(\cos k - 1) = 1 - 4D\frac{\Delta t}{\Delta x^2}\sin^2\left(\frac{k}{2}\right) \tag{11.37}$$

$$1 - 4D\frac{\Delta t}{\Delta x^2} \le g(k) \le 1 \tag{11.38}$$

hence stability requires

$$D\frac{\Delta t}{\Delta x^2} \le \frac{1}{2}. \tag{11.39}$$

---

[2]Differential equations which are higher order in time can be always brought to first order by introducing the time derivatives as additional variables.

11.2.3 Method of Lines

Alternatively time can be considered as a continuous variable. The discrete values of the function then are functions of time (so called lines)

$$f_m(t) \tag{11.40}$$

and a set of ordinary differential equations has to be solved. For instance for diffusion in one dimension (11.27) the equations

$$\frac{\mathrm{d}f_m}{\mathrm{d}t} = \frac{D}{h^2}(f_{m+1} + f_{m-1} - 2f_m) + S_m(t) \tag{11.41}$$

which can be written in matrix notation as

$$\frac{\mathrm{d}}{\mathrm{d}t}\begin{pmatrix} f_1 \\ f_1 \\ f_2 \\ \vdots \\ f_M \end{pmatrix} = \frac{D}{\Delta x^2}\begin{pmatrix} -2 & 1 & & & \\ 1 & -2 & 1 & & \\ & 1 & -2 & 1 & \\ & & \ddots & \ddots & \ddots \\ & & & 1 & -2 \end{pmatrix}\begin{pmatrix} f_1 \\ f_2 \\ f_3 \\ \vdots \\ f_M \end{pmatrix} + \begin{pmatrix} S_1 + \frac{D}{h^2}f_0 \\ S_2 \\ S_3 \\ \vdots \\ S_M + \frac{D}{h^2}f_{M+1} \end{pmatrix} \tag{11.42}$$

or briefly

$$\frac{\mathrm{d}}{\mathrm{d}t}\mathbf{f}(t) = A\mathbf{f}(t) + \mathbf{S}(t). \tag{11.43}$$

Several methods to integrate such a semi-discretized equation will be discussed in Chap. 12. If eigenvectors and eigenvalues of A are easy available, an eigenvector expansion can be used.

11.2.4 Eigenvector Expansion

A homogeneous system

$$\frac{\mathrm{d}}{\mathrm{d}t}\mathbf{f}(t) = A\mathbf{f}(t) \tag{11.44}$$

where the matrix A is obtained from discretizing the spatial derivatives, can be solved by an eigenvector expansion. From the eigenvalue problem

$$A\mathbf{f} = \lambda\mathbf{f} \tag{11.45}$$

we obtain the eigenvalues λ and eigenvectors $\mathbf{f}_\lambda$ which provide the particular solutions:

$$\mathbf{f}(t) = \mathrm{e}^{\lambda t}\mathbf{f}_\lambda \tag{11.46}$$

$$\frac{\mathrm{d}}{\mathrm{d}t}\left(\mathrm{e}^{\lambda t}\mathbf{f}_\lambda\right) = \lambda\left(\mathrm{e}^{\lambda t}\mathbf{f}_\lambda\right) = A\left(\mathrm{e}^{\lambda t}\mathbf{f}_\lambda\right). \tag{11.47}$$

These can be used to expand the general solution

$$\mathbf{f}(t) = \sum_\lambda C_\lambda e^{\lambda t} \mathbf{f}_\lambda. \tag{11.48}$$

The coefficients C_λ follow from the initial values by solving the linear equations

$$\mathbf{f}(t=0) = \sum_\lambda C_\lambda \mathbf{f}_\lambda. \tag{11.49}$$

If the differential equation is second order in time

$$\frac{d^2}{dt^2}\mathbf{f}(t) = A\mathbf{f}(t) \tag{11.50}$$

the particular solutions are

$$\mathbf{f}(t) = e^{\pm t\sqrt{\lambda}}\mathbf{f}_\lambda \tag{11.51}$$

$$\frac{d^2}{dt^2}\left(e^{\pm t\sqrt{\lambda}}\mathbf{f}_\lambda\right) = \lambda\left(e^{\pm t\sqrt{\lambda}}\mathbf{f}_\lambda\right) = A\left(e^{\pm t\sqrt{\lambda}}\mathbf{f}_\lambda\right) \tag{11.52}$$

and the eigenvector expansion is

$$\mathbf{f}(t) = \sum_\lambda \left(C_{\lambda+}e^{t\sqrt{\lambda}} + C_{\lambda-}e^{-t\sqrt{\lambda}}\right)\mathbf{f}_\lambda. \tag{11.53}$$

The coefficients $C_{\lambda\pm}$ follow from the initial amplitudes and velocities

$$\mathbf{f}(t=0) = \sum_\lambda (C_{\lambda+} + C_{\lambda-})\mathbf{f}_\lambda$$
$$\frac{d}{dt}\mathbf{f}(t=0) = \sum_\lambda \sqrt{\lambda}(C_{\lambda+} - C_{\lambda-})\mathbf{f}_\lambda. \tag{11.54}$$

For a first order inhomogeneous system

$$\frac{d}{dt}\mathbf{f}(t) = A\mathbf{f}(t) + \mathbf{S}(t) \tag{11.55}$$

the expansion coefficients have to be time dependent

$$\mathbf{f}(t) = \sum_\lambda C_\lambda(t)e^{\lambda t}\mathbf{f}_\lambda \tag{11.56}$$

and satisfy

$$\frac{d}{dt}\mathbf{f}(t) - A\mathbf{f}(t) = \sum_\lambda \frac{dC_\lambda}{dt}e^{\lambda t}\mathbf{f}_\lambda = \mathbf{S}(t). \tag{11.57}$$

After taking the scalar product with $\mathbf{f}_\mu$[3]

$$\frac{dC_\mu}{dt} = e^{-\mu t}\left(\mathbf{f}_\mu \mathbf{S}(t)\right) \tag{11.58}$$

---

[3]If A is not Hermitian we have to distinguish left- and right-eigenvectors.

can be solved by a simple time integration. For a second order system

$$\frac{d^2}{dt^2}\mathbf{f}(t) = A\mathbf{f}(t) + \mathbf{S}(t) \tag{11.59}$$

we introduce the first time derivative as a new variable

$$\mathbf{g} = \frac{d}{dt}\mathbf{f} \tag{11.60}$$

to obtain a first order system of double dimension

$$\frac{d}{dt}\begin{pmatrix}\mathbf{f}\\\mathbf{g}\end{pmatrix} = \begin{pmatrix}0 & 1\\A & 0\end{pmatrix}\begin{pmatrix}\mathbf{f}\\\mathbf{g}\end{pmatrix} + \begin{pmatrix}\mathbf{S}\\0\end{pmatrix} \tag{11.61}$$

where eigenvectors and eigenvalues can be found from those of A (11.45)

$$\begin{pmatrix}0 & 1\\A & 0\end{pmatrix}\begin{pmatrix}\mathbf{f}_\lambda\\\pm\sqrt{\lambda}\mathbf{f}_\lambda\end{pmatrix} = \begin{pmatrix}\pm\sqrt{\lambda}\mathbf{f}_\lambda\\\lambda\mathbf{f}_\lambda\end{pmatrix} = \pm\sqrt{\lambda}\begin{pmatrix}\mathbf{f}_\lambda\\\pm\sqrt{\lambda}\mathbf{f}_\lambda\end{pmatrix} \tag{11.62}$$

$$\left(\pm\sqrt{\lambda}\mathbf{f}_\lambda^T \ \ \mathbf{f}_\lambda^T\right)\begin{pmatrix}0 & 1\\A & 0\end{pmatrix} = \left(\lambda\mathbf{f}_\lambda^T \ \pm\sqrt{\lambda}\mathbf{f}_\lambda^T\right) = \pm\sqrt{\lambda}\left(\pm\sqrt{\lambda}\mathbf{f}_\lambda^T \ \ \mathbf{f}_\lambda^T\right). \tag{11.63}$$

Insertion of

$$\sum_\lambda C_{\lambda+}e^{\sqrt{\lambda}t}\begin{pmatrix}\mathbf{f}_\lambda\\\sqrt{\lambda}\mathbf{f}_\lambda\end{pmatrix} + C_{\lambda-}e^{-\sqrt{\lambda}t}\begin{pmatrix}\mathbf{f}_\lambda\\-\sqrt{\lambda}\mathbf{f}_\lambda\end{pmatrix}$$

gives

$$\sum_\lambda \frac{dC_{\lambda+}}{dt}e^{\sqrt{\lambda}t}\begin{pmatrix}\mathbf{f}_\lambda\\\sqrt{\lambda}\mathbf{f}_\lambda\end{pmatrix} + \frac{dC_{\lambda-}}{dt}e^{\sqrt{\lambda}t}\begin{pmatrix}\mathbf{f}_\lambda\\-\sqrt{\lambda}\mathbf{f}_\lambda\end{pmatrix} = \begin{pmatrix}\mathbf{S}(t)\\0\end{pmatrix} \tag{11.64}$$

and taking the scalar product with one of the left-eigenvectors we end up with

$$\frac{dC_{\lambda+}}{dt} = \frac{1}{2}(\mathbf{f}_\lambda\mathbf{S}(t))e^{-\sqrt{\lambda}t} \tag{11.65}$$

$$\frac{dC_{\lambda-}}{dt} = -\frac{1}{2}(\mathbf{f}_\lambda\mathbf{S}(t))e^{\sqrt{\lambda}t}. \tag{11.66}$$

11.3 Finite Volumes

Whereas the finite differences method uses function values

$$f_{i,j,k} = f(x_i, y_j, z_k) \tag{11.67}$$

at the grid points

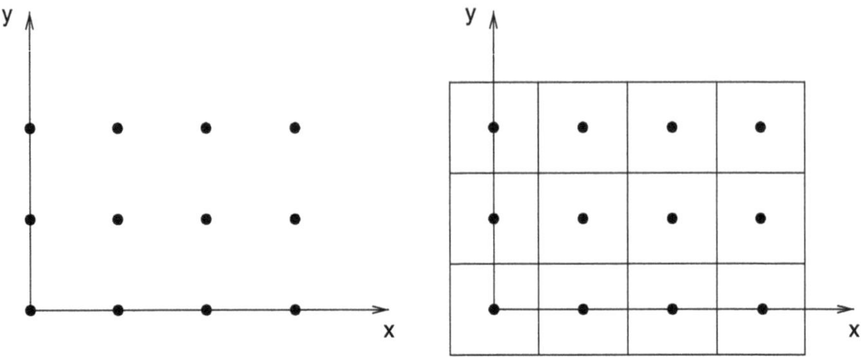

Fig. 11.1 (Finite volume method) The domain V is divided into small control volumes V_r, in the simplest case cubes around the grid points $\mathbf{r}_{ijk}$

$$\mathbf{r}_{ijk} = (x_i, y_j, z_k), \tag{11.68}$$

the finite volume method (FVM) [79] averages function values and derivatives over small control volumes V_r which are disjoint and span the domain V (Fig. 11.1)

$$V = \bigcup_r V_r \qquad V_r \cap V_{r'} = \emptyset \quad \forall r \neq r'. \tag{11.69}$$

The averages are

$$\overline{f}_r = \frac{1}{V_r} \int_{V_r} dV \, f(\mathbf{r}) \tag{11.70}$$

or in the simple case of cubic control volumes of equal size h^3

$$\overline{f}_{ijk} = \frac{1}{h^3} \int_{x_i-h/2}^{x_i+h/2} dx \int_{y_j-h/2}^{y_j+h/2} dy \int_{z_k-h/2}^{z_k+h/2} dz \, f(x, y, z). \tag{11.71}$$

Such average values have to be related to discrete function values by numerical integration (Chap. 4). The midpoint rule (4.17), for instance replaces the average by the central value

$$\overline{f}_{ijk} = f(x_i, y_j, z_k) + O(h^2) \tag{11.72}$$

whereas the trapezoidal rule (4.13) implies the average over the eight corners of the cube

$$\overline{f}_{ijk} = \frac{1}{8} \sum_{m,n,p=\pm 1} f(x_{i+m/2}, y_{j+n/2}, z_{k+p/2}) + O(h^2). \tag{11.73}$$

In (11.73) the function values refer to a dual grid [79] centered around the vertices of the original grid (11.68) (Fig. 11.2),

$$\mathbf{r}_{i+1/2, j+1/2, k+1/2} = \left(x_i + \frac{h}{2}, y_j + \frac{h}{2}, z_k + \frac{h}{2} \right). \tag{11.74}$$

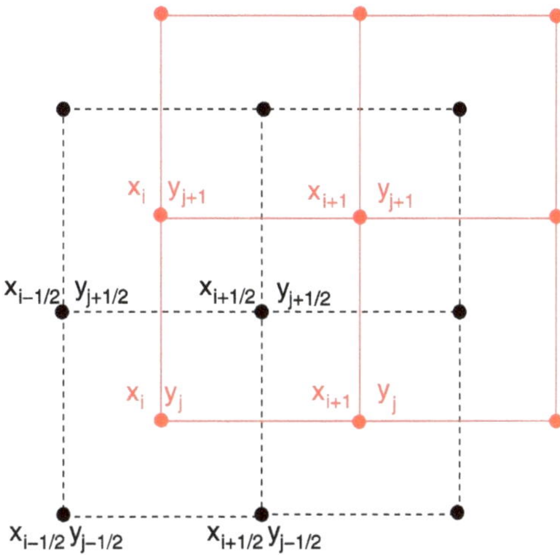

Fig. 11.2 (Dual grid) The dual grid (*black*) is centered around the vertices of the original grid (*red*)

The average gradient can be rewritten using the generalized Stokes' theorem as

$$\overline{\operatorname{grad} f_{ijk}} = \frac{1}{V}\int_{V_{ijk}} dV \, \operatorname{grad} f(\mathbf{r}) = \oint_{\partial V_{ijk}} f(\mathbf{r}) \, d\mathbf{A}. \tag{11.75}$$

For a cubic grid we have to integrate over the six faces of the control volume

$$\overline{\operatorname{grad} f_{ijk}} = \frac{1}{h^3}\begin{pmatrix} \int_{z_k-h/2}^{z_k+h/2} dz \int_{y_j-h/2}^{y_j+h/2} dy(f(x_i+\tfrac{h}{2},y,z)-f(x_i-\tfrac{h}{2},y,z)) \\ \int_{z_k-h/2}^{z_k+h/2} dz \int_{x_i-h/2}^{x_i+h/2} dx(f(x_i,y+\tfrac{h}{2},z)-f(x_i,y-\tfrac{h}{2},z)) \\ \int_{x_i-h/2}^{x_i+h/2} dx \int_{y_j-h/2}^{y_j+h/2} dy(f(x_i,y,z+\tfrac{h}{2})-f(x_i,y,z-\tfrac{h}{2})) \end{pmatrix}. \tag{11.76}$$

The integrals have to be evaluated numerically. Applying as the simplest approximation the midpoint rule (4.17)

$$\int_{x_i-h/2}^{x_i+h/2} dx \int_{y_j-h/2}^{y_j+h/2} dy \, f(x,y) = h^2\big(f(x_i,y_j)+O\big(h^2\big)\big) \tag{11.77}$$

this becomes

$$\overline{\operatorname{grad} f_{ijk}} = \frac{1}{h}\begin{pmatrix} f(x_i+\tfrac{h}{2},y_j,z_k)-f(x_i-\tfrac{h}{2},y_j,z_k) \\ f(x_i,y_j+\tfrac{h}{2},z_k)-f(x_i,y_j-\tfrac{h}{2},z_k) \\ f(x_i,y_j,z_k+\tfrac{h}{2})-f(x_i,y_j,z_k-\tfrac{h}{2}) \end{pmatrix} \tag{11.78}$$

which involves symmetric difference quotients. However, the function values in (11.78) refer neither to the original nor to the dual grid. Therefore we interpolate (Fig. 11.3)

$$f\left(x_i\pm\frac{h}{2},y_j,z_k\right) \approx \frac{1}{2}\big(f(x_i,y_j,z_k)+f(x_{i\pm1},y_j,z_k)\big) \tag{11.79}$$

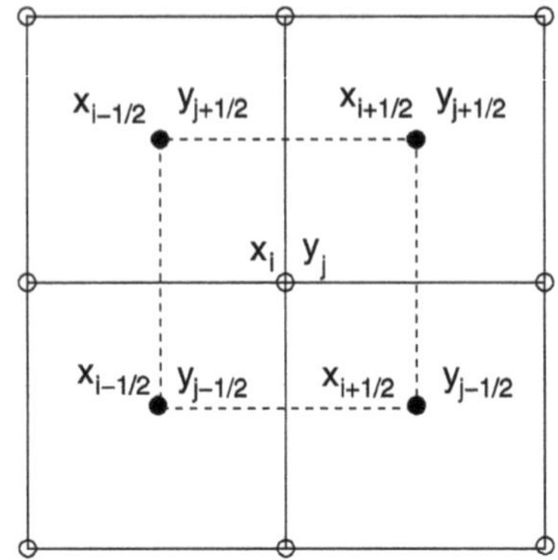

Fig. 11.3 (Interpolation between grid points) Interpolation is necessary to relate the averaged gradient (11.78) to the original or dual grid

$$\frac{1}{h}\left(f\left(x_i + \frac{h}{2}, y_j, z_k\right) - f\left(x_i - \frac{h}{2}, y_j, z_k\right)\right)$$

$$\approx \frac{1}{2h}\left(f(x_{i+1}, y_j, z_k) - f(x_{i-1}, y_j, z_k)\right) \tag{11.80}$$

or

$$f\left(x_i \pm \frac{h}{2}, y_j, z_k\right) \approx \frac{1}{4} \sum_{m,n=\pm 1} f\left(x_i \pm \frac{h}{2}, y_j + m\frac{h}{2}, z_k + n\frac{h}{2}\right). \tag{11.81}$$

The finite volume method is capable of treating discontinuities and is very flexible concerning the size and shape of the control volumes.

11.3.1 Discretization of fluxes

Integration of (11.10) over a control volume and application of Gauss' theorem gives the integral form of the conservation law

$$\frac{1}{V}\oint \mathbf{J}\, d\mathbf{A} + \frac{\partial}{\partial t}\frac{1}{V}\int f\, dV = \frac{1}{V}\int g\, dV \tag{11.82}$$

which involves the flux $\mathbf{J}$ of some property like particle concentration, mass, energy or momentum density or the flux of an electromagnetic field. The total flux through a control volume is given by the surface integral

$$\Phi = \oint_{\partial V} \mathbf{J}\, d\mathbf{A} \tag{11.83}$$

which in the special case of a cubic volume element of size h^3 becomes the sum over the six faces of the cube (Fig. 11.4)

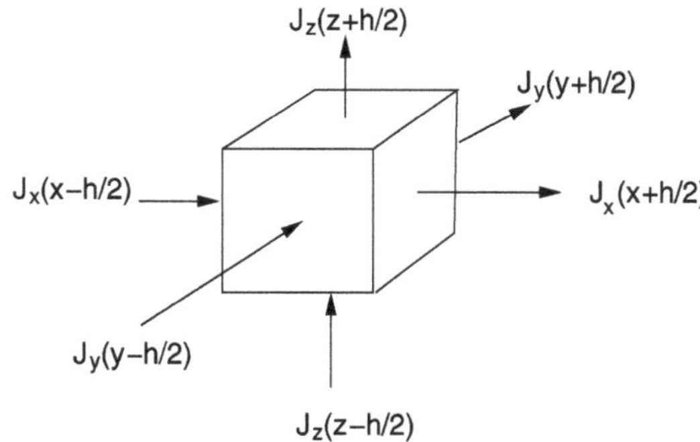

Fig. 11.4 Flux through a control volume

$$\Phi = \sum_{r=1}^{6} \int_{A_r} \mathbf{J}\, d\mathbf{A}$$

$$= \int_{x_i-h/2}^{x_i+h/2} dx \int_{y_j-h/2}^{y_j+h/2} dy \left(J_z\left(x, y, z_k + \frac{h}{2}\right) - J_z\left(x, y, z_k - \frac{h}{2}\right) \right)$$

$$+ \int_{x_i-h/2}^{x_i+h/2} dx \int_{z_k-h/2}^{z_k+h/2} dz \left(J_y\left(x, y_j + \frac{h}{2}, z\right) - J_z\left(x, y_j - \frac{h}{2}, z\right) \right)$$

$$+ \int_{z_k-h/2}^{z_k+h/2} dz \int_{y_j-h/2}^{y_j+h/2} dy \left(J_x\left(x_i + \frac{h}{2}, y, z\right) - J_z\left(x_i - \frac{h}{2}, y, z\right) \right).$$

$$(11.84)$$

The surface integral can be evaluated numerically (Chap. 4). Using the midpoint approximation (11.77) we obtain

$$\frac{1}{V}\Phi(x_i, y_j, z_k) = \frac{1}{h}\big(J_z(x_i, y_j, z_{k+1/2}) - J_z(x_i, y_j, z_{k-1/2})$$

$$+ J_y(x_i, y_{j+1/2}, z_k) - J_y(x_i, y_{j-1/2}, z_k)$$

$$+ J_x(x_{i+1/2}, y_j, z_k) - J_x(x_{i-1/2}, y_j, z_k)\big). \qquad (11.85)$$

The trapezoidal rule (4.13) introduces an average over the four corners (Fig. 11.3)

$$\int_{x_i-h/2}^{x_i+h/2} dx \int_{y_j-h/2}^{y_j+h/2} dy\, f(x, y)$$

$$= h^2\left(\frac{1}{4} \sum_{m,n=\pm 1} f(x_{i+m/2}, y_{j+n/2}) + O(h^2) \right) \qquad (11.86)$$

which replaces the flux values in (11.85) by the averages

$$J_x(x_{i\pm1/2}, y_j, z_k) = \frac{1}{4} \sum_{m=\pm1,n=\pm1} J_z(x_{i\pm1/2}, y_{j+m/2}, z_{k+n/2}) \qquad (11.87)$$

$$J_y(x_i, y_{j\pm1/2}, z_k) = \frac{1}{4} \sum_{m=\pm1,n=\pm1} J_z(x_{i+m/2}, y_{j\pm1/2}, z_{k+n/2}) \qquad (11.88)$$

$$J_z(x_i, y_j, z_{k\pm 1/2}) = \frac{1}{4} \sum_{m=\pm 1, n=\pm 1} J_z(x_{i+m/2}, y_{j+n/2}, z_{k\pm 1/2}). \quad (11.89)$$

One advantage of the finite volume method is that the flux is strictly conserved.

11.4 Weighted Residual Based Methods

A general method to discretize partial differential equations is to approximate the solution within a finite dimensional space of trial functions.[4] The partial differential equation is turned into a finite system of equations or a finite system of ordinary differential equations if time is treated as a continuous variable. This is the basis of spectral methods which make use of polynomials or Fourier series but also of the very successful finite element methods. Even finite difference methods and finite volume methods can be formulated as weighted residual based methods.

Consider a differential equation[5] on the domain V which is written symbolically with the differential operator $\mathcal{T}$

$$\mathcal{T}\big[u(\mathbf{r})\big] = f(\mathbf{r}) \quad \mathbf{r} \in V \tag{11.90}$$

and corresponding boundary conditions which are expressed with a boundary operator $\mathcal{B}$[6]

$$\mathcal{B}\big[u(\mathbf{r})\big] = g(\mathbf{r}) \quad \mathbf{r} \in \partial V. \tag{11.91}$$

The basic principle to obtain an approximate solution $\tilde{u}(\mathbf{r})$ is to choose a linear combination of expansion functions $N_i(\mathbf{r})\ i = 1 \ldots r$ as a trial function which fulfills the boundary conditions[7]

$$\tilde{u} = \sum_{i=1}^{r} u_i N_i(\mathbf{r}) \tag{11.92}$$

$$\mathcal{B}\big[\tilde{u}(\mathbf{r})\big] = g(\mathbf{r}). \tag{11.93}$$

In general (11.92) is not an exact solution and the residual

$$R(\mathbf{r}) = \mathcal{T}[\tilde{u}](\mathbf{r}) - f(\mathbf{r}) \tag{11.94}$$

will not be zero throughout the whole domain V. The function $\tilde{u}$ has to be determined such that the residual becomes "small" in a certain sense. To that end weight functions[8] $w_j\ j = 1 \ldots r$ are chosen to define the weighted residuals

---

[4] Also called expansion functions.

[5] Generalization to systems of equations is straightforward.

[6] One or more linear differential operators, usually a combination of the function and its first derivatives.

[7] This requirement can be replaced by additional equations for the u_i, for instance with the tau method [195].

[8] Also called test functions.

$$R_j(u_1 \ldots u_r) = \int dV\, w_j(\mathbf{r})\big(\mathcal{T}[\tilde{u}](\mathbf{r}) - f(\mathbf{r})\big). \tag{11.95}$$

The optimal parameters u_i are then obtained from the solution of the equations

$$R_j(u_1 \ldots u_r) = 0 \quad j = 1 \ldots r. \tag{11.96}$$

In the special case of a linear differential operator these equations are linear

$$\sum_{i=1}^{r} u_i \int dV\, w_j(\mathbf{r}) \mathcal{T}[N_i(\mathbf{r})] - \int dV\, w_j(\mathbf{r}) f(\mathbf{r}) = 0. \tag{11.97}$$

Several strategies are available to choose suitable weight functions.

11.4.1 Point Collocation Method

The collocation method uses the weight functions $w_j(\mathbf{r}) = \delta(\mathbf{r} - \mathbf{r}_j)$, with certain collocation points $\mathbf{r}_j \in V$. The approximation $\tilde{u}$ obeys the differential equation at the collocation points

$$0 = R_j = \mathcal{T}[\tilde{u}](\mathbf{r}_j) - f(\mathbf{r}_j) \tag{11.98}$$

and for a linear differential operator

$$0 = \sum_{i=1}^{r} u_i \mathcal{T}[N_i](\mathbf{r}_j) - f(\mathbf{r}_j). \tag{11.99}$$

The point collocation method is simple to use, especially for nonlinear problems. Instead of using trial functions satisfying the boundary conditions, extra collocation points on the boundary can be added (mixed collocation method).

11.4.2 Sub-domain Method

This approach uses weight functions which are the characteristic functions of a set of control volumes V_i which are disjoint and span the whole domain similar as for the finite volume method

$$V = \bigcup_j V_j \qquad V_j \cap V_{j'} = \emptyset \quad \forall j \neq j' \tag{11.100}$$

$$w_j(\mathbf{r}) = \begin{cases} 1 & \mathbf{r} \in V_j \\ 0 & \text{else.} \end{cases} \tag{11.101}$$

The residuals then are integrals over the control volumes and

$$0 = R_j = \int_{V_j} dV\, \big(\mathcal{T}[\tilde{u}](\mathbf{r}) - f(\mathbf{r})\big) \tag{11.102}$$

respectively

$$0 = \sum_i u_i \int_{V_j} dV\, \mathcal{T}[N_i](\mathbf{r}) - \int_{V_j} dV\, f(\mathbf{r}). \tag{11.103}$$

11.4.3 Least Squares Method

Least squares methods have become popular for first order systems of differential equations in computational fluid dynamics and computational electrodynamics [30, 140].

The L2-norm of the residual (11.94) is given by the integral

$$S = \int_V dV\, R(\mathbf{r})^2. \tag{11.104}$$

It is minimized by solving the equations

$$0 = \frac{\partial S}{\partial u_j} = 2 \int_V dV\, \frac{\partial R}{\partial u_j} R(\mathbf{r}) \tag{11.105}$$

which is equivalent to choosing the weight functions

$$w_j(\mathbf{r}) = \frac{\partial R}{\partial u_j} R(\mathbf{r}) = \frac{\partial}{\partial u_j} \mathcal{T}\left[\sum_i u_i N_i(\mathbf{r})\right] \tag{11.106}$$

or for a linear differential operator simply

$$w_j(\mathbf{r}) = \mathcal{T}[N_j(\mathbf{r})]. \tag{11.107}$$

Advantages of the least squares method are that boundary conditions can be incorporated into the residual and that S provides a measure for the quality of the solution which can be used for adaptive grid size control. On the other hand S involves a differential operator of higher order and therefore much smoother trial functions are necessary.

11.4.4 Galerkin Method

Galerkin's widely used method [87, 98] chooses the basis functions as weight functions

$$w_j(\mathbf{r}) = N_j(\mathbf{r}) \tag{11.108}$$

and solves the following system of equations

$$\int dV\, N_j(\mathbf{r}) \mathcal{T}\left[\sum_i u_i N_i(\mathbf{r})\right] - \int dV\, N_j(\mathbf{r}) f(\mathbf{r}) = 0 \tag{11.109}$$

or in the simpler linear case

$$\sum u_i \int_V dV\, N_j(\mathbf{r})\mathcal{T}(N_i(\mathbf{r})) = \int_V dV\, N_j(\mathbf{r}) f(\mathbf{r}, t). \tag{11.110}$$

11.5 Spectral and Pseudo-spectral Methods

Spectral methods use basis functions which are nonzero over the whole domain, the trial functions being mostly polynomials or Fourier sums [35]. They can be used to solve ordinary as well as partial differential equations. The combination of a spectral method with the point collocation method is also known as pseudo-spectral method.

11.5.1 Fourier Pseudo-spectral Methods

Linear differential operators become diagonal in Fourier space. Combination of Fourier series expansion and point collocation leads to equations involving a discrete Fourier transformation, which can be performed very efficiently with the Fast Fourier Transform methods.

For simplicity we consider only the one-dimensional case. We choose equidistant collocation points

$$x_m = m\Delta x \quad m = 0, 1 \ldots M - 1 \tag{11.111}$$

and expansion functions

$$N_j(x) = e^{ik_j x} \quad k_j = \frac{2\pi}{M\Delta x} j \quad j = 0, 1 \ldots M - 1. \tag{11.112}$$

For a linear differential operator

$$\mathcal{L}\left[e^{ik_j x}\right] = l(k_j) e^{ik_j x} \tag{11.113}$$

and the condition on the residual becomes

$$0 = R_m = \sum_{j=0}^{M-1} u_j l(k_j) e^{ik_j x_m} - f(x_m) \tag{11.114}$$

or

$$f(x_m) = \sum_{j=0}^{M-1} u_j l(k_j) e^{i2\pi m j/M} \tag{11.115}$$

which is nothing but a discrete Fourier back transformation (Sect. 7.2, (7.19)) which can be inverted to give

$$u_j l(k_j) = \frac{1}{N} \sum_{m=0}^{M-1} f(x_m) e^{-i2\pi m j/M}. \tag{11.116}$$

Instead of exponential expansion functions, sine and cosine functions can be used to satisfy certain boundary conditions, for instance to solve the Poisson equation within a cube (Sect. 17.1.2).

11.5.2 Example: Polynomial Approximation

Let us consider the initial value problem (Fig. 11.5)

$$\frac{d}{dx}u(x) - u(x) = 0 \quad u(0) = 1 \quad \text{for } 0 \le x \le 1 \tag{11.117}$$

with the well known solution

$$u(x) = e^x. \tag{11.118}$$

We choose a polynomial trial function with the proper initial value

$$\tilde{u}(x) = 1 + u_1 x + u_2 x^2. \tag{11.119}$$

The residual is

$$R(x) = u_1 + 2u_2 x - \left(1 + u_1 x + u_2 x^2\right)$$
$$= (u_1 - 1) + (2u_2 - u_1)x - u_2 x^2. \tag{11.120}$$

11.5.2.1 Point Collocation Method

For our example we need two collocation points to obtain two equations for the two unknowns $u_{1,2}$. We choose $x_1 = 0$, $x_2 = \frac{1}{2}$. Then we have to solve the equations

$$R(x_1) = u_1 - 1 = 0 \tag{11.121}$$
$$R(x_2) = \frac{1}{2}u_1 + \frac{3}{4}u_2 - 1 = 0 \tag{11.122}$$

which gives

$$u_1 = 1 \quad u_2 = \frac{2}{3} \tag{11.123}$$

$$u_c = 1 + x + \frac{2}{3}x^2. \tag{11.124}$$

11.5.2.2 Sub-domain Method

We need two sub-domains to obtain two equations for the two unknowns $u_{1,2}$. We choose $V_1 = \{x, 0 < x < \frac{1}{2}\}$, $V_2 = \{x, \frac{1}{2} < x < 1\}$. Integration gives

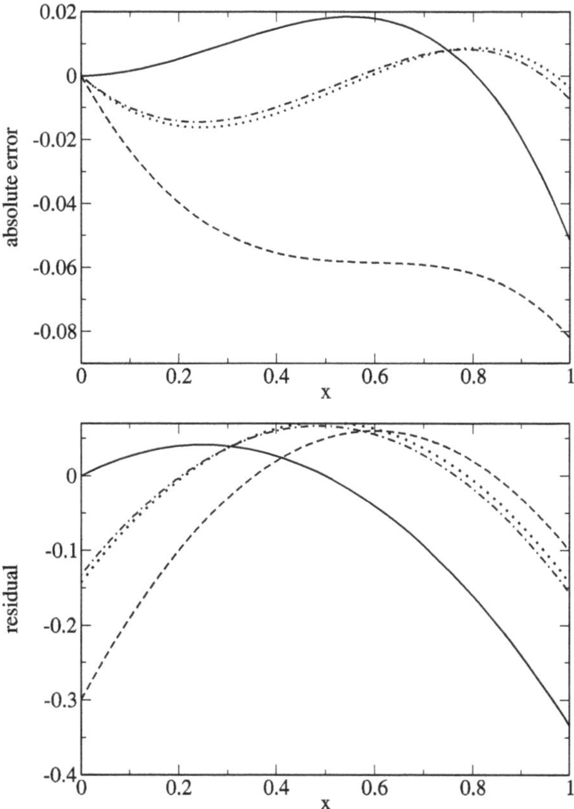

Fig. 11.5 (Approximate solution of a simple differential equation) The initial value problem $\frac{\mathrm{d}}{\mathrm{d}x}u(x) - u(x) = 0\ u(0) = 1$ for $0 \le x \le 1$ is approximately solved with a polynomial trial function $\tilde{u}(x) = 1 + u_1 x + u_2 x^2$. The parameters $u_{1,2}$ are optimized with the method of weighted residuals using point collocation (*full curve*), sub-domain collocation (*dotted curve*), Galerkin's method (*dashed curve*) and least squares (*dash-dotted curve*). The absolute error $\tilde{u}(x) - \mathrm{e}^x$ (*top*) and the residual $R(x) = \frac{\mathrm{d}}{\mathrm{d}x}\tilde{u}(x) - \tilde{u}(x) = (u_1 - 1) + (2u_2 - u_1)x - u_2 x^2$ both are smallest for the least squares and sub-domain collocation methods

$$R_1 = \frac{3}{8}u_1 + \frac{5}{24}u_2 - \frac{1}{2} = 0 \tag{11.125}$$

$$R_2 = \frac{1}{8}u_1 + \frac{11}{24}u_2 - \frac{1}{2} = 0 \tag{11.126}$$

$$u_1 = u_2 = \frac{6}{7} \tag{11.127}$$

$$u_{sdc} = 1 + \frac{6}{7}x + \frac{6}{7}x^2. \tag{11.128}$$

11.5.2.3 Galerkin Method

Galerkin's method uses the weight functions $w_1(x) = x$, $w_2(x) = x^2$. The equations

$$\int_0^1 dx\, w_1(x) R(x) = \frac{1}{6}u_1 + \frac{5}{12}u_2 - \frac{1}{2} = 0 \tag{11.129}$$

$$\int_0^1 dx\, w_2(x) R(x) = \frac{1}{12}u_1 + \frac{3}{10}u_2 - \frac{1}{3} = 0 \tag{11.130}$$

have the solution

$$u_1 = \frac{8}{11} \quad u_2 = \frac{10}{11} \tag{11.131}$$

$$u_G = 1 + \frac{8}{11}x + \frac{10}{11}x^2. \tag{11.132}$$

11.5.2.4 Least Squares Method

The integral of the squared residual

$$S = \int_0^1 dx\, R(x)^2 = 1 - u_1 - \frac{4}{3}u_2 + \frac{1}{3}u_1^2 + \frac{1}{2}u_1 u_2 + \frac{8}{15}u_2^2 \tag{11.133}$$

is minimized by solving

$$\frac{\partial S}{\partial u_1} = \frac{2}{3}u_1 + \frac{1}{2}u_2 - 1 = 0 \tag{11.134}$$

$$\frac{\partial S}{\partial u_2} = \frac{1}{2}u_1 + \frac{16}{15}u_2 - \frac{4}{3} = 0 \tag{11.135}$$

which gives

$$u_1 = \frac{72}{83} \quad u_2 = \frac{70}{83} \tag{11.136}$$

$$u_{LS} = 1 + \frac{72}{83}x + \frac{70}{83}x^2. \tag{11.137}$$

11.6 Finite Elements

The method of finite elements (FEM) is a very flexible method to discretize partial differential equations [84, 210]. It is rather dominant in a variety of engineering sciences. Usually the expansion functions N_i are chosen to have compact support. The integration volume is divided into disjoint sub-volumes

$$V = \bigcup_{i=1}^{r} V_i \qquad V_i \cap V_{i'} = \emptyset \forall i \neq i'. \tag{11.138}$$

The $N_i(\mathbf{x})$ are piecewise continuous polynomials which are nonzero only inside V_i and a few neighbor cells.

11.6.1 One-Dimensional Elements

In one dimension the domain is an interval $V = \{x; a \leq x \leq b\}$ and the sub-volumes are small sub-intervals $V_i = \{x; x_i \leq x \leq x_{i+1}\}$. The one-dimensional mesh is the

Fig. 11.6 (Finite elements in one dimension) The basis functions N_i are piecewise continuous polynomials and have compact support. In the simplest case they are composed of two linear functions over the sub-intervals $x_{i-1} \le x \le x_i$ and $x_i \le x \le x_{i+1}$

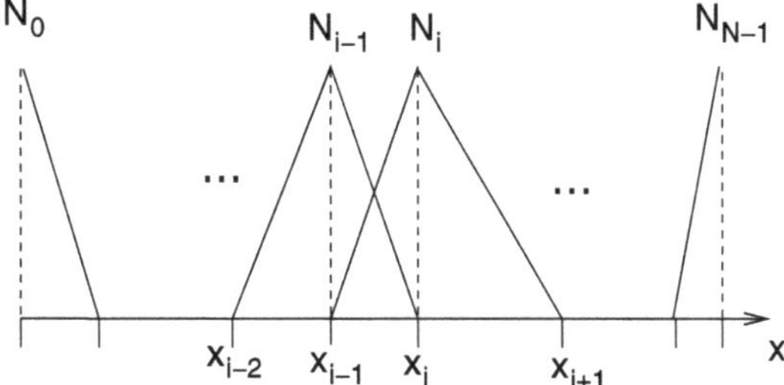

set of nodes $\{a = x_0, x_1 \ldots x_r = b\}$. Piecewise linear basis functions (Fig. 11.6) are in the 1-dimensional case given by

$$N_i(x) = \begin{cases} \frac{x_{i+1}-x}{x_{i+1}-x_i} & \text{for } x_i < x < x_{i+1} \\ \frac{x-x_{i-1}}{x_i-x_{i-1}} & \text{for } x_{i-1} < x < x_i \\ 0 & \text{else} \end{cases} \tag{11.139}$$

and the derivatives are (except at the nodes x_i)

$$N_i'(x) = \begin{cases} -\frac{1}{x_{i+1}-x_i} & \text{for } x_i < x < x_{i+1} \\ \frac{1}{x_i-x_{i-1}} & \text{for } x_{i-1} < x < x_i \\ 0 & \text{else.} \end{cases} \tag{11.140}$$

11.6.2 Two- and Three-Dimensional Elements

In two dimensions the mesh is defined by a finite number of points $(x_i, y_i) \in V$ (the nodes of the mesh). There is considerable freedom in the choice of these points and they need not be equally spaced.

11.6.2.1 Triangulation

The nodes can be regarded as forming the vertices of a triangulation[9] of the domain V (Fig. 11.7).

The piecewise linear basis function in one dimension (11.139) can be generalized to the two-dimensional case by constructing functions $N_i(x, y)$ which are zero at all nodes except (x_i, y_i)

$$N_i(x_j, y_j) = \delta_{i,j}. \tag{11.141}$$

---

[9]The triangulation is not determined uniquely by the nodes.

Fig. 11.7 (Triangulation of a two-dimensional domain) A two-dimensional mesh is defined by a set of node points which can be regarded to form the vertices of a triangulation

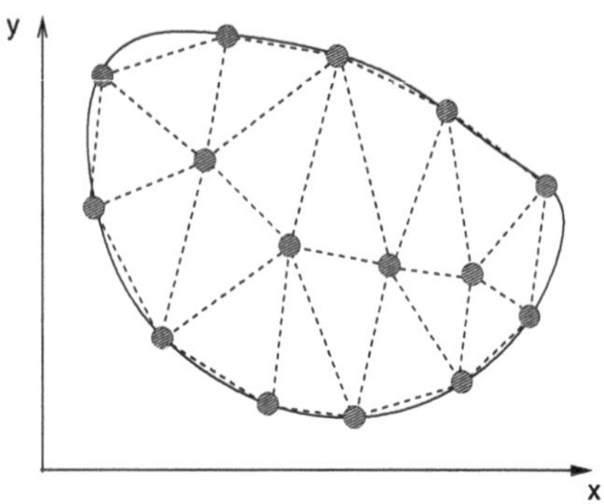

These functions are linear over each triangle which contains the vertex i and can be combined as the sum of small pyramids (Fig. 11.8). Let one of the triangles be denoted by its three vertices as T_{ijk}.[10] The corresponding linear function then is

$$n_{ijk}(x, y) = \alpha + \beta_x (x - x_i) + \beta_y (y - y_i) \tag{11.142}$$

where the coefficients follow from the conditions

$$n_{ijk}(x_i, y_i) = 1 \quad n_{ijk}(x_j, y_j) = n_{ijk}(x_k, y_k) = 0 \tag{11.143}$$

as

$$\alpha = 1 \quad \beta_x = \frac{y_j - y_k}{2A_{ijk}} \quad \beta_y = \frac{x_k - x_j}{2A_{ijk}} \tag{11.144}$$

with

$$A_{ijk} = \frac{1}{2} \det \begin{vmatrix} x_j - x_i & x_k - x_i \\ y_j - y_i & y_k - y_i \end{vmatrix} \tag{11.145}$$

which, apart from sign, is the area of the triangle T_{ijk}. The basis function N_i now is given by

$$N_i(x, y) = \begin{cases} n_{ijk}(x, y) & (x, y) \in T_{ijk} \\ 0 & \text{else.} \end{cases}$$

In three dimensions we consider tetrahedrons (Fig. 11.9) instead of triangles. The corresponding linear function of three arguments has the form

$$n_{i,j,k,l}(x, y, z) = \alpha + \beta_x (x - x_i) + \beta_y (y - y_i) + \beta_z (z - z_i) \tag{11.146}$$

and from the conditions $n_{i,j,k,l}(x_i, y_i, z_i) = 1$ and $n_{i,j,k,l} = 0$ on all other nodes we find (an algebra program is helpful at that point)

---

[10] The order of the indices does matter.

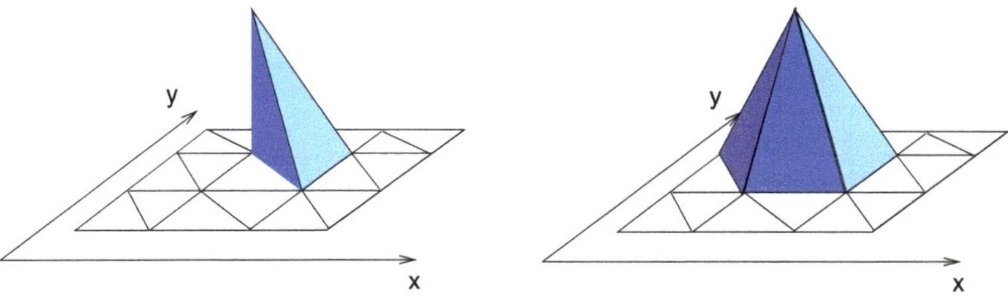

Fig. 11.8 (Finite elements in two dimensions) The simplest finite elements in two dimensions are piecewise linear functions $N_i(x, y)$ which are non-vanishing only at one node (x_i, y_i) (*right side*). They can be constructed from small pyramids built upon one of the triangles that contains this node (*left side*)

Fig. 11.9 (Tetrahedron) The tetrahedron is the three-dimensional case of a Euclidean simplex, i.e. the simplest polytope

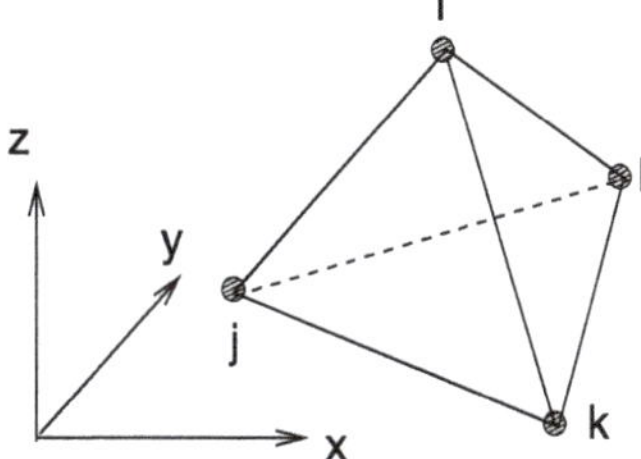

$$\alpha = 1$$

$$\beta_x = \frac{1}{6V_{ijkl}} \det \begin{vmatrix} y_k - y_j & y_l - y_j \\ z_k - z_j & z_l - z_j \end{vmatrix}$$

$$\beta_y = \frac{1}{6V_{ijkl}} \det \begin{vmatrix} z_k - z_j & z_l - z_j \\ x_k - x_j & x_l - x_j \end{vmatrix}$$

$$\beta_z = \frac{1}{6V_{ijkl}} \det \begin{vmatrix} x_k - x_j & x_l - x_j \\ y_k - y_j & y_l - y_j \end{vmatrix} \tag{11.147}$$

where V_{ijkl} is, apart from sign, the volume of the tetrahedron

$$V_{ijkl} = \frac{1}{6} \det \begin{vmatrix} x_j - x_i & x_k - x_i & x_l - x_i \\ y_j - y_i & y_k - y_i & y_l - y_i \\ z_j - z_i & z_k - z_i & z_l - z_i \end{vmatrix}. \tag{11.148}$$

11.6.2.2 Rectangular Elements

For a rectangular grid rectangular elements offer a practical alternative to triangles. Since equations for four nodes have to be fulfilled, the basic element needs four parameters, which is the case for a bilinear expression. Let us denote one of the rectangles which contains the vertex i as $R_{i,j,k,l}$. The other three edges are

Fig. 11.10 (Rectangular elements around one vertex) The basis function N_i is a bilinear function on each of the four rectangles containing the vertex (x_i, y_i)

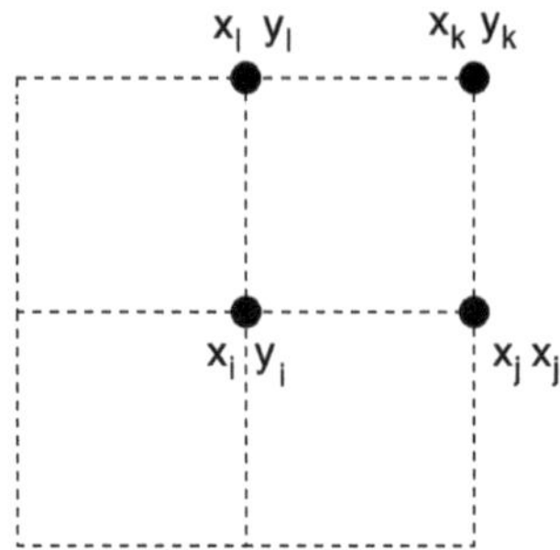

$$(x_j, y_j) = (x_i + b_x, y_i) \quad (x_k, y_k) = (x_i, y_i + b_y) \quad (x_l, y_l) = (x_i + b_x, y_i + b_y)$$

$$(11.149)$$

where $b_x = \pm h_x, b_y = \pm h_y$ corresponding to the four rectangles with the common vertex i (Fig. 11.10).

The bilinear function (Fig. 11.11) corresponding to R_{ijkl} is

$$n_{i,j,k,l}(x, y) = \alpha + \beta(x - x_i) + \gamma(y - y_i) + \eta(x - x_i)(y - y_i). \quad (11.150)$$

It has to fulfill

$$n_{i,j,k,l}(x_i, y_i) = 1 \quad n_{i,j,k,l}(x_j, y_j) = n_{i,j,k,l}(x_k, y_k) = n_{i,j,k,l}(x_l, y_l) = 0$$

$$(11.151)$$

from which we find

$$\alpha = 1 \quad \beta = -\frac{1}{b_x} \quad \gamma = -\frac{1}{b_y} \quad \eta = \frac{1}{b_x b_y} \quad (11.152)$$

$$n_{i,j,k,l}(x, y) = 1 - \frac{x - x_i}{b_x} - \frac{y - y_i}{b_y} + \frac{(x - x_i)}{b_x} \frac{(y - y_i)}{b_y}. \quad (11.153)$$

The basis function centered at node i then is

$$N_i(x, y) = \begin{cases} n_{i,j,k,l}(x, y) & (x, y) \in R_{i,j,k,l} \\ 0 & \text{else.} \end{cases} \quad (11.154)$$

Generalization to a three-dimensional grid is straightforward (Fig. 11.12). We denote one of the eight cuboids containing the node (x_i, y_i, z_i) as $C_{i,j_1...j_7}$ with $(x_{j_1}, y_{j_1}, z_{j_1}) = (x_i + b_x, y_i, z_i) \ldots (x_{j_7}, y_{j_7}, z_{j_7}) = (x_i + b_x, y_i + b_y, z_i + b_z)$. The corresponding trilinear function is

$$n_{i,j_1...j_7} = 1 - \frac{x - x_i}{b_x} - \frac{y - y_i}{b_y} - \frac{z - z_i}{b_z}$$

$$+ \frac{(x - x_i)}{b_x} \frac{(y - y_i)}{b_y} + \frac{(x - x_i)}{b_x} \frac{(z - z_i)}{b_z} + \frac{(z - z_i)}{b_z} \frac{(y - y_i)}{b_y}$$

$$- \frac{(x - x_i)}{b_x} \frac{(y - y_i)}{b_y} \frac{(z - z_i)}{b_z}. \quad (11.155)$$

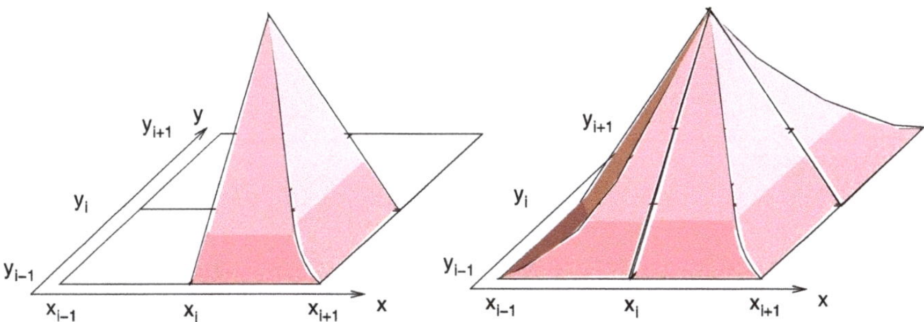

Fig. 11.11 (Bilinear elements on a rectangular grid) The basis functions $N_i(x, y)$ on a rectangular grid (*right side*) are piecewise bilinear functions (*left side*), which vanish at all nodes except (x_i, y_i) (*right side*)

Fig. 11.12
(Three-dimensional rectangular grid) The basis function N_i is trilinear on each of the eight cuboids containing the vertex i. It vanishes on all nodes except (x_i, y_i, z_i)

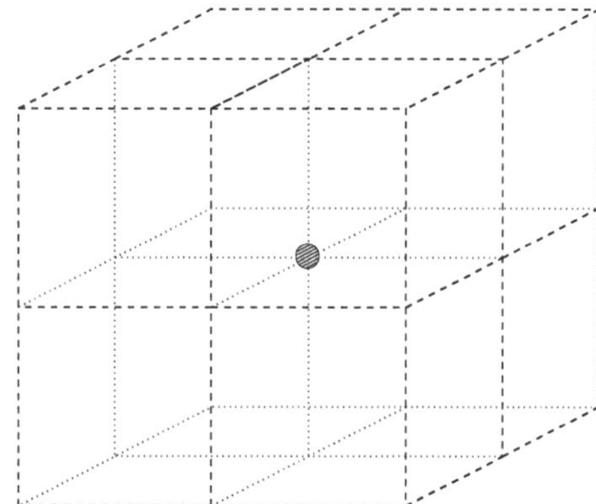

11.6.3 One-Dimensional Galerkin FEM

As an example we consider the one-dimensional linear differential equation (11.5)

$$\left(a\frac{\partial^2}{\partial x^2} + b\frac{\partial}{\partial x} + c\right)u(x) = f(x) \tag{11.156}$$

in the domain $0 \leq x \leq 1$ with boundary conditions

$$u(0) = u(1) = 0. \tag{11.157}$$

We use the basis functions from (11.139) on a one-dimensional grid with

$$x_{i+1} - x_i = h_i \tag{11.158}$$

and apply the Galerkin method [88]. The boundary conditions require

$$u_0 = u_{N-1} = 0. \tag{11.159}$$

The weighted residual is

$$0 = R_j = \sum_i u_i \int_0^1 dx\, N_j(x)\left(a\frac{\partial^2}{\partial x^2} + b\frac{\partial}{\partial x} + c\right)N_i(x) - \int_0^1 dx\, N_j(x)f(x).$$

$$(11.160)$$

First we integrate

$$\int_0^1 N_j(x)N_i(x)\,dx = \int_{x_{i-1}}^{x_{i+1}} N_j(x)N_i(x)\,dx = \begin{cases} \frac{h_i+h_{i-1}}{3} & j=i \\ \frac{h_i}{6} & j=i+1 \\ \frac{h_{i-1}}{6} & j=i-1 \\ 0 & |i-j|>1. \end{cases}$$

$$(11.161)$$

Integration of the first derivative gives

$$\int_0^1 dx\, N_j(x)N_i'(x) = \begin{cases} 0 & j=i \\ \frac{1}{2} & j=i-1 \\ -\frac{1}{2} & j=i+1 \\ 0 & \text{else.} \end{cases}$$

$$(11.162)$$

For the second derivative partial integration gives

$$\int_0^1 dx\, N_j(x)\frac{\partial^2}{\partial x^2}N_i(x)$$

$$= N_j(1)N_i'(1-\varepsilon) - N_j(0)N_i'(0+\varepsilon) - \int_0^1 dx\, N_j'(x)N_i'(x) \quad (11.163)$$

where the first two summands are zero due to the boundary conditions. Since N_i and N_i' are nonzero only for $x_{i-1} < x < x_{i+1}$ we find

$$\int_0^1 dx\, N_j(x)\frac{\partial^2}{\partial x^2}N_i(x) = -\int_{x_{i-1}}^{x_{i+1}} dx\, N_j'(x)N_i'(x)$$

$$= \begin{cases} \frac{1}{h_{i-1}} & j=i-1 \\ -\frac{1}{h_i} - \frac{1}{h_{i-1}} & i=j \\ \frac{1}{h_i} & j=i+1 \\ 0 & \text{else.} \end{cases}$$

$$(11.164)$$

Integration of the last term in (11.160) gives

$$\int_0^1 dx\, N_j(x)f(x) = \int_{x_{i-1}}^{x_{i+1}} dx\, N_j(x)f(x)$$

$$= \int_{x_{j-1}}^{x_j} dx\, \frac{x - x_{j-1}}{x_j - x_{j-1}}f(x)$$

$$+ \int_{x_j}^{x_{j+1}} dx\, \frac{x_{j+1} - x}{x_{j+1} - x_j}f(x).$$

$$(11.165)$$

Applying the trapezoidal rule[11] for both integrals we find

$$\int_{x_{j-1}}^{x_{j+1}} dx\, N_j(x) f(x) \approx f(x_j)\frac{h_j + h_{j-1}}{2}. \tag{11.166}$$

The discretized equation finally reads

$$a\left\{\frac{1}{h_{j-1}}u_{j-1} - \left(\frac{1}{h_j} + \frac{1}{h_{j-1}}\right)u_j + \frac{1}{h_j}u_{j+1}\right\}$$

$$+ b\left\{-\frac{1}{2}u_{j-1} + \frac{1}{2}u_{j+1}\right\}$$

$$+ c\left\{\frac{h_{j-1}}{6}u_{j-1} + \frac{h_j + h_{j-1}}{3}u_j + \frac{h_j}{6}u_{j+1}\right\}$$

$$= f(x_j)\frac{h_j + h_{j-1}}{2} \tag{11.167}$$

which can be written in matrix notation as

$$\mathbf{Au} = \mathbf{Bf} \tag{11.168}$$

where the matrix A is tridiagonal as a consequence of the compact support of the basis functions

$$A = a\begin{pmatrix} -\frac{1}{h_1} - \frac{1}{h_0},\ \frac{1}{h_1} & & & \\ & \ddots & & \\ & \frac{1}{h_{j-1}},\ -\frac{1}{h_j} - \frac{1}{h_{j-1}},\ \frac{1}{h_j} & & \\ & & \ddots & \\ & & \frac{1}{h_{N-3}},\ -\frac{1}{h_{N-2}} - \frac{1}{h_{N-3}} \end{pmatrix}$$

$$+ b\begin{pmatrix} 0,\ \frac{1}{2} & & \\ & \ddots & \\ -\frac{1}{2},\ 0,\ \frac{1}{2} & & \\ & \ddots & \\ & -\frac{1}{2},\ 0 \end{pmatrix}$$

$$\tag{11.169}$$

$$+ c\begin{pmatrix} \frac{(h_1+h_0)}{3},\ \frac{h_1}{6} & & \\ & \ddots & \\ \frac{h_{j-1}}{6},\ \frac{(h_j+h_{j-1})}{3},\ \frac{h_j}{6} & & \\ & \ddots & \\ & \frac{h_{N-3}}{6},\ \frac{(h_{N-2}+h_{N-3})}{3} \end{pmatrix}$$

---

[11]Higher accuracy can be achieved, for instance, by Gaussian integration.

$$B = \begin{pmatrix} \frac{h_0+h_1}{2} & & & & \\ & \ddots & & & \\ & & \frac{h_{j-1}+h_j}{2} & & \\ & & & \ddots & \\ & & & & \frac{h_{N-2}+h_{N-3}}{2} \end{pmatrix}.$$

For equally spaced nodes $h_i = h_{i-1} = h$ and after division by h (11.167) reduces to a system of equations where the derivatives are replaced by finite differences (11.20)

$$a\left\{ \frac{1}{h^2}u_{j-1} - \frac{2}{h^2}u_j + \frac{1}{h^2}u_{j+1} \right\}$$
$$+ b\left\{ -\frac{1}{2h}u_{j-1} + \frac{1}{2h}u_{j+1} \right\}$$
$$+ c\left\{ \frac{1}{6}u_{j-1} + \frac{2}{3}u_j + \frac{1}{6}u_{j+1} \right\}$$
$$= f(x_j) \tag{11.170}$$

and the function u is replaced by a certain average

$$\frac{1}{6}u_{j-1} + \frac{2}{3}u_j + \frac{1}{6}u_{j+1} = u_j + \frac{1}{6}(u_{j-1} - 2u_j + u_{j+1}). \tag{11.171}$$

The corresponding matrix in (11.169) is the so called mass matrix. Within the framework of the finite differences method the last expression equals

$$u_j + \frac{1}{6}(u_{j-1} - 2u_j + u_{j+1}) = u_j + \frac{h^2}{6}\left(\frac{d^2u}{dx^2}\right)_j + O(h^4) \tag{11.172}$$

hence replacing it by u_j (this is called mass lumping) introduces an error of the order $O(h^2)$.

11.7 Boundary Element Method

The boundary element method (BEM) [18, 276] is a method for linear partial differential equations which can be brought into boundary integral form[12] like Laplace's equation (Chap. 17)[13]

$$-\Delta \Phi(\mathbf{r}) = 0 \tag{11.173}$$

for which the fundamental solution

$$\Delta G(\mathbf{r}, \mathbf{r}') = -\delta(\mathbf{r} - \mathbf{r}')$$

---

[12] This is only possible if the fundamental solution or Green's function is available.

[13] The minus sign is traditionally used.

is given by

$$G(\mathbf{r} - \mathbf{r}') = \frac{1}{4\pi |\mathbf{r} - \mathbf{r}'|} \quad \text{in three dimensions} \tag{11.174}$$

$$G(\mathbf{r} - \mathbf{r}') = \frac{1}{2\pi} \ln \frac{1}{|\mathbf{r} - \mathbf{r}'|} \quad \text{in two dimensions.} \tag{11.175}$$

We apply Gauss's theorem to the expression [277]

$$\operatorname{div}\left[G(\mathbf{r} - \mathbf{r}') \operatorname{grad}(\Phi(\mathbf{r})) - \Phi(\mathbf{r}) \operatorname{grad}(G(\mathbf{r} - \mathbf{r}')) \right]$$
$$= -\Phi(\mathbf{r}) \Delta (G(\mathbf{r} - \mathbf{r}')). \tag{11.176}$$

Integration over a volume V gives

$$\oint_{\partial V} dA \left(G(\mathbf{r} - \mathbf{r}') \frac{\partial}{\partial n}(\Phi(\mathbf{r})) - \Phi(\mathbf{r}) \frac{\partial}{\partial n}(G(\mathbf{r} - \mathbf{r}')) \right)$$
$$= - \int_V dV \left(\Phi(\mathbf{r}) \Delta (G(\mathbf{r} - \mathbf{r}')) \right) = \Phi(\mathbf{r}'). \tag{11.177}$$

This integral equation determines the potential self-consistently by its value and normal derivative on the surface of the cavity. It can be solved numerically by dividing the surface into a finite number of boundary elements. The resulting system of linear equations often has smaller dimension than corresponding finite element approaches. However, the coefficient matrix is in general full and not necessarily symmetric.

Chapter 12
Equations of Motion

Simulation of a physical system means to calculate the time evolution of a model system in many cases. We consider a large class of models which can be described by a first order initial value problem

$$\frac{dY}{dt} = f\big(Y(t), t\big) \quad Y(t=0) = Y_0 \tag{12.1}$$

where Y is the state vector (possibly of very high dimension) which contains all information about the system. Our goal is to calculate the time evolution of the state vector $Y(t)$ numerically. For obvious reasons this can be done only for a finite number of values of t and we have to introduce a grid of discrete times t_n which for simplicity are assumed to be equally spaced:[1]

$$t_{n+1} = t_n + \Delta t. \tag{12.2}$$

Advancing time by one step involves the calculation of the integral

$$Y(t_{n+1}) - Y(t_n) = \int_{t_n}^{t_{n+1}} f\big(Y(t'), t'\big) dt' \tag{12.3}$$

which can be a formidable task since $f(Y(t), t)$ depends on time via the time dependence of all the elements of $Y(t)$. In this chapter we discuss several strategies for the time integration. The explicit Euler forward difference has low error order but is useful as a predictor step for implicit methods. A symmetric difference quotient is much more accurate. It can be used as the corrector step in combination with an explicit Euler predictor step and is often used for the time integration of partial differential equations. Methods with higher error order can be obtained from a Taylor series expansion, like the Nordsieck and Gear predictor-corrector methods which have been often applied in molecular dyamics calculations. Runge-Kutta methods are very important for ordinary differential equations. They are robust and allow an adaptive control of the step size. Very accurate results can be obtained for ordinary

---

[1] Control of the step width will be discussed later.

P.O.J. Scherer, *Computational Physics*, Graduate Texts in Physics, 207
DOI 10.1007/978-3-319-00401-3_12,
© Springer International Publishing Switzerland 2013

differential equations with extrapolation methods like the famous Gragg-Bulirsch-Stör method. If the solution is smooth enough, multistep methods are applicable, which use information from several points. Most known are Adams-Bashforth-Moulton methods and Gear methods (also known as backward differentiation methods), which are especially useful for stiff problems. The class of Verlet methods has been developed for molecular dynamics calculations. They are symplectic and time reversible and conserve energy over long trajectories.

12.1 The State Vector

The state of a classical N-particle system is given by the position in phase space, or equivalently by specifying position and velocity for all the N particles

$$Y = (\mathbf{r}_1, \mathbf{v}_1, \ldots, \mathbf{r}_N, \mathbf{v}_N). \tag{12.4}$$

The concept of a state vector is not restricted to a finite number of degrees of freedom. For instance a diffusive system can be described by the particle concentrations as a function of the coordinate, i.e. the elements of the state vector are now indexed by the continuous variable $\mathbf{x}$

$$Y = \big(c_1(\mathbf{x}) \cdots c_M(\mathbf{x})\big). \tag{12.5}$$

Similarly, a quantum particle moving in an external potential can be described by the amplitude of the wave function

$$Y = \big(\Psi(\mathbf{x})\big). \tag{12.6}$$

Numerical treatment of continuous systems is not feasible since even the ultimate high end computer can only handle a finite number of data in finite time. Therefore discretization is necessary (Chap. 11), by introducing a spatial mesh (Sects. 11.2, 11.3, 11.6), which in the simplest case means a grid of equally spaced points

$$\mathbf{x}_{ijk} = (ih, jh, kh) \quad i = 1 \cdots i_{\max}, \; j = 1 \cdots j_{\max}, \; k = 1 \cdots k_{\max} \tag{12.7}$$

$$Y = \big(c_1(\mathbf{x}_{ijk}) \ldots c_M(\mathbf{x}_{ijk})\big) \tag{12.8}$$

$$Y = \big(\Psi(\mathbf{x}_{ijk})\big) \tag{12.9}$$

or by expanding the continuous function with respect to a finite set of basis functions (Sect. 11.5). The elements of the state vector then are the expansion coefficients

$$|\Psi\rangle = \sum_{s=1}^{N} C_s |\Psi_s\rangle \tag{12.10}$$

$$Y = (C_1, \ldots, C_N). \tag{12.11}$$

If the density matrix formalism is used to take the average over a thermodynamic ensemble or to trace out the degrees of freedom of a heat bath, the state vector instead is composed of the elements of the density matrix

$$\rho = \sum_{s=1}^{N} \sum_{s'=1}^{N} \rho_{ss'} |\Psi_s\rangle \langle \Psi_{s'}| = \sum_{s=1}^{N} \sum_{s'=1}^{N} \overline{C_{s'}^* C_s} |\Psi_s\rangle \langle \Psi_{s'}| \tag{12.12}$$

$$Y = (\rho_{11} \cdots \rho_{1N}, \rho_{21} \cdots \rho_{2N}, \ldots, \rho_{N1} \cdots \rho_{NN}). \tag{12.13}$$

12.2 Time Evolution of the State Vector

We assume that all information about the system is included in the state vector. Then the simplest equation to describe the time evolution of the system gives the change of the state vector

$$\frac{dY}{dt} = f(Y, t) \tag{12.14}$$

as a function of the state vector (or more generally a functional in the case of a continuous system). Explicit time dependence has to be considered for instance to describe the influence of an external time dependent field.

Some examples will show the universality of this equation of motion:

- N-particle system

The motion of N interacting particles is described by

$$\frac{dY}{dt} = (\dot{\mathbf{r}}_1, \dot{\mathbf{v}}_1 \cdots) = (\mathbf{v}_1, \mathbf{a}_1 \cdots) \tag{12.15}$$

where the acceleration of a particle is given by the total force acting upon this particle and thus depends on all the coordinates and eventually time (velocity dependent forces could be also considered but are outside the scope of this book)

$$\mathbf{a}_i = \frac{\mathbf{F}_i(\mathbf{r}_1 \cdots \mathbf{r}_N, t)}{m_i}. \tag{12.16}$$

- Diffusion

Heat transport and other diffusive processes are described by the diffusion equation

$$\frac{\partial f}{\partial t} = D\Delta f + S(\mathbf{x}, t) \tag{12.17}$$

which in its simplest spatially discretized version for 1-dimensional diffusion reads

$$\frac{\partial f(\mathbf{x}_i)}{\partial t} = \frac{D}{\Delta x^2} \left(f(\mathbf{x}_{i+1}) + f(\mathbf{x}_{i-1}) - 2f(\mathbf{x}_i) \right) + S(\mathbf{x}_i, t). \tag{12.18}$$

- Waves

Consider the simple 1-dimensional wave equation

$$\frac{\partial^2 f}{\partial t^2} = c^2 \frac{\partial^2 f}{\partial x^2} \tag{12.19}$$

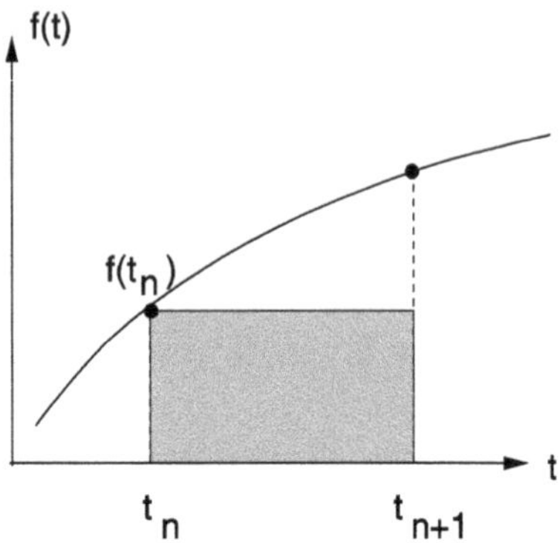

Fig. 12.1 Explicit Euler method

which by introducing the velocity $g(\mathbf{x}) = \frac{\partial}{\partial t} f(\mathbf{x})$ as an independent variable can be rewritten as

$$\frac{\partial}{\partial t}\big(f(\mathbf{x}), g(\mathbf{x})\big) = \left(g(\mathbf{x}), c^2 \frac{\partial^2}{\partial x^2} f(\mathbf{x})\right). \tag{12.20}$$

Discretization of space gives

$$\frac{\partial}{\partial t}\big(f(\mathbf{x}_i), g(\mathbf{x}_i)\big) = \left(g(\mathbf{x}_i), \frac{c^2}{\Delta x^2}\big(f(\mathbf{x}_{i+1}) + f(\mathbf{x}_{i-1}) - 2f(\mathbf{x}_i)\big)\right). \tag{12.21}$$

- Two-state quantum system

The Schrödinger equation for a two level system (for instance a spin-$\frac{1}{2}$ particle in a magnetic field) reads

$$\frac{d}{dt}\begin{pmatrix} C_1 \\ C_2 \end{pmatrix} = \begin{pmatrix} H_{11}(t) & H_{12}(t) \\ H_{21}(t) & H_{22}(t) \end{pmatrix}\begin{pmatrix} C_1 \\ C_2 \end{pmatrix}. \tag{12.22}$$

12.3 Explicit Forward Euler Method

The simplest method which is often discussed in elementary physics textbooks approximates the integrand by its value at the lower bound (Fig. 12.1):

$$Y(t_{n+1}) - Y(t_n) \approx f\big(Y(t_n), t_n\big)\Delta t. \tag{12.23}$$

The truncation error can be estimated from a Taylor series expansion

$$\begin{aligned}
Y(t_{n+1}) - Y(t_n) &= \Delta t \frac{dY}{dt}(t_n) + \frac{\Delta t^2}{2}\frac{d^2Y}{dt^2}(t_n) + \cdots \\
&= \Delta t f\big(Y(t_n), t_n\big) + O\big(\Delta t^2\big).
\end{aligned} \tag{12.24}$$

The explicit Euler method has several serious drawbacks

- Low error order

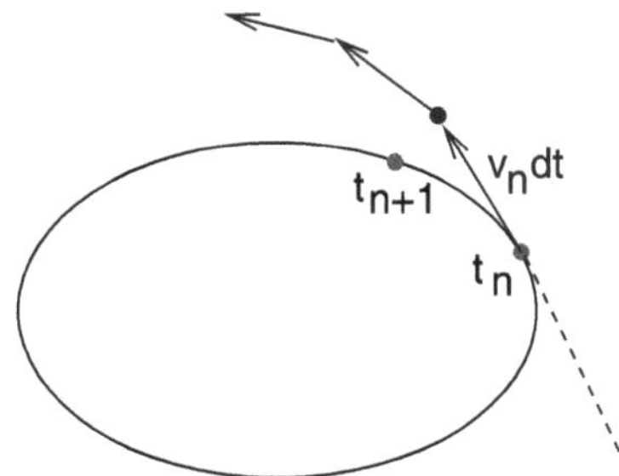

Fig. 12.2 Systematic errors of the Euler method

Suppose you want to integrate from the initial time t_0 to the final time $t_0 + T$. For a time step of Δt you have to perform $N = T/\Delta t$ steps. Assuming comparable error contributions from all steps the global error scales as $N\Delta t^2 = O(\Delta t)$. The error gets smaller as the time step is reduced but it may be necessary to use very small Δt to obtain meaningful results.

- Loss of orthogonality and normalization

The simple Euler method can produce systematic errors which are very inconvenient if you want, for instance, to calculate the orbits of a planetary system. This can be most easily seen from a very simple example. Try to integrate the following equation of motion (see Example 1.5 on page 12):

$$\frac{dz}{dt} = i\omega z. \tag{12.25}$$

The exact solution is obviously given by a circular orbit in the complex plane:

$$z = z_0 e^{i\omega t} \tag{12.26}$$

$$|z| = |z_0| = \text{const.} \tag{12.27}$$

Application of the Euler method gives

$$z(t_{n+1}) = z(t_n) + i\omega \Delta t\, z(t_n) = (1 + i\omega \Delta t) z(t_n) \tag{12.28}$$

and you find immediately

$$\left|z(t_n)\right| = \sqrt{1 + \omega^2 \Delta t^2}\left|z(t_{n-1})\right| = \left(1 + \omega^2 \Delta t^2\right)^{n/2}\left|z(t_0)\right| \tag{12.29}$$

which shows that the radius increases continually even for the smallest time step possible (Fig. 12.2).

The same kind of error appears if you solve the Schrödinger equation for a particle in an external potential or if you calculate the rotational motion of a rigid body. For the N-body system it leads to a violation of the conservation of phase space volume. This can introduce an additional sensitivity of the calculated results to the initial conditions. Consider a harmonic oscillator with the equation of motion

$$\frac{d}{dt}\begin{pmatrix} x(t) \\ v(t) \end{pmatrix} = \begin{pmatrix} v(t) \\ -\omega^2 x(t) \end{pmatrix}. \tag{12.30}$$

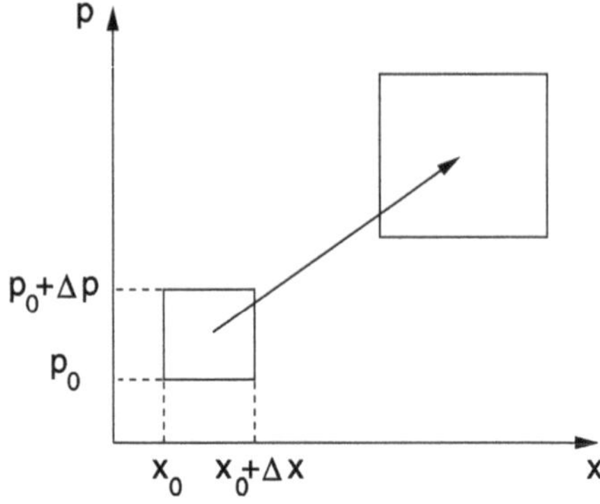

Fig. 12.3 Time evolution of the phase space volume

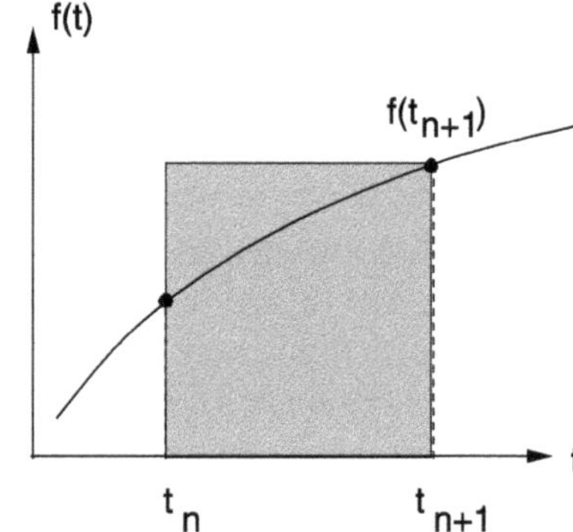

Fig. 12.4 Implicit backward Euler method

Application of the explicit Euler method gives

$$\begin{pmatrix} x(t + \Delta t) \\ v(t + \Delta t) \end{pmatrix} = \begin{pmatrix} x(t) \\ v(t) \end{pmatrix} + \begin{pmatrix} v(t) \\ -\omega^2 x(t) \end{pmatrix} \Delta t. \tag{12.31}$$

The change of the phase space volume (Fig. 12.3) is given by the Jacobi determinant

$$J = \left| \frac{\partial(x(t + \Delta t), v(t + \Delta t))}{\partial(x(t), v(t))} \right| = \left| \begin{matrix} 1 & \Delta t \\ -\omega^2 \Delta t & 1 \end{matrix} \right| = 1 + (\omega \Delta t)^2. \tag{12.32}$$

In this case the phase space volume increases continuously.

12.4 Implicit Backward Euler Method

Alternatively let us make a step backwards in time

$$Y(t_n) - Y(t_{n+1}) \approx -f\big(Y(t_{n+1}), t_{n+1}\big) \Delta t \tag{12.33}$$

which can be written as (Fig. 12.4)

$$Y(t_{n+1}) \approx Y(t_n) + f\big(Y(t_{n+1}), t_{n+1}\big) \Delta t. \tag{12.34}$$

Fig. 12.5 Improved Euler method

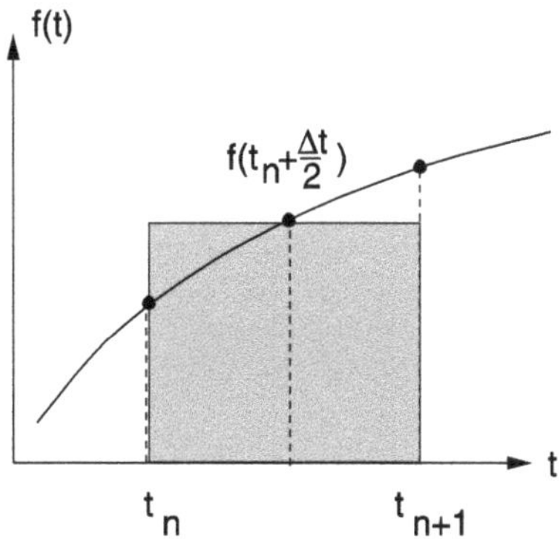

Taylor series expansion gives

$$Y(t_n) = Y(t_{n+1}) - \frac{\mathrm{d}}{\mathrm{d}t}Y(t_{n+1})\Delta t + \frac{\mathrm{d}^2}{\mathrm{d}t^2}Y(t_{n+1})\frac{\Delta t^2}{2} + \cdots \qquad (12.35)$$

which shows that the error order again is $O(\Delta t^2)$. The implicit method is sometimes used to avoid the inherent instability of the explicit method. For the examples in Sect. 12.3 it shows the opposite behavior. The radius of the circular orbit as well as the phase space volume decrease in time. The gradient at future time has to be estimated before an implicit step can be performed.

12.5 Improved Euler Methods

The quality of the approximation can be improved significantly by employing the midpoint rule (Fig. 12.5)

$$Y(t_{n+1}) - Y(t_n) \approx f\left(Y\left(t + \frac{\Delta t}{2}\right), t_n + \frac{\Delta t}{2}\right)\Delta t. \qquad (12.36)$$

The error is smaller by one order of Δt:

$$Y(t_n) + f\left(Y\left(t + \frac{\Delta t}{2}\right), t_n + \frac{\Delta t}{2}\right)\Delta t$$

$$= Y(t_n) + \left(\frac{\mathrm{d}Y}{\mathrm{d}t}(t_n) + \frac{\Delta t}{2}\frac{\mathrm{d}^2 Y}{\mathrm{d}t^2}(t_n) + \cdots\right)\Delta t$$

$$= Y(t_n + \Delta t) + O\left(\Delta t^3\right). \qquad (12.37)$$

The future value $Y(t + \frac{\Delta t}{2})$ can be obtained by two different approaches:

- Predictor-corrector method

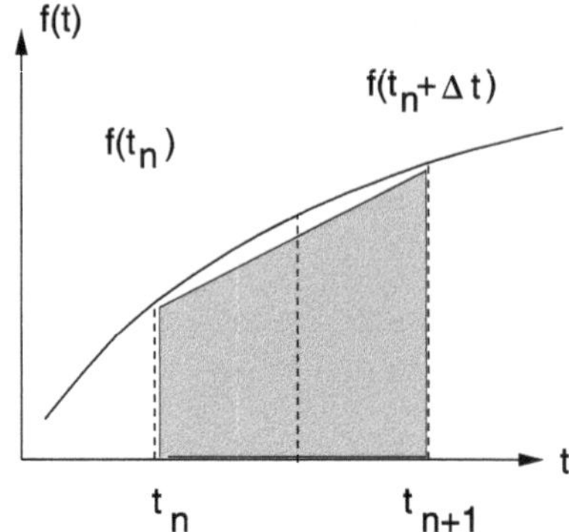

Fig. 12.6 Improved polygon (or Heun) method

Since $f(Y(t + \frac{\Delta t}{2}), t_n + \frac{\Delta t}{2})$ is multiplied with Δt, it is sufficient to use an approximation with lower error order. Even the explicit Euler step is sufficient. Together the following algorithm results:

$$\text{predictor step:} \quad Y^{(p)} = Y(t_n) + \frac{\Delta t}{2} f\big(Y(t_n), t_n\big)$$

$$\text{corrector step:} \quad Y(t_n + \Delta t) = Y(t_n) + \Delta t f\left(Y^{(p)}, t_n + \frac{\Delta t}{2}\right). \tag{12.38}$$

- Averaging (Heun method)

The average of $f(Y(t_n), t_n)$ and $f(Y(t_n + \Delta t), t + \Delta t)$ is another approximation to the midpoint value of comparable quality (Fig. 12.6).

Expansion around $t_n + \Delta t/2$ gives

$$\frac{1}{2}\big(f\big(Y(t_n), t_n\big) + f\big(Y(t_n + \Delta t), t + \Delta t\big)\big)$$

$$= f\left(Y\left(t_n + \frac{\Delta t}{2}\right), t_n + \frac{\Delta t}{2}\right) + O(\Delta t^2). \tag{12.39}$$

Inserting the average in (12.36) gives the following algorithm, which is also known as improved polygon method and corresponds to the trapezoidal rule for the integral (4.13) or to a combination of explicit and implicit Euler step:

$$Y(t_n + \Delta t) = Y(t_n) + \frac{\Delta t}{2}\big(f\big(Y(t_n), t_n\big) + f\big(Y(t_n + \Delta t), t + \Delta t\big)\big). \tag{12.40}$$

In the special case of a linear function $f(Y(t), t) = FY(t)$ (for instance rotational motion or diffusion) this can be solved formally by

$$Y(t_n + \Delta t) = \left(1 - \frac{\Delta t}{2}F\right)^{-1}\left(1 + \frac{\Delta t}{2}F\right)Y(t_n). \tag{12.41}$$

Numerically it is not necessary to perform the matrix inversion. Instead a linear system of equations is solved:

$$\left(1 - \frac{\Delta t}{2}F\right)Y(t_n + \Delta t) = \left(1 + \frac{\Delta t}{2}F\right)Y(t_n). \tag{12.42}$$

In certain cases the Heun method conserves the norm of the state vector, for instance if F has only imaginary eigenvalues (as for the 1-dimensional Schrödinger equation, see page 393).

In the general case a predictor step has to be made to estimate the state vector at $t_n + \Delta t$ before the Heun expression (12.40) can be evaluated:

$$Y^{(p)} = Y(t_n) + \Delta t f\big(Y(t_n), t_n\big). \tag{12.43}$$

12.6 Taylor Series Methods

Higher order methods can be obtained from a Taylor series expansion

$$Y(t_n + \Delta t) = Y(t_n) + \Delta t f\big(Y(t_n), t_n\big) + \frac{\Delta t^2}{2} \frac{\mathrm{d} f(Y(t_n), t_n)}{\mathrm{d} t} + \cdots. \tag{12.44}$$

The total time derivative can be expressed as

$$\frac{\mathrm{d} f}{\mathrm{d} t} = \frac{\partial f}{\partial Y} \frac{\mathrm{d} Y}{\mathrm{d} t} + \frac{\partial f}{\partial t} = f' f + \dot{f} \tag{12.45}$$

where the partial derivatives have been abbreviated in the usual way by $\frac{\partial f}{\partial t} = \dot{f}$ and $\frac{\partial f}{\partial Y} = f'$. Higher derivatives are given by

$$\frac{\mathrm{d}^2 f}{\mathrm{d} t^2} = f'' f^2 + f'^2 f + 2\dot{f}' f + \ddot{f} \tag{12.46}$$

$$\begin{aligned}
\frac{\mathrm{d}^3 f}{\mathrm{d} t^3} = {} & \frac{\partial^3 f}{\partial t^3} + f''' f^3 + 3 f'' f^2 + \ddot{f} f' + 3 f'' \dot{f} f \\
& + 3 \dot{f}' + 4 f'' f' f^2 + 5 \dot{f}' f' f + f'^3 f + f'^2 \dot{f}.
\end{aligned} \tag{12.47}$$

12.6.1 Nordsieck Predictor-Corrector Method

Nordsieck [184] determines an interpolating polynomial of degree m. As variables he uses the 0th to mth derivatives[2] evaluated at the current time t, for instance for $m = 5$ he uses the variables

$$Y(t) \tag{12.48}$$

$$g(t) = \frac{\mathrm{d}}{\mathrm{d} t} Y(t) \tag{12.49}$$

$$a(t) = \frac{\Delta t}{2} \frac{\mathrm{d}^2}{\mathrm{d} t^2} Y(t) \tag{12.50}$$

---

[2] In fact the derivatives of the interpolating polynomial which exist even if higher derivatives of f do not exist.

$$b(t) = \frac{\Delta t^2}{6} \frac{\mathrm{d}^3}{\mathrm{d}t^3} Y(t) \tag{12.51}$$

$$c(t) = \frac{\Delta t^3}{24} \frac{\mathrm{d}^4}{\mathrm{d}t^4} Y(t) \tag{12.52}$$

$$d(t) = \frac{\Delta t^4}{120} \frac{\mathrm{d}^5}{\mathrm{d}t^5} Y(t). \tag{12.53}$$

Taylor expansion gives approximate values at $t + \Delta t$

$$\begin{aligned}
Y(t + \Delta t) &= Y(t) + \Delta t \big[g(t) + a(t) + b(t) + c(t) + d(t) + e(t) \big] \\
&= Y^P(t + \Delta t) + e(t)\Delta t \tag{12.54}
\end{aligned}$$

$$\begin{aligned}
g(t + \Delta t) &= g(t) + 2a(t) + 3b(t) + 4c(t) + 5d(t) + 6e(t) \\
&= g^P(t + \Delta T) + 6e(t) \tag{12.55}
\end{aligned}$$

$$\begin{aligned}
a(t + \Delta t) &= a(t) + 3b(t) + 6c(t) + 10d(t) + 15e(t) \\
&= a^P(t + \Delta t) + 15e(t) \tag{12.56}
\end{aligned}$$

$$\begin{aligned}
b(t + \Delta t) &= b(t) + 4c(t) + 10d(t) + 20e(t) \\
&= b^P(t + \Delta t) + 20e(t) \tag{12.57}
\end{aligned}$$

$$c(t + \Delta t) = c(t) + 5d(t) + 15e(t) = c^P(t + \Delta t) + 15e(t) \tag{12.58}$$

$$d(t + \Delta t) = d(t) + 6e(t) = d^P(t + \Delta t) + 6e(t) \tag{12.59}$$

where the next term of the Taylor series $e(t) = \frac{\Delta t^5}{6!} \frac{\mathrm{d}^6}{\mathrm{d}t^6} Y(t)$ has been introduced as an approximation to the truncation error of the predicted values Y^P, g^P, etc. It can be estimated from the second equation

$$e = \frac{1}{6} \big[f\big(Y^P(t + \Delta t), t + \Delta t\big) - g^P(t + \Delta t) \big] = \frac{1}{6}\delta f. \tag{12.60}$$

This predictor-corrector method turns out to be rather unstable. However, stability can be achieved by slightly modifying the coefficients of the corrector step. Nordsieck suggested to use

$$Y(t + \Delta t) = Y^P(t + \Delta t) + \frac{95}{288}\delta f \tag{12.61}$$

$$a(t + \Delta t) = a^P(t + \Delta t) + \frac{25}{24}\delta f \tag{12.62}$$

$$b(t + \Delta t) = b^P(t + \Delta t) + \frac{35}{72}\delta f \tag{12.63}$$

$$c(t + \Delta t) = c^P(t + \Delta t) + \frac{5}{48}\delta f \tag{12.64}$$

$$d(t + \Delta t) = d^P(t + \Delta t) + \frac{1}{120}\delta f. \tag{12.65}$$

12.6.2 Gear Predictor-Corrector Methods

Gear [101] designed special methods for molecular dynamics simulations (Chap. 14) where Newton's law (12.15) has to be solved numerically. He uses again a truncated Taylor expansion for the predictor step

$$\mathbf{r}(t+\Delta t) = \mathbf{r}(t) + \mathbf{v}(t)\Delta t + \mathbf{a}(t)\frac{\Delta t^2}{2} + \dot{\mathbf{a}}(t)\frac{\Delta t^3}{6} + \ddot{\mathbf{a}}(t)\frac{\Delta t^4}{24} + \cdots \quad (12.66)$$

$$\mathbf{v}(t+\Delta t) = \mathbf{v}(t) + \mathbf{a}(t)\Delta t + \dot{\mathbf{a}}(t)\frac{\Delta t^2}{2} + \ddot{\mathbf{a}}(t)\frac{\Delta t^3}{6} + \cdots \quad (12.67)$$

$$\mathbf{a}(t+\Delta t) = \mathbf{a}(t) + \dot{\mathbf{a}}(t)\Delta t + \ddot{\mathbf{a}}(t)\frac{\Delta t^2}{2} + \cdots \quad (12.68)$$

$$\dot{\mathbf{a}}(t+\Delta t) = \dot{\mathbf{a}}(t) + \ddot{\mathbf{a}}(t)\Delta t + \cdots \quad (12.69)$$

$$\vdots$$

to calculate new coordinates etc. $\mathbf{r}^p_{n+1}, \mathbf{v}^p_{n+1}, \mathbf{a}^p_{n+1} \ldots$ (Fig. 12.7). The difference between the predicted acceleration and that calculated using the predicted coordinates

$$\delta\mathbf{a}_{n+1} = \mathbf{a}\left(\mathbf{r}^P_{n+1}, t+\Delta t\right) - \mathbf{a}^p_{n+1} \quad (12.70)$$

is then used as a measure of the error to correct the predicted values according to

$$\mathbf{r}_{n+1} = \mathbf{r}^p_{n+1} + c_1\delta\mathbf{a}_{n+1} \quad (12.71)$$

$$\mathbf{v}_{n+1} = \mathbf{v}^p_{n+1} + c_2\delta\mathbf{a}_{n+1} \quad (12.72)$$

$$\vdots$$

The coefficients c_i were determined to optimize stability and accuracy. For instance the fourth order Gear corrector reads

$$\mathbf{r}_{n+1} = \mathbf{r}^p_{n+1} + \frac{\Delta t^2}{12}\delta\mathbf{a}_{n+1} \quad (12.73)$$

$$\mathbf{v}_{n+1} = \mathbf{v}^p_{n+1} + \frac{5\Delta t}{12}\delta\mathbf{a}_{n+1} \quad (12.74)$$

$$\dot{\mathbf{a}}_{n+1} = \dot{\mathbf{a}}_n + \frac{1}{\Delta t}\delta\mathbf{a}_{n+1}. \quad (12.75)$$

Gear methods are generally not time reversible and show systematic energy drifts. A reversible symplectic predictor-corrector method has been presented recently by Martyna and Tuckerman [169].

12.7 Runge-Kutta Methods

If higher derivatives are not so easily available, they can be approximated by numerical differences. f is evaluated at several trial points and the results are combined to reproduce the Taylor series as close as possible [48].

Fig. 12.7 (Gear
predictor-corrector method)
The difference between
predicted acceleration $\mathbf{a}^p$ and
acceleration calculated for the
predicted coordinates $\mathbf{a}(\mathbf{r}^p)$ is
used as a measure of the error
to estimate the correction $\delta\mathbf{r}$

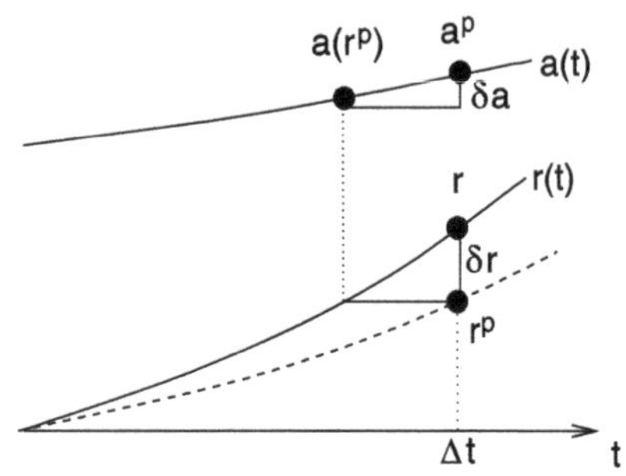

12.7.1 Second Order Runge-Kutta Method

Let us begin with two function values. As common in the literature we will denote
the function values as $K_1, K_2, \ldots$. From the gradient at time t_n

$$K_1 = f_n = f\big(Y(t_n), t_n\big) \tag{12.76}$$

we estimate the state vector at time $t_n + \Delta t$ as

$$Y(t_n + \Delta t) \approx \Delta t\, K_1. \tag{12.77}$$

The gradient at time $t_n + \Delta t$ is approximately

$$K_2 = f\big(Y(t_n) + \Delta t\, K_1, t_n + \Delta t\big) \tag{12.78}$$

which has the Taylor series expansion

$$K_2 = f_n + \big(\dot{f}_n + f'_n f_n\big)\Delta t + \cdots \tag{12.79}$$

and application of the trapezoidal rule (4.13) gives the 2nd order Runge-Kutta
method

$$Y_{n+1} = Y_n + \frac{\Delta t}{2}(K_1 + K_2) \tag{12.80}$$

which in fact coincides with the improved Euler or Heun method. Taylor series
expansion shows how the combination of K_1 and K_2 leads to an expression of higher
error order:

$$\begin{aligned}
Y_{n+1} &= Y_n + \frac{\Delta t}{2}\big(f_n + f_n + \big(\dot{f}_n + f'_n f_n\big)\Delta t + \cdots\big)\\
&= Y_n + f_n \Delta t + \frac{\mathrm{d} f_n}{\mathrm{d} t}\frac{\Delta t^2}{2} + \cdots .
\end{aligned} \tag{12.81}$$

12.7.2 Third Order Runge-Kutta Method

The accuracy can be further improved by calculating one additional function value
at mid-time. From (12.76) we estimate the gradient at mid-time by

$$K_2 = f\left(Y(t) + \frac{\Delta t}{2}K_1, t + \frac{\Delta t}{2}\right)$$

$$= f_n + \left(\dot{f}_n + f'_n f_n\right)\frac{\Delta t}{2} + \left(\ddot{f}_n + f''_n f_n^2 + 2\dot{f}'_n f_n\right)\frac{\Delta t^2}{8} + \cdots. \quad (12.82)$$

The gradient at time $t_n + \Delta t$ is then estimated as

$$K_3 = f\left(Y(t_n) + \Delta t(2K_2 - K_1), t_n + \Delta t\right)$$

$$= f_n + \dot{f}_n \Delta t + f'_n(2K_2 - K_1)\Delta t + \ddot{f}_n \frac{\Delta t^2}{2}$$

$$+ f''_n \frac{(2K_2 - K_1)^2 \Delta t^2}{2} + 2\dot{f}'_n \frac{(2K_2 - K_1)\Delta t^2}{2} + \cdots. \quad (12.83)$$

Inserting the expansion (12.82) gives the leading terms

$$K_3 = f_n + \left(\dot{f}_n + f'_n f_n\right)\Delta t + \left(2f'^2_n f_n + f''_n f_n^2 + \ddot{f}_n + 2f'_n \dot{f}_n + 2\dot{f}_n^2\right)\frac{\Delta t^2}{2} + \cdots. \quad (12.84)$$

Applying Simpson's rule (4.14) we combine the three gradients to get the 3rd order Runge-Kutta method

$$Y_{n+1} = Y(t_n) + \frac{\Delta t}{6}(K_1 + 4K_2 + K_3) \quad (12.85)$$

where the Taylor series

$$Y_{n+1} = Y(t_n) + \frac{\Delta t}{6}\Big(6f_n + 3\left(\dot{f}_n + f_n f'_n\right)\Delta t$$

$$+ \left(f'^2_n f_n + f''_n f_n^2 + 2\dot{f}'_n f_n + \ddot{f}_n + \dot{f}_n \ddot{f}'_n\right)\Delta t^2 + \cdots\Big)$$

$$= Y(t_n + \Delta t) + O\left(\Delta t^4\right) \quad (12.86)$$

recovers the exact Taylor series (12.44) including terms of order $O(\Delta t^3)$.

12.7.3 Fourth Order Runge-Kutta Method

The 4th order Runge-Kutta method (RK4) is often used because of its robustness and accuracy. It uses two different approximations for the midpoint

$$K_1 = f\left(Y(t_n), t_n\right)$$

$$K_2 = f\left(Y(t_n) + \frac{K_1}{2}\Delta t, t_n + \frac{\Delta t}{2}\right)$$

$$K_3 = f\left(Y(t_n) + \frac{K_2}{2}\Delta t, t_n + \frac{\Delta t}{2}\right)$$

$$K_4 = f\left(Y(t_n) + K_3\Delta t, t_n + \Delta t\right)$$

Fig. 12.8 Step doubling with
the fourth order Runge-Kutta
method

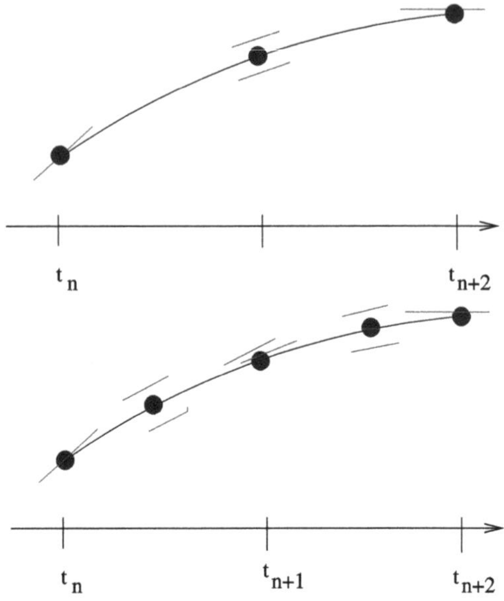

and Simpson's rule (4.14) to obtain

$$Y_{n+1} = Y(t_n) + \frac{\Delta t}{6}(K_1 + 2K_2 + 2K_3 + K_4) = Y(t_n + \Delta t) + O(\Delta t^5).$$

Expansion of the Taylor series is cumbersome but with the help of an algebra program one can easily check that the error is of order Δt^5.

12.8 Quality Control and Adaptive Step Size Control

For practical applications it is necessary to have an estimate for the local error and to adjust the step size properly. With the Runge-Kutta method this can be achieved by a step doubling procedure. We calculate y_{n+2} first by two steps Δt and then by one step $2\Delta t$. This needs 11 function evaluations as compared to 8 for the smaller step size only (Fig. 12.8). For the 4th order method we estimate the following errors:

$$\Delta\left(Y_{n+2}^{(\Delta t)}\right) = 2a\,\Delta t^5 \tag{12.87}$$

$$\Delta\left(Y_{n+2}^{(2\Delta t)}\right) = a(2\Delta t)^5. \tag{12.88}$$

The local error can be estimated from

$$\left|Y_{n+2}^{(\Delta t)} - Y_{n+2}^{(2\Delta t)}\right| = 30|a|\Delta t^5$$

$$\Delta\left(Y_{n+1}^{(\Delta t)}\right) = a\,\Delta t^5 = \frac{\left|Y_{n+2}^{(\Delta t)} - Y_{n+2}^{(2\Delta t)}\right|}{30}.$$

The step size Δt can now be adjusted to keep the local error within the desired limits.

12.9 Extrapolation Methods

Application of the extrapolation method to calculate the integral $\int_{t_n}^{t_{n+1}} f(t)\,dt$ produces very accurate results but can also be time consuming. The famous Gragg-Bulirsch-Stör method [244] starts from an explicit midpoint rule with a special starting procedure. The interval Δt is divided into a sequence of N sub-steps

$$h = \frac{\Delta t}{N}. \tag{12.89}$$

First a simple Euler step is performed

$$\begin{aligned} u_0 &= Y(t_n) \\ u_1 &= u_0 + h\,f(u_0, t_n) \end{aligned} \tag{12.90}$$

and then the midpoint rule is applied repeatedly to obtain

$$u_{j+1} = u_{j-1} + 2h\,f(u_j, t_n + jh) \quad j = 1, 2, \ldots, N-1. \tag{12.91}$$

Gragg [111] introduced a smoothing procedure to remove oscillations of the leading error term by defining

$$v_j = \frac{1}{4}u_{j-1} + \frac{1}{2}u_j + \frac{1}{4}u_{j+1}. \tag{12.92}$$

He showed that both approximations (12.91, 12.92) have an asymptotic expansion in powers of h^2 and are therefore well suited for an extrapolation method. The modified midpoint method can be summarized as follows:

$$\begin{aligned} u_0 &= Y(t_n) \\ u_1 &= u_0 + h\,f(u_0, t_n) \\ u_{j+1} &= u_{j-1} + 2hf(u_j, t_n + jh) \quad j = 1, 2, \ldots, N-1 \\ Y(t_n + \Delta t) &\approx \frac{1}{2}\big(u_N + u_{N-1} + hf(u_N, t_n + \Delta t)\big). \end{aligned} \tag{12.93}$$

The number of sub-steps N is increased according to a sequence like

$$N = 2, 4, 6, 8, 12, 16, 24, 32, 48, 64 \cdots \quad N_j = 2N_{j-2} \quad \text{Bulirsch-Stör sequence} \tag{12.94}$$

or

$$N = 2, 4, 6, 8, 10, 12 \cdots \quad N_j = 2j \quad \text{Deuflhard sequence.}$$

After each successive N is tried, a polynomial extrapolation is attempted. This extrapolation returns both the extrapolated values and an error estimate. If the error is still too large then N has to be increased further. A more detailed discussion can be found in [233, 234].

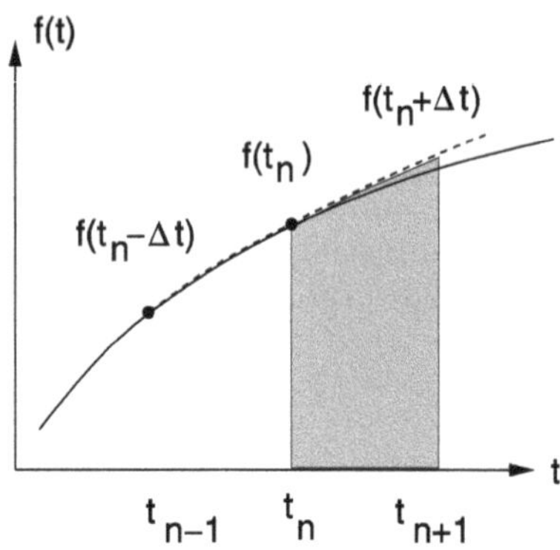

Fig. 12.9 Adams-Bashforth method

12.10 Linear Multistep Methods

All methods discussed so far evaluated one or more values of the gradient $f(Y(t), t)$ only within the interval $t_n \cdots t_n + \Delta t$. If the state vector changes sufficiently smooth then multistep methods can be applied. Linear multistep methods use a combination of function values Y_n and gradients f_n from several steps

$$Y_{n+1} = \sum_{j=1}^{k}(\alpha_j Y_{n-j+1} + \beta_j f_{n-j+1}\Delta t) + \beta_0 f_{n+1}\Delta t \qquad (12.95)$$

where the coefficients α, β are determined such, that a polynomial of certain order r is integrated exactly. The method is explicit if $\beta_0 = 0$ and implicit otherwise. Multistep methods have a small local error and need fewer function evaluations. On the other hand, they have to be combined with other methods (like Runge-Kutta) to start and end properly and it can be rather complicated to change the step size during the calculation. Three families of linear multistep methods are commonly used: explicit Adams-Bashforth methods, implicit Adams-Moulton methods and backward differentiation formulas (also known as Gear formulas [102]).

12.10.1 Adams-Bashforth Methods

The explicit Adams-Bashforth method of order r uses the gradients from the last $r - 1$ steps (Fig. 12.9) to obtain the polynomial

$$p(t_n) = f(Y_n, t_n) \cdots p(t_{n-r+1}) = f(Y_{n-r+1}, t_{n-r+1}) \qquad (12.96)$$

and to calculate the approximation

$$Y_{n+1} - Y_n \approx \int_{t_n}^{t_{n+1}} p(t)\, dt$$

which is generally a linear combination of $f_n \cdots f_{n-r+1}$. For example, the Adams-Bashforth formulas of order 2, 3, 4 are:

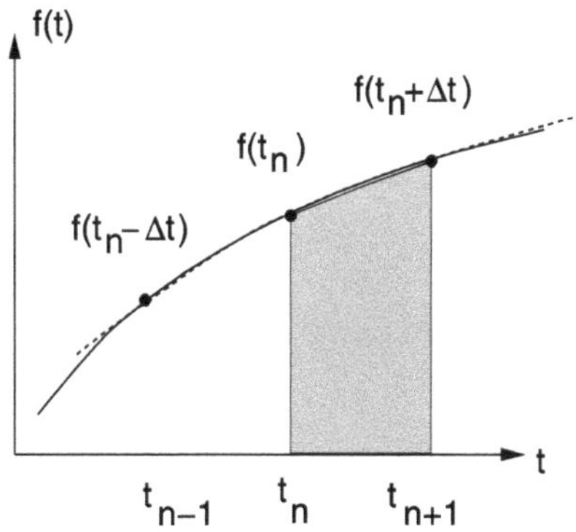

Fig. 12.10 Adams-Moulton method

$$Y_{n+1} - Y_n = \frac{\Delta t}{2}(3f_n - f_{n-1}) + O(\Delta t^3)$$

$$Y_{n+1} - Y_n = \frac{\Delta t}{12}(23f_n - 16f_{n-1} + 5f_{m-2}) + O(\Delta t^4) \tag{12.97}$$

$$Y_{n+1} - Y_n = \frac{\Delta t}{24}(55f_n - 59f_{n-1} + 37f_{n-2} - 9f_{n-3}) + O(\Delta t^5).$$

12.10.2 Adams-Moulton Methods

The implicit Adams-Moulton method also uses the yet not known value Y_{n+1} (Fig. 12.10) to obtain the polynomial

$$p(t_{n+1}) = f_{n+1} \cdots p(t_{n-r+2}) = f_{n-r+2}. \tag{12.98}$$

The corresponding Adams-Moulton formulas of order 2 to 4 are:

$$Y_{n+1} - Y_n = \frac{\Delta t}{2}(f_{n+1} + f_n) + O(\Delta t^3)$$

$$Y_{n+1} - Y_n = \frac{\Delta t}{12}(5f_{n+1} + 8f_n - f_{n-1}) + O(\Delta t^4) \tag{12.99}$$

$$Y_{n+1} - Y_n = \frac{\Delta t}{24}(9f_{n+1} + 19f_n - 5f_{n-1} + f_{n-2}) + O(\Delta t^5). \tag{12.100}$$

12.10.3 Backward Differentiation (Gear) Methods

Gear methods [102] are implicit and usually combined with a modified Newton method. They make use of previous function values $Y_n, Y_{n-1} \ldots$ and the gradient f_{n+1} at time $t + \Delta t$. Only methods of order $r \leq 6$ are stable and useful. The general formula (12.95) is

$$Y_{n+1} = \sum_{j=1}^{r} \alpha_j Y_{n-j+1} + \beta_0 f_{n+1} \Delta t. \tag{12.101}$$

For $r = 1$ this becomes

$$Y_{n+1} = \alpha_1 Y_n + \beta_0 f_1 \Delta t \tag{12.102}$$

and all linear polynomials

$$p = p_0 + p_1(t - t_n), \qquad \frac{dp}{dt} = p_1 \tag{12.103}$$

are integrated exactly if

$$p_0 + p_1 \Delta t = \alpha_1 p_0 + \beta_0 p_1 \tag{12.104}$$

which is the case for

$$\alpha_1 = 1, \quad \beta_0 = \Delta t. \tag{12.105}$$

Hence the first order Gear method is

$$Y_{n+1} = Y_n + f_{n+1} \Delta t + O(\Delta t^2) \tag{12.106}$$

which coincides with the implicit Euler method. The higher order stable Gear methods are given by

$$r = 2: \quad Y_{n+1} = \frac{4}{3} Y_n - \frac{1}{3} Y_{n-1} + \frac{2}{3} f_{n+1} \Delta t + O(\Delta t^3) \tag{12.107}$$

$$r = 3: \quad Y_{n+1} = \frac{18}{11} Y_n - \frac{9}{11} Y_{n-1} + \frac{2}{11} Y_{n-2} + \frac{6}{11} f_{n+1} \Delta t$$
$$+ O(\Delta t^4) \tag{12.108}$$

$$r = 4: \quad Y_{n+1} = \frac{48}{25} Y_n - \frac{36}{25} Y_{n-1} + \frac{16}{25} Y_{n-2}$$
$$- \frac{3}{25} Y_{n-3} + \frac{12}{25} f_{n+1} \Delta t + O(\Delta t^5) \tag{12.109}$$

$$r = 5: \quad Y_{n+1} = \frac{300}{137} Y_n - \frac{300}{137} Y_{n-1} + \frac{200}{137} Y_{n-2} - \frac{75}{137} Y_{n-3}$$
$$+ \frac{12}{137} Y_{n-4} + \frac{60}{137} f_{n+1} \Delta t + O(\Delta t^6) \tag{12.110}$$

$$r = 6: \quad Y_{n+1} = \frac{120}{49} Y_n - \frac{150}{49} Y_{n-1} + \frac{400}{147} Y_{n-2} - \frac{75}{49} Y_{n-3} + \frac{24}{49} Y_{n-4}$$
$$- \frac{10}{147} Y_{n-5} + \frac{20}{49} f_{n+1} \Delta t + O(\Delta t^7). \tag{12.111}$$

This class of algorithms is useful also for stiff problems (differential equations with strongly varying eigenvalues).

12.10.4 Predictor-Corrector Methods

The Adams-Bashforth-Moulton method combines the explicit method as a predictor step to calculate an estimate y_{n+1}^p with a corrector step using the implicit method of same order. The general class of linear multistep predictor-corrector methods [100] uses a predictor step

$$Y_{n+1}^{(0)} = \sum_{j=1}^{k} \left(\alpha_j^{(p)} Y_{n-j+1} + \beta_j^{(p)} f_{n-j+1} \Delta t \right) \tag{12.112}$$

which is corrected using the formula

$$Y_{n+1}^{(1)} = \sum_{j=1}^{k} \left(\alpha_j^{(c)} Y_{n-j+1} + \beta_j^{(c)} f_{n-j+1} \Delta t \right) + \beta_0 f\left(Y_{n+1}^{(0)}, t_{n+1} \right) \Delta t \tag{12.113}$$

and further iterations

$$Y_{n+1}^{(m+1)} = Y_{n+1}^{(m)} - \beta_0 \left[f\left(Y_{n+1}^{(m-1)}, t_{n+1} \right) - f\left(Y_{n+1}^{(m)}, t_{n+1} \right) \right] \Delta t$$
$$m = 1 \ldots M - 1 \tag{12.114}$$
$$Y_{n+1} = Y_{n+1}^{(M)}, \qquad \dot{Y}_{n+1} = f\left(Y_{n+1}^{(M-1)}, t_{n+1} \right). \tag{12.115}$$

The coefficients α, β have to be determined to optimize accuracy and stability.

12.11 Verlet Methods

For classical molecular dynamics simulations it is necessary to calculate very long trajectories. Here a family of symplectic methods often is used which conserve the phase space volume [3, 116, 193, 255, 257, 265]. The equations of motion of a classical interacting N-body system are

$$m_i \ddot{\mathbf{x}}_i = F_i \tag{12.116}$$

where the force acting on atom i can be calculated once a specific force field is chosen. Let us write these equations as a system of first order differential equations

$$\begin{pmatrix} \dot{\mathbf{x}}_i \\ \dot{\mathbf{v}}_i \end{pmatrix} = \begin{pmatrix} \mathbf{v}_i \\ \mathbf{a}_i \end{pmatrix} \tag{12.117}$$

where $\mathbf{x}(t)$ and $\mathbf{v}(t)$ are functions of time and the forces $m\mathbf{a}(\mathbf{x}(t))$ are functions of the time dependent coordinates.

12.11.1 Liouville Equation

We rewrite (12.117) as

$$\begin{pmatrix} \dot{\mathbf{x}} \\ \dot{\mathbf{v}} \end{pmatrix} = \mathcal{L} \begin{pmatrix} \mathbf{x} \\ \mathbf{v} \end{pmatrix} \tag{12.118}$$

where the Liouville operator $\mathcal{L}$ acts on the vector containing all coordinates and velocities:

$$\mathcal{L}\begin{pmatrix}\mathbf{x}\\\mathbf{v}\end{pmatrix} = \left(\mathbf{v}\frac{\partial}{\partial\mathbf{x}} + \mathbf{a}\frac{\partial}{\partial\mathbf{v}}\right)\begin{pmatrix}\mathbf{x}\\\mathbf{v}\end{pmatrix}. \tag{12.119}$$

The Liouville equation (12.118) can be formally solved by

$$\begin{pmatrix}\mathbf{x}(t)\\\mathbf{v}(t)\end{pmatrix} = e^{\mathcal{L}t}\begin{pmatrix}\mathbf{x}(0)\\\mathbf{v}(0)\end{pmatrix}. \tag{12.120}$$

For a better understanding let us evaluate the first members of the Taylor series of the exponential:

$$\mathcal{L}\begin{pmatrix}\mathbf{x}\\\mathbf{v}\end{pmatrix} = \left(\mathbf{v}\frac{\partial}{\partial\mathbf{x}} + \mathbf{a}\frac{\partial}{\partial\mathbf{v}}\right)\begin{pmatrix}\mathbf{x}\\\mathbf{v}\end{pmatrix} = \begin{pmatrix}\mathbf{v}\\\mathbf{a}\end{pmatrix} \tag{12.121}$$

$$\mathcal{L}^2\begin{pmatrix}\mathbf{x}\\\mathbf{v}\end{pmatrix} = \left(\mathbf{v}\frac{\partial}{\partial\mathbf{x}} + \mathbf{a}\frac{\partial}{\partial\mathbf{v}}\right)\begin{pmatrix}\mathbf{v}\\\mathbf{a(x)}\end{pmatrix} = \begin{pmatrix}\mathbf{a}\\\mathbf{v}\frac{\partial}{\partial\mathbf{x}}\mathbf{a}\end{pmatrix} \tag{12.122}$$

$$\mathcal{L}^3\begin{pmatrix}\mathbf{x}\\\mathbf{v}\end{pmatrix} = \left(\mathbf{v}\frac{\partial}{\partial\mathbf{x}} + \mathbf{a}\frac{\partial}{\partial\mathbf{v}}\right)\begin{pmatrix}\mathbf{a}\\\mathbf{v}\frac{\partial}{\partial\mathbf{x}}\mathbf{a}\end{pmatrix} = \begin{pmatrix}\mathbf{v}\frac{\partial}{\partial\mathbf{x}}\mathbf{a}\\\mathbf{a}\frac{\partial}{\partial\mathbf{x}}\mathbf{a} + \mathbf{v}\mathbf{v}\frac{\partial}{\partial\mathbf{x}}\frac{\partial}{\partial\mathbf{x}}\mathbf{a}\end{pmatrix}. \tag{12.123}$$

But since

$$\frac{\mathrm{d}}{\mathrm{d}t}\mathbf{a}\big(\mathbf{x}(t)\big) = \mathbf{v}\frac{\partial}{\partial\mathbf{x}}\mathbf{a} \tag{12.124}$$

$$\frac{\mathrm{d}^2}{\mathrm{d}t^2}\mathbf{a}\big(\mathbf{x}(t)\big) = \frac{\mathrm{d}}{\mathrm{d}t}\left(\mathbf{v}\frac{\partial}{\partial\mathbf{x}}\mathbf{a}\right) = \mathbf{a}\frac{\partial}{\partial\mathbf{x}}\mathbf{a} + \mathbf{v}\mathbf{v}\frac{\partial}{\partial\mathbf{x}}\frac{\partial}{\partial\mathbf{x}}\mathbf{a} \tag{12.125}$$

we recover

$$\left(1 + t\mathcal{L} + \frac{1}{2}t^2\mathcal{L}^2 + \frac{1}{6}t^3\mathcal{L}^3 + \cdots\right)\begin{pmatrix}\mathbf{x}\\\mathbf{v}\end{pmatrix} = \begin{pmatrix}\mathbf{x} + \mathbf{v}t + \frac{1}{2}t^2\mathbf{a} + \frac{1}{6}t^3\dot{\mathbf{a}} + \cdots\\\mathbf{v} + \mathbf{a}t + \frac{1}{2}t^2\dot{\mathbf{a}} + \frac{1}{6}t^3\ddot{\mathbf{a}} + \cdots\end{pmatrix}. \tag{12.126}$$

12.11.2 Split-Operator Approximation

We introduce a small time step $\Delta t = t/N$ and write

$$e^{\mathcal{L}t} = \left(e^{\mathcal{L}\Delta t}\right)^N. \tag{12.127}$$

For the small time step Δt the split-operator approximation can be used which approximately factorizes the exponential operator. For example, write the Liouville operator as the sum of two terms

$$\mathcal{L}_A = \mathbf{v}\frac{\partial}{\partial\mathbf{x}} \qquad \mathcal{L}_B = \mathbf{a}\frac{\partial}{\partial\mathbf{v}}$$

and make the approximation

$$e^{\mathcal{L}\Delta t} = e^{\mathcal{L}_A\Delta t}e^{\mathcal{L}_B\Delta t} + \cdots. \tag{12.128}$$

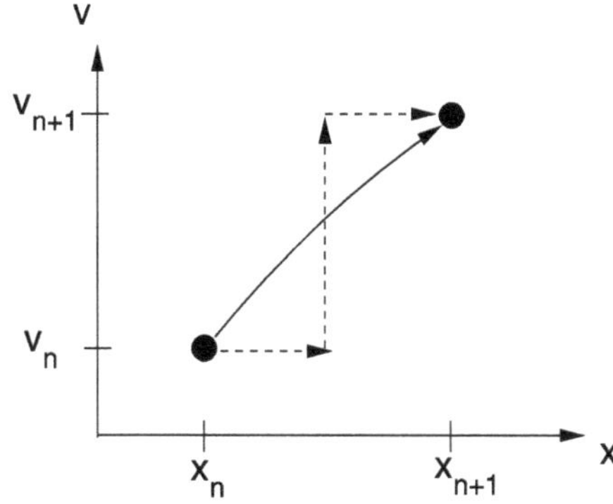

Fig. 12.11 (Position Verlet method) The exact integration path is approximated by two half-steps with constant velocities and one step with constant coordinates

Each of the two factors simply shifts positions or velocities

$$e^{\mathcal{L}_A \Delta t} \begin{pmatrix} \mathbf{x} \\ \mathbf{v} \end{pmatrix} = \begin{pmatrix} \mathbf{x} + \mathbf{v}\Delta t \\ \mathbf{v} \end{pmatrix} \qquad e^{\mathcal{L}_B \Delta t} \begin{pmatrix} \mathbf{x} \\ \mathbf{v} \end{pmatrix} = \begin{pmatrix} \mathbf{x} \\ \mathbf{v} + \mathbf{a}\Delta t \end{pmatrix} \qquad (12.129)$$

since these two steps correspond to either motion with constant velocities or constant coordinates and forces.

12.11.3 Position Verlet Method

Often the following approximation is used which is symmetrical in time

$$e^{\mathcal{L}\Delta t} = e^{\mathcal{L}_A \Delta t/2} e^{\mathcal{L}_B \Delta t} e^{\mathcal{L}_A \Delta t/2} + \cdots . \qquad (12.130)$$

The corresponding algorithm is the so called position Verlet method (Fig. 12.11):

$$\mathbf{x}_{n+\frac{1}{2}} = \mathbf{x}_n + \mathbf{v}_n \frac{\Delta t}{2} \qquad (12.131)$$

$$\mathbf{v}_{n+1} = \mathbf{v}_n + \mathbf{a}_{n+\frac{1}{2}} \Delta t = \mathbf{v}(t_n + \Delta t) + O(\Delta t^3) \qquad (12.132)$$

$$\mathbf{x}_{n+1} = \mathbf{x}_{n+\frac{1}{2}} + \mathbf{v}_{n+1} \frac{\Delta t}{2}$$

$$= \mathbf{x}_n + \frac{\mathbf{v}_n + \mathbf{v}_{n+1}}{2} \Delta t = \mathbf{x}(t_n + \Delta t) + O(\Delta t^3). \qquad (12.133)$$

12.11.4 Velocity Verlet Method

If we exchange operators A and B we have

$$e^{\mathcal{L}\Delta t} = e^{\mathcal{L}_B \Delta t/2} e^{\mathcal{L}_A \Delta t} e^{\mathcal{L}_B \Delta t/2} + \cdots \qquad (12.134)$$

which produces the velocity Verlet algorithm (Fig. 12.12):

Fig. 12.12 (Velocity Verlet method) The exact integration path is approximated by two half-steps with constant coordinates and one step with constant velocities

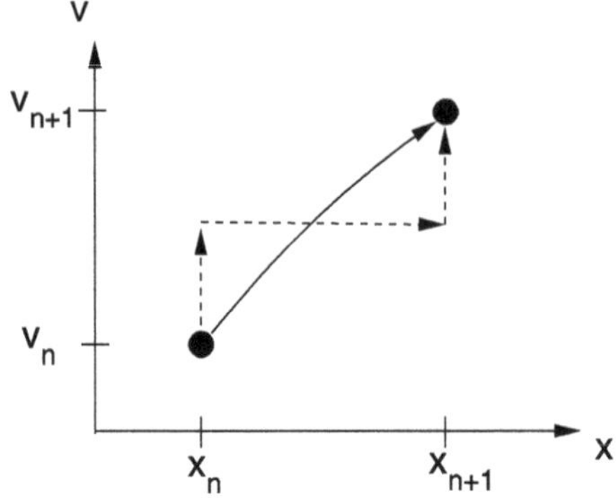

$$\mathbf{v}_{n+\frac{1}{2}} = \mathbf{v}_n + \mathbf{a}_n \frac{\Delta t}{2} \tag{12.135}$$

$$\mathbf{x}_{n+1} = \mathbf{x}_n + \mathbf{v}_{n+\frac{1}{2}}\Delta t = \mathbf{x}_n + \mathbf{v}_n \Delta t + \mathbf{a}_n \frac{\Delta t^2}{2}$$

$$= \mathbf{x}(t_n + \Delta t) + O\left(\Delta t^3\right) \tag{12.136}$$

$$\mathbf{v}_{n+1} = \mathbf{v}_{n+\frac{1}{2}} + \mathbf{a}_{n+1}\frac{\Delta t}{2} = \mathbf{v}_n + \frac{\mathbf{a}_n + \mathbf{a}_{n+1}}{2}\Delta t$$

$$= \mathbf{v}(t_n + \Delta t) + O\left(\Delta t^3\right). \tag{12.137}$$

12.11.5 Störmer-Verlet Method

The velocity Verlet method is equivalent to Störmer's version [208] of the Verlet method which is a two step method given by

$$\mathbf{x}_{n+1} = 2\mathbf{x}_n - \mathbf{x}_{n-1} + \mathbf{a}_n \Delta t^2 \tag{12.138}$$

$$\mathbf{v}_n = \frac{\mathbf{x}_{n+1} - \mathbf{x}_{n-1}}{2\Delta t}. \tag{12.139}$$

To show the equivalence we add two consecutive position vectors

$$\mathbf{x}_{n+2} + \mathbf{x}_{n+1} = 2\mathbf{x}_{n+1} + 2\mathbf{x}_n - \mathbf{x}_n - \mathbf{x}_{n-1} + (\mathbf{a}_{n+1} + \mathbf{a}_n)\Delta t^2 \tag{12.140}$$

which simplifies to

$$\mathbf{x}_{n+2} - \mathbf{x}_n - (\mathbf{x}_{n+1} - \mathbf{x}_n) = (\mathbf{a}_{n+1} + \mathbf{a}_n)\Delta t^2. \tag{12.141}$$

This can be expressed as the difference of two consecutive velocities:

$$2(\mathbf{v}_{n+1} - \mathbf{v}_n) = (\mathbf{a}_{n+1} + \mathbf{a}_n)\Delta t. \tag{12.142}$$

Now we substitute

$$\mathbf{x}_{n-1} = \mathbf{x}_{n+1} - 2\mathbf{v}_n \Delta t \tag{12.143}$$

to get

$$\mathbf{x}_{n+1} = 2\mathbf{x}_n - \mathbf{x}_{n+1} + 2\mathbf{v}_n \Delta t + \mathbf{a}_n \Delta t^2 \tag{12.144}$$

which simplifies to

$$\mathbf{x}_{n+1} = \mathbf{x}_n + \mathbf{v}_n \Delta t + \frac{\mathbf{a}_n}{2} \Delta t^2. \tag{12.145}$$

Thus the equations of the velocity Verlet algorithm have been recovered. However, since the Verlet method is a 2-step method, the choice of initial values is important. The Störmer-Verlet method starts from two coordinate sets x_0, x_1. The first step is

$$\mathbf{x}_2 = 2\mathbf{x}_1 - \mathbf{x}_0 + a_1 \Delta t^2 \tag{12.146}$$

$$\mathbf{v}_1 = \frac{\mathbf{x}_2 - \mathbf{x}_0}{2\Delta t} = \frac{\mathbf{x}_1 - \mathbf{x}_0}{\Delta t} + \frac{\mathbf{a}_1}{2} \Delta t^2. \tag{12.147}$$

The velocity Verlet method, on the other hand, starts from one set of coordinates and velocities $\mathbf{x}_1, \mathbf{v}_1$. Here the first step is

$$\mathbf{x}_2 = \mathbf{x}_1 + \mathbf{v}_1 \Delta t + \mathbf{a}_1 \frac{\Delta t^2}{2} \tag{12.148}$$

$$\mathbf{v}_2 = \mathbf{v}_1 + \frac{\mathbf{a}_1 + \mathbf{a}_2}{2} \Delta t. \tag{12.149}$$

The two methods give the same resulting trajectory if we choose

$$\mathbf{x}_0 = \mathbf{x}_1 - \mathbf{v}_1 \Delta t + \frac{\mathbf{a}_1}{2} \Delta t^2. \tag{12.150}$$

If, on the other hand, $\mathbf{x}_0$ is known with higher precision, the local error order of Störmer's algorithm changes as can be seen from addition of the two Taylor series

$$\mathbf{x}(t_n + \Delta t) = \mathbf{x}_n + \mathbf{v}_n \Delta t + \frac{\mathbf{a}_n}{2} \Delta t^2 + \frac{\dot{\mathbf{a}}_n}{6} \Delta t^3 + \cdots \tag{12.151}$$

$$\mathbf{x}(t_n - \Delta t) = \mathbf{x}_n - \mathbf{v}_n \Delta t + \frac{\mathbf{a}_n}{2} \Delta t^2 - \frac{\dot{\mathbf{a}}_n}{6} \Delta t^3 + \cdots \tag{12.152}$$

which gives

$$\mathbf{x}(t_n + \Delta t) = 2\mathbf{x}(t_n) - \mathbf{x}(t_n - \Delta t) + \mathbf{a}_n \Delta t^2 + O\left(\Delta t^4\right) \tag{12.153}$$

$$\frac{\mathbf{x}(t_n + \Delta t) - \mathbf{x}(t_n - \Delta t)}{2\Delta t} = \mathbf{v}_n + O\left(\Delta t^2\right). \tag{12.154}$$

12.11.6 Error Accumulation for the Störmer-Verlet Method

Equation (12.153) gives only the local error of one single step. Assume the start values $\mathbf{x}_0$ and $\mathbf{x}_1$ are exact. The next value $\mathbf{x}_2$ has an error with the leading term $\Delta x_2 = \alpha \Delta t^4$. If the trajectory is sufficiently smooth and the time step not too large

the coefficient α will vary only slowly and the error of the next few iterations is
given by

$$\Delta x_3 = 2\Delta x_2 - \Delta x_1 = 2\alpha \Delta t^4$$
$$\Delta x_4 = 2\Delta x_3 - \Delta x_2 = 3\alpha \Delta t^4$$
$$\vdots$$
$$\Delta x_{n+1} = n\alpha \Delta t^4. \tag{12.155}$$

This shows that the effective error order of the Störmer-Verlet method is only
$O(\Delta t^3)$ similar to the velocity Verlet method.

12.11.7 Beeman's Method

Beeman and Schofield [17, 229] introduced a method which is very similar to the
Störmer-Verlet method but calculates the velocities with higher precision. This is
important if, for instance, the kinetic energy has to be calculated. Starting from the
Taylor series

$$\mathbf{x}_{n+1} = \mathbf{x}_n + \mathbf{v}_n \Delta t + \mathbf{a}_n \frac{\Delta t^2}{2} + \dot{\mathbf{a}}_n \frac{\Delta t^3}{6} + \ddot{\mathbf{a}}_n \frac{\Delta t^4}{24} + \cdots \tag{12.156}$$

the derivative of the acceleration is approximated by a backward difference

$$\mathbf{x}_{n+1} = \mathbf{x}_n + \mathbf{v}_n \Delta t + \mathbf{a}_n \frac{\Delta t^2}{2} + \frac{\mathbf{a}_n - \mathbf{a}_{n-1}}{\Delta t} \frac{\Delta t^3}{6} + O(\Delta t^4)$$
$$= \mathbf{x}_n + \mathbf{v}_n \Delta t + \frac{4\mathbf{a}_n - \mathbf{a}_{n-1}}{6} \Delta t^2 + O(\Delta t^4). \tag{12.157}$$

This equation can be used as an explicit step to update the coordinates or as a predictor step in combination with the implicit corrector step

$$\mathbf{x}_{n+1} = \mathbf{x}_n + \mathbf{v}_n \Delta t + \mathbf{a}_n \frac{\Delta t^2}{2} + \frac{\mathbf{a}_{n+1} - \mathbf{a}_n}{\Delta t} \frac{\Delta t^3}{6} + O(\Delta t^4)$$
$$= \mathbf{x}_n + \mathbf{v}_n \Delta t + \frac{\mathbf{a}_{n+1} + 2\mathbf{a}_n}{6} \Delta t^2 + O(\Delta t^4) \tag{12.158}$$

which can be applied repeatedly (usually two iterations are sufficient). Similarly, the
Taylor series of the velocity is approximated by

$$\mathbf{v}_{n+1} = \mathbf{v}_n + \mathbf{a}_n \Delta t + \dot{\mathbf{a}}_n \frac{\Delta t^2}{2} + \ddot{\mathbf{a}}_n \frac{\Delta t^3}{6} + \cdots$$
$$= \mathbf{v}_n + \mathbf{a}_n \Delta t + \left(\frac{\mathbf{a}_{n+1} - \mathbf{a}_n}{\Delta t} + O(\Delta t) \right) \frac{\Delta t^2}{2} + \cdots$$
$$= \mathbf{v}_n + \frac{\mathbf{a}_{n+1} + \mathbf{a}_n}{2} \Delta t + O(\Delta t^3). \tag{12.159}$$

Inserting the velocity from (12.158) we obtain the corrector step for the velocity

$$\mathbf{v}_{n+1} = \frac{\mathbf{x}_{n+1} - \mathbf{x}_n}{\Delta t} - \frac{\mathbf{a}_{n+1} + 2\mathbf{a}_n}{6}\Delta t + \frac{\mathbf{a}_{n+1} + \mathbf{a}_n}{2}\Delta t + O\left(\Delta t^3\right)$$

$$= \frac{\mathbf{x}_{n+1} - \mathbf{x}_n}{\Delta t} + \frac{2\mathbf{a}_{n+1} + \mathbf{a}_n}{6}\Delta t + O\left(\Delta t^3\right). \tag{12.160}$$

In combination with (12.157) this can be replaced by

$$\mathbf{v}_{n+1} = \mathbf{v}_n + \frac{4\mathbf{a}_n - \mathbf{a}_{n-1}}{6}\Delta t + \frac{2\mathbf{a}_{n+1} + \mathbf{a}_n}{6}\Delta t + O\left(\Delta t^3\right)$$

$$= \mathbf{v}_n + \frac{2\mathbf{a}_{n+1} + 5\mathbf{a}_n - \mathbf{a}_{n-1}}{6}\Delta t + O\left(\Delta t^3\right). \tag{12.161}$$

Together, (12.157) and (12.161) provide an explicit method which is usually understood as Beeman's method. Inserting the velocity (12.160) from the previous step

$$\mathbf{v}_n = \frac{\mathbf{x}_n - \mathbf{x}_{n-1}}{\Delta t} + \frac{2\mathbf{a}_n + \mathbf{a}_{n-1}}{6}\Delta t + O\left(\Delta t^3\right) \tag{12.162}$$

into (12.157) gives

$$\mathbf{x}_{n+1} = 2\mathbf{x}_n - \mathbf{x}_{n-1} + \mathbf{a}_n \Delta t^2 + O\left(\Delta t^4\right) \tag{12.163}$$

which coincides with the Störmer-Verlet method (12.138). We conclude that Beeman's method should produce the same trajectory as the Störmer-Verlet method if numerical errors can be neglected and comparable initial values are used. In fact, the Störmer-Verlet method may suffer from numerical extinction and Beeman's method provides a numerically more favorable alternative.

12.11.8 The Leapfrog Method

Closely related to the Verlet methods is the so called leapfrog method [116]. It uses the simple decomposition

$$e^{\mathcal{L}\Delta t} \approx e^{\mathcal{L}_A \Delta t} e^{\mathcal{L}_B \Delta t} \tag{12.164}$$

but introduces two different time grids for coordinates and velocities which are shifted by $\Delta t/2$ (Fig. 12.13).

The leapfrog algorithm is given by

$$\mathbf{v}_{n+\frac{1}{2}} = \mathbf{v}_{n-\frac{1}{2}} + \mathbf{a}_n \Delta t \tag{12.165}$$

$$\mathbf{x}_{n+1} = \mathbf{x}_n + \mathbf{v}_{n+\frac{1}{2}} \Delta t. \tag{12.166}$$

Due to the shifted arguments the order of the method is increased as can be seen from the Taylor series:

$$\mathbf{x}(t_n) + \left(\mathbf{v}(t_n) + \frac{\Delta t}{2}\mathbf{a}(t_n) + \cdots\right)\Delta t = \mathbf{x}(t_n + \Delta t) + O\left(\Delta t^3\right) \tag{12.167}$$

$$\mathbf{v}\left(t_n + \frac{\Delta t}{2}\right) - \mathbf{v}\left(t_n - \frac{\Delta t}{2}\right) = \mathbf{a}(t_n)\Delta t + O\left(\Delta t^3\right). \tag{12.168}$$

Fig. 12.13 (Leapfrog method) The exact integration path is approximated by one step with constant coordinates and one step with constant velocities. Two different grids are used for coordinates and velocities which are shifted by $\Delta t/2$

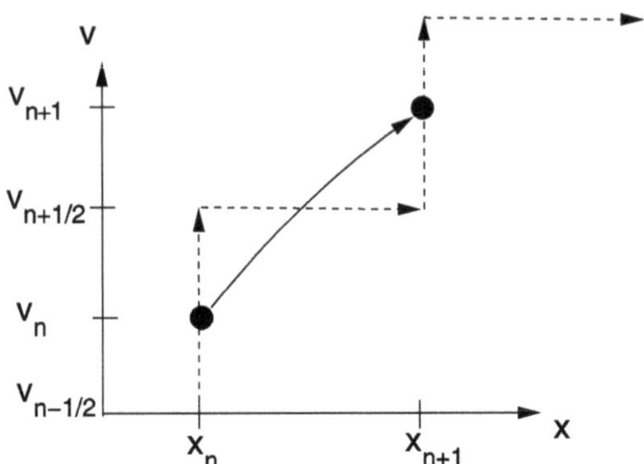

One disadvantage of the leapfrog method is that some additional effort is necessary if the velocities are needed. The simple expression

$$\mathbf{v}(t_n) = \frac{1}{2}\left(\mathbf{v}\left(t_n - \frac{\Delta t}{2}\right) + \mathbf{v}\left(t_n + \frac{\Delta t}{2}\right)\right) + O(\Delta t^2) \qquad (12.169)$$

is of lower error order than (12.168).

12.12 Problems

Problem 12.1 (Circular orbits)   In this computer experiment we consider a mass point moving in a central field. The equation of motion can be written as the following system of first order equations:

$$\begin{pmatrix} \dot{x} \\ \dot{y} \\ \dot{v}_x \\ \dot{v}_y \end{pmatrix} = \begin{pmatrix} 0 & 0 & 1 & 0 \\ 0 & 0 & 0 & 1 \\ -\dfrac{1}{(x^2+y^2)^{\frac{3}{2}}} & 0 & 0 & 0 \\ 0 & -\dfrac{1}{(x^2+y^2)^{\frac{3}{2}}} & 0 & 0 \end{pmatrix} \begin{pmatrix} x \\ y \\ v_x \\ v_y \end{pmatrix}. \qquad (12.170)$$

For initial values

$$\begin{pmatrix} x \\ y \\ v_x \\ v_y \end{pmatrix} = \begin{pmatrix} 1 \\ 0 \\ 0 \\ 1 \end{pmatrix} \qquad (12.171)$$

the exact solution is given by

$$x = \cos t \quad y = \sin t. \qquad (12.172)$$

The following methods are used to calculate the position $x(t)$, $y(t)$ and the energy

$$E_{tot} = E_{kin} + E_{pot} = \frac{1}{2}(v_x^2 + v_y^2) - \frac{1}{\sqrt{x^2 + y^2}}. \qquad (12.173)$$

- The explicit Euler method (12.3)

$$
\begin{aligned}
x(t_{n+1}) &= x(t_n) + v_x(t_n)\Delta t \\
y(t_{n+1}) &= y(t_n) + v_y(t_n)\Delta t \\
v_x(t_{n+1}) &= v_x(t_n) - \frac{x(t_n)}{R(t_n)^3}\Delta t \\
v_y(t_{n+1}) &= v_y(t_n) - \frac{y(t_n)}{R(t_n)^3}\Delta t.
\end{aligned}
\tag{12.174}
$$

- The 2nd order Runge-Kutta method (12.7.1)

which consists of the predictor step

$$
x(t_n + \Delta t/2) = x(t_n) + \frac{\Delta t}{2}v_x(t_n)
\tag{12.175}
$$

$$
y(t_n + \Delta t/2) = y(t_n) + \frac{\Delta t}{2}v_y(t_n)
\tag{12.176}
$$

$$
v_x(t_n + \Delta t/2) = v_x(t_n) - \frac{\Delta t}{2}\frac{x(t_n)}{R(t_n)^3}
\tag{12.177}
$$

$$
v_y(t_n + \Delta t/2) = v_y(t_n) - \frac{\Delta t}{2}\frac{y(t_n)}{R(t_n)^3}
\tag{12.178}
$$

and the corrector step

$$
x(t_{n+1}) = x(t_n) + \Delta t\, v_x(t_n + \Delta t/2)
\tag{12.179}
$$

$$
y(t_{n+1}) = y(t_n) + \Delta t\, v_y(t_n + \Delta t/2)
\tag{12.180}
$$

$$
v_x(t_{n+1}) = v_x(t_n) - \Delta t\frac{x(t_n + \Delta t/2)}{R^3(t_n + \Delta t/2)}
\tag{12.181}
$$

$$
v_y(t_{n+1}) = v_y(t_n) - \Delta t\frac{y(t_n + \Delta t/2)}{R^3(t_n + \Delta t/2)}.
\tag{12.182}
$$

- The fourth order Runge-Kutta method (12.7.3)
- The Verlet method (12.11.5)

$$
x(t_{n+1}) = x(t_n) + \big(x(t_n) - x(t_{n-1})\big) - \Delta t\frac{x(t_n)}{R^3(t_n)}
\tag{12.183}
$$

$$
y(t_{n+1}) = y(t_n) + \big(y(t_n) - y(t_{n-1})\big) - \Delta t\frac{y(t_n)}{R^3(t_n)}
\tag{12.184}
$$

$$
v_x(t_n) = \frac{x(t_{n+1}) - x(t_{n-1})}{2\Delta t} = \frac{x(t_n) - x(t_{n-1})}{\Delta t} - \frac{\Delta t}{2}\frac{x(t_n)}{R^3(t_n)}
\tag{12.185}
$$

$$
v_y(t_n) = \frac{y(t_{n+1}) - y(t_{n-1})}{2\Delta t} = \frac{y(t_n) - y(t_{n-1})}{\Delta t} - \frac{\Delta t}{2}\frac{y(t_n)}{R^3(t_n)}.
\tag{12.186}
$$

To start the Verlet method we need additional coordinates at time $-\Delta t$ which can be chosen from the exact solution or from the approximation

$$x(t_{-1}) = x(t_0) - \Delta t\, v_x(t_0) - \frac{\Delta t^2}{2}\frac{x(t_0)}{R^3(t_0)} \tag{12.187}$$

$$y(t_{-1}) = y(t_0) - \Delta t\, v_y(t_0) - \frac{\Delta t^2}{2}\frac{y(t_0)}{R^3(t_0)}. \tag{12.188}$$

- The leapfrog method (12.11.8)

$$x(t_{n+1}) = x(t_n) + v_x(t_{n+\frac{1}{2}})\Delta t \tag{12.189}$$

$$y(t_{n+1}) = y(t_n) + v_y(t_{n+\frac{1}{2}})\Delta t \tag{12.190}$$

$$v_x(t_{n+\frac{1}{2}}) = v_x(t_{n-\frac{1}{2}}) - \frac{x(t_n)}{R(t_n)^3}\Delta t \tag{12.191}$$

$$v_y(t_{n+\frac{1}{2}}) = v_y(t_{n-\frac{1}{2}}) - \frac{y(t_n)}{R(t_n)^3}\Delta t \tag{12.192}$$

where the velocity at time t_n is calculated from

$$v_x(t_n) = v_x(t_{n+\frac{1}{2}}) - \frac{\Delta t}{2}\frac{x(t_{n+1})}{R^3(t_{n+1})} \tag{12.193}$$

$$v_y(t_n) = v_y(t_{n+\frac{1}{2}}) - \frac{\Delta t}{2}\frac{y(t_{n+1})}{R^3(t_{n+1})}. \tag{12.194}$$

To start the leapfrog method we need the velocity at time $t_{-\frac{1}{2}}$ which can be taken from the exact solution or from

$$v_x(t_{-\frac{1}{2}}) = v_x(t_0) - \frac{\Delta t}{2}\frac{x(t_0)}{R^3(t_0)} \tag{12.195}$$

$$v_y(t_{-\frac{1}{2}}) = v_y(t_0) - \frac{\Delta t}{2}\frac{y(t_0)}{R^3(t_0)}. \tag{12.196}$$

Compare the conservation of energy for the different methods as a function of the time step Δt. Study the influence of the initial values for leapfrog and Verlet methods.

Problem 12.2 (*N*-body system) In this computer experiment we simulate the motion of three mass points under the influence of gravity. Initial coordinates and velocities as well as the masses can be varied. The equations of motion are solved with the 4th order Runge-Kutta method with quality control for different step sizes. The local integration error is estimated using the step doubling method. Try to simulate a planet with a moon moving round a sun!

Problem 12.3 (Adams-Bashforth method) In this computer experiment we simulate a circular orbit with the Adams-Bashforth method of order $2 \cdots 7$. The absolute error at time T

$$\Delta(T) = \left|x(T) - \cos(T)\right| + \left|y(t) - \sin(T)\right| + \left|v_x(T) + \sin(T)\right|$$
$$+ \left|v_y(T) - \cos(T)\right| \tag{12.197}$$

is shown as a function of the time step Δt in a log-log plot. From the slope

$$s = \frac{d(\log_{10}(\Delta))}{d(\log_{10}(\Delta t))} \tag{12.198}$$

the leading error order s can be determined. For very small step sizes rounding errors become dominating which leads to an increase $\Delta \sim (\Delta t)^{-1}$.

Determine maximum precision and optimal step size for different orders of the method. Compare with the explicit Euler method.

Part II
Simulation of Classical and Quantum Systems

Chapter 13
Rotational Motion

An asymmetric top under the influence of time dependent external forces is a rather complicated subject in mechanics. Efficient methods to describe the rotational motion are important as well in astrophysics as in molecular physics. The orientation of a rigid body relative to the laboratory system can be described by a 3×3 matrix. Instead of solving nine equations for all its components, the rotation matrix can be parametrized by the four real components of a quaternion. Euler angles use the minimum necessary number of three parameters but have numerical disadvantages. Care has to be taken to conserve the orthogonality of the rotation matrix. Omelyan's implicit quaternion method is very efficient and conserves orthogonality exactly. In computer experiments we compare different explicit and implicit methods for a free rotor, we simulate a rotor in an external field and the collision of two rotating molecules.

13.1 Transformation to a Body Fixed Coordinate System

Let us define a rigid body as a set of mass points m_i with fixed relative orientation (described by distances and angles).

The position of m_i in the laboratory coordinate system CS will be denoted by $\mathbf{r}_i$. The position of the center of mass (COM) of the rigid body is

$$\mathbf{R} = \frac{1}{\sum_i m_i} \sum_i m_i \mathbf{r}_i \tag{13.1}$$

and the position of m_i within the COM coordinate system CS_c (Fig. 13.1) is $\boldsymbol{\rho}_i$:

$$\mathbf{r}_i = \mathbf{R} + \boldsymbol{\rho}_i. \tag{13.2}$$

Let us define a body fixed coordinate system CS_{cb}, where the position $\boldsymbol{\rho}_{ib}$ of m_i is time independent $\frac{d}{dt}\boldsymbol{\rho}_{ib} = 0$. $\boldsymbol{\rho}_i$ and $\boldsymbol{\rho}_{ib}$ are connected by a linear vector function

$$\boldsymbol{\rho}_i = A \boldsymbol{\rho}_{ib} \tag{13.3}$$

P.O.J. Scherer, *Computational Physics*, Graduate Texts in Physics,
DOI 10.1007/978-3-319-00401-3_13,
© Springer International Publishing Switzerland 2013

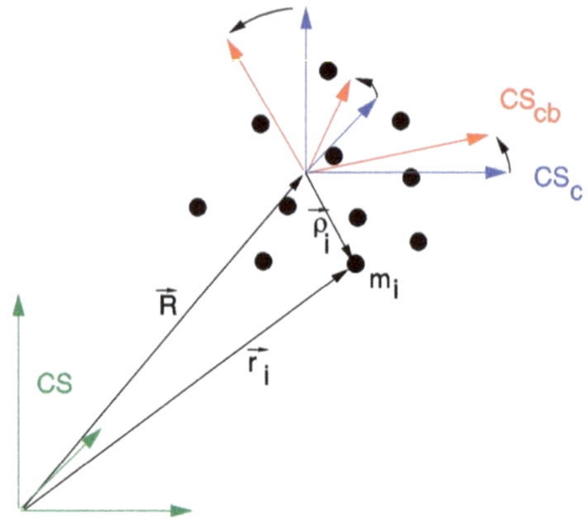

Fig. 13.1 (Coordinate systems) Three coordinate systems will be used: The laboratory system CS, the center of mass system CS_c and the body fixed system CS_{cb}

where A is a 3×3 matrix

$$A = \begin{pmatrix} a_{11} & a_{12} & a_{13} \\ a_{21} & a_{22} & a_{23} \\ a_{31} & a_{32} & a_{33} \end{pmatrix}. \tag{13.4}$$

13.2 Properties of the Rotation Matrix

Rotation conserves the length of ρ:[1]

$$\rho^T \rho = (A\rho)^T (A\rho) = \rho^T A^T A \rho. \tag{13.5}$$

Consider the matrix

$$M = A^T A - 1 \tag{13.6}$$

for which

$$\rho^T M \rho = 0 \tag{13.7}$$

holds for all vectors ρ. Let us choose the unit vector in x-direction:

$$\rho = \begin{pmatrix} 1 \\ 0 \\ 0 \end{pmatrix}.$$

Then we have

$$0 = \begin{pmatrix} 1 & 0 & 0 \end{pmatrix} \begin{pmatrix} M_{11} & M_{12} & M_{13} \\ M_{21} & M_{22} & M_{23} \\ M_{31} & M_{32} & M_{33} \end{pmatrix} \begin{pmatrix} 1 \\ 0 \\ 0 \end{pmatrix} = M_{11}. \tag{13.8}$$

---

[1] $\rho^T \rho$ denotes the scalar product of two vectors whereas $\rho \rho^T$ is the outer or matrix product.

Similarly by choosing a unit vector in y or z direction we find $M_{22} = M_{33} = 0$.

Now choose $\rho = \begin{pmatrix} 1 \\ 1 \\ 0 \end{pmatrix}$:

$$0 = \begin{pmatrix} 1 & 1 & 0 \end{pmatrix} \begin{pmatrix} M_{11} & M_{12} & M_{13} \\ M_{21} & M_{22} & M_{23} \\ M_{31} & M_{32} & M_{33} \end{pmatrix} \begin{pmatrix} 1 \\ 1 \\ 0 \end{pmatrix}$$

$$= \begin{pmatrix} 1 & 1 & 0 \end{pmatrix} \begin{pmatrix} M_{11} + M_{12} \\ M_{21} + M_{22} \\ M_{31} + M_{32} \end{pmatrix} = M_{11} + M_{22} + M_{12} + M_{21}. \tag{13.9}$$

Since the diagonal elements vanish we have $M_{12} = -M_{21}$. With

$$\rho = \begin{pmatrix} 1 \\ 0 \\ 1 \end{pmatrix}, \qquad \rho = \begin{pmatrix} 0 \\ 1 \\ 1 \end{pmatrix}$$

we find $M_{13} = -M_{31}$ and $M_{23} = -M_{32}$, hence M is skew symmetric and has three independent components

$$M = -M^T = \begin{pmatrix} 0 & M_{12} & M_{13} \\ -M_{12} & 0 & M_{23} \\ -M_{13} & -M_{23} & 0 \end{pmatrix}. \tag{13.10}$$

Inserting (13.6) we have

$$\left(A^T A - 1\right) = -\left(A^T A - 1\right)^T = -\left(A^T A - 1\right) \tag{13.11}$$

which shows that $A^T A = 1$ or equivalently $A^T = A^{-1}$. Hence $(\det(A))^2 = 1$ and A is an orthogonal matrix. For a pure rotation without reflection only $\det(A) = +1$ is possible.

From

$$\mathbf{r}_i = \mathbf{R} + A\rho_{ib} \tag{13.12}$$

we calculate the velocity

$$\frac{\mathrm{d}\mathbf{r}_i}{\mathrm{d}t} = \frac{\mathrm{d}\mathbf{R}}{\mathrm{d}t} + \frac{\mathrm{d}A}{\mathrm{d}t}\rho_{ib} + A\frac{\mathrm{d}\rho_{ib}}{\mathrm{d}t} \tag{13.13}$$

but since ρ_{ib} is constant by definition, the last summand vanishes

$$\dot{\mathbf{r}}_i = \dot{\mathbf{R}} + \dot{A}\rho_{ib} = \dot{\mathbf{R}} + \dot{A}A^{-1}\rho_i \tag{13.14}$$

and in the center of mass system we have

$$\frac{\mathrm{d}}{\mathrm{d}t}\rho_i = \dot{A}A^{-1}\rho_i = W\rho_i \tag{13.15}$$

Fig. 13.2 Infinitesimal rotation

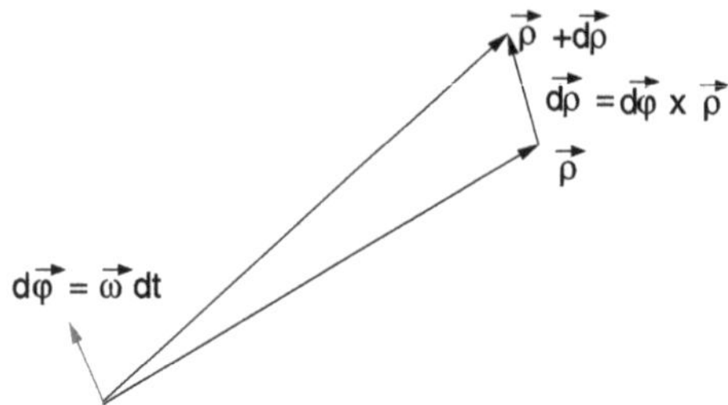

with the matrix

$$W = \dot{A}A^{-1}. \tag{13.16}$$

13.3 Properties of W, Connection with the Vector of Angular Velocity

Since rotation does not change the length of ρ_i, we have

$$0 = \frac{\mathrm{d}}{\mathrm{d}t}|\rho_i|^2 \to 0 = \rho_i \frac{\mathrm{d}}{\mathrm{d}t}\rho_i = \rho_i(W\rho_i) \tag{13.17}$$

or in matrix notation

$$0 = \rho_i^T W \rho_i. \tag{13.18}$$

This holds for arbitrary ρ_i. Hence W is skew symmetric and has three independent components

$$W = \begin{pmatrix} 0 & W_{12} & W_{13} \\ -W_{12} & 0 & W_{23} \\ -W_{13} & -W_{23} & 0 \end{pmatrix}. \tag{13.19}$$

Now consider an infinitesimal rotation by the angle $d\varphi$ (Fig. 13.2).

Then we have (the index i is suppressed)

$$d\rho = \frac{\mathrm{d}\rho}{\mathrm{d}t}dt = \begin{pmatrix} 0 & W_{12} & W_{13} \\ -W_{12} & 0 & W_{23} \\ -W_{13} & -W_{23} & 0 \end{pmatrix}\begin{pmatrix} \rho_1 \\ \rho_2 \\ \rho_3 \end{pmatrix}dt = \begin{pmatrix} W_{12}\rho_2 + W_{13}\rho_3 \\ -W_{12}\rho_1 + W_{23}\rho_3 \\ -W_{13}\rho_1 - W_{23}\rho_2 \end{pmatrix}dt \tag{13.20}$$

which can be written as a cross product:

$$d\rho = d\varphi \times \rho \tag{13.21}$$

with

$$d\boldsymbol{\varphi} = \begin{pmatrix} -W_{23}dt \\ W_{13}dt \\ -W_{12}dt \end{pmatrix}. \tag{13.22}$$

But this can be expressed in terms of the angular velocity $\boldsymbol{\omega}$ as

$$d\boldsymbol{\varphi} = \boldsymbol{\omega}dt \tag{13.23}$$

and finally we have

$$d\boldsymbol{\varphi} = \boldsymbol{\omega}dt = \begin{pmatrix} \omega_1 \\ \omega_2 \\ \omega_3 \end{pmatrix} dt \quad W = \begin{pmatrix} 0 & -\omega_3 & \omega_2 \\ \omega_3 & 0 & -\omega_1 \\ -\omega_2 & \omega_1 & 0 \end{pmatrix} \tag{13.24}$$

and the more common form of the equation of motion

$$\frac{d}{dt}\boldsymbol{\rho} = W\boldsymbol{\rho} = \boldsymbol{\omega} \times \boldsymbol{\rho}. \tag{13.25}$$

Example (Rotation around the z-axis) For constant angular velocity $\boldsymbol{\omega}$ the equation of motion

$$\frac{d}{dt}\boldsymbol{\rho} = W\boldsymbol{\rho} \tag{13.26}$$

has the formal solution

$$\boldsymbol{\rho} = e^{Wt}\boldsymbol{\rho}(0) = A(t)\boldsymbol{\rho}(0). \tag{13.27}$$

The angular velocity vector for rotation around the z-axis is

$$\boldsymbol{\omega} = \begin{pmatrix} 0 \\ 0 \\ \omega_3 \end{pmatrix} \tag{13.28}$$

and

$$W = \begin{pmatrix} 0 & -\omega_3 & 0 \\ \omega_3 & 0 & 0 \\ 0 & 0 & 0 \end{pmatrix}. \tag{13.29}$$

Higher powers of W can be easily calculated since

$$W^2 = \begin{pmatrix} -\omega_3^2 & 0 & 0 \\ 0 & -\omega_3^2 & 0 \\ 0 & 0 & 0 \end{pmatrix} \tag{13.30}$$

$$W^3 = -\omega_3^2 \begin{pmatrix} 0 & -\omega_3 & 0 \\ \omega_3 & 0 & 0 \\ 0 & 0 & 0 \end{pmatrix} \tag{13.31}$$

etc., and the rotation matrix is obtained from the Taylor series

$$
A(t) = e^{Wt} = 1 + Wt + \frac{1}{2}W^2 t^2 + \frac{1}{6}W^3 t^3 + \cdots
$$

$$
= 1 + \begin{pmatrix} \omega_3^2 t^2 & 0 & 0 \\ 0 & \omega_3^2 t^2 & 0 \\ 0 & 0 & 0 \end{pmatrix} \left(-\frac{1}{2} + \frac{\omega_3^2 t^2}{24} + \cdots \right)
$$

$$
+ \begin{pmatrix} 0 & -\omega_3 t & 0 \\ \omega_3 t & 0 & 0 \\ 0 & 0 & 0 \end{pmatrix} \left(1 - \frac{\omega_3^2 t^2}{6} + \cdots \right)
$$

$$
= \begin{pmatrix} \cos(\omega_3 t) & -\sin(\omega_3 t) & \\ \sin(\omega_3 t) & \cos(\omega_3 t) & \\ & & 1 \end{pmatrix}.
\tag{13.32}
$$

13.4 Transformation Properties of the Angular Velocity

Now imagine we are sitting on the rigid body and observe a mass point moving outside. Its position in the laboratory system is $\mathbf{r}_1$. In the body fixed system we observe it at

$$
\boldsymbol{\rho}_{1b} = A^{-1}(\mathbf{r}_1 - \mathbf{R})
\tag{13.33}
$$

and its velocity in the body fixed system is

$$
\dot{\boldsymbol{\rho}}_{1b} = A^{-1}(\dot{\mathbf{r}}_1 - \dot{\mathbf{R}}) + \frac{\mathrm{d}A^{-1}}{\mathrm{d}t}(\mathbf{r}_1 - \mathbf{R}).
\tag{13.34}
$$

The time derivative of the inverse matrix follows from

$$
0 = \frac{\mathrm{d}}{\mathrm{d}t}\left(A^{-1}A \right) = A^{-1}\dot{A} + \frac{\mathrm{d}A^{-1}}{\mathrm{d}t}A
\tag{13.35}
$$

$$
\frac{\mathrm{d}A^{-1}}{\mathrm{d}t} = -A^{-1}\dot{A}A^{-1} = -A^{-1}W
\tag{13.36}
$$

and hence

$$
\frac{\mathrm{d}A^{-1}}{\mathrm{d}t}(\mathbf{r}_1 - \mathbf{R}) = -A^{-1}W(\mathbf{r}_1 - \mathbf{R}).
\tag{13.37}
$$

Now we rewrite this using the angular velocity as it is observed in the body fixed system

$$
-A^{-1}W(\mathbf{r}_1 - \mathbf{R}) = -W_b A^{-1}(\mathbf{r}_1 - \mathbf{R}) = -W_b \boldsymbol{\rho}_{1b} = -\boldsymbol{\omega}_b \times \boldsymbol{\rho}_{1b}
\tag{13.38}
$$

where W transforms as like a rank-2 tensor

$$W_b = A^{-1} W A. \tag{13.39}$$

From this equation the transformation properties of $\boldsymbol{\omega}$ can be derived. We consider only rotation around one axis explicitly, since a general rotation matrix can always be written as a product of three rotations around different axes. For instance, rotation around the z-axis gives:

$$
\begin{aligned}
W_b &= \begin{pmatrix} 0 & -\omega_{b3} & \omega_{b2} \\ \omega_{b3} & 0 & -\omega_{b1} \\ -\omega_{b2} & \omega_{b1} & 0 \end{pmatrix} \\[1em]
&= \begin{pmatrix} \cos\varphi & \sin\varphi & 0 \\ -\sin\varphi & \cos\varphi & 0 \\ 0 & 0 & 1 \end{pmatrix} \begin{pmatrix} 0 & -\omega_3 & \omega_2 \\ \omega_3 & 0 & -\omega_1 \\ -\omega_2 & \omega_1 & 0 \end{pmatrix} \begin{pmatrix} \cos\varphi & -\sin\varphi & 0 \\ \sin\varphi & \cos\varphi & 0 \\ 0 & 0 & 1 \end{pmatrix} \\[1em]
&= \begin{pmatrix} 0 & -\omega_3 & \omega_2\cos\varphi - \omega_1\sin\varphi \\ \omega_3 & 0 & -(\omega_1\cos\varphi + \omega_2\sin\varphi) \\ -(\omega_2\cos\varphi - \omega_1\sin\varphi) & \omega_1\cos\varphi + \omega_2\sin\varphi & 0 \end{pmatrix}
\end{aligned}
\tag{13.40}
$$

which shows that

$$
\begin{pmatrix} \omega_{1b} \\ \omega_{2b} \\ \omega_{3b} \end{pmatrix} = \begin{pmatrix} \cos\varphi & \sin\varphi & 0 \\ -\sin\varphi & \cos\varphi & 0 \\ 0 & 0 & 1 \end{pmatrix} \begin{pmatrix} \omega_1 \\ \omega_2 \\ \omega_3 \end{pmatrix} = A^{-1}\boldsymbol{\omega}
\tag{13.41}
$$

i.e. $\boldsymbol{\omega}$ transforms like a vector under rotations. However, there is a subtle difference considering general coordinate transformations involving reflections. For example, under reflection at the xy-plane W is transformed according to

$$
\begin{aligned}
W_b &= \begin{pmatrix} 1 & 0 & 0 \\ 0 & 1 & 0 \\ 0 & 0 & -1 \end{pmatrix} \begin{pmatrix} 0 & -\omega_3 & \omega_2 \\ \omega_3 & 0 & -\omega_1 \\ -\omega_2 & \omega_1 & 0 \end{pmatrix} \begin{pmatrix} 1 & 0 & 0 \\ 0 & 1 & 0 \\ 0 & 0 & -1 \end{pmatrix} \\[1em]
&= \begin{pmatrix} 0 & -\omega_3 & -\omega_2 \\ \omega_3 & 0 & \omega_1 \\ \omega_2 & -\omega_1 & 0 \end{pmatrix}
\end{aligned}
\tag{13.42}
$$

and the transformed angular velocity vector is

$$
\begin{pmatrix} \omega_{1b} \\ \omega_{2b} \\ \omega_{3b} \end{pmatrix} = - \begin{pmatrix} 1 & 0 & 0 \\ 0 & 1 & 0 \\ 0 & 0 & -1 \end{pmatrix} \begin{pmatrix} \omega_1 \\ \omega_2 \\ \omega_3 \end{pmatrix}.
\tag{13.43}
$$

This is characteristic of a so called axial or pseudo-vector. Under a general coordinate transformation it transforms as

$$\boldsymbol{\omega}_b = \det(A) A \boldsymbol{\omega}. \tag{13.44}$$

13.5 Momentum and Angular Momentum

The total momentum is

$$\mathbf{P} = \sum_i m_i \dot{\mathbf{r}}_i = \sum_i m_i \dot{\mathbf{R}} = M\dot{\mathbf{R}} \tag{13.45}$$

since by definition we have $\sum_i m_i \boldsymbol{\rho}_i = 0$.

The total angular momentum can be decomposed into the contribution of the center of mass motion and the contribution relative to the center of mass

$$\mathbf{L} = \sum_i m_i \mathbf{r}_i \times \dot{\mathbf{r}}_i = M\mathbf{R} \times \dot{\mathbf{R}} + \sum_i m_i \boldsymbol{\rho}_i \times \dot{\boldsymbol{\rho}}_i = \mathbf{L}_{COM} + \mathbf{L}_{int}. \tag{13.46}$$

The second contribution is

$$\mathbf{L}_{int} = \sum_i m_i \boldsymbol{\rho}_i \times (\boldsymbol{\omega} \times \boldsymbol{\rho}_i) = \sum_i m_i \left(\boldsymbol{\omega}\rho_i^2 - \boldsymbol{\rho}_i(\boldsymbol{\rho}_i\boldsymbol{\omega})\right). \tag{13.47}$$

This is a linear vector function of $\boldsymbol{\omega}$, which can be expressed simpler by introducing the tensor of inertia

$$I = \sum_i m_i \rho_i^2 \mathbf{1} - m_i \boldsymbol{\rho}_i \boldsymbol{\rho}_i^T \tag{13.48}$$

or component-wise

$$I_{m,n} = \sum_i m_i \rho_i^2 \delta_{m,n} - m_i \rho_{i,m} \rho_{i,n} \tag{13.49}$$

as

$$\mathbf{L}_{int} = I\boldsymbol{\omega}. \tag{13.50}$$

13.6 Equations of Motion of a Rigid Body

Let $\mathbf{F}_i$ be an external force acting on m_i. Then the equation of motion for the center of mass is

$$\frac{d^2}{dt^2} \sum_i m_i \mathbf{r}_i = M\ddot{\mathbf{R}} = \sum_i \mathbf{F}_i = \mathbf{F}_{ext}. \tag{13.51}$$

If there is no total external force $\mathbf{F}_{ext}$, the center of mass moves with constant velocity

$$\mathbf{R} = \mathbf{R}_0 + \mathbf{V}(t - t_0). \tag{13.52}$$

The time derivative of the angular momentum equals the total external torque

$$\frac{d}{dt}\mathbf{L} = \frac{d}{dt} \sum_i m_i \mathbf{r}_i \times \dot{\mathbf{r}}_i = \sum_i m_i \mathbf{r}_i \times \ddot{\mathbf{r}}_i = \sum_i \mathbf{r}_i \times \mathbf{F}_i = \sum_i \mathbf{N}_i = \mathbf{N}_{ext} \tag{13.53}$$

which can be decomposed into

$$\mathbf{N}_{ext} = \mathbf{R} \times \mathbf{F}_{ext} + \sum_i \boldsymbol{\rho}_i \times \mathbf{F}_i. \tag{13.54}$$

With the decomposition of the angular momentum

$$\frac{d}{dt}\mathbf{L} = \frac{d}{dt}\mathbf{L}_{COM} + \frac{d}{dt}\mathbf{L}_{int} \tag{13.55}$$

we have two separate equations for the two contributions:

$$\frac{d}{dt}\mathbf{L}_{COM} = \frac{d}{dt}M\mathbf{R} \times \dot{\mathbf{R}} = M\mathbf{R} \times \ddot{\mathbf{R}} = \mathbf{R} \times \mathbf{F}_{ext} \tag{13.56}$$

$$\frac{d}{dt}\mathbf{L}_{int} = \sum_i \boldsymbol{\rho}_i \times \mathbf{F}_i = \mathbf{N}_{ext} - \mathbf{R} \times \mathbf{F}_{ext} = \mathbf{N}_{int}. \tag{13.57}$$

13.7 Moments of Inertia

The angular momentum (13.50) is

$$\mathbf{L}_{rot} = I\boldsymbol{\omega} = AA^{-1}IAA^{-1}\boldsymbol{\omega} = AI_b\boldsymbol{\omega}_b \tag{13.58}$$

where the tensor of inertia in the body fixed system is

$$I_b = A^{-1}IA = A^{-1}\left(\sum_i m_i\boldsymbol{\rho}_i^T\rho_i - m_i\boldsymbol{\rho}_i\boldsymbol{\rho}_i^T\right)A$$

$$= \sum_i m_iA^T\boldsymbol{\rho}_i^T\rho_iA - m_iA^T\boldsymbol{\rho}_i\boldsymbol{\rho}_i^TA$$

$$= \sum_i m_i\rho_{ib}^2 - m_i\boldsymbol{\rho}_{ib}\boldsymbol{\rho}_{ib}^T. \tag{13.59}$$

Since I_b does not depend on time (by definition of the body fixed system) we will use the principal axes of I_b as the axes of the body fixed system. Then I_b takes the simple form

$$I_b = \begin{pmatrix} I_1 & 0 & 0 \\ 0 & I_2 & 0 \\ 0 & 0 & I_3 \end{pmatrix} \tag{13.60}$$

with the principle moments of inertia $I_{1,2,3}$.

13.8 Equations of Motion for a Rotor

The following equations describe pure rotation of a rigid body:

$$\frac{d}{dt}A = WA = AW_b \tag{13.61}$$

$$\frac{d}{dt}\mathbf{L}_{int} = \mathbf{N}_{int} \tag{13.62}$$

$$W = \begin{pmatrix} 0 & -\omega_3 & \omega_2 \\ \omega_3 & 0 & -\omega_1 \\ -\omega_2 & \omega_1 & 0 \end{pmatrix} \qquad W_{ij} = -\varepsilon_{ijk}\omega_k \tag{13.63}$$

$$\mathbf{L}_{int} = A\mathbf{L}_{int,b} = I\omega = AI_b\omega_b \tag{13.64}$$

$$\omega_b = I_b^{-1}\mathbf{L}_{int,b} = \begin{pmatrix} I_1^{-1} & 0 & 0 \\ 0 & I_2^{-1} & 0 \\ 0 & 0 & I_3^{-1} \end{pmatrix}\mathbf{L}_{int,b} \quad \omega = A\omega_b \tag{13.65}$$

$$I_b = \text{const.} \tag{13.66}$$

13.9 Explicit Methods

Equation (13.61) for the rotation matrix and (13.62) for the angular momentum have to be solved by a suitable algorithm. The simplest integrator is the explicit Euler method (Fig. 13.3) [241]:

$$A(t + \Delta t) = A(t) + A(t)W_b(t)\Delta t + O(\Delta t^2) \tag{13.67}$$

$$\mathbf{L}_{int}(t + \Delta t) = \mathbf{L}_{int}(t) + \mathbf{N}_{int}(t)\Delta t + O(\Delta t^2). \tag{13.68}$$

Expanding the Taylor series of $A(t)$ to second order we have the second order approximation (Fig. 13.3)

$$A(t + \Delta t) = A(t) + A(t)W_b(t)\Delta t + \frac{1}{2}\left(A(t)W_b^2(t) + A(t)\dot{W}_b(t)\right)\Delta t^2 + O(\Delta t^3). \tag{13.69}$$

A corresponding second order expression for the angular momentum involves the time derivative of the forces and is usually not practicable.

The time derivative of W can be expressed via the time derivative of the angular velocity which can be calculated as follows:

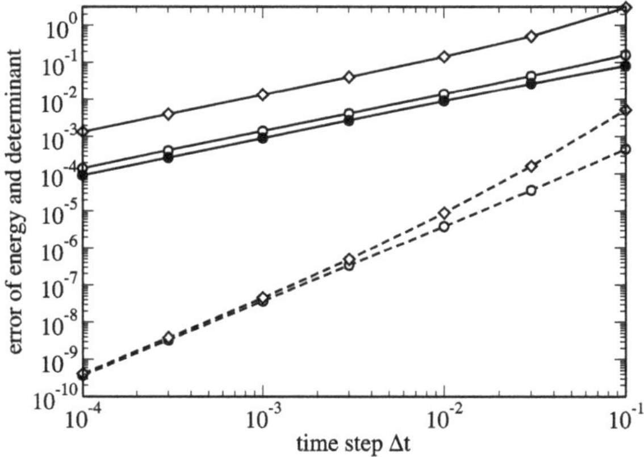

Fig. 13.3 (Global error of the explicit methods) The equations of a free rotor (13.8) are solved using the explicit first order (*full curves*) and second order (*dashed curves*) method. The deviations $|\det(A) - 1|$ (*diamonds*) and $|E_{kin} - E_{kin}(0)|$ (*circles*) at $t = 10$ are shown as a function of the time step Δt. *Full circles* show the energy deviation of the first order method with reorthogonalization. The principal moments of inertia are $I_b = \mathrm{diag}(1, 2, 3)$ and the initial angular momentum is $\mathbf{L} = (1, 1, 1)$. See also Problem 13.1

$$\frac{\mathrm{d}}{\mathrm{d}t}\boldsymbol{\omega}_b = \frac{\mathrm{d}}{\mathrm{d}t}\left(I_b^{-1}A^{-1}\mathbf{L}_{int}\right) = I_b^{-1}\left(\frac{\mathrm{d}}{\mathrm{d}t}A^{-1}\right)\mathbf{L}_{int} + I_b^{-1}A^{-1}\mathbf{N}_{int}$$

$$= I_b^{-1}\left(-A^{-1}W\right)\mathbf{L}_{int} + I_b^{-1}A^{-1}\mathbf{N}_{int}$$

$$= -I_b^{-1}W_b\mathbf{L}_{int,b} + I_b^{-1}\mathbf{N}_{int,b}. \tag{13.70}$$

Alternatively, in the laboratory system

$$\frac{\mathrm{d}}{\mathrm{d}t}\boldsymbol{\omega} = \frac{\mathrm{d}}{\mathrm{d}t}(A\boldsymbol{\omega}_b) = WA\boldsymbol{\omega}_b - AI_b^{-1}A^{-1}W\mathbf{L}_{int} + AI_b^{-1}A^{-1}\mathbf{N}_{int}$$

$$= AI_b^{-1}A(\mathbf{N}_{int} - W\mathbf{L}_{int}) \tag{13.71}$$

where the first summand vanishes due to

$$WA\boldsymbol{\omega}_b = AW_b\boldsymbol{\omega}_b = A\boldsymbol{\omega}_b \times \boldsymbol{\omega}_b = 0. \tag{13.72}$$

Substituting the angular momentum we have

$$\frac{\mathrm{d}}{\mathrm{d}t}\boldsymbol{\omega}_b = I_b^{-1}\mathbf{N}_{int,b} - I_b^{-1}W_bI_b\boldsymbol{\omega}_b \tag{13.73}$$

which reads in components:

$$\begin{pmatrix} \dot{\omega}_{b1} \\ \dot{\omega}_{b2} \\ \dot{\omega}_{b3} \end{pmatrix} = \begin{pmatrix} I_{b1}^{-1} N_{b1} \\ I_{b2}^{-1} N_{b2} \\ I_{b3}^{-1} N_{b3} \end{pmatrix}$$

$$- \begin{pmatrix} I_{b1}^{-1} & & \\ & I_{b2}^{-1} & \\ & & I_{b3}^{-1} \end{pmatrix} \begin{pmatrix} 0 & -\omega_{b3} & \omega_{b2} \\ \omega_{b3} & 0 & -\omega_{b1} \\ -\omega_{b2} & \omega_{b1} & 0 \end{pmatrix} \begin{pmatrix} I_{b1}\omega_{b1} \\ I_{b2}\omega_{b2} \\ I_{b3}\omega_{b3} \end{pmatrix}.$$

$$(13.74)$$

Evaluation of the product gives a set of equations which are well known as Euler's equations:

$$\dot{\omega}_{b1} = \frac{I_{b2} - I_{b3}}{I_{b1}} \omega_{b2}\omega_{b3} + \frac{N_{b1}}{I_{b1}}$$

$$\dot{\omega}_{b2} = \frac{I_{b3} - I_{b1}}{I_{b2}} \omega_{b3}\omega_{b1} + \frac{N_{b2}}{I_{b2}} \qquad (13.75)$$

$$\dot{\omega}_{b3} = \frac{I_{b1} - I_{b2}}{I_{b3}} \omega_{b1}\omega_{b2} + \frac{N_{b3}}{I_{b3}}.$$

13.10 Loss of Orthogonality

The simple methods above do not conserve the orthogonality of A. This is an effect of higher order but the error can accumulate quickly. Consider the determinant of A. For the simple explicit Euler scheme we have

$$\det(A + \Delta A) = \det(A + W A \Delta t) = \det A \det(1 + W \Delta t)$$

$$= \det A \left(1 + \omega^2 \Delta t^2\right). \qquad (13.76)$$

The error is of order Δt^2, but the determinant will continuously increase, i.e. the rigid body will explode. For the second order integrator we find

$$\det(A + \Delta A) = \det\left(A + W A \Delta t + \frac{\Delta t^2}{2} \left(W^2 A + \dot{W} A\right) \right)$$

$$= \det A \det\left(1 + W \Delta t + \frac{\Delta t^2}{2} \left(W^2 + \dot{W}\right) \right). \qquad (13.77)$$

This can be simplified to give

$$\det(A + \Delta A) = \det A \left(1 + \dot{\omega}\omega \Delta t^3 + \cdots \right). \qquad (13.78)$$

The second order method behaves somewhat better since the product of angular velocity and acceleration can change in time. To assure that A remains a rotation matrix we must introduce constraints or reorthogonalize A at least after some steps (for instance every time when $|\det(A) - 1|$ gets larger than a certain threshold). The following method with a symmetric correction matrix is a very useful alternative [127]. The non-singular square matrix A can be decomposed into the product of an orthonormal matrix $\widetilde{A}$ and a positive semi-definite matrix S

$$A = \widetilde{A}S \tag{13.79}$$

with the positive definite square root of the symmetric matrix $A^T A$

$$S = \left(A^T A\right)^{1/2} \tag{13.80}$$

and

$$\widetilde{A} = AS^{-1} = A\left(A^T A\right)^{-1/2} \tag{13.81}$$

which is orthonormal as can be seen from

$$\widetilde{A}^T \widetilde{A} = \left(S^{-1}\right)^T A^T A S^{-1} = S^{-1}S^2 S^{-1} = 1. \tag{13.82}$$

Since the deviation of A from orthogonality is small, we make the approximations

$$S = 1 + s \tag{13.83}$$

$$A^T A = S^2 \approx 1 + 2s \tag{13.84}$$

$$s \approx \frac{A^T A - 1}{2} \tag{13.85}$$

$$S^{-1} \approx 1 - s \approx 1 + \frac{1 - A^T A}{2} + \cdots \tag{13.86}$$

which can be easily evaluated.

13.11 Implicit Method

The quality of the method can be significantly improved by taking the time derivative at midstep (Fig. 13.4) (12.5):

$$A(t + \Delta t) = A(t) + A\left(t + \frac{\Delta t}{2}\right)W\left(t + \frac{\Delta t}{2}\right)\Delta t + \cdots \tag{13.87}$$

$$\mathbf{L}_{int}(t + \Delta t) = \mathbf{L}_{int}(t) + \mathbf{N}_{int}\left(t + \frac{\Delta t}{2}\right)\Delta t + \cdots . \tag{13.88}$$

Taylor series expansion gives

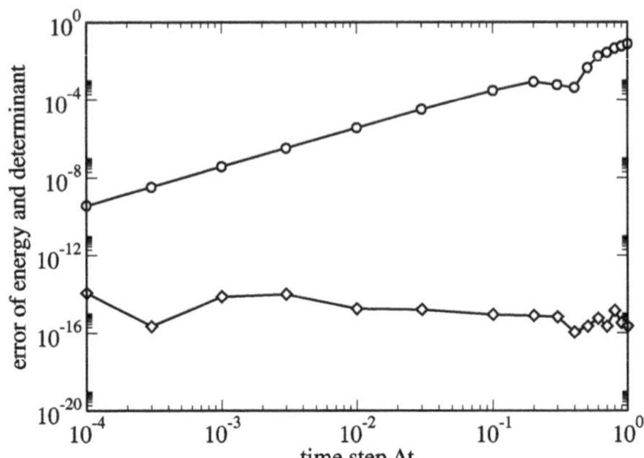

Fig. 13.4 (Global error of the implicit method) The equations of a free rotor (13.8) are solved using the implicit method. The deviations $|\det(A) - 1|$ (*diamonds*) and $|E_{kin} - E_{kin}(0)|$ (*circles*) at $t = 10$ are shown as a function of the time step Δt. Initial conditions as in Fig. 13.3. See also Problem 13.1

$$A\left(t + \frac{\Delta t}{2}\right) W\left(t + \frac{\Delta t}{2}\right) \Delta t$$

$$= A(t)W(t)\Delta t + \dot{A}(t)W(t)\frac{\Delta t^2}{2} + A(t)\dot{W}(t)\frac{\Delta t^2}{2} + O\left(\Delta t^3\right) \quad (13.89)$$

$$= A(t)W(t)\Delta t + \left(A(t)W^2(t) + A(t)\dot{W}(t)\right)\frac{\Delta t^2}{2} + O\left(\Delta t^3\right) \quad (13.90)$$

which has the same error order as the explicit second order method. The matrix $A(t + \frac{\Delta t}{2})$ at mid-time can be approximated by

$$\frac{1}{2}\left(A(t) + A(t + \Delta t)\right) = A\left(t + \frac{\Delta t}{2}\right) + \frac{\Delta t^2}{4}\ddot{A}\left(t + \frac{\Delta t}{2}\right) + \cdots$$

$$= A\left(t + \frac{\Delta t}{2}\right) + O\left(\Delta t^2\right) \quad (13.91)$$

which does not change the error order of the implicit integrator which now becomes

$$A(t + \Delta t) = A(t) + \frac{1}{2}\left(A(t) + A(t + \Delta t)\right)W\left(t + \frac{\Delta t}{2}\right)\Delta t + O\left(\Delta t^3\right). \quad (13.92)$$

This equation can be formally solved by

$$A(t + \Delta t) = A(t)\left(1 + \frac{\Delta t}{2}W\left(t + \frac{\Delta t}{2}\right)\right)\left(1 - \frac{\Delta t}{2}W\left(t + \frac{\Delta t}{2}\right)\right)^{-1}$$

$$= A(t)T_b\left(\frac{\Delta t}{2}\right). \quad (13.93)$$

Alternatively, using angular velocities in the laboratory system we have the similar expression

$$A(t + \Delta t) = \left[1 - \frac{\Delta t}{2} W\left(t + \frac{\Delta t}{2}\right)\right]^{-1} \left[1 + \frac{\Delta t}{2} W\left(t + \frac{\Delta t}{2}\right)\right] A(t)$$

$$= T\left(\frac{\Delta t}{2}\right) A(t). \tag{13.94}$$

The angular velocities at midtime can be calculated with sufficient accuracy from

$$W\left(t + \frac{\Delta t}{2}\right) = W(t) + \frac{\Delta t}{2}\dot{W}(t) + O\left(\Delta t^2\right). \tag{13.95}$$

With the help of an algebra program we easily prove that

$$\det\left(1 + \frac{\Delta t}{2} W\right) = \det\left(1 - \frac{\Delta t}{2} W\right) = 1 + \frac{\omega^2 \Delta t^2}{4} \tag{13.96}$$

and therefore the determinant of the rotation matrix is conserved. The necessary matrix inversion can be easily done:

$$\left[1 - \frac{\Delta t}{2} W\right]^{-1}$$

$$= \begin{pmatrix} 1 + \frac{\omega_1^2 \Delta t^2}{4} & -\omega_3 \frac{\Delta t}{2} + \omega_1\omega_2 \frac{\Delta t^2}{4} & \omega_2 \frac{\Delta t}{2} + \omega_1\omega_3 \frac{\Delta t^2}{4} \\ \omega_3 \frac{\Delta t}{2} + \omega_1\omega_2 \frac{\Delta t^2}{4} & 1 + \frac{\omega_2^2 \Delta t^2}{4} & -\omega_1 \frac{\Delta t}{2} + \omega_2\omega_3 \frac{\Delta t^2}{4} \\ -\omega_2 \frac{\Delta t}{2} + \omega_1\omega_3 \frac{\Delta t^2}{4} & \omega_1 \frac{\Delta t}{2} + \omega_2\omega_3 \frac{\Delta t^2}{4} & 1 + \frac{\omega_3^2 \Delta t^2}{4} \end{pmatrix}$$

$$\times \frac{1}{1 + \omega^2 \frac{\Delta t^2}{4}}. \tag{13.97}$$

The matrix product is explicitly

$$T_b = \left[1 + \frac{\Delta t}{2} W_b\right]\left[1 - \frac{\Delta t}{2} W_b\right]^{-1}$$

$$= \begin{pmatrix} 1 + \frac{\omega_{b1}^2 - \omega_{b2}^2 - \omega_{b3}^2}{4} \Delta t^2 & -\omega_{b3} \Delta t + \omega_{b1}\omega_{b2} \frac{\Delta t^2}{2} & \omega_{b2} \Delta t + \omega_{b1}\omega_{b3} \frac{\Delta t^2}{2} \\ \omega_{b3} \Delta t + \omega_{b1}\omega_{b2} \frac{\Delta t^2}{2} & 1 + \frac{-\omega_{b1}^2 + \omega_{b2}^2 - \omega_{b3}^2}{4} \Delta t^2 & -\omega_{b1} \Delta t + \omega_{b2}\omega_{b3} \frac{\Delta t^2}{2} \\ -\omega_{b2} \Delta t + \omega_{b1}\omega_{b3} \frac{\Delta t^2}{2} & \omega_{b1} \Delta t + \omega_{b2}\omega_{b3} \frac{\Delta t^2}{2} & 1 + \frac{-\omega_{b1}^2 - \omega_{b2}^2 + \omega_{b3}^2}{4} \Delta t^2 \end{pmatrix}$$

$$\times \frac{1}{1 + \omega_b^2 \frac{\Delta t^2}{4}}. \tag{13.98}$$

With the help of an algebra program it can be proved that this matrix is even orthogonal

$$T_b^T T_b = 1 \tag{13.99}$$

and hence the orthonormality of A is conserved. The approximation for the angular momentum

$$\mathbf{L}_{int}(t) + \mathbf{N}_{int}\left(t + \frac{\Delta t}{2}\right)\Delta t$$

$$= \mathbf{L}_{int}(t) + \mathbf{N}_{int}(t)\Delta t + \dot{\mathbf{N}}_{int}(t)\frac{\Delta t^2}{2} + \cdots = \mathbf{L}_{int}(t + \Delta t) + O\left(\Delta t^3\right) \tag{13.100}$$

can be used in an implicit way

$$\mathbf{L}_{int}(t + \Delta t) = \mathbf{L}_{int}(t) + \frac{\mathbf{N}_{int}(t + \Delta t) + \mathbf{N}_{int}(t)}{2}\Delta t + O\left(\Delta t^3\right). \tag{13.101}$$

Alternatively Euler's equations can be used in the form [190, 191]

$$\omega_{b1}\left(t + \frac{\Delta t}{2}\right) = \omega_{b1}\left(t - \frac{\Delta t}{2}\right) + \frac{I_{b2} - I_{b3}}{I_{b1}}\omega_{b2}(t)\omega_{b3}(t)\Delta t + \frac{N_{b1}}{I_{b1}}\Delta t \text{ etc.} \tag{13.102}$$

where the product $\omega_{b2}(t)\omega_{b3}(t)$ is approximated by

$$\omega_{b2}(t)\omega_{b3}(t) = \frac{1}{2}\left[\omega_{b2}\left(t - \frac{\Delta t}{2}\right)\omega_{b3}\left(t - \frac{\Delta t}{2}\right) + \omega_{b2}\left(t + \frac{\Delta t}{2}\right)\omega_{b3}\left(t + \frac{\Delta t}{2}\right)\right]. \tag{13.103}$$

$\omega_{b1}(t + \frac{\Delta t}{2})$ is determined by iterative solution of the last two equations. Starting with $\omega_{b1}(t - \frac{\Delta t}{2})$ convergence is achieved after few iterations.

Example (Free symmetric rotor) For the special case of a free symmetric rotor ($I_{b2} = I_{b3}$, $\mathbf{N}_{int} = 0$) Euler's equations simplify to:

$$\dot{\omega}_{b1} = 0 \tag{13.104}$$

$$\dot{\omega}_{b2} = \frac{I_{b2(3)} - I_{b1}}{I_{b2(3)}}\omega_{b1}\omega_{b3} = \lambda\omega_{b3} \tag{13.105}$$

$$\dot{\omega}_{b3} = \frac{I_{b1} - I_{b2(3)}}{I_{b2(3)}}\omega_{b1}\omega_{b2} = -\lambda\omega_{b2} \tag{13.106}$$

$$\lambda = \frac{I_{b2(3)} - I_{b1}}{I_{b2(3)}}\omega_{b1}. \tag{13.107}$$

Coupled equations of this type appear often in physics. The solution can be found using a complex quantity

$$\Omega = \omega_{b2} + i\omega_{b3} \tag{13.108}$$

which obeys the simple differential equation

$$\dot{\Omega} = \dot{\omega}_{b2} + i\dot{\omega}_{b3} = -i(i\lambda\omega_{b3} + \lambda\omega_{b2}) = -i\lambda\Omega \qquad (13.109)$$

with the solution

$$\Omega = \Omega_0 e^{-i\lambda t}. \qquad (13.110)$$

Finally

$$\boldsymbol{\omega}_b = \begin{pmatrix} \omega_{b1}(0) \\ \Re(\Omega_0 e^{-i\lambda t}) \\ \Im(\Omega_0 e^{-i\lambda t}) \end{pmatrix} = \begin{pmatrix} \omega_{b1}(0) \\ \omega_{b2}(0)\cos(\lambda t) + \omega_{b3}(0)\sin(\lambda t) \\ \omega_{b3}(0)\cos(\lambda t) - \omega_{b2}(0)\sin(\lambda t) \end{pmatrix} \qquad (13.111)$$

i.e. $\boldsymbol{\omega}_b$ rotates around the 1-axis with frequency λ.

13.12 Kinetic Energy of a Rotor

The kinetic energy of the rotor is

$$E_{kin} = \sum_i \frac{m_i}{2}\dot{r}_i^2 = \sum_i \frac{m_i}{2}(\dot{\mathbf{R}} + \dot{A}\boldsymbol{\rho}_{ib})^2$$

$$= \sum_i \frac{m_i}{2}(\dot{R}^T + \boldsymbol{\rho}_{ib}^T\dot{A}^T)(\dot{R} + \dot{A}\boldsymbol{\rho}_{ib})$$

$$= \frac{M}{2}\dot{R}^2 + \sum_i \frac{m_i}{2}\boldsymbol{\rho}_{ib}^T\dot{A}^T\dot{A}\boldsymbol{\rho}_{ib}. \qquad (13.112)$$

The second part is the contribution of the rotational motion. It can be written as

$$E_{rot} = \sum_i \frac{m_i}{2}\boldsymbol{\rho}_{ib}^T W_b^T A^T A W_b\boldsymbol{\rho}_{ib} = -\sum_i \frac{m_i}{2}\boldsymbol{\rho}_{ib}^T W_b^2\boldsymbol{\rho}_{ib} = \frac{1}{2}\omega_b^T I_b\omega_b \qquad (13.113)$$

since

$$-W_b^2 = \begin{pmatrix} \omega_{b3}^2 + \omega_{b2}^2 & -\omega_{b1}\omega_{b2} & -\omega_{b1}\omega_{b3} \\ -\omega_{b1}\omega_{b2} & \omega_{b1}^2 + \omega_{b3}^2 & -\omega_{b2}\omega_{b3} \\ -\omega_{b1}\omega_{b3} & -\omega_{b2}\omega_{b3} & \omega_{b1}^2 + \omega_{b2}^2 \end{pmatrix} = \omega_b^2 - \boldsymbol{\omega}_b\boldsymbol{\omega}_b^T. \qquad (13.114)$$

13.13 Parametrization by Euler Angles

So far we had to solve equations for all 9 components of the rotation matrix. But there are six constraints since the column vectors of the matrix have to be orthonormalized. Therefore the matrix can be parametrized with less than 9 variables. In fact it is sufficient to use only three variables. This can be achieved by splitting the full rotation into three rotations around different axis. Most common are Euler angles defined by the orthogonal matrix [106]

$$\begin{pmatrix} \cos\psi\cos\phi - \cos\theta\sin\phi\sin\psi & -\sin\psi\cos\phi - \cos\theta\sin\phi\cos\psi & \sin\theta\sin\phi \\ \cos\psi\sin\phi + \cos\theta\cos\phi\sin\psi & -\sin\psi\sin\phi + \cos\theta\cos\phi\cos\psi & -\sin\theta\cos\phi \\ \sin\theta\sin\psi & \sin\theta\cos\psi & \cos\theta \end{pmatrix}$$

(13.115)

obeying the equations

$$\dot{\phi} = \omega_x \frac{\sin\phi\cos\theta}{\sin\theta} + \omega_y \frac{\cos\phi\cos\theta}{\sin\theta} + \omega_z \tag{13.116}$$

$$\dot{\theta} = \omega_x \cos\phi + \omega_y \sin\phi \tag{13.117}$$

$$\dot{\psi} = \omega_x \frac{\sin\phi}{\sin\theta} - \omega_y \frac{\cos\phi}{\sin\theta}. \tag{13.118}$$

Different versions of Euler angles can be found in the literature, together with the closely related cardanic angles. For all of them a $\sin\theta$ appears in the denominator which causes numerical instabilities at the poles. One possible solution to this problem is to switch between two different coordinate systems.

13.14 Cayley-Klein Parameters, Quaternions, Euler Parameters

There exists another parametrization of the rotation matrix which is very suitable for numerical calculations. It is connected with the algebra of the so called quaternions. The vector space of the complex 2×2 matrices can be spanned using Pauli matrices by

$$1 = \begin{pmatrix} 1 & 0 \\ 0 & 1 \end{pmatrix} \quad \sigma_x = \begin{pmatrix} 0 & 1 \\ 1 & 0 \end{pmatrix} \quad \sigma_y = \begin{pmatrix} 0 & -i \\ i & 0 \end{pmatrix} \quad \sigma_z = \begin{pmatrix} 1 & 0 \\ 0 & -1 \end{pmatrix}. \tag{13.119}$$

Any complex 2×2 matrix can be written as a linear combination

$$c_0 1 + \mathbf{c}\boldsymbol{\sigma}. \tag{13.120}$$

Accordingly any vector $\mathbf{x} \in R^3$ can be mapped onto a complex 2×2 matrix:

$$\mathbf{x} \to P = \begin{pmatrix} z & x - iy \\ x + iy & -z \end{pmatrix}. \tag{13.121}$$

Rotation of the coordinate system leads to the transformation

$$P' = QPQ^\dagger \tag{13.122}$$

where

$$Q = \begin{pmatrix} \alpha & \beta \\ \gamma & \delta \end{pmatrix} \tag{13.123}$$

is a complex 2×2 rotation matrix. Invariance of the length ($|\mathbf{x}| = \sqrt{-\det(P)}$) under rotation implies that Q must be unitary, i.e. $Q^\dagger = Q^{-1}$ and its determinant must be 1. Explicitly

$$Q^\dagger = \begin{pmatrix} \alpha^* & \gamma^* \\ \beta^* & \delta^* \end{pmatrix} = Q^{-1} = \frac{1}{\alpha\delta - \beta\gamma} \begin{pmatrix} \delta & -\beta \\ -\gamma & \alpha \end{pmatrix} \tag{13.124}$$

and Q has the form

$$Q = \begin{pmatrix} \alpha & \beta \\ -\beta^* & \alpha^* \end{pmatrix} \quad \text{with } |\alpha|^2 + |\beta|^2 = 1. \tag{13.125}$$

Setting $x_\pm = x \pm iy$, the transformed matrix has the same form as P:

$$QPQ^\dagger = \begin{pmatrix} \alpha^*\beta x_+ + \beta^*\alpha x_- + (|\alpha|^2 - |\beta|^2)z & -\beta^2 x_+ + \alpha^2 x_- - 2\alpha\beta z \\ \alpha^{*2}x_+ - \beta^{*2}x_- - 2\alpha^*\beta^* z & -\alpha^*\beta x_+ - \alpha\beta^* x_- - (|\alpha|^2 - |\beta|^2)z \end{pmatrix}$$

$$= \begin{pmatrix} z' & x'_- \\ x'_+ & -z' \end{pmatrix}. \tag{13.126}$$

From comparison we find the transformed vector components:

$$x' = \frac{1}{2}(x'_+ + x'_-) = \frac{1}{2}(\alpha^{*2} - \beta^2)x_+ + \frac{1}{2}(\alpha^2 - \beta^{*2})x_- - (\alpha\beta + \alpha^*\beta^*)z$$

$$= \frac{\alpha^{*2} + \alpha^2 - \beta^{*2} - \beta^2}{2}x + \frac{i(\alpha^{*2} - \alpha^2 + \beta^{*2} - \beta^2)}{2}y$$

$$- (\alpha\beta + \alpha^*\beta^*)z \tag{13.127}$$

$$y' = \frac{1}{2i}(x'_+ - x'_-) = \frac{1}{2i}(\alpha^{*2} + \beta^2)x_+ + \frac{1}{2i}(-\beta^{*2} - \alpha^2)x_- + \frac{1}{i}(-\alpha^*\beta^* + \alpha\beta)z$$

$$= \frac{\alpha^{*2} - \alpha^2 - \beta^{*2} + \beta^2}{2i}x + \frac{\alpha^{*2} + \alpha^2 + \beta^{*2} + \beta^2}{2}y$$

$$+ i(\alpha^*\beta^* - \alpha\beta)z \tag{13.128}$$

$$z' = (\alpha^*\beta + \alpha\beta^*)x + i(\alpha^*\beta - \alpha\beta^*)y + (|\alpha|^2 - |\beta|^2)z. \tag{13.129}$$

This gives us the rotation matrix in terms of the Cayley-Klein parameters α and β:

$$A = \begin{pmatrix} \dfrac{\alpha^{*2}+\alpha^2-\beta^{*2}-\beta^2}{2} & \dfrac{i(\alpha^{*2}-\alpha^2+\beta^{*2}-\beta^2)}{2} & -(\alpha\beta + \alpha^*\beta^*) \\ \dfrac{\alpha^{*2}-\alpha^2-\beta^{*2}+\beta^2}{2i} & \dfrac{\alpha^{*2}+\alpha^2+\beta^{*2}+\beta^2}{2} & \dfrac{1}{i}(-\alpha^*\beta^* + \alpha\beta) \\ (\alpha^*\beta + \alpha\beta^*) & i(\alpha^*\beta - \alpha\beta^*) & (|\alpha|^2 - |\beta|^2) \end{pmatrix}. \tag{13.130}$$

For practical calculations one often prefers to have four real parameters instead of two complex ones. The so called Euler parameters q_0, q_1, q_2, q_3 are defined by

$$\alpha = q_0 + iq_3 \qquad \beta = q_2 + iq_1. \tag{13.131}$$

Now the matrix Q

$$Q = \begin{pmatrix} q_0 + iq_3 & q_2 + iq_1 \\ -q_2 + iq_1 & q_0 - iq_3 \end{pmatrix} = q_0 1 + iq_1\sigma_x + iq_2\sigma_y + iq_3\sigma_z \tag{13.132}$$

becomes a so called quaternion which is a linear combination of the four matrices

$$U = 1 \qquad I = i\sigma_z \qquad J = i\sigma_y \qquad K = i\sigma_x \tag{13.133}$$

which obey the following multiplication rules:

$$I^2 = J^2 = K^2 = -U$$
$$IJ = -JI = K$$
$$JK = -KJ = I \tag{13.134}$$
$$KI = -IK = J.$$

In terms of Euler parameters the rotation matrix reads

$$A = \begin{pmatrix} q_0^2 + q_1^2 - q_2^2 - q_3^2 & 2(q_1q_2 + q_0q_3) & 2(q_1q_3 - q_0q_2) \\ 2(q_1q_2 - q_0q_3) & q_0^2 - q_1^2 + q_2^2 - q_3^2 & 2(q_2q_3 + q_0q_1) \\ 2(q_1q_3 + q_0q_2) & 2(q_2q_3 - q_0q_1) & q_0^2 - q_1^2 - q_2^2 + q_3^2 \end{pmatrix} \tag{13.135}$$

and from the equation $\dot{A} = WA$ we derive the equation of motion for the quaternion

$$\begin{pmatrix} \dot{q}_0 \\ \dot{q}_1 \\ \dot{q}_2 \\ \dot{q}_3 \end{pmatrix} = \frac{1}{2} \begin{pmatrix} 0 & \omega_1 & \omega_2 & \omega_3 \\ -\omega_1 & 0 & -\omega_3 & \omega_2 \\ -\omega_2 & \omega_3 & 0 & -\omega_1 \\ -\omega_3 & -\omega_2 & \omega_1 & 0 \end{pmatrix} \begin{pmatrix} q_0 \\ q_1 \\ q_2 \\ q_3 \end{pmatrix} \tag{13.136}$$

or from $\dot{A} = AW_b$ the alternative equation

$$\begin{pmatrix} \dot{q}_0 \\ \dot{q}_1 \\ \dot{q}_2 \\ \dot{q}_3 \end{pmatrix} = \frac{1}{2} \begin{pmatrix} 0 & \omega_{1b} & \omega_{2b} & \omega_{3b} \\ -\omega_{1b} & 0 & \omega_{3b} & -\omega_{2b} \\ -\omega_{2b} & -\omega_{3b} & 0 & \omega_{1b} \\ -\omega_{3b} & \omega_{2b} & -\omega_{1b} & 0 \end{pmatrix} \begin{pmatrix} q_0 \\ q_1 \\ q_2 \\ q_3 \end{pmatrix}. \tag{13.137}$$

Both of these equations can be written briefly in the form

$$\dot{\mathbf{q}} = \widetilde{W}\mathbf{q}. \tag{13.138}$$

Example (Rotation around the z-axis) Rotation around the z-axis corresponds to the quaternion with Euler parameters

$$\mathbf{q} = \begin{pmatrix} \cos\frac{\omega t}{2} \\ 0 \\ 0 \\ -\sin\frac{\omega t}{2} \end{pmatrix} \tag{13.139}$$

as can be seen from the rotation matrix

$$
A = \begin{pmatrix} (\cos\frac{\omega t}{2})^2 - (\sin\frac{\omega t}{2})^2 & -2\cos\frac{\omega t}{2}\sin\frac{\omega t}{2} & 0 \\ 2\cos\frac{\omega t}{2}\sin\frac{\omega t}{2} & (\cos\frac{\omega t}{2})^2 - (\sin\frac{\omega t}{2})^2 & 0 \\ 0 & 0 & (\cos\frac{\omega t}{2})^2 + (\sin\frac{\omega t}{2})^2 \end{pmatrix}
$$

$$
= \begin{pmatrix} \cos\omega t & -\sin\omega t & 0 \\ \sin\omega t & \cos\omega t & 0 \\ 0 & 0 & 1 \end{pmatrix}. \tag{13.140}
$$

The time derivative of $\mathbf{q}$ obeys the equation

$$
\dot{\mathbf{q}} = \frac{1}{2}\begin{pmatrix} 0 & 0 & 0 & \omega \\ 0 & 0 & -\omega & 0 \\ 0 & \omega & 0 & 0 \\ -\omega & 0 & 0 & 0 \end{pmatrix}\begin{pmatrix} \cos\frac{\omega t}{2} \\ 0 \\ 0 \\ -\sin\frac{\omega t}{2} \end{pmatrix} = \begin{pmatrix} -\frac{\omega}{2}\sin\omega t \\ 0 \\ 0 \\ -\frac{\omega}{2}\cos\omega t \end{pmatrix}. \tag{13.141}
$$

After a rotation by 2π the quaternion changes its sign, i.e. $\mathbf{q}$ and $-\mathbf{q}$ parametrize the same rotation matrix!

13.15 Solving the Equations of Motion with Quaternions

As with the matrix method we can obtain a simple first or second order algorithm from the Taylor series expansion

$$
\mathbf{q}(t+\Delta t) = \mathbf{q}(t) + \widetilde{W}(t)\mathbf{q}(t)\Delta t + \left(\dot{\widetilde{W}}(t) + \widetilde{W}^2(t)\right)\mathbf{q}(t)\frac{\Delta t^2}{2} + \cdots. \tag{13.142}
$$

Now only one constraint remains, which is the conservation of the norm of the quaternion. This can be taken into account by rescaling the quaternion whenever its norm deviates too much from unity.

It is also possible to use Omelyan's [192] method:

$$
\mathbf{q}(t+\Delta t) = \mathbf{q}(t) + \widetilde{W}\left(t + \frac{\Delta t}{2}\right)\frac{1}{2}\left(\mathbf{q}(t) + \mathbf{q}(t+\Delta t)\right) \tag{13.143}
$$

gives

$$
\mathbf{q}(t+\Delta t) = \left(1 - \frac{\Delta t}{2}\widetilde{W}\right)^{-1}\left(1 + \frac{\Delta t}{2}\widetilde{W}\right)\mathbf{q}(t) \tag{13.144}
$$

where the inverse matrix is

$$
\left(1 - \frac{\Delta t}{2}\widetilde{W}\right)^{-1} = \frac{1}{1 + \omega^2\frac{\Delta t^2}{16}}\left(1 + \frac{\Delta t}{2}\widetilde{W}\right) \tag{13.145}
$$

Fig. 13.5 Free asymmetric rotor

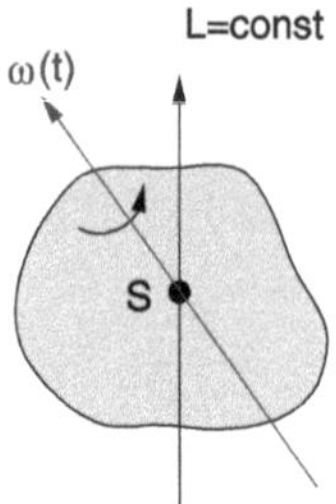

and the matrix product

$$\left(1 - \frac{\Delta t}{2}\widetilde{W}\right)^{-1}\left(1 + \frac{\Delta t}{2}\widetilde{W}\right) = \frac{1 - \omega^2\frac{\Delta t^2}{16}}{1 + \omega^2\frac{\Delta t^2}{16}} + \frac{\Delta t}{1 + \omega^2\frac{\Delta t^2}{16}}\widetilde{W}. \tag{13.146}$$

This method conserves the norm of the quaternion and works quite well.

13.16 Problems

Problem 13.1 (Free rotor, Fig. 13.5) In this computer experiment we compare different methods for a free rotor (Sect. 13.8):

- explicit first order method (13.67)

$$A(t + \Delta t) = A(t) + A(t)W_b(t)\Delta t + O\left(\Delta t^2\right) \tag{13.147}$$

- explicit second order method (13.69)

$$A(t + \Delta t) = A(t) + A(t)W_b(t)\Delta t + \frac{1}{2}\left(A(t)W_b^2(t) + A(t)\dot{W}_b(t)\right)\Delta t^2 + O\left(\Delta t^3\right) \tag{13.148}$$

- implicit second order method (13.93)

$$A(t + \Delta t) = A(t)\left(1 + \frac{\Delta t}{2}W\left(t + \frac{\Delta t}{2}\right)\right)\left(1 - \frac{\Delta t}{2}W\left(t + \frac{\Delta t}{2}\right)\right)^{-1} + O\left(\Delta t^3\right). \tag{13.149}$$

The explicit methods can be combined with reorthogonalization according to (13.79) or with the Gram-Schmidt method. Reorthogonalization threshold and time step can be varied and the error of kinetic energy and determinant are plotted as a function of the total simulation time.

Problem 13.2 (Rotor in a field, Fig. 13.6) In this computer experiment we simulate a molecule with a permanent dipole moment in a homogeneous electric field **E**. We neglect vibrations and describe the molecule as a rigid body consisting of nuclei

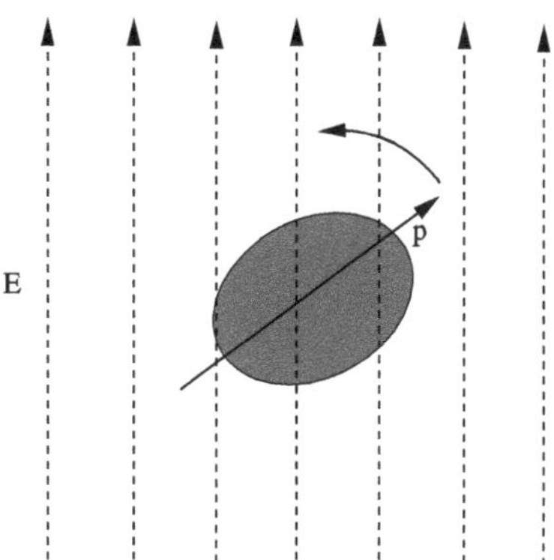

Fig. 13.6 Rotor in an electric field

with masses m_i and partial charges Q_i. The total charge is $\sum_i Q_i = 0$. The dipole moment is

$$\mathbf{p} = \sum_i Q_i \mathbf{r}_i \tag{13.150}$$

and external force and torque are

$$\mathbf{F}_{ext} = \sum_i Q_i \mathbf{E} = 0 \tag{13.151}$$

$$\mathbf{N}_{ext} = \sum_i Q_i \mathbf{r}_i \times \mathbf{E} = \mathbf{p} \times \mathbf{E}. \tag{13.152}$$

The angular momentum changes according to

$$\frac{d}{\Delta t}\mathbf{L}_{int} = \mathbf{p} \times \mathbf{E} \tag{13.153}$$

where the dipole moment is constant in the body fixed system. We use the implicit integrator for the rotation matrix (13.93) and the equation

$$\dot{\boldsymbol{\omega}}_b(t) = -I_b^{-1} W_b(t)\mathbf{L}_{int,b}(t) + I_b^{-1} A^{-1}(t)\big(\mathbf{p}(t) \times \mathbf{E}\big) \tag{13.154}$$

to solve the equations of motion numerically. Obviously the component of the angular momentum parallel to the field is constant. The potential energy is

$$U = -\sum_i Q_i \mathbf{E}\mathbf{r}_i = -\mathbf{p}\mathbf{E}. \tag{13.155}$$

Problem 13.3 (Molecular collision)   This computer experiment simulates the collision of two rigid methane molecules (Fig. 13.7). The equations of motion are solved with the implicit quaternion method (13.143) and the velocity Verlet method (12.11.4). The two molecules interact by a standard 6–12 Lennard-Jones potential (14.24) [3]. For comparison the attractive r^{-6} part can be switched off. The initial

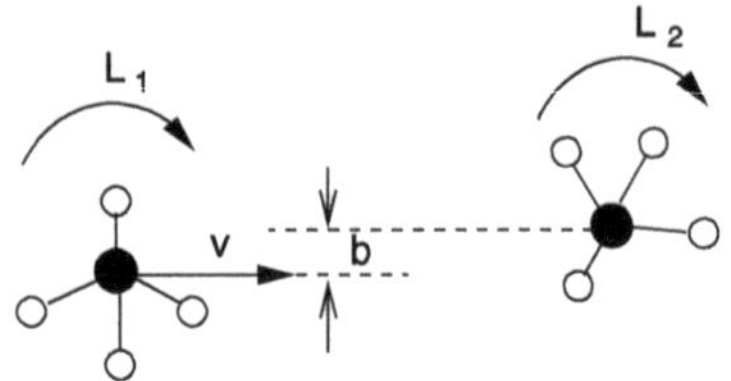

Fig. 13.7 Molecular collision

angular momenta as well as the initial velocity v and collision parameter b can be varied. Total energy and momentum are monitored and the decomposition of the total energy into translational, rotational and potential energy are plotted as a function of time.

Study the exchange of momentum and angular momentum and the transfer of energy between translational and rotational degrees of freedom.

Chapter 14
Molecular Mechanics

Classical molecular mechanics simulations have become a very valuable tool for the investigation of atomic and molecular systems [97, 115, 157, 212, 225], mainly in the area of materials science and molecular biophysics. Based on the Born-Oppenheimer separation which assumes that the electrons move much faster than the nuclei, nuclear motion is described quantum mechanically by the Hamiltonian

$$H = \left[T^{Nuc} + U\left(\mathbf{r}_j^{Nuc}\right) \right]. \tag{14.1}$$

Molecular mechanics uses the corresponding classical energy function

$$T^{Nuc} + U\left(\mathbf{r}_j^{Nuc}\right) = \sum_j \frac{(p_j^{Nuc})^2}{2m_j} + U\left(\mathbf{r}_j^{Nuc}\right) \tag{14.2}$$

which treats the atoms as mass points interacting by classical forces

$$\mathbf{F}_i = -\operatorname{grad}_{\mathbf{r}_i} U\left(\mathbf{r}_j^{Nuc}\right). \tag{14.3}$$

Stable structures, i.e. local minima of the potential energy can be found by the methods discussed in Chap. 6. Small amplitude motions around an equilibrium geometry are described by a harmonic normal mode analysis. Molecular dynamics (MD) simulations solve the classical equations of motion

$$m_i \frac{d^2 \mathbf{r}_i}{dt^2} = \mathbf{F}_i = -\operatorname{grad}_{\mathbf{r}_i} U \tag{14.4}$$

numerically.

The potential energy function $U(\mathbf{r}_j^{Nuc})$ can be calculated with simplified quantum methods for not too large systems [50, 152]. Classical MD simulations for larger molecules use empirical force fields, which approximate the potential energy surface of the electronic ground state. They are able to describe structural and conformational changes but not chemical reactions which usually involve more than one electronic state. Among the most popular classical force fields are AMBER [63], CHARMM [163] and GROMOS [58, 262].

In this chapter we discuss the most important interaction terms, which are conveniently expressed in internal coordinates, i.e. bond lengths, bond angles and dihedral angles. We derive expressions for the gradients of the force field with respect

P.O.J. Scherer, *Computational Physics*, Graduate Texts in Physics,
DOI 10.1007/978-3-319-00401-3_14,
© Springer International Publishing Switzerland 2013

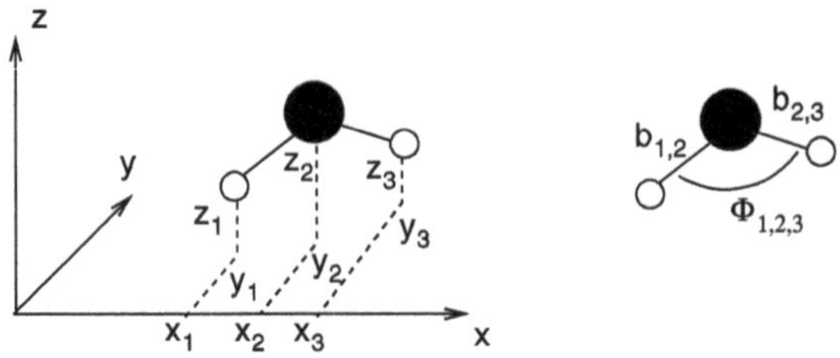

Fig. 14.1 (Molecular coordinates) Cartesian coordinates (*left*) are used to solve the equations of motion whereas the potential energy is more conveniently formulated in internal coordinates (*right*)

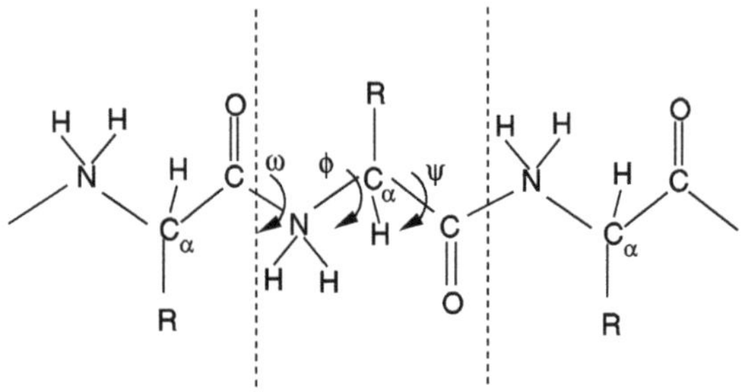

Fig. 14.2 (Conformation of a protein) The relative orientation of two successive protein residues can be described by three angles (Ψ, Φ, ω)

to Cartesian coordinates. In a computer experiment we simulate a glycine dipeptide and demonstrate the principles of energy minimization, normal mode analysis and dynamics simulation.

14.1 Atomic Coordinates

The most natural coordinates for the simulation of molecules are the Cartesian co-ordinates (Fig. 14.1) of the atoms,

$$\mathbf{r}_i = (x_i, y_i, z_i) \tag{14.5}$$

which can be collected into a $3N$-dimensional vector

$$(\xi_1, \xi_2 \cdots \xi_{3N}) = (x_1, y_1, z_1, x_2 \cdots x_N, y_N, z_N). \tag{14.6}$$

The second derivatives of the Cartesian coordinates appear directly in the equations of motion (14.4)

$$m_r \ddot{\xi}_r = F_r \quad r = 1 \cdots 3N. \tag{14.7}$$

Cartesian coordinates have no direct relation to the structural properties of molecules. For instance a protein is a long chain of atoms (the so called backbone) with additional side groups (Fig. 14.2).

The protein structure can be described more intuitively with the help of atomic distances and angles. Internal coordinates are (Fig. 14.3) distances between two bonded atoms (bond lengths)

$$b_{ij} = |\mathbf{r}_{ij}| = |\mathbf{r}_i - \mathbf{r}_j|, \tag{14.8}$$

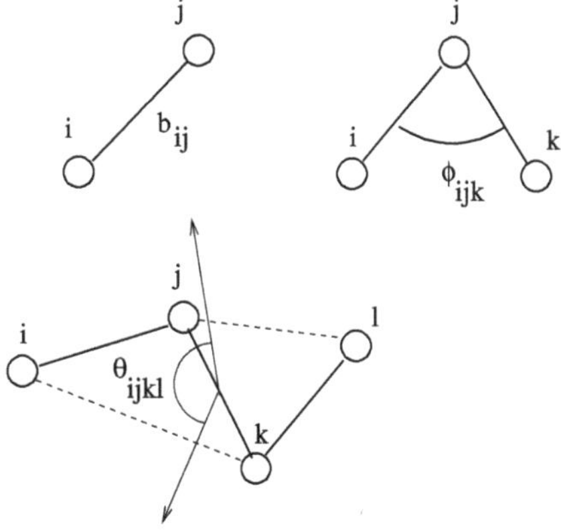

Fig. 14.3 (Internal coordinates) The structure of a molecule can be described by bond lengths, bond angles and dihedral angles

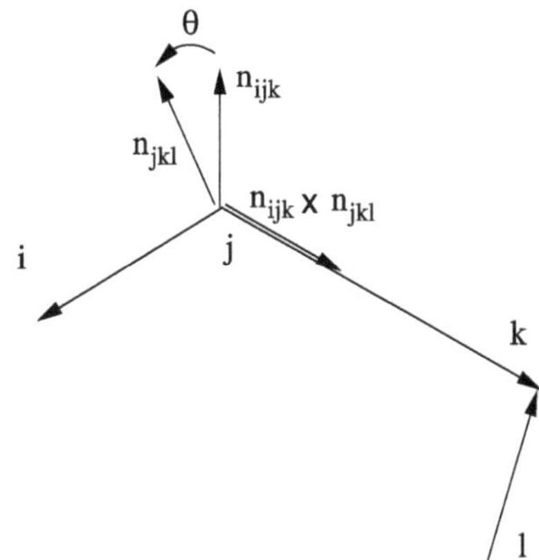

Fig. 14.4 Dihedral angle

angles between two bonds (bond angles)

$$\phi_{ijk} = \arccos\left(\frac{\mathbf{r}_{ij}\mathbf{r}_{kj}}{|\mathbf{r}_{ij}||\mathbf{r}_{kj}|}\right) \tag{14.9}$$

and dihedral angles which describe the planarity and torsions of the molecule. A dihedral angle (Fig. 14.4) is the angle between two planes which are defined by three bonds

$$\theta_{ijkl} = \text{sign}(\theta_{ijkl})\arccos(\mathbf{n}_{ijk}\mathbf{n}_{jkl}) \tag{14.10}$$

$$\mathbf{n}_{ijk} = \frac{\mathbf{r}_{ij} \times \mathbf{r}_{kj}}{|\mathbf{r}_{ij} \times \mathbf{r}_{kj}|} \quad \mathbf{n}_{jkl} = \frac{\mathbf{r}_{kj} \times \mathbf{r}_{kl}}{|\mathbf{r}_{kj} \times \mathbf{r}_{kl}|} \tag{14.11}$$

where the conventional sign of the dihedral angle [136] is determined by

$$\text{sign}\,\theta_{ijkl} = \text{sign}\big(\mathbf{r}_{kj}(\mathbf{n}_{ijk} \times \mathbf{n}_{jkl})\big). \tag{14.12}$$

Internal coordinates are very convenient for the formulation of a force field. On the other hand, the kinetic energy (14.2) becomes complicated if expressed in internal coordinates. Therefore both kinds of coordinates are used in molecular dynamics

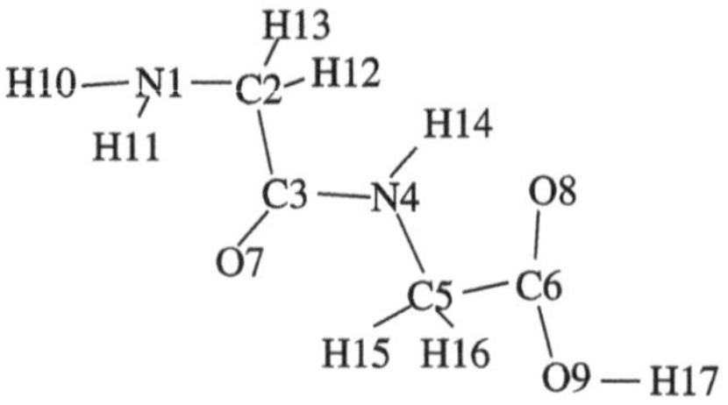

Fig. 14.5 (Glycine dipeptide model) The glycine dipeptide is the simplest model for a peptide. It is simulated in Problem 14.1. Optimized internal coordinates are shown in Table 14.1

calculations. The internal coordinates are usually arranged in Z-matrix form. Each line corresponds to one atom i and shows its position relative to three atoms j, k, l in terms of the bond length b_{ij}, the bond angle ϕ_{ijk} and the dihedral angle θ_{ijkl} (Fig. 14.5 and Table 14.1).

14.2 Force Fields

Classical force fields are usually constructed as an additive combination of many interaction terms. Generally these can be divided into intramolecular contributions

Table 14.1 (Z-matrix) The optimized values of the internal coordinates from Problem 14.1 are shown in Z-matrix form. Except for the first three atoms the position of atom i is given by its distance b_{ij} to atom j, the bond angle ϕ_{ijk} and the dihedral angle θ_{ijkl}

Number i	Label	j	k	l	Bond length (Å) b_{ij}	Bond angle ϕ_{ijk}	Dihedral θ_{ijkl}
1	N1						
2	C2	1			1.45		
3	C3	2	1		1.53	108.6	
4	N4	3	2	1	1.35	115.0	160.7
5	C5	4	3	2	1.44	122.3	−152.3
6	C6	5	4	3	1.51	108.7	−153.1
7	O7	3	2	1	1.23	121.4	−26.3
8	O8	6	5	4	1.21	124.4	123.7
9	O9	6	5	4	1.34	111.5	−56.5
10	H10	1	2	3	1.02	108.7	−67.6
11	H11	1	2	3	1.02	108.7	49.3
12	H12	2	3	4	1.10	109.4	−76.8
13	H13	2	3	4	1.10	109.4	38.3
14	H14	4	3	2	1.02	123.1	27.5
15	H15	5	4	3	1.10	111.2	−32.5
16	H16	5	4	3	1.10	111.1	86.3
17	H17	9	6	5	0.97	106.9	−147.4

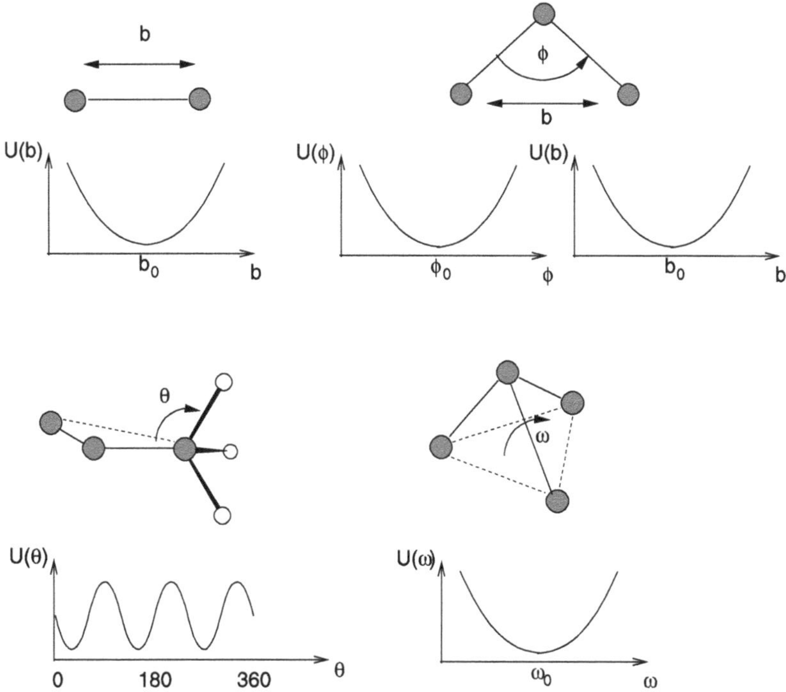

Fig. 14.6 Intramolecular forces

U_{bonded} which determine the configuration and motion of a single molecule and intermolecular contributions $U_{non\text{-}bonded}$ describing interactions between different atoms or molecules

$$U = U_{bonded} + U_{non\text{-}bonded}. \tag{14.13}$$

14.2.1 Intramolecular Forces

The most important intramolecular forces depend on the deviation of bond lengths, bond angles and dihedral angles from their equilibrium values. For simplicity a sum of independent terms is used as for the CHARMM force field [39, 40, 163]

$$U_{intra} = \sum U_{ij}^{bond} + \sum U_{ijk}^{angle} + \sum U_{ijk}^{UB} + \sum U_{ijkl}^{dihedral} + \sum U_{ijkl}^{improper}. \tag{14.14}$$

The forces are derived from potential functions which are in the simplest case approximated by harmonic oscillator parabolas (Fig. 14.6), like the bond stretching energy

$$U_{ij}^{bond} = \frac{1}{2}k_{ij}\left(b_{ij} - b_{ij}^0\right)^2 \tag{14.15}$$

Table 14.2 (Atom types of the glycine dipeptide) Atom types for glycine oligopeptides according to Bautista and Seminario [15]. The atoms are classified by element and bonding environment. Atoms of the same atom type are considered equivalent

Atom type	Atoms
C	C3,
C_1	C2, C5
C_2	C6
N	N4
N_2	N1
O	O7
O_1	O9
O_2	O8
H	H14
H_1	H12, H13, H15, H16
H_2	H17
H_3	H10, H11

angle bending terms

$$U_{ijk}^{angle} = \frac{1}{2} k_{ijk} \left(\phi_{ijk} - \phi_{ijk}^0 \right)^2 \tag{14.16}$$

together with the Urey-Bradly correction

$$U_{ijk}^{UB} = \frac{1}{2} k_{ijk} \left(b_{ik} - b_{ik}^0 \right)^2 \tag{14.17}$$

and "improper dihedral" terms which are used to keep planarity

$$U_{ijkl}^{improper} = \frac{1}{2} k_{ijkl} \left(\theta_{ijkl} - \theta_{ijkl}^0 \right)^2. \tag{14.18}$$

Torsional energy contributions are often described by a cosine function[1]

$$U_{ijkl}^{dihedral} = k_{ijkl} \left(1 - \cos\left(m\theta_{ijkl} - \theta_{ijkl}^0 \right) \right) \tag{14.19}$$

where $m = 1, 2, 3, 4, 6$ describes the symmetry. For instance $m = 3$ for the three equivalent hydrogen atoms of a methyl group. In most cases the phase shift $\theta_{ijkl}^0 = 0$ or $\theta_{ijkl}^0 = \pi$. Then the dihedral potential can be expanded as a polynomial of $\cos\theta$, for instance

$$m = 1: \quad U_{ijkl}^{dihedral} = k(1 \pm \cos\theta_{ijkl}) \tag{14.20}$$

$$m = 2: \quad U_{ijkl}^{dihedral} = k \pm k\left(1 - 2(\cos\theta_{ijkl})^2\right) \tag{14.21}$$

$$m = 3: \quad U_{ijkl}^{dihedral} = k\left(1 \pm 3\cos\theta_{ijkl} \mp 4(\cos\theta_{ijkl})^3\right). \tag{14.22}$$

For more general θ_{ijkl}^0 the torsional potential can be written as a polynomial of $\cos\theta_{ijkl}$ and $\sin\theta_{ijkl}$.

---

[1]Some force-fields like Desmond [33] or UFF [213] use a more general sum $k \sum_{m=0}^{M} c_m \cos(m\theta - \theta^0)$.

Table 14.3 (Bond stretching parameters) Equilibrium bond lengths (Å) and force constants (kcal mol^{-1} Å^{-2}) for the glycine dipeptide from [15]

Bond type	b^0	k	Bonds
$r_{C,N}$	1.346	1296.3	C3-N4
$r_{C1,N}$	1.438	935.5	N4-C5
$r_{C1,N2}$	1.452	887.7	N1-C2
$r_{C2,C1}$	1.510	818.9	C5-C6
$r_{C,C1}$	1.528	767.9	C2-C3
$r_{C2,O2}$	1.211	2154.5	C6-O8
$r_{C,O}$	1.229	1945.7	C3-O7
$r_{C2,O1}$	1.339	1162.1	C6-O9
$r_{N,H}$	1.016	1132.4	N4-H14
$r_{N2,H3}$	1.020	1104.5	N1-H10, N1-H11
$r_{C1,H1}$	1.098	900.0	C2-H12, C2-H13, C5-H15, C5-H16
$r_{O1,H2}$	0.974	1214.6	O9-H17

The atoms are classified by element and bonding environment. Atoms of the same atom type are considered equivalent and the parameters transferable (for an example see Tables 14.2, 14.3, 14.4).

14.2.2 Intermolecular Interactions

Interactions between non-bonded atoms

$$U_{non\text{-}bonded} = U^{Coul} + U^{vdW} \tag{14.23}$$

include the Coulomb interaction and the weak attractive van der Waals forces which are usually combined with a repulsive force at short distances to account for the Pauli principle. Very often a sum of pairwise Lennard-Jones potentials is used (Fig. 14.7) [3]

$$U^{vdw} = \sum_{A \neq B} \sum_{i \in A, j \in B} U_{i,j}^{vdw} = \sum_{A \neq B} \sum_{ij} 4\varepsilon_{ij} \left(\frac{\sigma_{ij}^{12}}{r_{ij}^{12}} - \frac{\sigma_{ij}^{6}}{r_{ij}^{6}} \right). \tag{14.24}$$

The charge distribution of a molecular system can be described by a set of multipoles at the position of the nuclei, the bond centers and further positions (lone pairs for example). Such distributed multipoles can be calculated quantum chemically for not too large molecules. In the simplest models only partial charges are taken into account giving the Coulomb energy as a sum of atom-atom interactions

$$U^{Coul} = \sum_{A \neq B} \sum_{i \in A, j \in B} \frac{q_i q_j}{4\pi \varepsilon_0 r_{ij}}. \tag{14.25}$$

Table 14.4 (Bond angle parameters) Equilibrium bond angles (deg) and force constants (kcal mol^{-1} rad^{-2}) for the glycine dipeptide from [15]

Angle type	ϕ^0	k	Angles
$\phi_{N,C,C1}$	115.0	160.0	C2-C3-N4
$\phi_{C1,N,C}$	122.3	160.1	C3-N4-C5
$\phi_{C1,C2,O1}$	111.5	156.0	C5-C6-O9
$\phi_{C1,C2,O2}$	124.4	123.8	C5-C6-O8
$\phi_{C1,C,O}$	121.4	127.5	C2-C3-O7
$\phi_{O2,C2,O1}$	124.1	146.5	O8-C6-O9
$\phi_{N,C,O}$	123.2	132.7	N4-C3-O7
$\phi_{C,C1,H1}$	110.1	74.6	H12-C2-C3, H13-C2-C3
$\phi_{C2,C1,H1}$	109.4	69.6	H16-C5-C6, H15-C5-C6
$\phi_{C,N,H}$	123.1	72.0	C3-N4-H14
$\phi_{C1,N,H}$	114.6	68.3	C5-N4-H14
$\phi_{C1,N2,H3}$	108.7	71.7	H10-N1-C2, H11-N1-C2
$\phi_{H1,C1,H1}$	106.6	48.3	H13-C2-H12, H15-C5-H16
$\phi_{H3,N2,H3}$	107.7	45.2	H10-N1-H11
$\phi_{C,C1,N2}$	109.0	139.8	N1-C2-C3
$\phi_{C2,C1,N}$	108.6	129.0	N4-C5-C6
$\phi_{C2,O1,H2}$	106.9	72.0	H17-O9-C6
$\phi_{N,C1,H1}$	111.1	73.3	H15-C5-N4, H16-C5-N4
$\phi_{N2,C1,H1}$	112.6	80.1	H13-C2-N1, H12-C2-N1

More sophisticated force fields include higher charge multipoles and polarization effects.

14.3 Gradients

The equations of motion are usually solved in Cartesian coordinates and the gradients of the potential are needed in Cartesian coordinates. Since the potential depends only on relative position vectors $\mathbf{r}_{ij}$, the gradient with respect to a certain atom position $\mathbf{r}_k$ can be calculated from

$$\mathrm{grad}_{\mathbf{r}_k} = \sum_{i<j} (\delta_{ik} - \delta_{jk})\,\mathrm{grad}_{\mathbf{r}_{ij}}. \tag{14.26}$$

Therefore it is sufficient to calculate gradients with respect to the difference vectors. Numerically efficient methods to calculate first and second derivatives of many force field terms are given in [156, 259, 260]. The simplest potential terms depend only on the distance of two atoms. For instance bond stretching terms, Lennard-Jones and Coulomb energies have the form

$$U_{ij} = U(r_{ij}) = U(|\mathbf{r}_{ij}|) \tag{14.27}$$

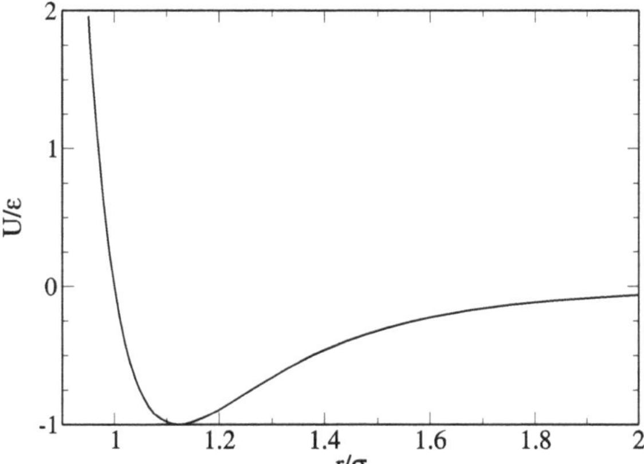

Fig. 14.7 (Lennard-Jones potential) The 6–12 potential (14.24) has its minimum at $r_{\min} = \sqrt[6]{2}\sigma \approx 1.12\sigma$ with $U_{\min} = -\epsilon$

where the gradient is

$$\operatorname{grad}_{\mathbf{r}_{ij}} U_{ij} = \frac{\Delta U}{\Delta r}\frac{\mathbf{r}_{ij}}{|\mathbf{r}_{ij}|}. \tag{14.28}$$

The most important gradients of this kind are

$$\operatorname{grad}_{\mathbf{r}_{ij}} U_{i,j}^{bond} = k\left(r_{ij} - b^0\right)\frac{\mathbf{r}_{ij}}{r_{ij}} = k\left(1 - \frac{b^0}{r_{ij}}\right)\mathbf{r}_{ij} \tag{14.29}$$

$$\operatorname{grad}_{\mathbf{r}_{ij}} U_{i,j}^{vdw} = 24\varepsilon_{ij}\left(-2\frac{\sigma_{ij}^{12}}{r_{ij}^{14}} + \frac{\sigma_{ij}^{6}}{r_{ij}^{8}}\right)\mathbf{r}_{ij} \tag{14.30}$$

$$\operatorname{grad}_{\mathbf{r}_{ij}} U_{ij}^{Coul} = -\frac{q_i q_j}{4\pi \varepsilon_0 r_{ij}^3}\mathbf{r}_{ij}. \tag{14.31}$$

The gradient of the harmonic bond angle potential is

$$\operatorname{grad}_{\mathbf{r}} U_{i,j,k}^{angle} = k\left(\phi_{ijk} - \phi^0\right)\operatorname{grad}_{\mathbf{r}} \phi_{ijk} \tag{14.32}$$

where the gradient of the angle can be calculated from the gradient of its cosine

$$\operatorname{grad}_{\mathbf{r}_{ij}} \phi_{ijk} = -\frac{1}{\sin\phi_{ijk}}\operatorname{grad}_{\mathbf{r}_{ij}} \cos\phi_{ijk} = -\frac{1}{\sin\phi_{ijk}}\left(\frac{\mathbf{r}_{kj}}{|\mathbf{r}_{ij}||\mathbf{r}_{kj}|} - \frac{\mathbf{r}_{ij}\mathbf{r}_{kj}}{|\mathbf{r}_{ij}|^3|\mathbf{r}_{kj}|}\mathbf{r}_{ij}\right)$$

$$= -\frac{1}{\sin\phi_{ijk}}\left(\frac{\mathbf{r}_{kj}}{|\mathbf{r}_{ij}||\mathbf{r}_{kj}|} - \frac{\cos\phi_{ijk}}{|\mathbf{r}_{ij}|^2}\mathbf{r}_{ij}\right) \tag{14.33}$$

$$\operatorname{grad}_{\mathbf{r}_{kj}} \phi_{ijk} = -\frac{1}{\sin\phi_{ijk}}\left(\frac{\mathbf{r}_{ij}}{|\mathbf{r}_{ij}||\mathbf{r}_{kj}|} - \frac{\cos\phi_{ijk}}{|\mathbf{r}_{kj}|^2}\mathbf{r}_{kj}\right). \tag{14.34}$$

In principle, the sine function in the denominator could lead to numerical problems which can be avoided by treating angles close to 0 or π separately or using a function of $\cos\phi_{ijk}$ like the trigonometric potential

$$U_{ijk}^{angle} = \frac{1}{2}k_{ijk}\left(\cos\phi_{ijk} - \cos\phi_{ijk}^0\right)^2 \tag{14.35}$$

instead [5, 213, 224]. Alternatively, the gradient of ϕ can be brought to a form which is free of singularities by expressing the sine in the denominator by a cosine [59]

$$
\begin{aligned}
\mathrm{grad}_{\mathbf{r}_{ij}}\,\phi_{ijk} &= -\frac{1}{\sqrt{r_{ij}^2 r_{kj}^2 (1 - \cos^2 \phi_{ijk})}}\left(\mathbf{r}_{kj} - \frac{\mathbf{r}_{ij}\mathbf{r}_{kj}}{r_{ij}^2}\mathbf{r}_{ij}\right) \\
&= -\frac{r_{ij}^2 \mathbf{r}_{kj} - (\mathbf{r}_{ij}\mathbf{r}_{kj})\mathbf{r}_{ij}}{r_{ij}\sqrt{(r_{ij}^2 \mathbf{r}_{kj} - (\mathbf{r}_{ij}\mathbf{r}_{kj})\mathbf{r}_{ij})^2}} \\
&= -\frac{1}{r_{ij}}\frac{\mathbf{r}_{ij} \times (\mathbf{r}_{kj} \times \mathbf{r}_{ij})}{|\mathbf{r}_{ij} \times (\mathbf{r}_{kj} \times \mathbf{r}_{ij})|}
\end{aligned}
\tag{14.36}
$$

and similarly

$$
\mathrm{grad}_{\mathbf{r}_{kj}}\,\phi_{ijk} = -\frac{1}{r_{kj}}\frac{\mathbf{r}_{kj} \times (\mathbf{r}_{ij} \times \mathbf{r}_{kj})}{|\mathbf{r}_{kj} \times (\mathbf{r}_{ij} \times \mathbf{r}_{kj})|}.
\tag{14.37}
$$

Gradients of the dihedral potential are most easily calculated for $\theta_{ijkl}^0 = 0$ or π. In that case, the dihedral potential is a polynomial of $\cos\theta_{ijkl}$ only (14.20)–(14.22) and

$$
\mathrm{grad}_{\mathbf{r}}\,U_{ijkl}^{dihedral} = \frac{\mathrm{d}U_{ijkl}^{dihedral}}{\mathrm{d}\cos\theta_{ijkl}}\,\mathrm{grad}_{\mathbf{r}}\cos\theta_{ijkl}
\tag{14.38}
$$

whereas in the general case $0 < \theta_{ijkl} < \pi$ application of the chain rule gives

$$
\mathrm{grad}\,U_{ijkl}^{dihedral} = mk_{ijkl}\sin\!\left(m\!\left(\theta_{ijkl} - \theta^0\right)\right)\mathrm{grad}\,\theta_{ijkl}.
\tag{14.39}
$$

If this is evaluated with the help of

$$
\mathrm{grad}\,\theta_{ijkl} = -\frac{1}{\sin\theta_{ijkl}}\,\mathrm{grad}\cos\theta_{ijkl}
\tag{14.40}
$$

singularities appear for $\theta = 0$ and π. The same is the case for the gradients of the harmonic improper potential

$$
\mathrm{grad}\,U_{ijkl}^{improper} = k\!\left(\theta_{ijkl} - \theta_{ijkl}^0\right)\mathrm{grad}\,\theta_{ijkl}.
\tag{14.41}
$$

Again, one possibility which has been often used, is to treat angles close to 0 or π separately [39]. However, the gradient of the angle θ_{ijkl} can be calculated directly, which is much more efficient [19].

The gradient of the cosine follows from application of the product rule

$$
\mathrm{grad}\cos\theta = \mathrm{grad}\!\left(\frac{\mathbf{r}_{ij} \times \mathbf{r}_{kj}}{|\mathbf{r}_{ij} \times \mathbf{r}_{kj}|}\frac{\mathbf{r}_{kj} \times \mathbf{r}_{kl}}{|\mathbf{r}_{kj} \times \mathbf{r}_{kl}|}\right).
\tag{14.42}
$$

First we derive the differentiation rule

$$
\begin{aligned}
\mathrm{grad}_{\mathbf{a}}\!\left[(\mathbf{a} \times \mathbf{b})(\mathbf{c} \times \mathbf{d})\right] &= \mathrm{grad}_{\mathbf{a}}\!\left[(\mathbf{ac})(\mathbf{bd}) - (\mathbf{ad})(\mathbf{bc})\right] \\
&= \mathbf{c}(\mathbf{bd}) - \mathbf{d}(\mathbf{bc}) = \mathbf{b} \times (\mathbf{c} \times \mathbf{d})
\end{aligned}
\tag{14.43}
$$

which helps us to find

$$\text{grad}_{r_{ij}}(\mathbf{r}_{ij} \times \mathbf{r}_{kj})(\mathbf{r}_{kj} \times \mathbf{r}_{kl}) = \mathbf{r}_{kj} \times (\mathbf{r}_{kj} \times \mathbf{r}_{kl}) \tag{14.44}$$

$$\text{grad}_{r_{kl}}(\mathbf{r}_{ij} \times \mathbf{r}_{kj})(\mathbf{r}_{kj} \times \mathbf{r}_{kl}) = \mathbf{r}_{kj} \times (\mathbf{r}_{kj} \times \mathbf{r}_{ij}) \tag{14.45}$$

$$\text{grad}_{r_{kj}}(\mathbf{r}_{ij} \times \mathbf{r}_{kj})(\mathbf{r}_{kj} \times \mathbf{r}_{kl}) = \mathbf{r}_{kl} \times (\mathbf{r}_{ij} \times \mathbf{r}_{kj}) + \mathbf{r}_{ij} \times (\mathbf{r}_{kl} \times \mathbf{r}_{kj}) \tag{14.46}$$

and

$$\text{grad}_{r_{ij}} \frac{1}{|\mathbf{r}_{ij} \times \mathbf{r}_{kj}|} = -\frac{\mathbf{r}_{kj} \times (\mathbf{r}_{ij} \times \mathbf{r}_{kj})}{|\mathbf{r}_{ij} \times \mathbf{r}_{kj}|^3} \tag{14.47}$$

$$\text{grad}_{r_{kj}} \frac{1}{|\mathbf{r}_{ij} \times \mathbf{r}_{kj}|} = -\frac{\mathbf{r}_{ij} \times (\mathbf{r}_{kj} \times \mathbf{r}_{ij})}{|\mathbf{r}_{ij} \times \mathbf{r}_{kj}|^3} \tag{14.48}$$

$$\text{grad}_{r_{kj}} \frac{1}{|\mathbf{r}_{kj} \times \mathbf{r}_{kl}|} = -\frac{\mathbf{r}_{kl} \times (\mathbf{r}_{kj} \times \mathbf{r}_{kl})}{|\mathbf{r}_{kj} \times \mathbf{r}_{kl}|^3} \tag{14.49}$$

$$\text{grad}_{r_{kl}} \frac{1}{|\mathbf{r}_{kj} \times \mathbf{r}_{kl}|} = -\frac{\mathbf{r}_{kj} \times (\mathbf{r}_{kl} \times \mathbf{r}_{kj})}{|\mathbf{r}_{kj} \times \mathbf{r}_{kl}|^3}. \tag{14.50}$$

Finally we collect terms to obtain the gradients of the cosine [59]

$$\begin{aligned}
\text{grad}_{\mathbf{r}_{ij}} \cos\theta_{ijkl} &= \frac{\mathbf{r}_{kj} \times (\mathbf{r}_{kj} \times \mathbf{r}_{kl})}{|\mathbf{r}_{ij} \times \mathbf{r}_{kj}||\mathbf{r}_{kj} \times \mathbf{r}_{kl}|} - \frac{\mathbf{r}_{kj} \times (\mathbf{r}_{ij} \times \mathbf{r}_{kj})}{|\mathbf{r}_{ij} \times \mathbf{r}_{kj}|^2} \cos\theta_{ijkl} \\
&= \frac{\mathbf{r}_{kj}}{|\mathbf{r}_{ij} \times \mathbf{r}_{kj}|} \times (\mathbf{n}_{jkl} - \mathbf{n}_{ijk} \cos\theta) \\
&= \frac{\mathbf{r}_{kj}}{|\mathbf{r}_{ij} \times \mathbf{r}_{kj}|} \times \left(\mathbf{n}_{jkl} - \mathbf{n}_{ijk}(\mathbf{n}_{jkl}\mathbf{n}_{ijk})\right) \\
&= \frac{\mathbf{r}_{kj}}{|\mathbf{r}_{ij} \times \mathbf{r}_{kj}|} \times \left(\mathbf{n}_{ijk} \times (\mathbf{n}_{jkl} \times \mathbf{n}_{ijk})\right) \\
&= \frac{\mathbf{r}_{kj}}{|\mathbf{r}_{ij} \times \mathbf{r}_{kj}|} \times \left(\mathbf{n}_{ijk} \times \frac{1}{r_{kj}}(-\mathbf{r}_{kj}) \sin\theta\right) \\
&= \frac{\sin\theta}{r_{kj}} \frac{\mathbf{r}_{kj}}{|\mathbf{r}_{ij} \times \mathbf{r}_{kj}|} \times \left(\mathbf{n}_{ijk} \times (-\mathbf{r}_{kj})\right) \\
&= \frac{\sin\theta}{r_{kj}} \frac{1}{|\mathbf{r}_{ij} \times \mathbf{r}_{kj}|}(-\mathbf{n}_{ijk} r_{kj}^2) \\
&= -r_{kj} \sin\theta \frac{\mathbf{n}_{ijk}}{|\mathbf{r}_{ij} \times \mathbf{r}_{kj}|} \tag{14.51}
\end{aligned}$$

$$\begin{aligned}
\text{grad}_{\mathbf{r}_{kl}} \cos\theta_{ijkl} &= \frac{\mathbf{r}_{kj} \times (\mathbf{r}_{kj} \times \mathbf{r}_{ij})}{|\mathbf{r}_{ij} \times \mathbf{r}_{kj}||\mathbf{r}_{kj} \times \mathbf{r}_{kl}|} - \frac{\mathbf{r}_{kj} \times (\mathbf{r}_{kl} \times \mathbf{r}_{kj})}{|\mathbf{r}_{kj} \times \mathbf{r}_{kl}|^2} \cos\theta_{ijkl} \\
&= \frac{\mathbf{r}_{kj}}{|\mathbf{r}_{kj} \times \mathbf{r}_{kl}|} \times (-\mathbf{n}_{ijk} + \mathbf{n}_{jkl} \cos\theta) \\
&= \frac{\mathbf{r}_{kj}}{|\mathbf{r}_{kj} \times \mathbf{r}_{kl}|} \times \left(-\mathbf{n}_{ijk} + \mathbf{n}_{jkl}(\mathbf{n}_{ijk}\mathbf{n}_{jkl})\right) \\
&= -\frac{\mathbf{r}_{kj}}{|\mathbf{r}_{kj} \times \mathbf{r}_{kl}|} \times \left(\mathbf{n}_{jkl} \times (\mathbf{n}_{ijk} \times \mathbf{n}_{jkl})\right)
\end{aligned}$$

$$= -\frac{\mathbf{r}_{kj}}{|\mathbf{r}_{kj} \times \mathbf{r}_{kl}|} \times \left(\mathbf{n}_{jkl} \times \left(\frac{\mathbf{r}_{kj}}{r_{kj}} \sin\theta\right)\right)$$

$$= -\frac{\sin\theta}{r_{kj}|\mathbf{r}_{kj} \times \mathbf{r}_{kl}|} \mathbf{r}_{kj} \times (\mathbf{n}_{jkl} \times \mathbf{r}_{kj})$$

$$= -\frac{r_{kj}\sin\theta}{|\mathbf{r}_{kj} \times \mathbf{r}_{kl}|} \mathbf{n}_{jkl} \tag{14.52}$$

$$\operatorname{grad}_{\mathbf{r}_{kj}} \cos\theta_{ijkl}$$

$$= \frac{\mathbf{r}_{kl} \times (\mathbf{r}_{ij} \times \mathbf{r}_{kj}) + \mathbf{r}_{ij} \times (\mathbf{r}_{kl} \times \mathbf{r}_{kj})}{|\mathbf{r}_{ij} \times \mathbf{r}_{kj}||\mathbf{r}_{kj} \times \mathbf{r}_{kl}|}$$

$$- \frac{\mathbf{r}_{ij} \times (\mathbf{r}_{kj} \times \mathbf{r}_{ij})}{|\mathbf{r}_{ij} \times \mathbf{r}_{kj}|^2} \cos\theta - \frac{\mathbf{r}_{kl} \times (\mathbf{r}_{kj} \times \mathbf{r}_{kl})}{|\mathbf{r}_{kj} \times \mathbf{r}_{kl}|^2} \cos\theta$$

$$= \frac{\mathbf{r}_{ij}}{|\mathbf{r}_{ij} \times \mathbf{r}_{kj}|} \times (-\mathbf{n}_{jkl} + \mathbf{n}_{ijk}\cos\theta) + \frac{\mathbf{r}_{kl}}{|\mathbf{r}_{kj} \times \mathbf{r}_{kl}|} \times (\mathbf{n}_{ijk} - \mathbf{n}_{jkl}\cos\theta)$$

$$= -\frac{\mathbf{r}_{ij}}{|\mathbf{r}_{ij} \times \mathbf{r}_{kj}|} \times (\mathbf{n}_{ijk} \times (\mathbf{n}_{jkl} \times \mathbf{n}_{ijk})) + \frac{\mathbf{r}_{kl}}{|\mathbf{r}_{kj} \times \mathbf{r}_{kl}|} (\mathbf{n}_{jkl} \times (\mathbf{n}_{ijk} \times \mathbf{n}_{jkl}))$$

$$= \frac{\mathbf{r}_{ij}}{|\mathbf{r}_{ij} \times \mathbf{r}_{kj}|} \times \left(\mathbf{n}_{ijk} \times \left(\frac{\mathbf{r}_{kj}}{r_{kj}} \sin\theta\right)\right) + \frac{\mathbf{r}_{kl}}{|\mathbf{r}_{kj} \times \mathbf{r}_{kl}|} \left(\mathbf{n}_{jkl} \times \left(\frac{\mathbf{r}_{kj}}{r_{kj}} \sin\theta\right)\right)$$

$$= \frac{\sin\theta}{r_{kj}} \frac{1}{|\mathbf{r}_{ij} \times \mathbf{r}_{kj}|} \mathbf{r}_{ij} \times (\mathbf{n}_{ijk} \times \mathbf{r}_{kj}) + \frac{\sin\theta}{r_{kj}} \frac{1}{|\mathbf{r}_{kj} \times \mathbf{r}_{kl}|} \mathbf{r}_{kl} \times (\mathbf{n}_{jkl} \times \mathbf{r}_{kj})$$

$$= \frac{\sin\theta}{r_{kj}} \frac{\mathbf{n}_{ijk}(\mathbf{r}_{ij}\mathbf{r}_{kj})}{|\mathbf{r}_{ij} \times \mathbf{r}_{kj}|} + \frac{\sin\theta}{r_{kj}} \frac{\mathbf{n}_{jkl}(\mathbf{r}_{kl}\mathbf{r}_{kj})}{|\mathbf{r}_{kj} \times \mathbf{r}_{kl}|}$$

$$= -\frac{\mathbf{r}_{ij}\mathbf{r}_{kj}}{r_{kj}^2} \operatorname{grad}_{ij} \cos\theta - \frac{\mathbf{r}_{kl}\mathbf{r}_{kj}}{r_{kj}^2} \operatorname{grad}_{kl} \cos\theta. \tag{14.53}$$

14.4 Normal Mode Analysis

The nuclear motion around an equilibrium configuration can be approximately described as the combination of independent harmonic normal modes. Equilibrium configurations can be found with the methods discussed in Sect. 6.2. The convergence is usually rather slow (Fig. 14.8) except for the full Newton-Raphson method, which needs the calculation and inversion of the Hessian matrix.

14.4.1 Harmonic Approximation

At an equilibrium configuration

$$\xi_i = \xi_i^{eq} \tag{14.54}$$

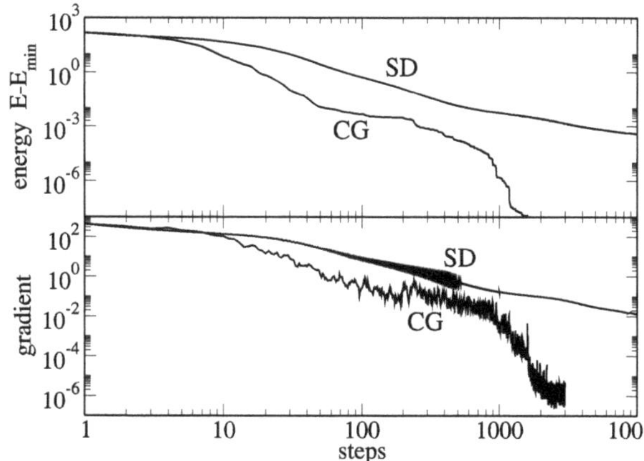

Fig. 14.8 (Convergence of energy and gradient) The energy of the glycine dipeptide is minimized with the methods of steepest descent and conjugate gradients

the gradient of the potential energy vanishes. For small deviations from the equilibrium

$$\zeta_i = \xi_i - \xi_i^{eq} \tag{14.55}$$

approximation by a truncated Taylor series gives

$$U(\zeta_1 \cdots \zeta_{3N}) = U_0 + \frac{1}{2} \sum_{i,j} \frac{\partial^2 U}{\partial \zeta_i \partial \zeta_j} \zeta_i \zeta_j + \cdots \approx U_0 + \frac{1}{2} \sum_{i,j} H_{i,j} \zeta_i \zeta_j \tag{14.56}$$

and the equations of motion are approximately

$$m_i \ddot{\zeta}_i = -\frac{\partial}{\partial \zeta_i} U = -\sum_j H_{i,j} \zeta_j . \tag{14.57}$$

Assuming periodic oscillations

$$\zeta_i = \zeta_i^0 e^{i\omega t} \tag{14.58}$$

we have

$$m_i \omega^2 \zeta_i^0 = \sum_j H_{ij} \zeta_j^0 . \tag{14.59}$$

If mass weighted coordinates are used, defined as

$$\tau_i = \sqrt{m_i} \zeta_i \tag{14.60}$$

this becomes an ordinary eigenvalue problem

$$\omega^2 \tau_i^0 = \sum_j \frac{H_{ij}}{\sqrt{m_i m_j}} \tau_j^0 . \tag{14.61}$$

The eigenvectors $\mathbf{u}_r$ of the symmetric matrix

$$\tilde{H}_{ij} = \frac{H_{ij}}{\sqrt{m_i m_j}} \tag{14.62}$$

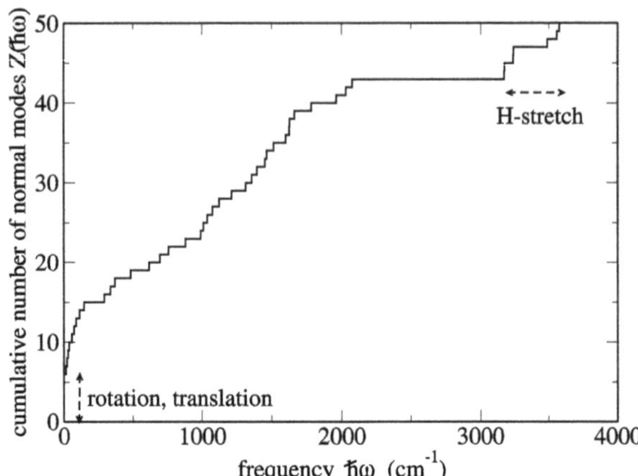

Fig. 14.9 (Normal mode distribution for the dipeptide model) The cumulative distribution (Sect. 8.1.2) of normal mode frequencies is shown for the glycine dipeptide. Translations and rotations of the molecule correspond to the lowest 6 frequencies which are close to zero. The highest frequencies between $3100\ \mathrm{cm}^{-1}$ and $3600\ \mathrm{cm}^{-1}$ correspond to the stretching modes of the 8 hydrogen atoms

are the solutions of

$$\sum_j \tilde{H}_{ij} u_{jr} = \lambda_r u_{ir} \tag{14.63}$$

and satisfy (14.61)

$$\omega^2 u_{ir} = \sum \tilde{H}_{ij} u_{jr} = \lambda_r u_{ir} \tag{14.64}$$

with normal mode frequencies

$$\omega_r = \sqrt{\lambda_r}. \tag{14.65}$$

Finally, the Cartesian coordinates are linear combinations of all normal modes

$$\zeta_i = \sum_r C_r \frac{u_{ir}}{\sqrt{m_i}} e^{i\omega_r t}. \tag{14.66}$$

In a true local energy minimum the Hessian matrix H_{ij} is positive definite and all frequencies are real valued. The six lowest frequencies are close to zero and correspond to translations and rotations of the whole system (Fig. 14.9).

14.5 Problems

Problem 14.1 (Simulation of a glycine dipeptide) In this computer experiment a glycine dipeptide (Fig. 14.5) is simulated. Parameters for bond stretching (Table 14.3) and bond angle (Table 14.4) terms have been derived from quantum calculations by Bautista and Seminario [15].

- Torsional potential terms (Table 14.5) can be added to make the structure more rigid. This is especially important for the O9-H17, N4-H14 and N1-H10 bonds, which rotate almost freely without torsional potentials.

Table 14.5 (Torsional potential terms) Torsional potential terms $V_{ijkl} = k_{ijkl}(1 - \cos(\theta_{ijkl} - \theta^0_{ijkl}))$, which can be added to the forcefield. Minimum angles are from the optimized structure without torsional terms (14.1). The barrier height of $2k_{ijkl} = 2$ kcal/mol is only a guessed value

i	j	k	l	θ^0_{ijkl}	k_{ijkl}	Backbone
10	1	2	3	-67.6	1.0	
14	4	3	2	27.5	1.0	
17	9	6	5	-147.4	1.0	
4	3	2	1	160.7	1.0	Ψ
5	4	3	2	-152.3	1.0	ω
6	5	4	3	-153.1	1.0	Φ
8	6	5	4	123.7	1.0	
9	6	5	4	-56.5	1.0	
15	5	4	3	-32.5	1.0	
16	5	4	3	86.3	1.0	
7	3	2	1	-26.3	1.0	

- The energy can be minimized with the methods of steepest descent or conjugate gradients.
- A normal mode analysis can be performed (the Hessian matrix is calculated by numerical differentiation). The r-th normal mode can be visualized by modulating the coordinates periodically according to

$$\xi_i = \xi_i^{eq} + C_r \frac{u_{ir}}{\sqrt{m_i}} \cos \omega_r t. \tag{14.67}$$

- The motion of the atoms can be simulated with the Verlet method. You can stretch the O9-H17 or N4-H14 bond and observe, how the excitation spreads over the molecule.

Chapter 15
Thermodynamic Systems

An important application for computer simulations is the calculation of thermodynamic averages in an equilibrium system. We discuss two different examples:

In the first case the classical equations of motion are solved for a system of particles interacting pairwise by Lennard-Jones forces (Lennard-Jones fluid). The thermodynamic average is taken along the trajectory, i.e. over the calculated coordinates at different times $\mathbf{r}_i(t_n)$. We evaluate the pair distance distribution function

$$g(R) = \frac{1}{N^2 - N} \left\langle \sum_{i \neq j} \delta(r_{ij} - R) \right\rangle, \tag{15.1}$$

the velocity auto-correlation function

$$C(t) = \left\langle \mathbf{v}(t_0)\mathbf{v}(t) \right\rangle \tag{15.2}$$

and the mean square displacement

$$\Delta x^2 = \left\langle \left(\mathbf{x}(t) - \mathbf{x}(t_0)\right)^2 \right\rangle. \tag{15.3}$$

In the second case the Metropolis method is applied to a one- or two-dimensional system of interacting spins (Ising model). The thermodynamic average is taken over a set of random configurations $\mathbf{q}^{(n)}$. We study the average magnetization

$$\langle M \rangle = \mu \langle S \rangle \tag{15.4}$$

in a magnetic field and the phase transition to the ferromagnetic state.

15.1 Simulation of a Lennard-Jones Fluid

The Lennard-Jones fluid is a simple model of a realistic atomic fluid. It has been studied by computer simulations since Verlet's early work [118, 265] and serves as a test case for the theoretical description of liquids [142, 182] and the liquid-gas [270] and liquid-solid phase transitions [146, 176].

P.O.J. Scherer, *Computational Physics*, Graduate Texts in Physics,
DOI 10.1007/978-3-319-00401-3_15,
© Springer International Publishing Switzerland 2013

In the following we describe a simple computer model of 125 interacting particles[1] without internal degrees of freedom (see problems section). The force on atom i is given by the gradient of the pairwise Lennard-Jones potential (14.24)

$$\mathbf{F}_i = \sum_{j \neq i} \mathbf{F}_{ij} = -4\varepsilon \sum_{j \neq i} \nabla_i \left(\frac{\sigma^{12}}{r_{ij}^{12}} - \frac{\sigma^6}{r_{ij}^6} \right)$$

$$= 4\varepsilon \sum_{j \neq i} \left(\frac{12\sigma^{12}}{r_{ij}^{14}} - \frac{6\sigma^6}{r_{ij}^8} \right) (\mathbf{r}_i - \mathbf{r}_j). \tag{15.5}$$

We use argon parameters $m = 6.69 \times 10^{-26}$ kg, $\varepsilon = 1.654 \times 10^{-21}$ J, $\sigma = 3.405 \times 10^{-10}$ m [3]. After introduction of reduced units for length $\mathbf{r}^* = \frac{1}{\sigma}\mathbf{r}$, energy $E^* = \frac{1}{\varepsilon}E$ and time $t^* = \sqrt{\varepsilon/m\sigma^2}\, t$, the potential energy

$$U^* = \sum_{ij} 4 \left(\frac{1}{r_{ij}^{*12}} - \frac{1}{r_{ij}^{*6}} \right) \tag{15.6}$$

and the equation of motion

$$\frac{\mathrm{d}^2}{\mathrm{d}t^{*2}}\mathbf{r}_i^* = 4 \sum_{j \neq i} \left(\frac{12}{r_{ij}^{*14}} - \frac{6}{r_{ij}^{*8}} \right) (\mathbf{r}_i^* - \mathbf{r}_j^*) \tag{15.7}$$

become universal expressions, i.e. there exists only one universal Lennard-Jones system. To reduce computer time, usually the 6–12 potential is modified at larger distances which can influence the simulation results [239]. In our model a simple cutoff of potential and forces at $r_{max} = 10$ Å is used.

15.1.1 Integration of the Equations of Motion

The equations of motion are integrated with the Verlet algorithm (Sect. 12.11.5)

$$\Delta\mathbf{r}_i = \mathbf{r}_i(t) - \mathbf{r}_i(t - \Delta t) + \frac{\mathbf{F}_i(t)}{m}\Delta t^2 \tag{15.8}$$

$$\mathbf{r}_i(t + \Delta t) = \mathbf{r}_i(t) + \Delta\mathbf{r}_i + O(\Delta t^4). \tag{15.9}$$

We use a higher order expression for the velocities to improve the accuracy of the calculated kinetic energy

$$\mathbf{v}_{i+1} = \frac{\Delta\mathbf{r}_i}{\Delta t} + \frac{5\mathbf{F}_i(t) - 2\mathbf{F}_i(t - \Delta t)}{6m}\Delta t + O(\Delta t^3). \tag{15.10}$$

---

[1] This small number of particles allows a graphical representation of the system during the simulation.

Fig. 15.1 Reflecting walls

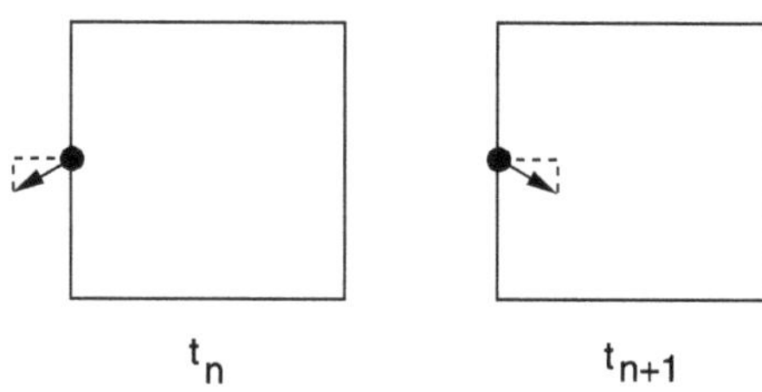

15.1.2 Boundary Conditions and Average Pressure

Molecular dynamics simulations often involve periodic boundary conditions to reduce finite size effects. Here we employ an alternative method which simulates a box with elastic walls. This allows us to calculate explicitly the pressure on the walls of the box.

The atoms are kept in the cube by reflecting walls, i.e. whenever an atom passes a face of the cube, the normal component of the velocity vector is changed in sign (Fig. 15.1). Thus the kinetic energy is conserved but a momentum of $m\,\Delta v = 2mv_\perp$ is transferred to the wall. The average momentum change per time can be interpreted as a force acting upon the wall

$$F_\perp = \left\langle \frac{\sum_{refl.} 2mv_\perp}{dt} \right\rangle. \tag{15.11}$$

The pressure p is given by

$$p = \frac{1}{6L^2}\left\langle \frac{\sum_{walls} \sum_{refl.} 2mv_\perp}{dt} \right\rangle. \tag{15.12}$$

With the Verlet algorithm the reflection can be realized by exchanging the values of the corresponding coordinate at times t_n and t_{n-1}.

15.1.3 Initial Conditions and Average Temperature

At the very beginning the $N = 125$ atoms are distributed over equally spaced lattice points within the cube. Velocities are randomly distributed according to a Gaussian distribution for each Cartesian component v_μ

$$f(v_\mu) = \sqrt{\frac{m}{2\pi k_B T}}\, e^{-mv_\mu^2/2k_B T} \tag{15.13}$$

corresponding to a Maxwell speed distribution

$$f(|v|) = \left(\frac{m}{2\pi k_B T}\right)^{3/2} 4\pi v^2 e^{-mv^2/2k_B T}. \tag{15.14}$$

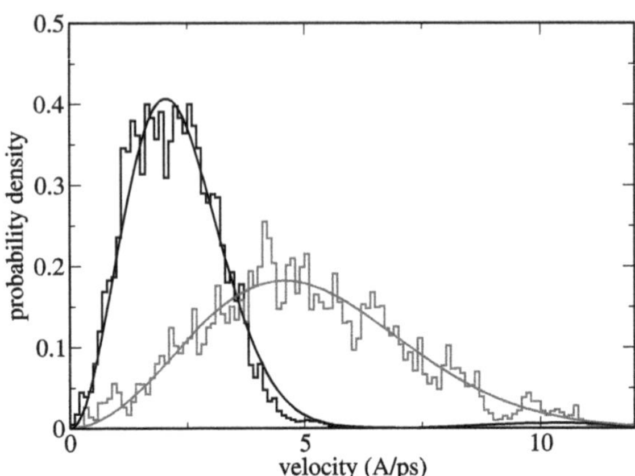

Fig. 15.2 (Velocity distribution) The velocity distribution is shown for $T = 100$ K and $T = 500$ K (*histograms*) and compared to the Maxwell speed distribution (*solid curves*)

Assuming thermal equilibrium, the effective temperature is calculated from the kinetic energy

$$k_B T = \frac{2}{3N} E_{kin}.$$ (15.15)

The desired temperature T_o is established by the rescaling procedure

$$\mathbf{v}_i \rightarrow \mathbf{v}_i \sqrt{\frac{k_B T_o}{k_B T_{actual}}}$$ (15.16)

which is applied repeatedly during an equilibration run. The velocity distribution $f(|v|)$ can be monitored. It approaches quickly a stationary Maxwell distribution (Fig. 15.2).

A smoother method to control temperature is the Berendsen thermostat algorithm [21]

$$\mathbf{v}_i \rightarrow \mathbf{v}_i \sqrt{1 + \frac{\Delta t}{\tau_{therm}} \frac{k T_o - k T_{actual}}{k T_{actual}}}$$ (15.17)

where τ_{therm} is a suitable relaxation time (for instance $\tau_{therm} = 20\Delta t$). This method can be used also during the simulation. However, it does not generate the trajectory of a true canonical ensemble. If this is necessary, more complicated methods have to be used [128].

15.1.4 Analysis of the Results

After an initial equilibration phase the system is simulated at constant energy (NVE simulation) or at constant temperature (NVT) simulation with the Berendsen thermostat method. Several static and dynamic properties can be determined.

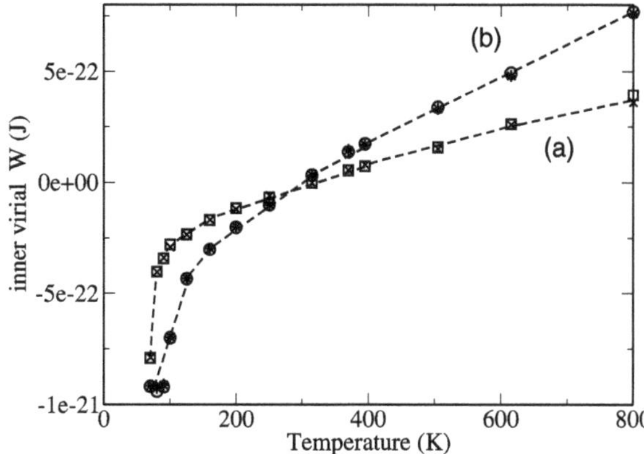

Fig. 15.3 (Inner virial) The inner virial W (15.20, *crosses* and *stars*) is compared to $pV - k_B T$ (*squares* and *circles*) for two values of the particle density $N/V = 10^{-3}\ \text{Å}^{-1}$ (*a*) and $1.95 \times 10^{-3}\ \text{Å}^{-1}$ (*b*), corresponding to reduced densities $n^* = \sigma^3 N/V$ of 0.040 and 0.077

15.1.4.1 Deviation from the Ideal Gas Behavior

A dilute gas is approximately ideal with

$$pV = Nk_B T. \tag{15.18}$$

For a real gas the interaction between the particles has to be taken into account. From the equipartition theorem it can be found that[2]

$$pV = Nk_B T + W \tag{15.19}$$

with the inner virial (Fig. 15.3)

$$W = \left\langle \frac{1}{3} \sum_i \mathbf{r}_i \mathbf{F}_i \right\rangle \tag{15.20}$$

which can be expanded as a power series of the number density $n = N/V$ [231] to give

$$pV = Nk_B T \left(1 + b(T)\frac{N}{V} + c(T)\left(\frac{N}{V}\right)^2 + \cdots \right). \tag{15.21}$$

The virial coefficient $b(T)$ can be calculated exactly for the Lennard-Jones gas [231]:

$$b(T) = -\frac{2\pi}{3}\sigma^3 \sum_{j=0}^{\infty} \frac{2^{j-3/2}}{j!} \Gamma\left(\frac{2j-1}{4}\right) \left(\frac{\varepsilon}{k_B T}\right)^{(j/2+1/4)}. \tag{15.22}$$

For comparison we calculate the quantity

$$\frac{V}{N}\left(\frac{pV}{Nk_B T} - 1\right) \tag{15.23}$$

which for small values of the particle density $n = N/V$ correlates well (Fig. 15.4) with expression (15.22).

---

[2]MD simulations with periodic boundary conditions use this equation to calculate the pressure.

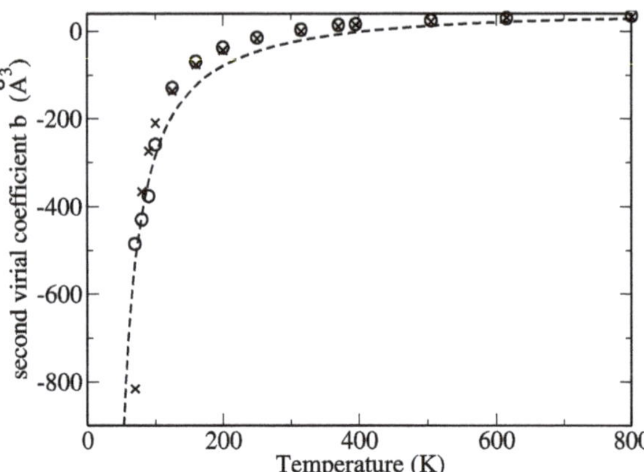

Fig. 15.4 (Second virial coefficient) The value of $\frac{V}{N}(\frac{pV}{k_BT} - 1)$ is shown for two values of the particle density $N/V = 10^{-3}$ Å (*crosses*) and 1.95×10^{-3} Å^{-1} (*circles*) and compared to the exact second virial coefficient b (*dashed curve*) (15.22)

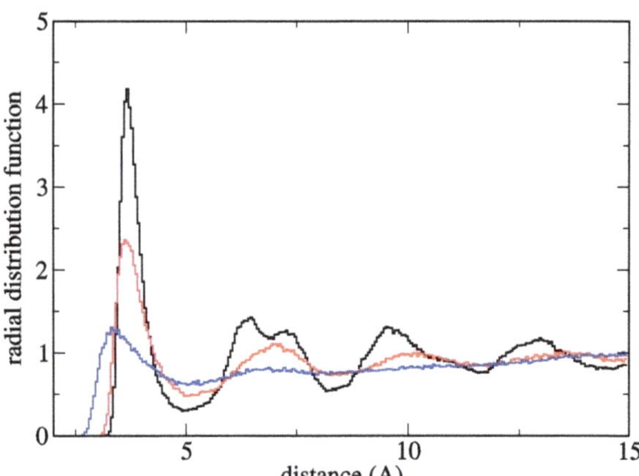

Fig. 15.5 (Radial pair distribution) The normalized radial distribution function $g(R)/g_{ideal}(R)$ is evaluated for $kT = 35$ K, 100 K, 1000 K and a density of $n = 0.025\,A^{-3}$ corresponding to a reduced density $n^* = \sigma^3 N/V$ of 1.0. At this density the Lennard-Jones system shows a liquid-solid transition at a temperature of ca. 180 K [146]

15.1.4.2 Structural Order

A convenient measure for structural order [78] is the radial pair distribution function (Fig. 15.5)

$$g(R) = \left\langle \frac{1}{N(N-1)} \sum_{i \neq j} \delta(r_{ij} - R) \right\rangle = \frac{P(R < r_{ij} < R + dR)}{dR} \qquad (15.24)$$

which is usually normalized with respect to an ideal gas, for which

$$g_{ideal}(R) = 4\pi n R^2 dR. \qquad (15.25)$$

For small distances $g(R)/g_{ideal}(R)$ vanishes due to the strong repulsive force. It peaks at the distance of nearest neighbors and approaches unity at very large distances. In the condensed phase additional maxima appear showing the degree of short (liquid) and long range (solid) order.

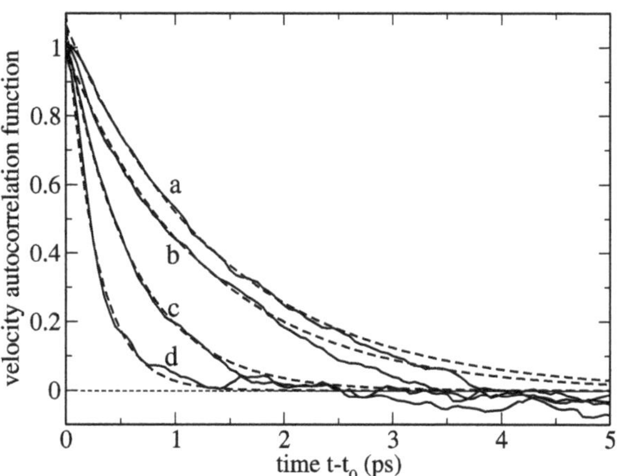

Fig. 15.6 (Velocity auto-correlation function) The Lennard-Jones system is simulated for $k_B T = 200$ K and different values of the density $n^* = 0.12$ (*a*), 0.18 (*b*), 0.32 (*c*), 0.62 (*d*). The velocity auto-correlation function (*full curves*) is averaged over 20 trajectories and fitted by an exponential function (*dashed curves*)

Equation (15.25) is not valid for our small model system without periodic boundary conditions. Therefore g_{ideal} was calculated numerically to normalize the results shown in Fig. 15.5.

15.1.4.3 Ballistic and Diffusive Motion

The velocity auto-correlation function (Fig. 15.6)

$$C(t) = \langle \mathbf{v}(t)\mathbf{v}(t_0)\rangle \tag{15.26}$$

decays as a function of the delay time $t - t_0$ due to collisions of the particles. In a stationary state it does not depend on the initial time t_0. Integration leads to the mean square displacement (Fig. 15.6)

$$\Delta x^2(t) = \langle (\mathbf{x}(t) - \mathbf{x}(t_0))^2\rangle. \tag{15.27}$$

In the absence of collisions the mean square displacement grows with $(t - t_0)^2$, representing a ballistic type of motion. Collisions lead to a diffusive kind of motion where the mean square displacement grows only linearly with time. The transition between this two types of motion can be analyzed within the model of Brownian motion [219] where the collisions are replaced by a fluctuating random force $\Gamma(t)$ and a damping constant γ.

The equation of motion in one dimension is

$$\dot{v} + \gamma v = \Gamma(t) \tag{15.28}$$

with

$$\langle \Gamma(t)\rangle = 0 \tag{15.29}$$

$$\langle \Gamma(t)\Gamma(t')\rangle = \frac{2\gamma k_B T}{m}\delta(t - t'). \tag{15.30}$$

The velocity correlation decays exponentially

$$\langle v(t)v(t_0)\rangle = \frac{k_BT}{m}e^{-\gamma|t-t_0|} \tag{15.31}$$

with the average velocity square given by

$$\langle v^2\rangle = C(t_0) = \frac{k_BT}{m} = \frac{\langle E_{kin}\rangle}{\frac{m}{2}} \tag{15.32}$$

and the integral of the correlation function equals

$$\int_{t_0}^{\infty} C(t)\,dt = \frac{k_BT}{\gamma m}. \tag{15.33}$$

The average of Δx^2 is

$$\langle (x(t)-x(t_0))^2\rangle = \frac{2k_BT}{m\gamma}(t-t_0) - \frac{2k_BT}{m\gamma^2}\left(1-e^{-\gamma(t-t_0)}\right). \tag{15.34}$$

For small time differences $t-t_0$ the motion is ballistic with the thermal velocity

$$\langle (x(t)-x(t_0))^2\rangle \approx \frac{k_BT}{m}(t-t_0)^2 = \langle v^2\rangle(t-t_0)^2. \tag{15.35}$$

For large time differences diffusive motion emerges with

$$\langle (x(t)-x(t_0))^2\rangle \approx \frac{2k_BT}{m\gamma}(t-t_0) = 2D(t-t_0) \tag{15.36}$$

with the diffusion constant given by the Einstein relation

$$D = \frac{k_BT}{m\gamma}. \tag{15.37}$$

For a three-dimensional simulation the Cartesian components of the position or velocity vector add up independently. The diffusion coefficient can be determined from

$$D = \frac{1}{6}\lim_{t\to\infty}\frac{\langle (x(t)-x(t_0))^2\rangle}{t-t_0} \tag{15.38}$$

or, alternatively from (15.33) [3]

$$D = \frac{1}{3}\int_{t_0}^{\infty}\langle \mathbf{v}(t)\mathbf{v}(t_0)\rangle\,dt. \tag{15.39}$$

This equation is more generally valid also outside the Brownian limit (Green-Kubo formula). The Brownian model represents the simulation data quite well at low particle densities (Figs. 15.6, 15.7). For higher densities the velocity auto-correlation function shows a very rapid decay followed by a more or less structured tail [3, 159, 256].

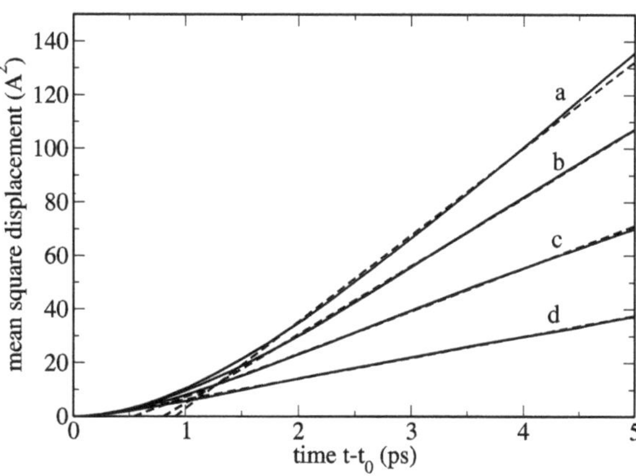

Fig. 15.7 (Mean square displacement) The Lennard-Jones system is simulated for $k_B T = 200$ K and different values of the density $n^* = 0.12$ (*a*), 0.18 (*b*), 0.32 (*c*), 0.62 (*d*). The mean square displacement (*full curves*) is averaged over 20 trajectories and fitted by a linear function (*dashed lines*) for $t - t_0 > 1.5$ ps

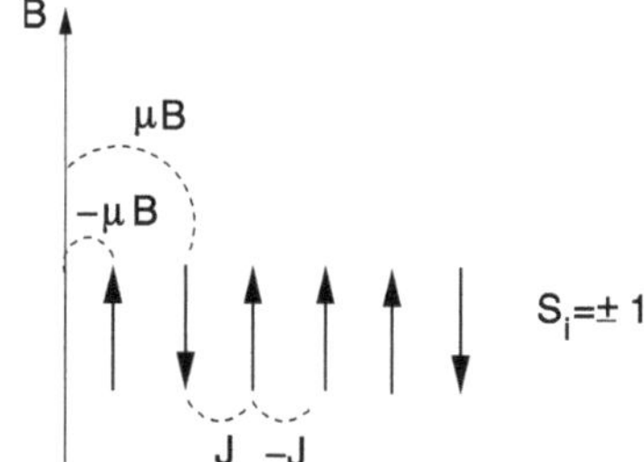

Fig. 15.8 (Ising model) N spins can be up or down. The interaction with the magnetic field is $-\mu B S_i$, the interaction between nearest neighbors is $-J S_i S_j$

15.2 Monte Carlo Simulation

The basic principles of Monte Carlo simulations are discussed in Chap. 8. Here we will apply the Metropolis algorithm to simulate the Ising model in one or two dimensions. The Ising model [26, 134] is primarily a model for the phase transition of a ferromagnetic system. However, it has further applications for instance for a polymer under the influence of an external force or protonation equilibria in proteins.

15.2.1 One-Dimensional Ising Model

We consider a chain consisting of N spins which can be either up ($S_i = 1$) or down ($S_i = -1$). The total energy in a magnetic field is (Fig. 15.8)

$$H = -MB = -B \sum_{i=1}^{N} \mu S_i \tag{15.40}$$

and the average magnetic moment of one spin is

$$\langle M \rangle = \mu \frac{e^{\mu B/kT} - e^{-\mu B/kT}}{e^{\mu B/kT} + e^{-\mu B/kT}} = \mu \tanh\left(\frac{\mu B}{kT}\right). \tag{15.41}$$

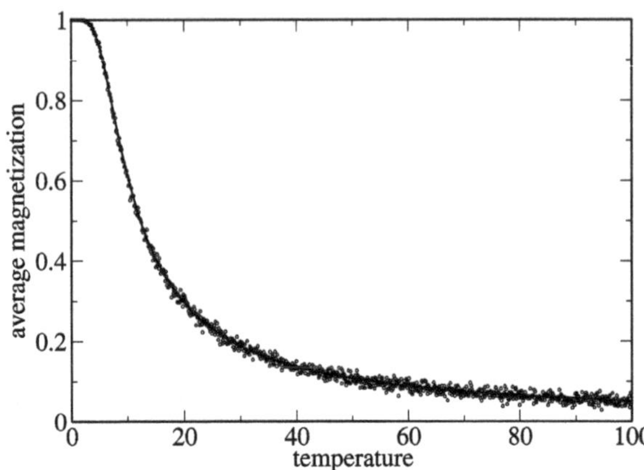

Fig. 15.9 (Numerical simulation of the 1-dimensional Ising model) The average magnetization per spin is calculated from a MC simulation (*circles*) and compared to the exact solution (15.43). Parameters are $\mu B = -5$ and $J = -2$

If interaction between neighboring spins is included the energy of a configuration $(S_1 \cdots S_N)$ becomes

$$H = -\mu B \sum_{i=1}^{N} S_i - J \sum_{i=1}^{N-1} S_i S_{i+1}.$$ (15.42)

The 1-dimensional model can be solved analytically [231]. In the limit $N \to \infty$ the magnetization is

$$\langle M \rangle = \mu \frac{\sinh(\frac{\mu B}{kT})}{\sqrt{\sinh^2(\frac{\mu B}{kT}) + e^{4J/kT}}}.$$ (15.43)

The numerical simulation (Fig. 15.9) starts either with the ordered state $S_i = 1$ or with a random configuration. New configurations are generated with the Metropolis method as follows:

- flip one randomly chosen spin S_i[3] and calculate the energy change due to the change $\Delta S_i = (-S_i) - S_i = -2S_i$

$$\Delta E = -\mu B \Delta S_i - J \Delta S_i (S_{i+1} + S_{i-1}) = 2\mu B S_i + 2J S_i (S_{i+1} + S_{i-1})$$ (15.44)

- if $\Delta E < 0$ then accept the flip, otherwise accept it with a probability of $P = e^{-\Delta E/kT}$

As a simple example consider $N = 3$ spins which have 8 possible configurations. The probabilities of the trial step $T_{i \to j}$ are shown in Table 15.1.

The table is symmetric and all configurations are connected.

---

[3]Or try one spin after the other.

Table 15.1 transition probabilities for a 3-spin system ($p = 1/3$)

	$+++$	$++-$	$+-+$	$+--$	$-++$	$-+-$	$--+$	$---$
$+++$	0	p	p	0	p	0	0	0
$++-$	p	0	0	p	0	p	0	0
$+-+$	p	0	0	p	0	0	p	0
$+--$	0	p	p	0	0	0	0	p
$-++$	p	0	0	0	0	p	p	0
$-+-$	0	p	0	0	p	0	0	p
$--+$	0	0	p	0	p	0	0	p
$---$	0	0	0	p	0	p	p	0

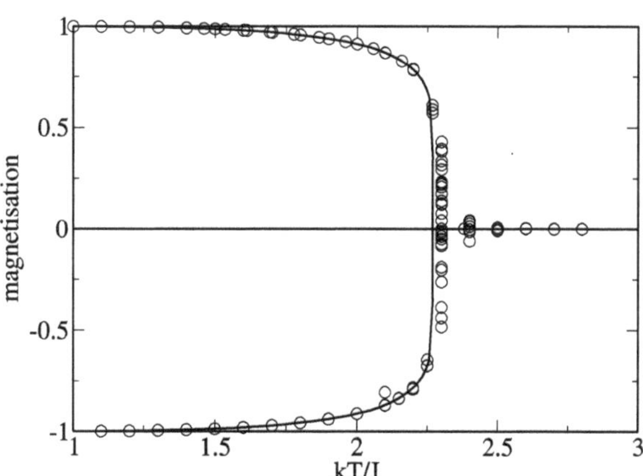

Fig. 15.10 (Numerical simulation of the 2-dimensional Ising model) The average magnetization per spin is calculated for $B = 0$ from a MC simulation (*circles*) and compared to (15.46)

15.2.2 Two-Dimensional Ising Model

For dimension $d > 1$ the Ising model behaves qualitatively different as a phase transition appears. For $B = 0$ (Fig. 15.10) the 2-dimensional Ising-model with 4 nearest neighbors can be solved analytically [171, 194]. The magnetization disappears above the critical temperature T_c, which is given by

$$\frac{J}{kT_c} = -\frac{1}{2}\ln(\sqrt{2}-1) \approx \frac{1}{2.27}. \tag{15.45}$$

Below T_c the average magnetization is given by

$$\langle M \rangle = \left(1 - \frac{1}{\sinh^4(\frac{2J}{kT})}\right)^{\frac{1}{8}}. \tag{15.46}$$

Fig. 15.11 Two state model

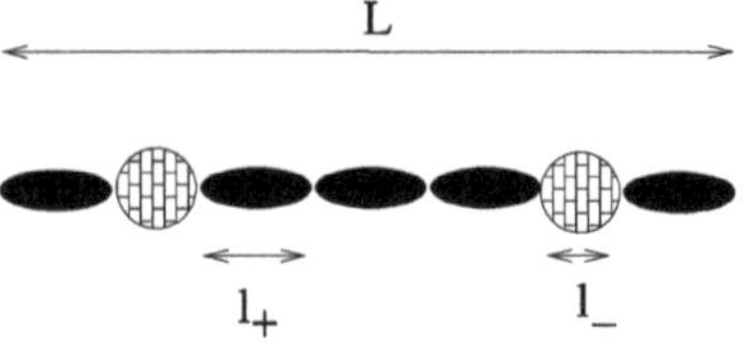

15.3 Problems

Problem 15.1 (Lennard-Jones fluid) In this computer experiment a Lennard-Jones fluid is simulated. The pressure is calculated from the average transfer of momentum (15.12) and compared with expression (15.19).

- Equilibrate the system and observe how the distribution of squared velocities approaches a Maxwell distribution.
- Equilibrate the system for different values of temperature and volume and investigate the relation between pV/N and kT.
- observe the radial distribution function for different values of temperature and densities. Try to locate phase transitions.
- determine the decay time of the velocity correlation function and compare with the behavior of the mean square displacement which shows a transition from ballistic to diffusive motion.

Problem 15.2 (One-dimensional Ising model) In this computer experiment we simulate a linear chain of $N = 500$ spins with periodic boundaries and interaction between nearest neighbors only. We go along the chain and try to flip one spin after the other according to the Metropolis method.

After trying to flip the last spin S_N the total magnetization

$$M = \sum_{i=1}^{N} S_i \tag{15.47}$$

is calculated. It is averaged over 500 such cycles and then compared graphically with the analytical solution for the infinite chain (15.43). Temperature and magnetic field can be varied.

Problem 15.3 (Two-state model for a polymer) Consider a polymer (Fig. 15.11) consisting of N units which can be in two states $S_i = +1$ or $S_i = -1$ with corresponding lengths l_+ and l_-. The interaction between neighboring units takes one of the values w_{++}, w_{+-}, w_{--}. Under the influence of an external force κ the energy of the polymer is

$$E = -\kappa \sum_i l(S_i) + \sum_i w(S_i, S_{i+1}). \tag{15.48}$$

This model is isomorphic to the one-dimensional Ising model,

$$E = -\kappa N \frac{l_- + l_+}{2} - \kappa \frac{l_+ - l_-}{2} \sum S_i \tag{15.49}$$

$$+ \sum \left(w_{+-} + \frac{w_{++} - w_{+-}}{2} S_i + \frac{w_{+-} - w_{--}}{2} S_{i+1} \right.$$

$$\left. + \frac{w_{++} + w_{--} - 2w_{+-}}{2} S_i S_{i+1} \right) \tag{15.50}$$

$$= \kappa N \frac{l_- + l_+}{2} + N w_{+-}$$

$$- \kappa \frac{l_+ - l_-}{2} M + \frac{w_{++} - w_{--}}{2} M$$

$$+ \frac{w_{++} + w_{--} - 2w_{+-}}{2} \sum S_i S_{i+1}. \tag{15.51}$$

Comparison with (15.42) shows the correspondence

$$-J = \frac{w_{++} + w_{--} - 2w_{+-}}{2} \tag{15.52}$$

$$-\mu B = -\kappa \frac{l_+ - l_-}{2} + \frac{w_{++} - w_{--}}{2} \tag{15.53}$$

$$L = \sum l(S_i) = N \frac{l_+ + l_-}{2} + \frac{l_+ - l_-}{2} M. \tag{15.54}$$

In this computer experiment we simulate a linear chain of $N = 20$ units with periodic boundaries and nearest neighbor interaction as in the previous problem.

The fluctuations of the chain conformation are shown graphically and the magnetization of the isomorphic Ising model is compared with the analytical expression for the infinite system (15.43). Temperature and magnetic field can be varied as well as the coupling J. For negative J the anti-ferromagnetic state becomes stable at low magnetic field strengths.

Problem 15.4 (Two-dimensional Ising model) In this computer experiment a 200 × 200 square lattice with periodic boundaries and interaction with the 4 nearest neighbors is simulated. The fluctuations of the spins can be observed. At low temperatures ordered domains with parallel spin appear. The average magnetization is compared with the analytical expression for the infinite system (15.46).

Chapter 16
Random Walk and Brownian Motion

Random walk processes are an important class of stochastic processes. They have many applications in physics, computer science, ecology, economics and other fields. A random walk [199] is a sequence of successive random steps. In this chapter we study Markovian [166, 167][1] discrete time[2] models. In one dimension the position of the walker after n steps approaches a Gaussian distribution, which does not depend on the distribution of the single steps. This follows from the central limit theorem and can be checked in a computer experiment. A 3-dimensional random walk provides a simple statistical model for the configuration of a biopolymer, the so called freely jointed chain model. In a computer experiment we generate random structures and calculate the gyration tensor, an experimentally observable quantity,which gives information on the shape of a polymer. Simulation of the dynamics is simplified if the fixed length segments of the freely jointed chain are replaced by Hookean springs. This is utilized in a computer experiment to study the dependence of the polymer extension on an applied external force (this effect is known as entropic elasticity). The random motion of a heavy particle in a bath of light particles, known as Brownian motion, can be described by Langevin dynamics, which replace the collisions with the light particles by an average friction force proportional to the velocity and a randomly fluctuating force with zero mean and infinitely short correlation time. In a computer experiment we study Brownian motion in a harmonic potential.

16.1 Markovian Discrete Time Models

The time evolution of a system is described in terms of an N-dimensional vector $\mathbf{r}(t)$, which can be for instance the position of a molecule in a liquid, or the price

---

[1] Different steps are independent.

[2] A special case of the more general continuous time random walk with a waiting time distribution of $P(\tau) = \delta(\tau - \Delta t)$.

P.O.J. Scherer, *Computational Physics*, Graduate Texts in Physics,
DOI 10.1007/978-3-319-00401-3_16,

Fig. 16.1 Discrete time random walk

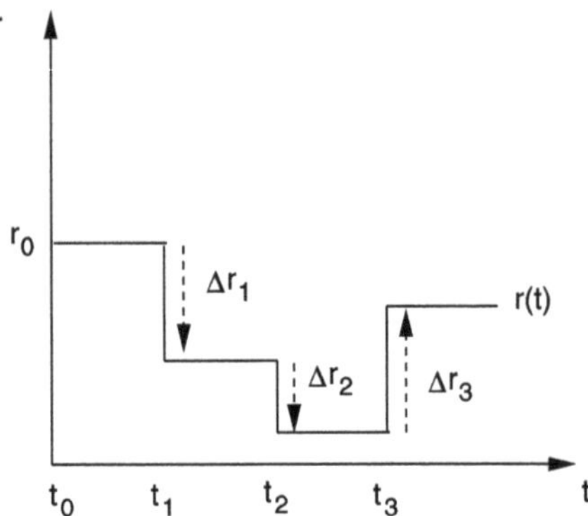

of a fluctuating stock. At discrete times $t_n = n\Delta t$ the position changes suddenly (Fig. 16.1)

$$\mathbf{r}(t_{n+1}) = \mathbf{r}(t_n) + \Delta\mathbf{r}_n \tag{16.1}$$

where the steps are distributed according to the probability distribution[3]

$$P(\Delta\mathbf{r}_n = \mathbf{b}) = f(\mathbf{b}). \tag{16.2}$$

The probability of reaching the position $\mathbf{R}$ after $n + 1$ steps obeys the equation

$$P_{n+1}(\mathbf{R}) = P\big(\mathbf{r}(t_{n+1}) = \mathbf{R}\big)$$
$$= \int d^N\mathbf{b}\, P_n(\mathbf{R} - \mathbf{b}) f(\mathbf{b}). \tag{16.3}$$

16.2 Random Walk in One Dimension

Consider a random walk in one dimension. We apply the central limit theorem to calculate the probability distribution of the position r_n after n steps. The first two moments and the standard deviation of the step distribution are

$$\overline{b} = \int db\, b f(b) \quad \overline{b^2} = \int db\, b^2 f(b) \quad \sigma_b = \sqrt{\overline{b^2} - \overline{b}^2}. \tag{16.4}$$

Hence the normalized quantity

$$\xi_i = \frac{\Delta x_i - \overline{b}}{\sigma_b} \tag{16.5}$$

is a random variable with zero average and unit standard deviation. The distribution function of the new random variable

---

[3] General random walk processes are characterized by a distribution function $P(\mathbf{R}, \mathbf{R}')$. Here we consider only correlated processes for which $P(\mathbf{R}, \mathbf{R}') = P(\mathbf{R}' - \mathbf{R})$.

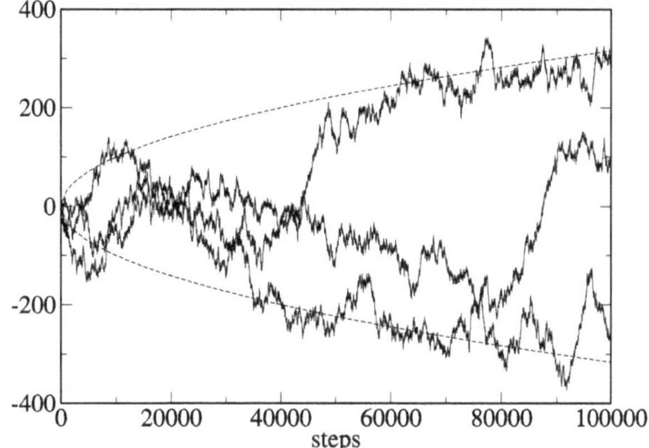

Fig. 16.2 (Random walk with constant step size) The figure shows the position r_n for three different 1-dimensional random walks with step size $\Delta x = \pm 1$. The *dashed curves* show the width $\pm\sigma = \pm\sqrt{n}$ of the Gaussian approximation (16.8)

$$\eta_n = \frac{\xi_1 + \xi_2 + \cdots + \xi_n}{\sqrt{n}} = \frac{r_n - n\overline{b}}{\sigma_b \sqrt{n}} \tag{16.6}$$

approaches a normal distribution for large n

$$f(\eta_n) \to \frac{1}{\sqrt{2\pi}} e^{-\eta_n^2/2} \tag{16.7}$$

and finally from

$$f(r_n)dr_n = f(\eta_n)d\eta_n = f(\eta_n)\frac{dr_n}{\sigma_b \sqrt{n}}$$

we have

$$f(r_n) = \frac{1}{\sqrt{2\pi n}\sigma_b} \exp\left\{-\frac{(r_n - n\overline{b})^2}{2n\sigma_b^2}\right\}. \tag{16.8}$$

The position of the walker after n steps obeys approximately a Gaussian distribution centered at $\overline{r}_n = n\overline{b}$ with a standard deviation of

$$\sigma_{r_n} = \sqrt{n}\sigma_b. \tag{16.9}$$

16.2.1 Random Walk with Constant Step Size

In the following we consider the classical example of a 1-dimensional random walk process with constant step size. At time t_n the walker takes a step of length Δx to the left with probability p or to the right with probability $q = 1 - p$ (Figs. 16.2, 16.3).

The corresponding step size distribution function is

$$f(b) = p\delta(b + \Delta x) + q\delta(b - \Delta x) \tag{16.10}$$

Fig. 16.3 Random walk with constant step size

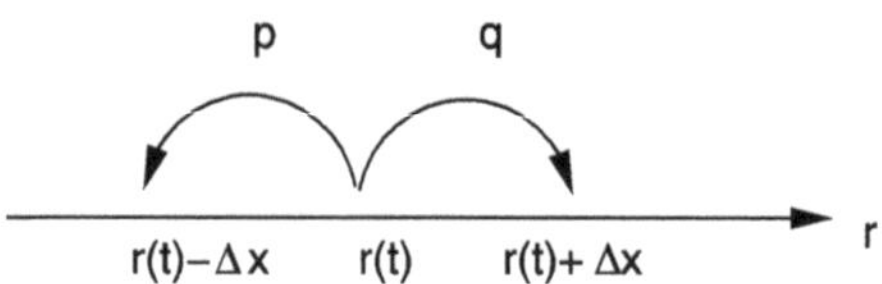

with the first two moments

$$\overline{b} = (q - p)\Delta x \quad \overline{b^2} = \Delta x^2. \tag{16.11}$$

Let the walker start at $r(t_0) = 0$. The probability $P_n(m)$ of reaching position $m\Delta x$ after n steps obeys the recursion

$$P_{n+1}(m) = pP_n(m+1) + qP_n(m-1) \tag{16.12}$$

which obviously leads to a binomial distribution. From the expansion of

$$(p+q)^n = \sum \binom{n}{m} p^m q^{n-m} \tag{16.13}$$

we see that

$$P_n(n-2m) = \binom{n}{m} p^m q^{n-m} \tag{16.14}$$

or after substitution $m' = n - 2m = -n, -n+2 \cdots n-2, n$:

$$P_n(m') = \binom{n}{(n-m')/2} p^{(n-m')/2} q^{(n+m')/2}. \tag{16.15}$$

Since the steps are uncorrelated we easily find the first two moments

$$\overline{r}_n = \sum_{i=1}^{n} \overline{\Delta x_i} = n\overline{b} = n\Delta x(q-p) \tag{16.16}$$

and

$$\overline{r_n^2} = \overline{\left(\sum_{i=1}^{n} \Delta x_i\right)^2} = \overline{\sum_{i,j=1}^{n} \Delta x_i \Delta x_j} = \sum_{i=1}^{n} \overline{(\Delta x_i)^2} = n\overline{b^2} = n\Delta x^2. \tag{16.17}$$

16.3 The Freely Jointed Chain

We consider a simple statistical model for the conformation of a biopolymer like DNA or a protein.

The polymer is modeled by a 3-dimensional chain consisting of M units with constant bond length and arbitrary relative orientation (Fig. 16.4). The configuration can be described by a point in a $3(M+1)$-dimensional space which is reached after M steps $\Delta \mathbf{r}_i = \mathbf{b}_i$ of a 3-dimensional random walk with constant step size

$$\mathbf{r}_M = \mathbf{r}_0 + \sum_{i=1}^{M} \mathbf{b}_i. \tag{16.18}$$

Fig. 16.4 Freely jointed
chain with constant bond
length b

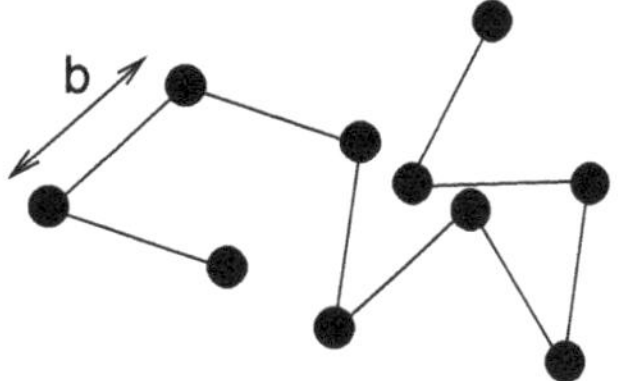

16.3.1 Basic Statistic Properties

The M bond vectors

$$\mathbf{b}_i = \mathbf{r}_i - \mathbf{r}_{i-1} \tag{16.19}$$

have a fixed length $|\mathbf{b}_i| = b$ and are oriented randomly. The first two moments are

$$\overline{\mathbf{b}_i} = 0 \quad \overline{\mathbf{b}_i^2} = b^2. \tag{16.20}$$

Since different units are independent

$$\overline{\mathbf{b}_i \mathbf{b}_j} = \delta_{i,j} b^2. \tag{16.21}$$

Obviously the relative position of segment j

$$\mathbf{R}_j = \mathbf{r}_j - \mathbf{r}_0 = \sum_{i=1}^{j} \mathbf{b}_i$$

has zero mean

$$\overline{\mathbf{R}_j} = \sum_{i=1}^{j} \overline{\mathbf{b}_i} = 0 \tag{16.22}$$

and its second moment is

$$\overline{R_j^2} = \overline{\left(\sum_{i=1}^{j} \mathbf{b}_i \sum_{k=1}^{j} \mathbf{b}_k \right)} = \sum_{i,k=1}^{j} \overline{\mathbf{b}_i \mathbf{b}_k} = jb^2. \tag{16.23}$$

For the end to end distance (Fig. 16.5)

$$\mathbf{R}_M = \mathbf{r}_M - \mathbf{r}_0 = \sum_{i=1}^{M} \mathbf{b}_i \tag{16.24}$$

this gives

$$\overline{\mathbf{R}_M} = 0, \quad \overline{R_M^2} = Mb^2. \tag{16.25}$$

Let us apply the central limit theorem for large M. For the x coordinate of the end
to end vector we have

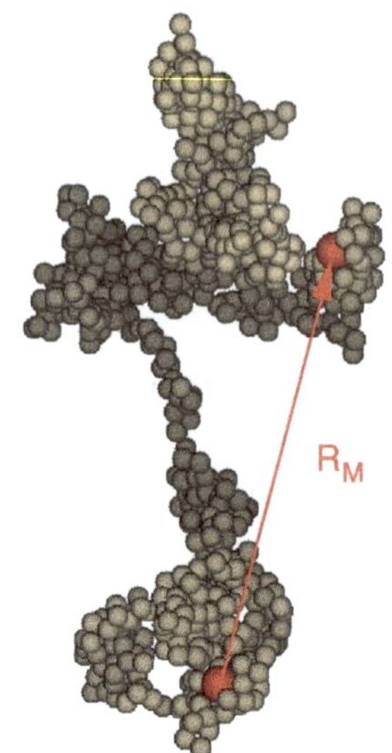

Fig. 16.5 (Freely jointed chain) The figure shows a random 3-dimensional structure with 1000 segments visualized as balls (Molden graphics [223])

$$X = \sum_{i=1}^{M} \mathbf{b}_i \mathbf{e}_x = b \sum_i \cos \theta_i . \tag{16.26}$$

With the help of the averages[4]

$$\overline{\cos \theta_i} = \frac{1}{4\pi} \int_0^{2\pi} d\phi \int_0^{\pi} \cos \theta \sin \theta \, d\theta = 0 \tag{16.27}$$

$$\overline{(\cos \theta_i)^2} = \frac{1}{4\pi} \int_0^{2\pi} d\phi \int_0^{\pi} \cos^2 \theta \sin \theta d\theta = \frac{1}{3} \tag{16.28}$$

we find that the scaled difference

$$\xi_i = \sqrt{3} \cos \theta_i \tag{16.29}$$

has zero mean and unit variance and therefore the sum

$$\widetilde{X} = \frac{\sqrt{3}}{b\sqrt{M}} X = \sqrt{\frac{3}{M}} \sum_{i=1}^{M} \cos \theta_i \tag{16.30}$$

converges to a normal distribution:

$$P(\widetilde{X}) = \frac{1}{\sqrt{2\pi}} \exp\left\{ -\frac{\widetilde{X}^2}{2} \right\}. \tag{16.31}$$

---

[4]For a 1-dimensional polymer $\overline{\cos \theta_i} = 0$ and $\overline{(\cos \theta_i)^2} = 1$. In two dimensions $\overline{\cos \theta_i} = \frac{1}{\pi} \int_0^{\pi} \cos \theta \, d\theta = 0$ and $\overline{(\cos \theta_i)^2} = \frac{1}{\pi} \int_0^{\pi} \cos^2 \theta \, d\theta = \frac{1}{2}$. To include these cases the factor 3 in the exponent of (16.33) should be replaced by the dimension d.

Hence

$$P(X) = \frac{1}{\sqrt{2\pi}} \frac{\sqrt{3}}{b\sqrt{M}} \exp\left\{-\frac{3}{2Mb^2}X^2\right\} \tag{16.32}$$

and finally in 3 dimensions

$$P(\mathbf{R}_M) = P(X)\,P(Y)\,P(Z)$$
$$= \frac{\sqrt{27}}{b^3\sqrt{(2\pi M)^3}} \exp\left\{-\frac{3}{2Mb^2}\mathbf{R}_M^2\right\}. \tag{16.33}$$

16.3.2 Gyration Tensor

For the center of mass

$$\mathbf{R}_c = \frac{1}{M}\sum_{i=1}^{M}\mathbf{R}_i$$

we find

$$\overline{\mathbf{R}_c} = 0 \quad \overline{R_c^2} = \frac{1}{M^2}\sum_{i,j}\overline{\mathbf{R}_i\mathbf{R}_j}$$

and since

$$\overline{\mathbf{R}_i\mathbf{R}_j} = \min(i,j)b^2$$

we have

$$\overline{R_c^2} = \frac{b^2}{M^2}\left(2\sum_{i=1}^{M}i(M-i+1)-\sum_{i=1}^{M}i\right) = \frac{b^2}{M^2}\left(\frac{M^3}{3}+\frac{M^2}{2}+\frac{M}{6}\right) \approx \frac{Mb^2}{3}.$$

The gyration radius [170] is generally defined by

$$R_g^2 = \frac{1}{M}\sum_{i=1}^{M}\overline{(\mathbf{R}_i-\mathbf{R}_c)^2}$$
$$= \frac{1}{M}\sum_{i=1}^{M}\left(\overline{R_i^2}+\overline{R_c^2}-2\frac{1}{M}\sum_{j=1}^{M}\overline{\mathbf{R}_i\mathbf{R}_j}\right) = \frac{1}{M}\sum_{i}\overline{(R_i^2)}-\overline{R_c^2}$$
$$= b^2\frac{M+1}{2}-\frac{b^2}{M^2}\left(\frac{M^3}{3}+\frac{M^2}{2}+\frac{M}{6}\right) = b^2\left(\frac{M}{6}-\frac{1}{6M}\right) \approx \frac{Mb^2}{6}.$$

R_g can be also written as

$$R_g^2 = \left(\frac{1}{M}\sum_i\overline{R_i^2}-\frac{1}{M^2}\sum_{ij}\overline{\mathbf{R}_i\mathbf{R}_j}\right) = \frac{1}{2M^2}\sum_{i=1}^{M}\sum_{j=1}^{M}\overline{(\mathbf{R}_i-\mathbf{R}_j)^2}$$

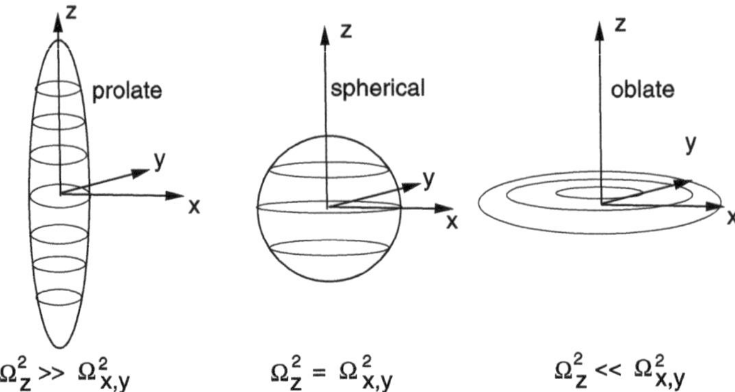

$$\Omega_z^2 \gg \Omega_{x,y}^2 \qquad\qquad \Omega_z^2 = \Omega_{x,y}^2 \qquad\qquad \Omega_z^2 \ll \Omega_{x,y}^2$$

Fig. 16.6 (Gyration tensor) The eigenvalues of the gyration tensor give information on the shape of the polymer. If the extension is larger (smaller) along one direction than in the perpendicular plane, one eigenvalue is larger (smaller) than the two other

Fig. 16.7 Polymer model with Hookean springs

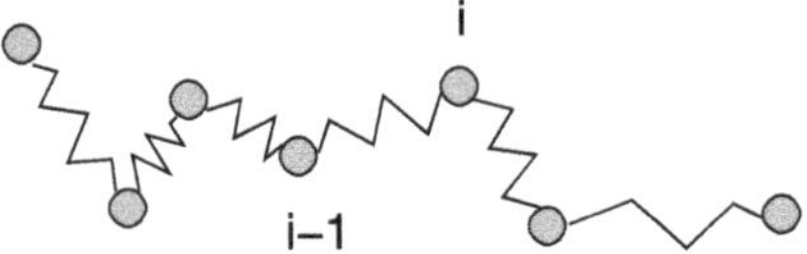

and can be experimentally measured with the help of scattering phenomena. It is related to the gyration tensor which is defined as

$$\Omega_g = \frac{1}{M} \sum_i \overline{(\mathbf{R}_i - \mathbf{R}_c)(\mathbf{R}_i - \mathbf{R}_c)^T}.$$

Its trace is

$$\mathrm{tr}(\Omega_g) = R_g^2$$

and its eigenvalues give us information about the shape of the polymer (Fig. 16.6).

16.3.3 Hookean Spring Model

Simulation of the dynamics of the freely jointed chain is complicated by the constraints which are implied by the constant chain length. Much simpler is the simulation of a model which treats the segments as Hookean springs (Fig. 16.7). In the limit of a large force constant the two models give equivalent results.

We assume that the segments are independent (self crossing is not avoided). Then for one segment the energy contribution is

$$E_i = \frac{f}{2}\left(|\mathbf{b}_i| - b\right)^2.$$

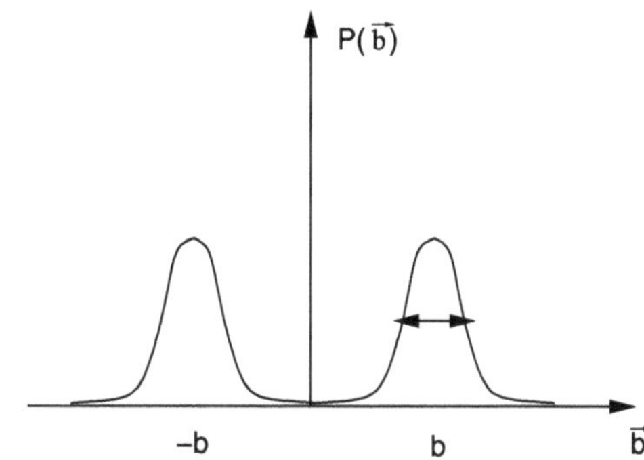

Fig. 16.8 (Distribution of bond vectors) The bond vector distribution for a 1-dimensional chain of springs has maxima at $\pm b$. For large force constants the width of the two peaks becomes small and the chain of springs resembles a freely jointed chain with constant bond length

If the fluctuations are small

$$\overline{\left|\,|\mathbf{b}_i| - b\,\right|} \ll b$$

then (Fig. 16.8)

$$\overline{|\mathbf{b}_i|} \approx b \quad \overline{\mathbf{b}_i^2} \approx b^2$$

and the freely jointed chain model (16.33) gives the entropy as a function of the end to end vector

$$S = -k_B \ln\big(P(\mathbf{R}_M)\big) = -k_B \ln\left(\frac{\sqrt{27}}{b^3 \sqrt{(2\pi M)^3}}\right) + \frac{3k_B}{2Mb^2}\mathbf{R}_M{}^2. \tag{16.34}$$

If one end of the polymer is fixed at $\mathbf{r}_0 = 0$ and a force κ is applied to the other end, the free energy is given by

$$F = TS - \kappa\mathbf{R}_M = \frac{3k_BT}{2Mb^2}\mathbf{R}_M^2 - \kappa\mathbf{R}_M + \text{const.} \tag{16.35}$$

In thermodynamic equilibrium the free energy is minimal, hence the average extension is

$$\overline{\mathbf{R}_M} = \frac{Mb^2}{3k_BT}\kappa. \tag{16.36}$$

This linear behavior is similar to a Hookean spring with an effective force constant

$$f_{\text{eff}} = \frac{Mb^2}{3k_BT} \tag{16.37}$$

and is only valid for small forces. For large forces the freely jointed chain asymptotically reaches its maximum length of $R_{M,\text{max}} = Mb$, whereas for the chain of springs $R_M \to M(b + \kappa/f)$.

16.4 Langevin Dynamics

A heavy particle moving in a bath of much smaller and lighter particles (for instance atoms and molecules of the air) shows what is known as Brownian motion

[41, 76, 77]. Due to collisions with the thermally moving bath particles it experiences a fluctuating force which drives the particle into a random motion. The French physicist Paul Langevin developed a model to describe this motion without including the light particles explicitly. The fluctuating force is divided into a macroscopic friction force proportional to the velocity

$$\mathbf{F}_{fr} = -\gamma\mathbf{v} \tag{16.38}$$

and a randomly fluctuating force with zero mean and infinitely short correlation time

$$\overline{\mathbf{F}_{rand}(t)} = 0 \quad \overline{\mathbf{F}_{rand}(t)\mathbf{F}_{rand}(t')} = \overline{\mathbf{F}_{rand}^2}\delta(t - t'). \tag{16.39}$$

The equations of motion for the heavy particle are

$$\begin{aligned} \frac{d}{dt}\mathbf{x} &= \mathbf{v} \\ \frac{d}{dt}\mathbf{v} &= -\gamma\mathbf{v} + \frac{1}{m}\mathbf{F}_{fr}(t) - \frac{1}{m}\nabla U(\mathbf{x}) \end{aligned} \tag{16.40}$$

with the macroscopic friction coefficient γ and the potential $U(\mathbf{x})$.

The behavior of the random force can be better understood if we introduce a time grid $t_{n+1} - t_n = \Delta t$ and take the limit $\Delta t \to 0$. We assume that the random force has a constant value during each interval

$$\mathbf{F}_{rand}(t) = \mathbf{F}_n \quad t_n \leq t < t_{n+1} \tag{16.41}$$

and that the values at different intervals are uncorrelated

$$\overline{\mathbf{F}_n\mathbf{F}_m} = \delta_{m,n}\overline{\mathbf{F}_n^2}. \tag{16.42}$$

The auto-correlation function then is given by

$$\overline{\mathbf{F}_{rand}(t)\mathbf{F}_{rand}(t')} = \begin{cases} 0 & \text{different intervals} \\ \overline{\mathbf{F}_n^2} & \text{same interval.} \end{cases} \tag{16.43}$$

Division by Δt gives a sequence of functions which converges to a delta function in the limit $\Delta t \to 0$

$$\frac{1}{\Delta t}\overline{\mathbf{F}_{rand}(t)\mathbf{F}_{rand}(t')} \to \overline{\mathbf{F}_n^2}\,\delta(t - t'). \tag{16.44}$$

Hence we find

$$\overline{\mathbf{F}_n^2} = \frac{1}{\Delta t}\overline{\mathbf{F}_{rand}^2}. \tag{16.45}$$

Within a short time interval $\Delta t \to 0$ the velocity changes by

$$\mathbf{v}(t_n + \Delta t) = \mathbf{v} - \gamma\mathbf{v}\Delta t - \frac{1}{m}\nabla U(\mathbf{x})\Delta t + \frac{1}{m}\mathbf{F}_n\Delta t + \cdots \tag{16.46}$$

and taking the square gives

$$\mathbf{v}^2(t_n + \Delta t) = \mathbf{v}^2 - 2\gamma\mathbf{v}^2\Delta t - \frac{2}{m}\mathbf{v}\nabla U(\mathbf{x})\Delta t + \frac{2}{m}\mathbf{v}\mathbf{F}_n\Delta t + \frac{\mathbf{F}_n^2}{m^2}(\Delta t)^2 + \cdots. \tag{16.47}$$

Hence for the total energy

$$E(t_n + \Delta t) = \frac{m}{2}\mathbf{v}^2(t_n + \Delta t) + U\big(\mathbf{x}(t_n + \Delta t)\big)$$
$$= \frac{m}{2}\mathbf{v}^2(t_n + \Delta t) + U(\mathbf{x}) + \mathbf{v}\nabla U(\mathbf{x})\Delta t + \cdots \qquad (16.48)$$

we have

$$E(t_n + \Delta t) = E(t_n) - m\gamma\mathbf{v}^2\Delta t + \mathbf{v}\mathbf{F}_n\Delta t + \frac{\mathbf{F}_n^2}{2m}(\Delta t)^2 + \cdots. \qquad (16.49)$$

On the average the total energy $\overline{E}$ should be constant and furthermore in d dimensions

$$\frac{m}{2}\overline{\mathbf{v}^2} = \frac{d}{2}k_B T. \qquad (16.50)$$

Therefore we conclude

$$m\gamma\overline{\mathbf{v}^2} = \frac{\Delta t}{2m}\overline{\mathbf{F}_n^2} = \frac{1}{2m}\overline{\mathbf{F}_{rand}^2} \qquad (16.51)$$

from which we obtain finally

$$\overline{\mathbf{F}_n^2} = \frac{2m\gamma d}{\Delta t}k_B T. \qquad (16.52)$$

16.5 Problems

Problem 16.1 (Random walk in one dimension) This program generates random walks with (a) fixed step length $\Delta x = \pm 1$ or (b) step length equally distributed over the interval $-\sqrt{3} < \Delta x < \sqrt{3}$. It also shows the variance, which for large number of walks approaches $\sigma = \sqrt{n}$. See also Fig. 16.2.

Problem 16.2 (Gyration tensor) The program calculates random walks with M steps of length b. The bond vectors are generated from M random points $\mathbf{e}_i$ on the unit sphere as $\mathbf{b}_i = b\mathbf{e}_i$. End to end distance, center of gravity and gyration radius are calculated and can be averaged over numerous random structures. The gyration tensor 16.3.2 is diagonalized and the ordered eigenvalues are averaged.

Problem 16.3 (Brownian motion in a harmonic potential) The program simulates a particle in a 1-dimensional harmonic potential

$$U(\mathbf{x}) = \frac{f}{2}x^2 - \kappa x \qquad (16.53)$$

where κ is an external force. We use the improved Euler method (12.36). First the coordinate and the velocity at mid time are estimated

$$\mathbf{x}\left(t_n + \frac{\Delta t}{2}\right) = \mathbf{x}(t_n) + \mathbf{v}(t_n)\frac{\Delta t}{2} \tag{16.54}$$

$$\mathbf{v}\left(t_n + \frac{\Delta t}{2}\right) = \mathbf{v}(t_n) - \gamma\mathbf{v}(t_n)\frac{\Delta t}{2} + \frac{\mathbf{F}_n}{m}\frac{\Delta t}{2} - \frac{f}{m}\mathbf{x}(t_n)\frac{\Delta t}{2} \tag{16.55}$$

where $\mathbf{F}_n$ is a random number obeying (16.52). Then the values at t_{n+1} are calculated as

$$\mathbf{x}(t_n + \Delta t) = \mathbf{x}(t_n) + \mathbf{v}\left(t_n + \frac{\Delta t}{2}\right)\Delta t \tag{16.56}$$

$$\mathbf{v}(t_n + \Delta t) = \mathbf{v}(t_n) - \gamma\mathbf{v}\left(t_n + \frac{\Delta t}{2}\right)\Delta t + \frac{\mathbf{F}_n}{m}\Delta t - \frac{f}{m}\mathbf{x}\left(t_n + \frac{\Delta t}{2}\right)\Delta t. \tag{16.57}$$

Problem 16.4 (Force extension relation) The program simulates a chain of springs (Sect. 16.3.3) with potential energy

$$U = \frac{f}{2}\sum\left(|\mathbf{b}_i| - b\right)^2 - \kappa\,\mathbf{R}_M. \tag{16.58}$$

The force can be varied and the extension along the force direction is averaged over numerous time steps.

Chapter 17
Electrostatics

The electrostatic potential $\Phi(\mathbf{r})$ of a charge distribution $\rho(\mathbf{r})$ is a solution[1] of Poisson's equation

$$\Delta\Phi(\mathbf{r}) = -\rho(\mathbf{r}) \tag{17.1}$$

which, for spatially varying dielectric constant $\varepsilon(\mathbf{r})$ becomes

$$\operatorname{div}\bigl(\varepsilon(\mathbf{r})\operatorname{grad}\Phi(\mathbf{r})\bigr) = -\rho(\mathbf{r}) \tag{17.2}$$

and, if mobile charges are taken into account, like for an electrolyte or semiconductor, turns into the Poisson-Boltzmann equation

$$\operatorname{div}\bigl(\varepsilon(\mathbf{r})\operatorname{grad}\Phi(\mathbf{r})\bigr) = -\rho_{fix}(\mathbf{r}) - \sum_i n_i^0 Z_i e\, e^{-Z_i e\Phi(\mathbf{r})/k_B T}. \tag{17.3}$$

In this chapter we discretize the Poisson and the linearized Poisson-Boltzmann equation by finite volume methods which are applicable even in case of discontinuous ε. We solve the discretized equations iteratively with the method of successive over-relaxation. The solvation energy of a charged sphere in a dielectric medium is calculated to compare the accuracy of several methods. This can be studied also in a computer experiment.

Since the Green's function is analytically available for the Poisson and Poisson-Boltzmann equations, alternatively the method of boundary elements can be applied, which can reduce the computer time, for instance for solvation models. A computer experiment simulates a point charge within a spherical cavity and calculates the solvation energy with the boundary element method.

17.1 Poisson Equation

From a combination of the basic equations of electrostatics

---

[1]The solution depends on the boundary conditions, which in the simplest case are given by $\lim_{|\mathbf{r}|\to\infty}\Phi(\mathbf{r}) = 0$.

P.O.J. Scherer, *Computational Physics*, Graduate Texts in Physics,   305
DOI 10.1007/978-3-319-00401-3_17,
© Springer International Publishing Switzerland 2013

Fig. 17.1 (Finite volume for the Poisson equation) The control volume is a *small cube* centered at a grid point (*full circle*)

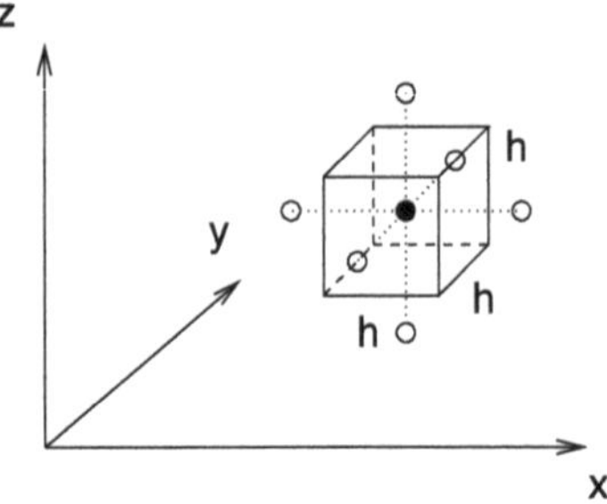

$$\operatorname{div} D(\mathbf{r}) = \rho(\mathbf{r}) \tag{17.4}$$

$$D(\mathbf{r}) = \varepsilon(\mathbf{r}) E(\mathbf{r}) \tag{17.5}$$

$$E(\mathbf{r}) = -\operatorname{grad} \Phi(\mathbf{r}) \tag{17.6}$$

the generalized Poisson equation is obtained

$$\operatorname{div}\big(\varepsilon(\mathbf{r}) \operatorname{grad} \Phi(\mathbf{r})\big) = -\rho(\mathbf{r}) \tag{17.7}$$

which can be written in integral form with the help of Gauss's theorem

$$\oint_{\partial V} d\mathbf{A}\, \operatorname{div}\big(\varepsilon(\mathbf{r}) \operatorname{grad} \Phi(\mathbf{r})\big) = \int_V dV\, \varepsilon(\mathbf{r}) \operatorname{grad} \Phi(\mathbf{r}) = -\int_V dV\, \rho(\mathbf{r}). \tag{17.8}$$

If $\varepsilon(\mathbf{r})$ is continuously differentiable, the product rule for differentiation gives

$$\varepsilon(\mathbf{r}) \Delta \Phi(\mathbf{r}) + \big(\operatorname{grad} \varepsilon(\mathbf{r})\big)\big(\operatorname{grad} \Phi(\mathbf{r})\big) = -\rho(\mathbf{r}) \tag{17.9}$$

which for constant ε simplifies to the Poisson equation

$$\Delta \Phi(\mathbf{r}) = -\frac{\rho(\mathbf{r})}{\varepsilon}. \tag{17.10}$$

17.1.1 Homogeneous Dielectric Medium

We begin with the simplest case of a dielectric medium with constant ε and solve (17.10) numerically. We use a finite volume method (Sect. 11.3) which corresponds to a finite element method with piecewise constant test functions. The integration volume is divided into small cubes V_{ijk} which are centered at the grid points (Fig. 17.1)

$$r_{ijk} = (hi, hj, hk). \tag{17.11}$$

Integration of (17.10) over the control volume V_{ijk} around $\mathbf{r}_{ijk}$ gives

$$\int_V dV\, \operatorname{div} \operatorname{grad} \Phi = \oint_{\partial V} \operatorname{grad} \Phi\, \mathbf{dA} = -\frac{1}{\varepsilon} \int_V dV\, \rho(\mathbf{r}) = -\frac{Q_{ijk}}{\varepsilon}. \tag{17.12}$$

Q_{ijk} is the total charge in the control volume. The flux integral is approximated by (11.85)

$$\oint_{\partial V} \operatorname{grad} \Phi \, \mathbf{dA} = -h^2 \bigg(\frac{\partial \Phi}{\partial x}(x_{i+1/2}, y_i, z_i) - \frac{\partial \Phi}{\partial x}(x_{i-1/2}, y_i, z_i)$$
$$+ \frac{\partial \Phi}{\partial y}(x_i y_{i+1/2}, z_i) - \frac{\partial \Phi}{\partial y}(x_i, y_{i-1/2}, z_i)$$
$$+ \frac{\partial \Phi}{\partial z}(x_i, y_i, z_{i+1/2}) - \frac{\partial \Phi}{\partial z}(x_i, y_i, z_{i-1/2}) \bigg). \tag{17.13}$$

The derivatives are approximated by symmetric differences

$$\oint_{\partial V} \operatorname{grad} \Phi \, \mathbf{dA} = -h\big(\Phi(x_{i+1}, y_i, z_i) - \Phi(x_i, y_i, z_i)$$
$$- \big(\Phi(x_i, y_i, z_i) - \Phi(x_{i-1}, y_i, z_i) \big)$$
$$+ \Phi(x_i, y_{i+1}, z_i) - \Phi(x_i, y_i, z_i)$$
$$- \big(\Phi(x_i, y_i, z_i) - \Phi(x_i, y_{i-1}, z_i) \big)$$
$$+ \Phi(x_i, y_i, z_{i+1}) - \Phi(x_i, y_i, z_i)$$
$$- \big(\Phi(x_i, y_i, z_i) - \Phi(x_i, y_i, z_{i-1}) \big) \big)$$
$$= -h\big(\Phi(x_{i-1}, y_i, z_i) + \Phi(x_{i+1}, y_i, z_i)$$
$$+ \Phi(x_i, y_{i-1}, z_i) + \Phi(x_i, y_{i+1}, z_i)$$
$$+ \Phi(x_i, y_i, z_{i-1}) + \Phi(x_i, y_i, z_{i+1})$$
$$- 6\Phi(x_i, y_i, z_{i1}) \big) \tag{17.14}$$

which coincides with the simplest discretization of the second derivatives (3.40). Finally we obtain the discretized Poisson equation in the more compact form

$$\sum_{s=1}^{6} \big(\Phi(r_{ijk} + d\mathbf{r}_s) - \Phi(r_{ijk}) \big) = -\frac{Q_{ijk}}{\varepsilon h} \tag{17.15}$$

which involves an average over the 6 neighboring cells

$$d\mathbf{r}_1 = (-h, 0, 0) \ldots d\mathbf{r}_6 = (0, 0, h). \tag{17.16}$$

17.1.2 Numerical Methods for the Poisson Equation

Equation (17.15) is a system of linear equations with very large dimension (for a grid with $100 \times 100 \times 100$ points the dimension of the matrix is $10^6 \times 10^6$!). Our computer experiments use the iterative method (5.5)

$$\Phi^{new}(r_{ijk}) = \frac{1}{6} \bigg(\sum_s \Phi^{old}(\mathbf{r}_{ijk} + d\mathbf{r}_s) + \frac{Q_{ijk}}{\varepsilon h} \bigg). \tag{17.17}$$

Jacobi's method ((5.110) on page 74) makes all the changes in one step whereas the Gauss-Seidel method ((5.113) on page 74) makes one change after the other. The chessboard (or black red method) divides the grid into two subgrids (with $i + j + k$ even or odd) which are treated subsequently. The vector $d\mathbf{r}_s$ connects points of different subgrids. Therefore it is not necessary to store intermediate values like for the Gauss-Seidel method.

Convergence can be improved with the method of successive over-relaxation (SOR, (5.117) on page 75) using a mixture of old and new values

$$\Phi^{new}(r_{ijk}) = (1 - \omega)\Phi^{old}(\mathbf{r}_{ijk}) + \omega\frac{1}{6}\left(\sum_s \Phi^{old}(\mathbf{r}_{ijk} + d\mathbf{r}_s) + \frac{Q_{ijk}}{\varepsilon h}\right)$$

(17.18)

with the relaxation parameter ω. For $1 < \omega < 2$ convergence is faster than for $\omega = 1$. The optimum choice of ω for the Poisson problem in any dimension is discussed in [279].

Convergence can be further improved by multigrid methods [120, 283]. Error components with short wavelengths are strongly damped during a few iterations whereas it takes a very large number of iterations to remove the long wavelength components. But here a coarser grid is sufficient and reduces computing time. After a few iterations a first approximation Φ_1 is obtained with the finite residual

$$r_1 = \Delta\Phi_1 + \frac{1}{\varepsilon}\rho.$$

(17.19)

Then more iterations on a coarser grid are made to find an approximate solution Φ_2 of the equation

$$\Delta\Phi = -r_1 = -\frac{1}{\varepsilon}\rho - \Delta\Phi_1.$$

(17.20)

The new residual is

$$r_2 = \Delta\Phi_2 + r_1.$$

(17.21)

Function values of Φ_2 on the finer grid are obtained by interpolation and finally the sum $\Phi_1 + \Phi_2$ provides an improved approximation to the solution since

$$\Delta(\Phi_1 + \Phi_2) = -\frac{1}{\varepsilon}\rho + r_1 + (r_2 - r_1) = -\frac{1}{\varepsilon}\rho + r_2.$$

(17.22)

This method can be extended to a hierarchy of many grids.

Alternatively, the Poisson equation can be solved non-iteratively with pseudospectral methods [238, 247]. For instance, if the boundary is the surface of a cube, eigenfunctions of the Laplacian are for homogeneous boundary conditions ($\Phi = 0$) given by

$$N_{\mathbf{k}}(\mathbf{r}) = \sin(k_x x)\sin(k_y y)\sin(k_z z)$$

(17.23)

and for no-flow boundary conditions ($\frac{\partial}{\partial n}\Phi = 0$) by

$$N_{\mathbf{k}}(\mathbf{r}) = \cos(k_x x)\cos(k_y y)\cos(k_z z)$$

(17.24)

which can be used as expansion functions for the potential

$$\Phi(\mathbf{r}) = \sum_{k_x,k_y,k_z} \Phi_{\mathbf{k}} N_{\mathbf{k}}(\mathbf{r}). \tag{17.25}$$

Introducing collocation points $\mathbf{r}_j$ the condition on the residual becomes

$$0 = \Delta\Phi(\mathbf{r}_j) + \frac{1}{\varepsilon}\rho(\mathbf{r}_j) = \sum_{k_x,k_y,k_z} k^2 \Phi_{\mathbf{k}} N_{\mathbf{k}}(\mathbf{r}_j) + \frac{1}{\varepsilon}\rho(\mathbf{r}_j) \tag{17.26}$$

which can be inverted with an inverse discrete sine transformation, (respectively an inverse discrete cosine transformation for no-flux boundary conditions) to obtain the Fourier components of the potential. Another discrete sine (or cosine) transformation gives the potential in real space.

17.1.3 Charged Sphere

As a simple example we consider a sphere of radius R with a homogeneous charge density of

$$\rho_0 = e \cdot \frac{3}{4\pi R^3}. \tag{17.27}$$

The exact potential is given by

$$\Phi(r) = \frac{e}{4\pi\varepsilon_0 R} + \frac{e}{8\pi\varepsilon_0 R}\left(1 - \frac{r^2}{R^2}\right) \quad \text{for } r < R \tag{17.28}$$

$$\Phi(r) = \frac{e}{4\pi\varepsilon_0 r} \quad \text{for } r > R.$$

The charge density (17.27) is discontinuous at the surface of the sphere. Integration over a control volume smears out this discontinuity which affects the potential values around the boundary (Fig. 17.2). Alternatively we could assign the value $\rho(\mathbf{r}_{ijk})$ which is either ρ_0 (17.27) or zero to each control volume which retains a sharp transition but changes the shape of the boundary surface and does not conserve the total charge. This approach was discussed in the first edition of this book in connection with a finite differences method. The most precise but also complicated method divides the control volumes at the boundary into two irregularly shaped parts [188, 189].

Initial guess as well as boundary values are taken from

$$\Phi_0(r) = \frac{e}{4\pi\varepsilon_0 \max(r, h)} \tag{17.29}$$

which provides proper boundary values but is far from the final solution inside the sphere. The interaction energy is given by (Sect. 17.5)

$$E_{int} = \frac{1}{2}\int_V \rho(\mathbf{r})\Phi(\mathbf{r})\, dV = \frac{3}{20}\frac{e^2}{\pi\varepsilon_0 R}. \tag{17.30}$$

Calculated potential (Fig. 17.3) and interaction energy (Figs. 17.4, 17.5) converge rapidly. The optimum relaxation parameter is around $\omega \approx 1.9$.

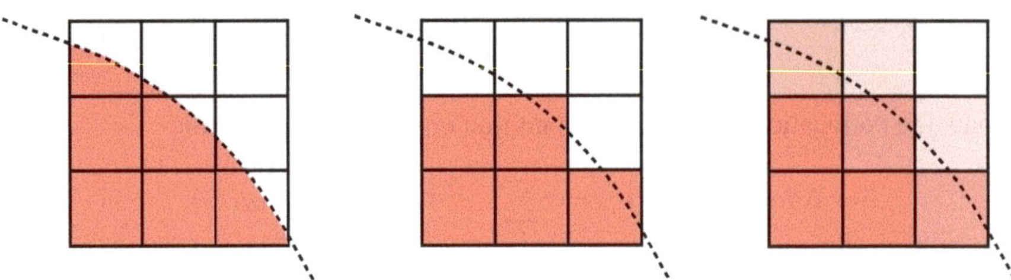

Fig. 17.2 (Discretization of the discontinuous charge density) *Left*: the most precise method divides the control volumes at the boundary into two irregularly shaped parts. *Middle*: assigning either the value ρ_0 or zero retains the discontinuity but changes the shape of the boundary. *Right*: averaging over a control volume smears out the discontinuous transition

Fig. 17.3 (Electrostatic potential of a charged sphere) A charged sphere is simulated with radius $R = 0.25$ Å and a homogeneous charge density $\rho = e \cdot 3/4\pi R^3$. The grid consists of 200^3 points with a spacing of $h = 0.025$ Å. The calculated potential (*circles*) is compared to the exact solution ((17.28), *solid curve*), the initial guess is shown by the *dashed line*

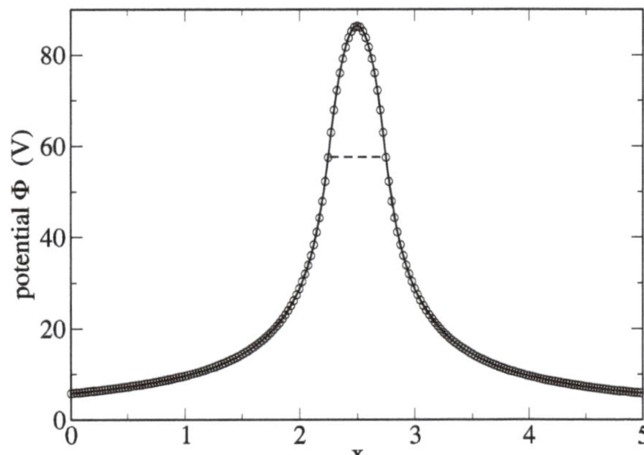

Fig. 17.4 (Influence of the relaxation parameter) The convergence of the interaction energy ((17.30), which has a value of 34.56 eV for this example) is studied as a function of the relaxation parameter ω. The optimum value is around $\omega \approx 1.9$. For $\omega > 2$ there is no convergence. The *dashed line* shows the exact value

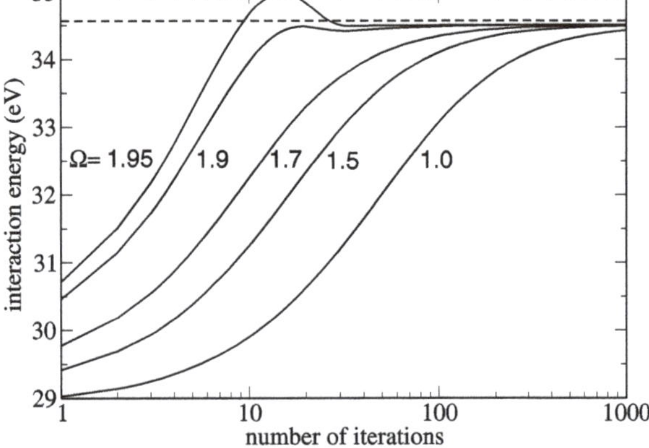

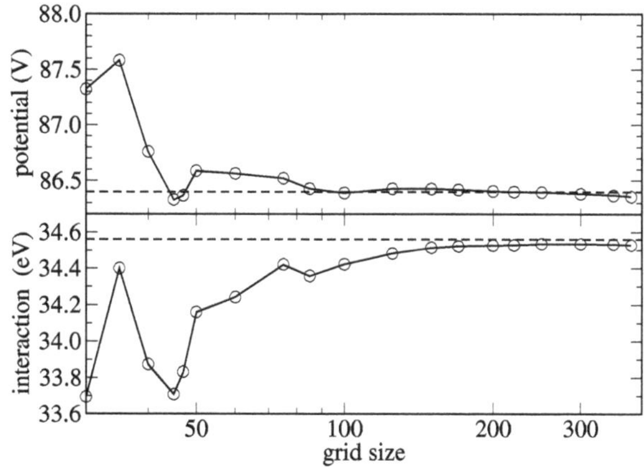

Fig. 17.5 (Influence of grid size) The convergence of the interaction energy (17.30) and the central potential value are studied as a function of grid size. The *dashed lines* show the exact values

17.1.4 Variable ε

In the framework of the finite volume method we take the average over a control volume to discretize ε^2 and Φ

$$\varepsilon_{ijk} = \overline{\varepsilon}(\mathbf{r}_{ijk}) = \frac{1}{h^3} \int_{V_{ijk}} dV\, \varepsilon(\mathbf{r}) \tag{17.31}$$

$$\Phi_{ijk} = \overline{\Phi}(\mathbf{r}_{ijk}) = \frac{1}{h^3} \int_{V_{ijk}} dV\, \Phi(\mathbf{r}). \tag{17.32}$$

Integration of (17.7) gives

$$\int_V dV\, \mathrm{div}\big(\varepsilon(\mathbf{r})\,\mathrm{grad}\,\Phi(\mathbf{r})\big) = \oint_{\partial V} \varepsilon(\mathbf{r})\,\mathrm{grad}\,\Phi\,\mathbf{dA} = -\int_V dV\, \rho(\mathbf{r}) = -Q_{ijk}. \tag{17.33}$$

The surface integral is

$$\oint_{\partial V} \mathbf{dA}\,\varepsilon\,\mathrm{grad}\,\Phi = \sum_{s\in faces} \int_{A_s} dA\, \varepsilon(\mathbf{r})\frac{\partial}{\partial n}\Phi. \tag{17.34}$$

Applying the midpoint rule (11.77) we find (Fig. 17.6)

$$\oint_{\partial V} \mathbf{dA}\varepsilon\,\mathrm{grad}\,\Phi \approx h^2 \sum_{r=1}^{6} \varepsilon\left(\mathbf{r}_{ijk} + \frac{1}{2}d\mathbf{r}_s\right)\frac{\partial}{\partial n}\Phi\left(\mathbf{r}_{ijk} + \frac{1}{2}d\mathbf{r}_s\right). \tag{17.35}$$

The potential Φ as well as the product $\varepsilon(\mathbf{r})\frac{\partial\Phi}{\partial n}$ are continuous, therefore we make the approximation [188]

---

[2]But see Sect. 17.1.5 for the case of discontinuous ε.

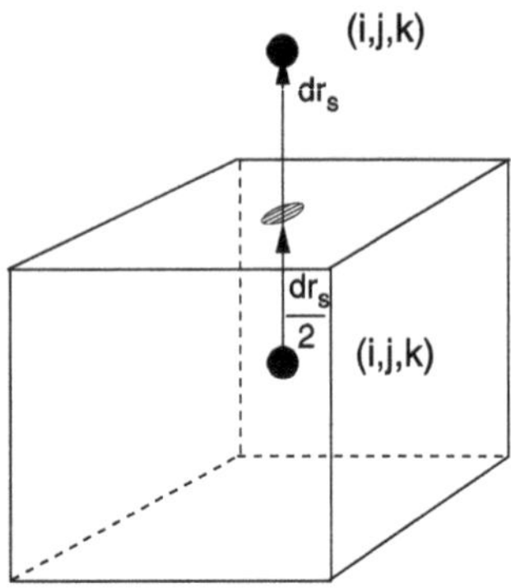

Fig. 17.6 Face center of the control volume

$$
\varepsilon\left(\mathbf{r}_{ijk}+\frac{1}{2}d\mathbf{r}_s\right)\frac{\partial\Phi}{\partial n}\left(\mathbf{r}_{ijk}+\frac{1}{2}d\mathbf{r}_s\right)
$$

$$
=\overline{\varepsilon}(\mathbf{r}_{ijk})\frac{\overline{\Phi}(\mathbf{r}_{ijk}+\frac{1}{2}d\mathbf{r}_s)-\overline{\Phi}(\mathbf{r}_{ijk})}{\frac{h}{2}}
$$

$$
=\overline{\varepsilon}(\mathbf{r}_{ijk}+d\mathbf{r}_s)\frac{\overline{\Phi}(\mathbf{r}_{ijk}+d\mathbf{r}_s)-\overline{\Phi}(\mathbf{r}_{ijk}+\frac{1}{2}d\mathbf{r}_s)}{\frac{h}{2}}. \tag{17.36}
$$

From this equation the unknown potential value on the face of the control volume $\overline{\Phi}(\mathbf{r}_{ijk}+\frac{1}{2}d\mathbf{r}_s)$ (Fig. 17.6) can be calculated

$$
\overline{\Phi}\left(\mathbf{r}_{ijk}+\frac{1}{2}d\mathbf{r}_s\right)=\frac{\overline{\varepsilon}(\mathbf{r}_{ijk})\overline{\Phi}(\mathbf{r}_{ijk})+\overline{\varepsilon}(\mathbf{r}_{ijk}+d\mathbf{r}_s)\overline{\Phi}(\mathbf{r}_{ijk}+d\mathbf{r}_s)}{\overline{\varepsilon}(\mathbf{r}_{ijk})+\overline{\varepsilon}(\mathbf{r}_{ijk}+d\mathbf{r}_s)} \tag{17.37}
$$

which gives

$$
\varepsilon\left(\mathbf{r}_{ijk}+\frac{1}{2}d\mathbf{r}_s\right)\frac{\partial}{\partial n}\Phi\left(\mathbf{r}_{ijk}+\frac{1}{2}d\mathbf{r}_s\right)
$$

$$
=\frac{2\overline{\varepsilon}(\mathbf{r}_{ijk})\overline{\varepsilon}(\mathbf{r}_{ijk}+d\mathbf{r}_s)}{\overline{\varepsilon}(\mathbf{r}_{ijk})+\overline{\varepsilon}(\mathbf{r}_{ijk}+d\mathbf{r}_s)}\frac{\overline{\Phi}(\mathbf{r}_{ijk}+d\mathbf{r}_s)-\overline{\Phi}(\mathbf{r}_{ijk})}{h}. \tag{17.38}
$$

Finally we obtain the discretized equation

$$
-Q_{ijk}=h\sum_{s=1}^{6}\frac{2\overline{\varepsilon}(\mathbf{r}_{ijk}+d\mathbf{r}_s)\overline{\varepsilon}(\mathbf{r}_{ijk})}{\overline{\varepsilon}(\mathbf{r}_{ijk}+d\mathbf{r}_s)+\overline{\varepsilon}(\mathbf{r}_{ijk})}\left(\overline{\Phi}(\mathbf{r}_{ijk}+d\mathbf{r}_s)-\overline{\Phi}(\mathbf{r}_{ijk})\right) \tag{17.39}
$$

which can be solved iteratively according to

$$
\Phi^{new}(\mathbf{r}_{ijk})=\frac{\sum\frac{2\varepsilon(\mathbf{r}_{ijk}+d\mathbf{r}_s)\varepsilon(\mathbf{r}_{ijk})}{\varepsilon(\mathbf{r}_{ijk}+d\mathbf{r}_s)+\varepsilon(\mathbf{r}_{ijk})}\Phi^{old}(\mathbf{r}_{ijk}+d\mathbf{r}_s)+\frac{Q_{ijk}}{h}}{\sum\frac{2\varepsilon(\mathbf{r}_{ijk}+d\mathbf{r}_s)\varepsilon(\mathbf{r}_{ijk})}{\varepsilon(\mathbf{r}_{ijk}+d\mathbf{r}_s)+\varepsilon(\mathbf{r}_{ijk})}}. \tag{17.40}
$$

Fig. 17.7 (Transition of ε) The discontinuous $\varepsilon(r)$ (*black line*) is averaged over the control volumes to obtain the discretized values ε_{ijk} (*full circles*). Equation (17.40) takes the harmonic average over two neighbor cells (*open circles*) and replaces the discontinuity by a smooth transition over a distance of about h

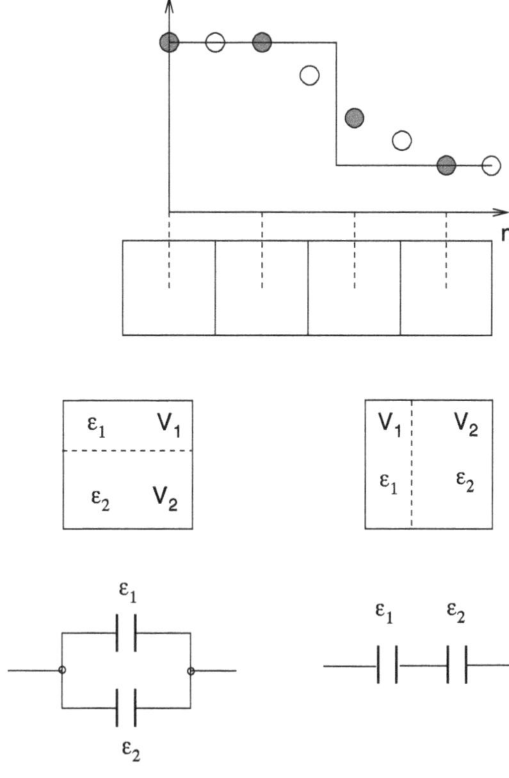

Fig. 17.8 Average of ε over a control volume

17.1.5 Discontinuous ε

For practical applications models are often used with piecewise constant ε. A simple example is the solvation of a charged molecule in a dielectric medium (Fig. 17.9). Here $\varepsilon = \varepsilon_0$ within the molecule and $\varepsilon = \varepsilon_0 \varepsilon_1$ within the medium. At the boundary ε is discontinuous. In (17.40) the discontinuity is replaced by a smooth transition between the two values of ε (Fig. 17.7).

If the discontinuity of ε is inside a control volume V_{ijk} then (17.31) takes the arithmetic average

$$\overline{\varepsilon}_{ijk} = V_{ijk}^{(1)}\varepsilon_1 + V_{ijk}^{(2)}\varepsilon_2 \tag{17.41}$$

which corresponds to the parallel connection of two capacities (Fig. 17.8). Depending on geometry, a serial connection may be more appropriate which corresponds to the weighted harmonic average

$$\overline{\varepsilon}_{ijk} = \frac{1}{V_{ijk}^{(1)}\varepsilon_1^{-1} + V_{ijk}^{(2)}\varepsilon_2^{-1}}. \tag{17.42}$$

Fig. 17.9 (Solvation of a charged sphere in a dielectric medium) Charge density and dielectric constant are discontinuous at the surface of the sphere

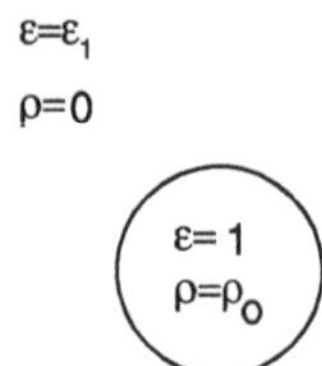

17.1.6 Solvation Energy of a Charged Sphere

We consider again a charged sphere, which is now embedded in a dielectric medium (Fig. 17.9) with relative dielectric constant ε_1.

For a spherically symmetrical problem (17.7) can be solved by application of Gauss's theorem

$$4\pi r^2 \varepsilon(r)\frac{d\Phi}{dr} = -4\pi \int_0^r \rho(r')r'^2\,dr' = -q(r) \tag{17.43}$$

$$\Phi(r) = -\int_0^r \frac{q(r)}{4\pi r^2 \varepsilon(r)} + \Phi(0). \tag{17.44}$$

For the charged sphere we find

$$q(r) = \begin{cases} Qr^3/R^3 & \text{for } r < R \\ Q & \text{for } r > R \end{cases} \tag{17.45}$$

$$\Phi(r) = -\frac{Q}{4\pi\varepsilon_0 R^3}\frac{r^2}{2} + \Phi(0) \quad \text{for } r < R \tag{17.46}$$

$$\Phi(r) = -\frac{Q}{8\pi\varepsilon_0 R} + \Phi(0) + \frac{Q}{4\pi\varepsilon_0\varepsilon_1}\left(\frac{1}{r} - \frac{1}{R}\right) \quad \text{for } r > R. \tag{17.47}$$

The constant $\Phi(0)$ is chosen to give vanishing potential at infinity

$$\Phi(0) = \frac{Q}{4\pi\varepsilon_0\varepsilon_1 R} + \frac{Q}{8\pi\varepsilon_0 R}. \tag{17.48}$$

The interaction energy is

$$E_{int} = \frac{1}{2}\int_0^R 4\pi r^2 dr\,\rho\Phi(r) = \frac{Q^2(5+\varepsilon_1)}{40\pi\varepsilon_0\varepsilon_1 R}. \tag{17.49}$$

Numerical results for $\varepsilon_1 = 4$ are shown in Fig. 17.10.

17.1.7 The Shifted Grid Method

An alternative approach uses a different grid (Fig. 17.11) for ε which is shifted by $h/2$ in all directions [44] or, more generally, a dual grid (11.74),

$$\varepsilon_{ijk} = \overline{\varepsilon}(\mathbf{r}_{i+1/2, j+1/2, k+1/2}). \tag{17.50}$$

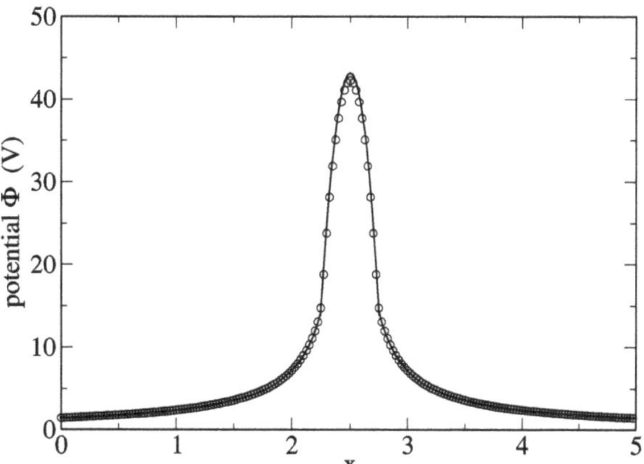

Fig. 17.10 (Charged sphere in a dielectric medium) Numerical results for $\varepsilon_1 = 4$ outside the sphere and 200^3 grid points (*circles*) are compared to the exact solution ((17.46), (17.47), *solid curves*)

The value of ε has to be averaged over four neighboring cells to obtain the discretized equation

$$
-\frac{Q_{ijk}}{h^2} = \sum_s \varepsilon(\mathbf{r}_{ijk} + d\mathbf{r}_s)\frac{\partial \Phi}{\partial n}(\mathbf{r}_{ijk} + d\mathbf{r}_s)
$$

$$
= \frac{\Phi_{i,j,k+1} - \Phi_{i,j,k}}{h}\frac{\varepsilon_{ijk} + \varepsilon_{i,j-1,k} + \varepsilon_{i-1,j,k} + \varepsilon_{i-1,j-1,k}}{4}
$$

$$
+ \frac{\Phi_{i,j,k-1} - \Phi_{i,j,k}}{h}\frac{\varepsilon_{ijk-1} + \varepsilon_{i,j-1,k-1} + \varepsilon_{i-1,j,k-1} + \varepsilon_{i-1,j-1,k-1}}{4}
$$

$$
+ \frac{\Phi_{i+1,j,k} - \Phi_{i,j,k}}{h}\frac{\varepsilon_{ijk} + \varepsilon_{i,j-1,k} + \varepsilon_{i,j,k-1} + \varepsilon_{i,j-1,k-1}}{4}
$$

$$
+ \frac{\Phi_{i-1,j,k} - \Phi_{i,j,k}}{h}\frac{\varepsilon_{i-1jk} + \varepsilon_{i-1,j-1,k} + \varepsilon_{i-1,j,k-1} + \varepsilon_{i-1,j-1,k-1}}{4}
$$

$$
+ \frac{\Phi_{i,j+1,k} - \Phi_{i,j,k}}{h}\frac{\varepsilon_{ijk} + \varepsilon_{i-1,j,k} + \varepsilon_{i,j,k-1} + \varepsilon_{i-1,j,k-1}}{4}
$$

$$
+ \frac{\Phi_{i,j-1,k} - \Phi_{i,j,k}}{h}\frac{\varepsilon_{ij-1k} + \varepsilon_{i-1,j-1,k} + \varepsilon_{i,j-1,k-1} + \varepsilon_{i-1,j-1,k-1}}{4}.
$$

$$
\tag{17.51}
$$

The shifted grid method is especially useful if ε changes at planar interfaces. Numerical results of several methods are compared in Fig. 17.12.

17.2 Poisson-Boltzmann Equation

Electrostatic interactions are very important in molecular physics. Bio-molecules are usually embedded in an environment which is polarizable and contains mobile charges (Na^+, K^+, Mg^{++}, $\mathrm{Cl}^- \cdots$).

We divide the charge density formally into a fixed and a mobile part

$$
\rho(\mathbf{r}) = \rho_{fix}(\mathbf{r}) + \rho_{mobile}(\mathbf{r}). \tag{17.52}
$$

Fig. 17.11 (Shifted grid method) A different grid is used for the discretization of ε which is shifted by $h/2$ in all directions

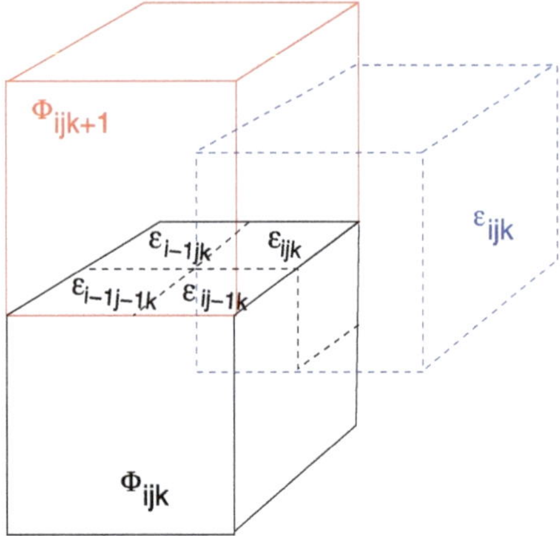

Fig. 17.12 (Comparison of numerical errors) The Coulomb interaction of a charged sphere is calculated with several methods for 100^3 grid points. *Circles*: ((17.40), ε averaged), *diamonds*: ((17.40), ε^{-1} averaged), *squares*: ((17.51), ε averaged), *triangles*: ((17.51), ε^{-1} averaged), *solid curve*: analytical solution (17.49)

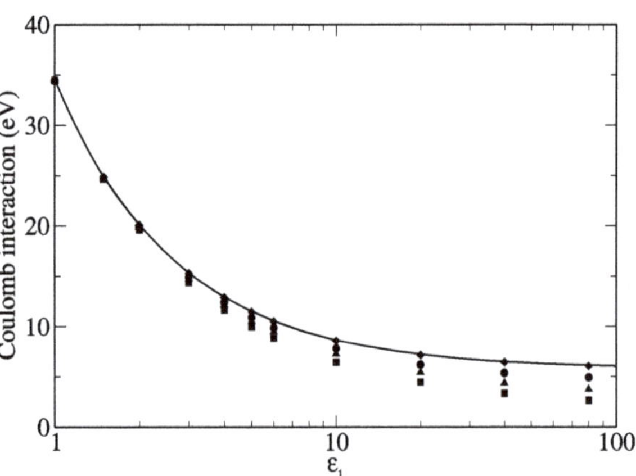

The fixed part represents, for instance, the charge distribution of a protein molecule which, neglecting polarization effects, is a given quantity and provides the inhomogeneity of the equation. The mobile part, on the other hand, represents the sum of all mobile charges (e is the elementary charge and Z_i the charge number of ion species i)

$$\rho_{mobile}(\mathbf{r}) = \sum_i Z_i e n_i(\mathbf{r}) \tag{17.53}$$

which move around until an equilibrium is reached which is determined by the mutual interaction of the ions. The famous Debye-Hückel [69] and Gouy-Chapman models [56, 110] assume that the electrostatic interaction

$$U(\mathbf{r}) = Z_i e \Phi(\mathbf{r}) \tag{17.54}$$

is dominant and the density of the ions n_i is given by a Boltzmann-distribution

$$n_i(\mathbf{r}) = n_i^{(0)} e^{-Z_i e \Phi(\mathbf{r})/k_B T} . \tag{17.55}$$

The potential $\Phi(\mathbf{r})$ has to be calculated in a self consistent way together with the density of mobile charges. The charge density of the free ions is

$$\rho_{mobile}(\mathbf{r}) = \sum_i n_i^{(0)} e Z_i e^{-Z_i e\Phi/k_B T} \tag{17.56}$$

and the Poisson equation (17.7) turns into the Poisson-Boltzmann equation [91]

$$\mathrm{div}\big(\varepsilon(\mathbf{r})\,\mathrm{grad}\,\Phi(\mathbf{r})\big) + \sum_i n_i^{(0)} e Z_i e^{-Z_i e\Phi/k_B T} = -\rho_{fix}(\mathbf{r}). \tag{17.57}$$

17.2.1 Linearization of the Poisson-Boltzmann Equation

For small ion concentrations the exponential can be expanded

$$e^{-Z_i e\Phi/kT} \approx 1 - \frac{Z_i e\Phi}{k_B T} + \frac{1}{2}\left(\frac{Z_i e\Phi}{k_B T}\right)^2 + \cdots. \tag{17.58}$$

For a neutral system

$$\sum_i n_i^{(0)} Z_i e = 0 \tag{17.59}$$

and the linearized Poisson-Boltzmann equation is obtained:

$$\mathrm{div}\big(\varepsilon(\mathbf{r})\,\mathrm{grad}\,\Phi(\mathbf{r})\big) - \sum_i n_i^{(0)} \frac{Z_i^2 e^2}{k_B T}\,\Phi(\mathbf{r}) = -\rho_{fix}. \tag{17.60}$$

With

$$\varepsilon(\mathbf{r}) = \varepsilon_0 \varepsilon_r(\mathbf{r}) \tag{17.61}$$

and the definition

$$\kappa(\mathbf{r})^2 = \frac{e^2}{\varepsilon_0 \varepsilon_r(\mathbf{r}) k_B T} \sum_i n_i^{(0)} Z_i^2 \tag{17.62}$$

we have finally

$$\mathrm{div}\big(\varepsilon_r(\mathbf{r})\,\mathrm{grad}\,\Phi(\mathbf{r})\big) - \varepsilon_r \kappa^2 \Phi = -\frac{1}{\varepsilon_0}\rho. \tag{17.63}$$

For a charged sphere with radius a embedded in a homogeneous medium the solution of (17.63) is given by

$$\Phi = \frac{A}{r} e^{-\kappa r} \qquad A = \frac{e}{4\pi \varepsilon_0 \varepsilon_r}\frac{e^{\kappa a}}{1+\kappa a}. \tag{17.64}$$

The potential is shielded by the ions. Its range is of the order $\lambda_{Debye} = 1/\kappa$ (the so-called Debye length).

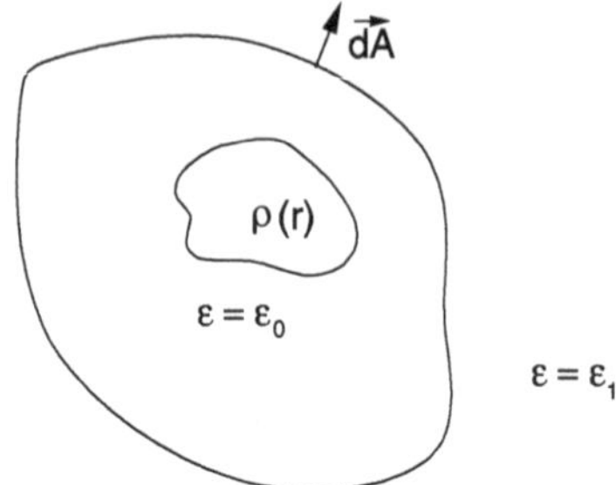

Fig. 17.13 Cavity in a
dielectric medium

17.2.2 Discretization of the Linearized Poisson-Boltzmann Equation

To solve (17.63) the discrete equation (17.39) is generalized to [181]

$$\sum_s \frac{2\varepsilon_r(\mathbf{r}_{ijk})\varepsilon_r(\mathbf{r}_{ijk}+d\mathbf{r}_s)}{\varepsilon_r(\mathbf{r}_{ijk})+\varepsilon_r(\mathbf{r}_{ijk}+d\mathbf{r}_s)}\left(\Phi(\mathbf{r}_{ijk}+d\mathbf{r}_s)-\Phi(\mathbf{r}_{ijk})\right)$$

$$-\varepsilon_r(\mathbf{r}_{ijk})\kappa^2(\mathbf{r}_{ijk})h^2\Phi(\mathbf{r}_{ijk})=-\frac{Q_{ijk}}{h\varepsilon_0}. \tag{17.65}$$

If ε is constant then we have to iterate

$$\Phi^{new}(\mathbf{r}_{ijk})=\frac{\frac{Q_{ijk}}{h\varepsilon_0\varepsilon_r}+\sum_s\Phi^{old}(\mathbf{r}_{ijk}+d\mathbf{r}_s)}{6+h^2\kappa^2(\mathbf{r}_{ijk})}. \tag{17.66}$$

17.3 Boundary Element Method for the Poisson Equation

Often continuum models are used to describe the solvation of a subsystem which is
treated with a high accuracy method. The polarization of the surrounding solvent or
protein is described by its dielectric constant ε and the subsystem is placed inside a
cavity with $\varepsilon = \varepsilon_0$ (Fig. 17.13). Instead of solving the Poisson equation for a large
solvent volume another kind of method is often used which replaces the polarization
of the medium by a distribution of charges over the boundary surface.

In the following we consider model systems which are composed of two spatial
regions:

- the outer region is filled with a dielectric medium (ε_1) and contains no free
  charges
- the inner region ("Cavity") contains a charge distribution $\rho(r)$ and its dielectric
  constant is $\varepsilon = \varepsilon_0$.

17.3.1 Integral Equations for the Potential

Starting from the Poisson equation

$$\mathrm{div}\big(\varepsilon(\mathbf{r})\,\mathrm{grad}\,\Phi(\mathbf{r})\big) = -\rho(\mathbf{r}) \tag{17.67}$$

we will derive some useful integral equations in the following. First we apply Gauss's theorem to the expression [277]

$$\mathrm{div}\big[G(\mathbf{r}-\mathbf{r}')\varepsilon(\mathbf{r})\,\mathrm{grad}\big(\Phi(\mathbf{r})\big) - \Phi(\mathbf{r})\varepsilon(\mathbf{r})\,\mathrm{grad}\big(G(\mathbf{r}-\mathbf{r}')\big)\big]$$
$$= -\rho(\mathbf{r})G(\mathbf{r}-\mathbf{r}') - \Phi(\mathbf{r})\varepsilon(\mathbf{r})\,\mathrm{div}\,\mathrm{grad}\big(G(\mathbf{r}-\mathbf{r}')\big)$$
$$- \Phi(\mathbf{r})\,\mathrm{grad}\,\varepsilon(\mathbf{r})\,\mathrm{grad}\big(G(\mathbf{r}-\mathbf{r}')\big) \tag{17.68}$$

with the yet undetermined function $G(\mathbf{r}-\mathbf{r}')$. Integration over a volume V gives

$$-\int_V dV\,\big(\rho(\mathbf{r})G(\mathbf{r}-\mathbf{r}') + \Phi(\mathbf{r})\varepsilon(\mathbf{r})\,\mathrm{div}\,\mathrm{grad}\big(G(\mathbf{r}-\mathbf{r}')\big)$$
$$+ \Phi(\mathbf{r})\,\mathrm{grad}\,\varepsilon(\mathbf{r})\,\mathrm{grad}\big(G(\mathbf{r}-\mathbf{r}')\big)\big)$$
$$= \oint_{\partial V} dA\,\left(G(\mathbf{r}-\mathbf{r}')\varepsilon(\mathbf{r})\frac{\partial}{\partial n}\big(\Phi(\mathbf{r})\big) - \Phi(\mathbf{r})\varepsilon(\mathbf{r})\frac{\partial}{\partial n}\big(G(\mathbf{r}-\mathbf{r}')\big)\right). \tag{17.69}$$

Now choose G as the fundamental solution of the Poisson equation

$$G_0(\mathbf{r}-\mathbf{r}') = -\frac{1}{4\pi\,|\mathbf{r}-\mathbf{r}'|} \tag{17.70}$$

which obeys

$$\mathrm{div}\,\mathrm{grad}\,G_0 = \delta(\mathbf{r}-\mathbf{r}') \tag{17.71}$$

to obtain the following integral equation for the potential:

$$\Phi(\mathbf{r}')\varepsilon(\mathbf{r}) = \int_V dV\,\frac{\rho(\mathbf{r})}{4\pi\,|\mathbf{r}-\mathbf{r}'|} + \frac{1}{4\pi}\int_V dV\,\Phi(\mathbf{r})\,\mathrm{grad}\,\varepsilon(\mathbf{r})\,\mathrm{grad}\left(\frac{1}{|\mathbf{r}-\mathbf{r}'|}\right)$$
$$-\frac{1}{4\pi}\oint_{\partial V} dA\,\left(\frac{1}{|\mathbf{r}-\mathbf{r}'|}\varepsilon(\mathbf{r})\frac{\partial}{\partial n}\big(\Phi(\mathbf{r})\big) + \Phi(\mathbf{r})\varepsilon(\mathbf{r})\frac{\partial}{\partial n}\left(\frac{1}{|\mathbf{r}-\mathbf{r}'|}\right)\right). \tag{17.72}$$

First consider as the integration volume a sphere with increasing radius. Then the surface integral vanishes for infinite radius ($\Phi \to 0$ at large distances) [277].

The gradient of $\varepsilon(\mathbf{r})$ is nonzero only on the boundary surface Fig. 17.14 of the cavity and with the limiting procedure ($d \to 0$)

$$\mathrm{grad}\,\varepsilon(\mathbf{r})\,dV = \mathbf{n}\frac{\varepsilon_1 - 1}{d}\varepsilon_0\,dV = dA\,\mathbf{n}(\varepsilon_1 - 1)\varepsilon_0$$

we obtain

$$\Phi(\mathbf{r}') = \frac{1}{\varepsilon(\mathbf{r}')}\int_{cav} dV\,\frac{\rho(\mathbf{r})}{4\pi\,|\mathbf{r}-\mathbf{r}'|}$$
$$+ \frac{(\varepsilon_1 - 1)\varepsilon_0}{4\pi\,\varepsilon(\mathbf{r}')}\oint_S dA\,\Phi(\mathbf{r})\frac{\partial}{\partial n}\frac{1}{|\mathbf{r}-\mathbf{r}'|}. \tag{17.73}$$

Fig. 17.14 Discontinuity at
the cavity boundary

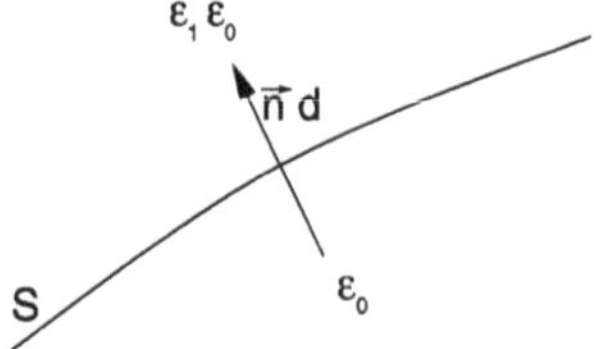

This equation allows to calculate the potential inside and outside the cavity from
the given charge density and the potential at the boundary.

Next we apply (17.72) to the cavity volume (where $\varepsilon = \varepsilon_0$) and obtain

$$\Phi_{in}(\mathbf{r}') = \int_V dV \frac{\rho(\mathbf{r})}{4\pi |\mathbf{r} - \mathbf{r}'| \varepsilon_0}$$
$$- \frac{1}{4\pi} \oint_S dA \left(\Phi_{in}(\mathbf{r}) \frac{\partial}{\partial n} \frac{1}{|\mathbf{r} - \mathbf{r}'|} - \frac{1}{|\mathbf{r} - \mathbf{r}'|} \frac{\partial}{\partial n} \Phi_{in}(\mathbf{r}) \right). \quad (17.74)$$

From comparison with (17.73) we have

$$\oint_S dA \frac{1}{|\mathbf{r} - \mathbf{r}'|} \frac{\partial}{\partial n} \Phi_{in}(\mathbf{r}) = \varepsilon_1 \oint_S dA\, \Phi_{in}(\mathbf{r}) \frac{\partial}{\partial n} \frac{1}{|\mathbf{r} - \mathbf{r}'|}$$

and the potential can be alternatively calculated from the values of its normal gradient at the boundary

$$\Phi(\mathbf{r}') = \frac{1}{\varepsilon(\mathbf{r}')} \int_{cav} dV \frac{\rho(\mathbf{r})}{4\pi |\mathbf{r} - \mathbf{r}'|} + \frac{(1 - \frac{1}{\varepsilon_1})\varepsilon_0}{4\pi \varepsilon(\mathbf{r}')} \oint_S dA \frac{1}{|\mathbf{r} - \mathbf{r}'|} \frac{\partial}{\partial n} \Phi_{in}(\mathbf{r}).$$
$$(17.75)$$

This equation can be interpreted as the potential generated by the charge density ρ
plus an additional surface charge density

$$\sigma(\mathbf{r}) = \left(1 - \frac{1}{\varepsilon_1} \right) \varepsilon_0 \frac{\partial}{\partial n} \Phi_{in}(\mathbf{r}). \quad (17.76)$$

Integration over the volume outside the cavity (where $\varepsilon = \varepsilon_1 \varepsilon_0$) gives the following
expression for the potential:

$$\Phi_{out}(\mathbf{r}') = \frac{1}{4\pi} \oint_S dA \left(\Phi_{out}(\mathbf{r}) \frac{\partial}{\partial n} \frac{1}{|\mathbf{r} - \mathbf{r}'|} - \frac{1}{|\mathbf{r} - \mathbf{r}'|} \frac{\partial}{\partial n} \Phi_{out}(\mathbf{r}) \right). \quad (17.77)$$

At the boundary the potential is continuous

$$\Phi_{out}(\mathbf{r}) = \Phi_{in}(\mathbf{r}) \quad \mathbf{r} \in A \quad (17.78)$$

whereas the normal derivative (hence the normal component of the electric field)
has a discontinuity

$$\varepsilon_1 \frac{\partial \Phi_{out}}{\partial n} = \frac{\partial \Phi_{in}}{\partial n}. \quad (17.79)$$

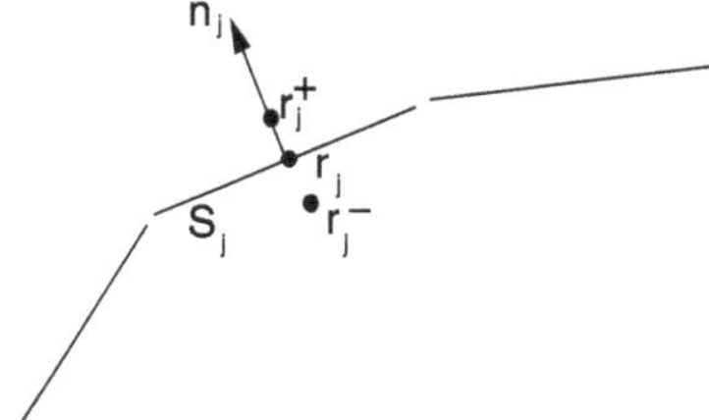

Fig. 17.15 Representation of the boundary by surface elements

17.3.2 Calculation of the Boundary Potential

For a numerical treatment the boundary surface is approximated by a finite set of small surface elements S_i, $i = 1 \ldots N$ centered at $\mathbf{r}_i$ with an area A_i and normal vector $\mathbf{n}_i$ (Fig. 17.15). (We assume planar elements in the following, the curvature leads to higher order corrections.)

The corresponding values of the potential and its normal derivative are denoted as $\Phi_i = \Phi(\mathbf{r}_i)$ and $\frac{\partial \Phi_i}{\partial n} = \mathbf{n}_i \operatorname{grad} \Phi(\mathbf{r}_i)$. At a point $\mathbf{r}_j^{\pm}$ close to the element S_j we obtain the following approximate equations:

$$
\Phi_{in}\left(\mathbf{r}_j^-\right) = \int_V dV \, \frac{\rho(\mathbf{r})}{4\pi \, |\mathbf{r} - \mathbf{r}_j^-| \varepsilon_0}
$$

$$
- \frac{1}{4\pi} \sum_i \Phi_i \oint_{S_i} dA \, \frac{\partial}{\partial n} \frac{1}{|\mathbf{r} - \mathbf{r}_j^-|} + \frac{1}{4\pi} \sum_i \frac{\partial \Phi_{i,in}}{\partial n} \oint_{S_i} dA \, \frac{1}{|\mathbf{r} - \mathbf{r}_j^-|}
$$

$$(17.80)$$

$$
\Phi_{out}\left(\mathbf{r}_j^+\right) = \frac{1}{4\pi} \sum_i \Phi_i \oint_{S_i} dA \, \frac{\partial}{\partial n} \frac{1}{|\mathbf{r} - \mathbf{r}_j^+|} - \frac{1}{4\pi} \sum_i \frac{\partial \Phi_{i,out}}{\partial n} \oint_{S_i} dA \, \frac{1}{|\mathbf{r} - \mathbf{r}_j^+|}.
$$

$$(17.81)$$

These two equations can be combined to obtain a system of equations for the potential values only. To that end we approach the boundary symmetrically with $\mathbf{r}_i^{\pm} = \mathbf{r}_i \pm d\mathbf{n}_i$. Under this circumstance

$$
\oint_{S_i} dA \, \frac{1}{|\mathbf{r} - \mathbf{r}_j^+|} = \oint_{S_i} dA \, \frac{1}{|\mathbf{r} - \mathbf{r}_j^-|}
$$

$$
\oint_{S_i} dA \, \frac{\partial}{\partial n} \frac{1}{|\mathbf{r} - \mathbf{r}_i^+|} = - \oint_{S_i} dA \, \frac{\partial}{\partial n} \frac{1}{|\mathbf{r} - \mathbf{r}_i^-|}
$$

$$(17.82)$$

$$
\oint_{S_i} dA \, \frac{\partial}{\partial n} \frac{1}{|\mathbf{r} - \mathbf{r}_j^+|} = \oint_{S_i} dA \, \frac{\partial}{\partial n} \frac{1}{|\mathbf{r} - \mathbf{r}_j^-|} \qquad j \neq i
$$

and we find

$$
(1 + \varepsilon_1)\Phi_j = \int_V dV \frac{\rho(\mathbf{r})}{4\pi \varepsilon_0 |\mathbf{r} - \mathbf{r}_j|}
$$

$$
- \frac{1}{4\pi} \sum_{i \neq j} (1 - \varepsilon_1)\Phi_i \oint_{S_i} dA \, \frac{\partial}{\partial n} \frac{1}{|\mathbf{r} - \mathbf{r}_j^-|}
$$

Fig. 17.16 Projection of the
surface element

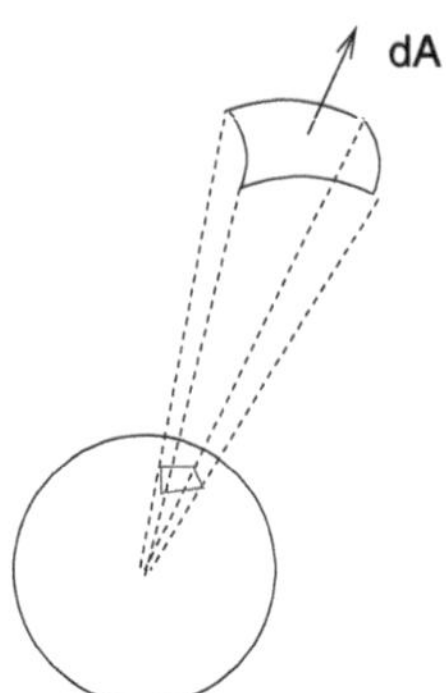

$$-\frac{1}{4\pi}(1+\varepsilon_1)\Phi_j \oint_{S_j} dA \,\frac{\partial}{\partial n}\frac{1}{|\mathbf{r}-\mathbf{r}_j^-|}. \tag{17.83}$$

The integrals for $i \neq j$ can be approximated by

$$\oint_{S_i} dA \,\frac{\partial}{\partial n}\frac{1}{|\mathbf{r}-\mathbf{r}_j^-|} = A_i\,\mathbf{n}_i\,\mathrm{grad}_i\,\frac{1}{|\mathbf{r}_i-\mathbf{r}_j|}. \tag{17.84}$$

The second integral has a simple geometrical interpretation (Fig. 17.16).

Since $\mathrm{grad}\,\frac{1}{|r-r'|} = -\frac{1}{|r-r'|^2}\frac{r-r'}{|r-r'|}$ the area element $d\mathbf{A}$ is projected onto a sphere
with unit radius. The integral $\oint_{S_j} d\mathbf{A}\,\mathrm{grad}_{r-}\frac{1}{|\mathbf{r_j}-\mathbf{r}_j^-|}$ is given by the solid angle of $\mathbf{S}_j$
with respect to r'. For $r' \to r_j$ from inside this is just minus half of the full space
angle of 4π. Thus we have

$$(1+\varepsilon_1)\Phi_j = \int_V dV\,\frac{\rho(\mathbf{r})}{4\pi|\mathbf{r}-\mathbf{r}_j|\varepsilon_0}$$
$$-\frac{1}{4\pi}\sum_{i\neq j}(1-\varepsilon_1)\Phi_i A_i\,\frac{\partial}{\partial n_i}\frac{1}{|\mathbf{r}_i-\mathbf{r}_j|}+\frac{1}{2}(1+\varepsilon_1)\Phi_j \tag{17.85}$$

or

$$\Phi_j = \frac{2}{1+\varepsilon_1}\int_V dV\,\frac{\rho(\mathbf{r})}{4\pi\varepsilon_0|\mathbf{r}-\mathbf{r}_j|}+\frac{1}{2\pi}\sum_{i\neq j}\frac{\varepsilon_1-1}{\varepsilon_1+1}\Phi_i A_i\,\frac{\partial}{\partial n_i}\frac{1}{|\mathbf{r}_i-\mathbf{r}_j|}. \tag{17.86}$$

This system of equations can be used to calculate the potential on the boundary. The
potential inside the cavity is then given by (17.73). Numerical stability is improved
by a related method which considers the potential gradient along the boundary. Taking the normal derivative

$$\frac{\partial}{\partial n_j} = \mathbf{n}_j\,\mathrm{grad}_{r_j\pm} \tag{17.87}$$

of (17.80), (17.81) gives

$$\frac{\partial}{\partial n_j}\Phi_{in}\left(\mathbf{r}_j^-\right) = \frac{\partial}{\partial n_j}\int_V dV \frac{\rho(\mathbf{r})}{4\pi\,|\mathbf{r}-\mathbf{r}_j^-|\,\varepsilon_0}$$

$$-\frac{1}{4\pi}\sum_i \Phi_i \oint_{S_i} dA\,\frac{\partial^2}{\partial n\,\partial n_j}\frac{1}{|\mathbf{r}-\mathbf{r}_j^-|}$$

$$+\frac{1}{4\pi}\sum_i \frac{\partial\Phi_{i,in}}{\partial n}\oint_{S_i} dA\,\frac{\partial}{\partial n_j}\frac{1}{|\mathbf{r}-\mathbf{r}_j^-|} \tag{17.88}$$

$$\frac{\partial}{\partial n_j}\Phi_{out}\left(\mathbf{r}_j^+\right) = \frac{1}{4\pi}\sum_i \Phi_i \oint_{S_i} dA\,\frac{\partial^2}{\partial n\,\partial n_j}\frac{1}{|\mathbf{r}-\mathbf{r}_j^+|}$$

$$-\frac{1}{4\pi}\sum_i \frac{\partial\Phi_{i,out}}{\partial n}\oint_{S_i} dA\,\frac{\partial}{\partial n_j}\frac{1}{|\mathbf{r}-\mathbf{r}_j^+|}. \tag{17.89}$$

In addition to (17.82) we have now

$$\oint_{S_i} dA\,\frac{\partial^2}{\partial n\,\partial n_j}\frac{1}{|\mathbf{r}-\mathbf{r}_j^-|} = \oint_{S_i} dA\,\frac{\partial^2}{\partial n\,\partial n_j}\frac{1}{|\mathbf{r}-\mathbf{r}_j^+|} \tag{17.90}$$

and the sum of the two equations gives

$$\left(1+\frac{1}{\varepsilon_1}\right)\frac{\partial}{\partial n_j}\Phi_{in,j}$$

$$=\frac{\partial}{\partial n_j}\left(\int_V dV\,\frac{\rho(\mathbf{r})}{4\pi\varepsilon_0|\mathbf{r}-\mathbf{r}_j|} + \frac{1-\frac{1}{\varepsilon_1}}{4\pi}\sum_{i\neq j} A_i \frac{\partial\Phi_{i,in}}{\partial n}\frac{1}{|\mathbf{r}_i-\mathbf{r}_j|}\right)$$

$$+\frac{1+\frac{1}{\varepsilon_1}}{2\pi}\frac{\partial\Phi_{j,in}}{\partial n} \tag{17.91}$$

or finally

$$\frac{\partial}{\partial n_j}\Phi_{in,j} = \frac{2\varepsilon_1}{\varepsilon_1+1}\frac{\partial}{\partial n_j}\int_V dV\,\frac{\rho(\mathbf{r})}{4\pi\varepsilon_0|\mathbf{r}-\mathbf{r}_j|}$$

$$+2\frac{\varepsilon_1-1}{\varepsilon_1+1}\sum_{i\neq j} A_i \frac{\partial\Phi_{i,in}}{\partial n}\frac{\partial}{\partial n_j}\frac{1}{|\mathbf{r}_i-\mathbf{r}_j|}. \tag{17.92}$$

In terms of the surface charge density this reads:

$$\sigma_j' = 2\varepsilon_0\frac{(1-\varepsilon_1)}{(1+\varepsilon_1)}\left(-\mathbf{n}_j\,\mathrm{grad}\int dV\,\frac{\rho(\mathbf{r})}{4\pi\varepsilon_0|\mathbf{r}-\mathbf{r}'|} + \frac{1}{4\pi\varepsilon_0}\sum_{i\neq j}\sigma_i' A_i \frac{\mathbf{n}_j(\mathbf{r}_j-\mathbf{r}_i)}{|\mathbf{r}_i-\mathbf{r}_j|^3}\right). \tag{17.93}$$

This system of linear equations can be solved directly or iteratively (a simple damping scheme $\sigma_m' \to \omega\sigma_m' + (1-\omega)\sigma_{m,old}'$ with $\omega \approx 0.6$ helps to get rid of oscillations). From the surface charges $\sigma_i A_i$ the potential is obtained with the help of (17.75).

17.4 Boundary Element Method for the Linearized Poisson-Boltzmann Equation

We consider now a cavity within an electrolyte. The fundamental solution of the linear Poisson-Boltzmann equation (17.63)

$$G_\kappa\left(\mathbf{r} - \mathbf{r}'\right) = -\frac{e^{-\kappa|\mathbf{r}-\mathbf{r}'|}}{4\pi|\mathbf{r} - \mathbf{r}'|} \tag{17.94}$$

obeys

$$\operatorname{div}\operatorname{grad} G_\kappa\left(\mathbf{r} - \mathbf{r}'\right) - \kappa^2 G_\kappa\left(\mathbf{r} - \mathbf{r}'\right) = \delta\left(\mathbf{r} - \mathbf{r}'\right). \tag{17.95}$$

Inserting into Green's theorem (17.69) we obtain the potential outside the cavity

$$\Phi_{out}(\mathbf{r}') = -\oint_S dA\left(\Phi_{out}(\mathbf{r})\frac{\partial}{\partial n}G_\kappa\left(\mathbf{r} - \mathbf{r}'\right) - G_\kappa\left(\mathbf{r} - \mathbf{r}'\right)\frac{\partial}{\partial n}\Phi_{out}(\mathbf{r})\right) \tag{17.96}$$

which can be combined with (17.74), (17.79) to give the following equations [32]

$$(1 + \varepsilon_1)\Phi\left(\mathbf{r}'\right) = \oint_S dA\left[\Phi(\mathbf{r})\frac{\partial}{\partial n}(G_0 - \varepsilon_1 G_\kappa) - (G_0 - G_\kappa)\frac{\partial}{\partial n}\Phi_{in}(\mathbf{r})\right]$$
$$+ \int_{cav}\frac{\rho(\mathbf{r})}{4\pi\varepsilon_0|\mathbf{r} - \mathbf{r}'|}dV \tag{17.97}$$

$$(1 + \varepsilon_1)\frac{\partial}{\partial n'}\Phi_{in}\left(\mathbf{r}'\right) = \oint_S dA\,\Phi(\mathbf{r})\frac{\partial^2}{\partial n\partial n'}(G_0 - G_\kappa)$$
$$- \oint_S dA\,\frac{\partial}{\partial n}\Phi_{in}(\mathbf{r})\frac{\partial}{\partial n'}\left(G_0 - \frac{1}{\varepsilon_1}G_k\right)$$
$$+ \frac{\partial}{\partial n'}\int_{cav}\frac{\rho(\mathbf{r})}{4\pi\varepsilon|\mathbf{r} - \mathbf{r}'|}dV. \tag{17.98}$$

For a set of discrete boundary elements the following equations determine the values of the potential and its normal derivative at the boundary:

$$\frac{1 + \varepsilon_1}{2}\Phi_j = \sum_{i\neq j}\Phi_i\oint dA\frac{\partial}{\partial n}(G_0 - \varepsilon_1 G_\kappa) - \sum_{i\neq j}\frac{\partial}{\partial n}\Phi_{i,in}\oint dA\,(G_0 - G_\kappa)$$
$$+ \int\frac{\rho(\mathbf{r})}{4\pi\varepsilon_0|\mathbf{r} - \mathbf{r}_i|}dV \tag{17.99}$$

$$\frac{1 + \varepsilon_1}{2}\frac{\partial}{\partial n'}\Phi_{i,in} = \sum_{i\neq j}\Phi_i\oint dA\,\frac{\partial^2}{\partial n\partial n'}(G_0 - G_\kappa)$$
$$- \sum_{i\neq j}\frac{\partial}{\partial n}\Phi_{i,in}\oint dA\frac{\partial}{\partial n'}\left(G_0 - \frac{1}{\varepsilon_1}G_k\right)$$
$$+ \frac{\partial}{\partial n'}\int\frac{\rho(\mathbf{r})}{4\pi\varepsilon|\mathbf{r} - \mathbf{r}_i|}dV. \tag{17.100}$$

The situation is much more involved than for the simpler Poisson equation (with $\kappa = 0$) since the calculation of many integrals including such with singularities is necessary [32, 143].

17.5 Electrostatic Interaction Energy (Onsager Model)

A very important quantity in molecular physics is the electrostatic interaction of a molecule and the surrounding solvent [11, 237]. We calculate it by taking a small part of the charge distribution from infinite distance ($\Phi(r \to \infty) = 0$) into the cavity. The charge distribution thereby changes from $\lambda\rho(r)$ to $(\lambda + d\lambda)\rho(r)$ with $0 \le \lambda \le 1$. The corresponding energy change is

$$
dE = \int d\lambda \cdot \rho(r)\Phi_\lambda(r)\,dV
$$
$$
= \int d\lambda \cdot \rho(r)\left(\sum_n \frac{\sigma_n(\lambda)A_n}{4\pi\varepsilon_0|r - r_n|} + \int \frac{\lambda\rho(r')}{4\pi\varepsilon_0|r - r'|}\,dV'\right)dV.
$$
$$
(17.101)
$$

Multiplication of (17.93) by a factor of λ shows that the surface charges $\lambda\sigma_n$ are the solution corresponding to the charge density $\lambda\rho(r)$. It follows that $\sigma_n(\lambda) = \lambda\sigma_n$ and hence

$$
dE = \lambda\,d\lambda \int \rho(r)\left(\sum_n \frac{\sigma_n A_n}{4\pi\varepsilon_0|r - r_n|} + \frac{\rho(r')}{4\pi\varepsilon_0|r - r'|}\,dV'\right). \quad (17.102)
$$

The second summand is the self energy of the charge distribution which does not depend on the medium. The first summand vanishes without a polarizable medium and gives the interaction energy. Hence we have the final expression

$$
E_{int} = \int dE = \int_0^1 \lambda\,d\lambda \int \rho(r) \sum_n \frac{\sigma_n A_n}{4\pi\varepsilon_0|r - r_n|}\,dV
$$
$$
= \sum_n \sigma_n A_n \int \frac{\rho(r)}{8\pi\varepsilon_0|r - r_n|}\,dV. \quad (17.103)
$$

For the special case of a spherical cavity with radius a an analytical solution by a multipole expansion is available [148]

$$
E_{int} = -\frac{1}{8\pi\varepsilon_0} \sum_l \sum_{m=-l}^{l} \frac{(l+1)(\varepsilon_1 - 1)}{[l + \varepsilon_1(l+1)]a^{2l+1}} M_l^m M_l^m \quad (17.104)
$$

with the multipole moments

$$
M_l^m = \int \rho(r,\theta,\varphi)\sqrt{\frac{4\pi}{2l+1}}\,r^l Y_l^m(\theta,\varphi)\,dV. \quad (17.105)
$$

The first two terms of this series are:

Fig. 17.17 Surface charges

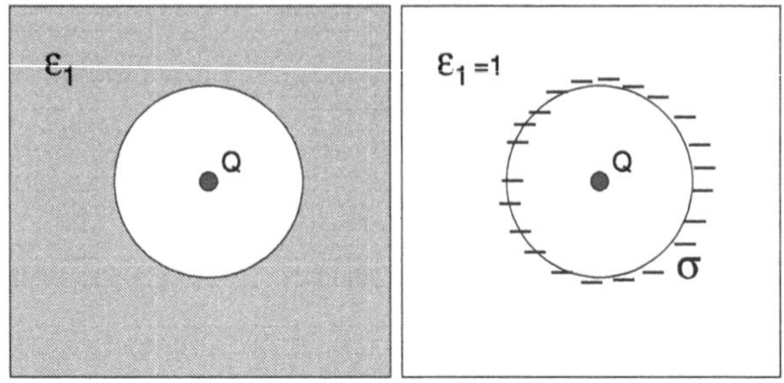

$$E_{int}^{(0)} = -\frac{1}{8\pi\varepsilon_0}\frac{\varepsilon_1 - 1}{\varepsilon_1 a}M_0^0 M_0^0 = -\frac{1}{8\pi\varepsilon_0}\left(1 - \frac{1}{\varepsilon_1}\right)\frac{Q^2}{a} \tag{17.106}$$

$$E_{int}^{(1)} = -\frac{1}{8\pi\varepsilon_0}\frac{2(\varepsilon_1 - 1)}{(1 + 2\varepsilon_1)a^3}\left(M_1^{-1}M_1^{-1} + M_1^0 M_1^0 + M_1^1 M_1^1\right)$$

$$= -\frac{1}{8\pi\varepsilon_0}\frac{2(\varepsilon_1 - 1)}{1 + 2\varepsilon_1}\frac{\mu^2}{a^3}. \tag{17.107}$$

17.5.1 Example: Point Charge in a Spherical Cavity

Consider a point charge Q in the center of a spherical cavity of radius R (Fig. 17.17).
The dielectric constant is given by

$$\varepsilon = \begin{cases} \varepsilon_0 & r < R \\ \varepsilon_1\varepsilon_0 & r > R. \end{cases} \tag{17.108}$$

Electric field and potential are inside the cavity

$$E = \frac{Q}{4\pi\varepsilon_0 r^2} \qquad \Phi = \frac{Q}{4\pi\varepsilon_0 r} + \frac{Q}{4\pi\varepsilon_0 R}\left(\frac{1}{\varepsilon_1} - 1\right) \tag{17.109}$$

and outside

$$E = \frac{Q}{4\pi\varepsilon_1\varepsilon_0 r^2} \qquad \Phi = \frac{Q}{4\pi\varepsilon_1\varepsilon_0 r} \quad r > R \tag{17.110}$$

which in terms of the surface charge density σ is

$$E = \frac{Q + 4\pi R^2\sigma}{4\pi\varepsilon_0 r^2} \quad r > R \tag{17.111}$$

with the total surface charge

$$4\pi R^2\sigma = Q\left(\frac{1}{\varepsilon_1} - 1\right). \tag{17.112}$$

The solvation energy (17.103) is given by

$$E_{int} = \frac{Q^2}{8\pi\varepsilon_0}\left(\frac{1}{\varepsilon_1} - 1\right) \tag{17.113}$$

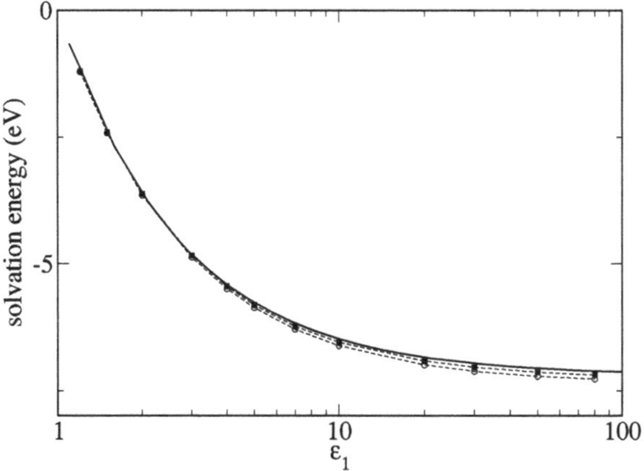

Fig. 17.18 (Solvation energy with the boundary element method) A spherical cavity is simulated with radius $a = 1$ Å which contains a point charge in its center. The solvation energy is calculated with 25×25 (*circles*) and 50×50 (*squares*) surface elements of equal size. The exact expression (17.106) is shown by the *solid curve*

which is the first term (17.106) of the multipole expansion. Figure 17.18 shows numerical results.

17.6 Problems

Problem 17.1 (Linearized Poisson-Boltzmann equation) This computer experiment simulates a homogeneously charged sphere in a dielectric medium (Fig. 17.19). The electrostatic potential is calculated from the linearized Poisson-Boltzmann equation (17.65) on a cubic grid of up to 100^3 points. The potential $\Phi(x)$ is shown along a line through the center together with a log-log plot of the maximum change per iteration

$$\left| \Phi^{(n+1)}(\mathbf{r}) - \Phi^{(n)}(\mathbf{r}) \right| \tag{17.114}$$

as a measure of convergence.

Explore the dependence of convergence on

- the initial values which can be chosen either $\Phi(\mathbf{r}) = 0$ or from the analytical solution

$$\Phi(\mathbf{r}) = \begin{cases} \dfrac{Q}{8\pi\varepsilon\varepsilon_0 a} \dfrac{2+\varepsilon(1+\kappa a)}{1+\kappa a} - \dfrac{Q}{8\pi\varepsilon_0 a^3} r^2 & \text{for } r < a \\[2ex] \dfrac{Qe^{-\kappa(r-a)}}{4\pi\varepsilon_0\varepsilon(\kappa a+1)r} & \text{for } r > a \end{cases} \tag{17.115}$$

- the relaxation parameter ω for different combinations of ε and κ
- the resolution of the grid

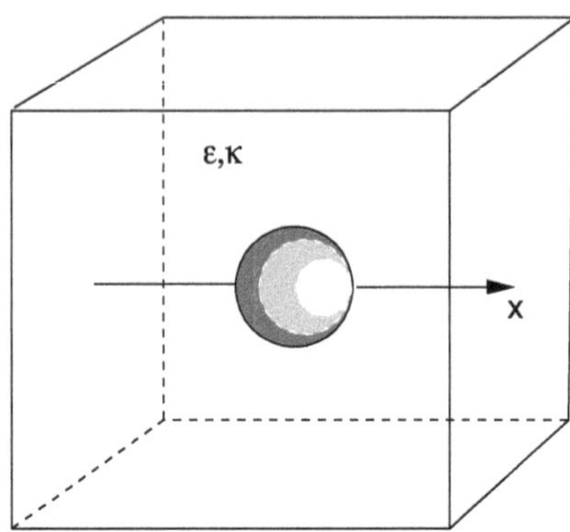

Fig. 17.19 Charged sphere in a dielectric medium

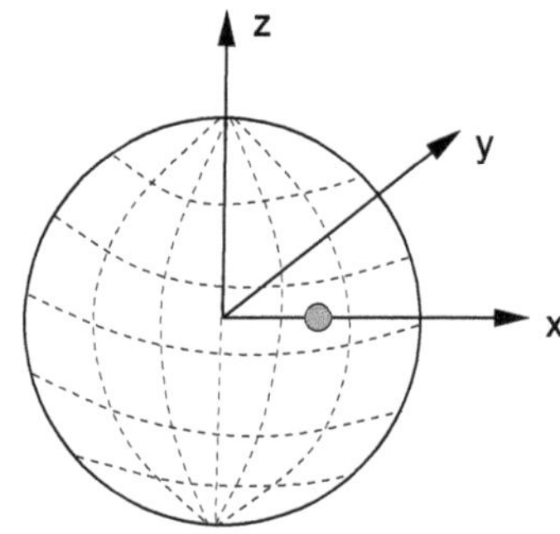

Fig. 17.20 Point charge inside a spherical cavity

Problem 17.2 (Boundary element method) In this computer experiment the solvation energy of a point charge within a spherical cavity (Fig. 17.20) is calculated with the boundary element method (17.93).

The calculated solvation energy is compared to the analytical value from (17.104)

$$E_{solv} = \frac{Q^2}{8\pi\varepsilon_0 R} \sum_{n=1}^{\infty} \frac{s^{2n}}{R^{2n}} \frac{(\varepsilon_1 - \varepsilon_2)(n+1)}{n\varepsilon_1 + (n+1)\varepsilon_2} \qquad (17.116)$$

where R is the cavity radius and s is the distance of the charge from the center of the cavity.

Explore the dependence of accuracy and convergence on

- the damping parameter ω
- the number of surface elements ($6 \times 6 \cdots 42 \times 42$) which can be chosen either as $d\phi\, d\theta$ or $d\phi\, d\cos\theta$ (equal areas)
- the position of the charge

Chapter 18
Waves

Waves are oscillations that move in space and time and are able to transport energy from one point to another. Quantum mechanical wavefunctions are discussed in Chap. 21. In this chapter we simulate classical waves which are, for instance, important in acoustics and electrodynamics. We use the method of finite differences to discretize the wave equation in one spatial dimension

$$\frac{\partial^2}{\partial t^2} f(t, x) = c^2 \frac{\partial^2}{\partial x^2} f(t, x). \tag{18.1}$$

Numerical solutions are obtained by an eigenvector expansion using trigonometric functions or by time integration. Accuracy and stability of different methods are compared. The wave function is second order in time and can be integrated directly with a two-step method. Alternatively, it can be converted into a first order system of equations of double dimension. Here, the velocity appears explicitly and velocity dependent damping can be taken into account. Finally, the second order wave equation can be replaced by two coupled first order equations for two variables (like velocity and density in case of acoustic waves), which can be solved by quite general methods. We compare the leapfrog, Lax-Wendroff and Crank-Nicolson methods. Only the Crank-Nicolson method is stable for Courant numbers $\alpha > 1$. It is an implicit method and can be solved iteratively. In a series of computer experiments we simulate waves on a string. We study reflection at an open or fixed boundary and at the interface between two different media. We compare dispersion and damping for different methods.

18.1 Classical Waves

In classical physics there are two main types of waves:

Electromagnetic waves do not require a medium. They are oscillations of the electromagnetic field and propagate also in vacuum. As an example consider a plane wave

P.O.J. Scherer, *Computational Physics*, Graduate Texts in Physics,
DOI 10.1007/978-3-319-00401-3_18,

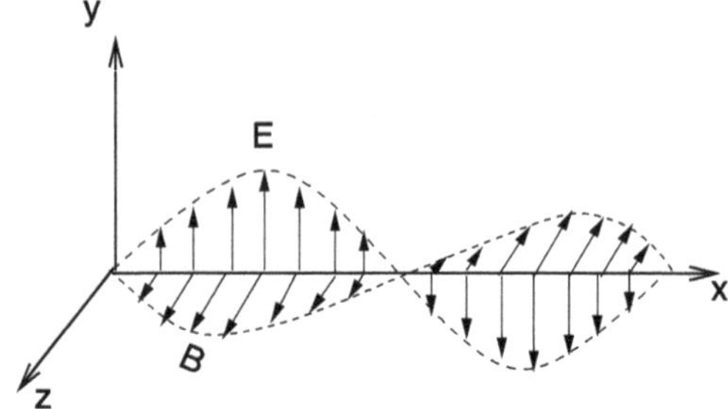

Fig. 18.1 Electromagnetic wave

which propagates in x-direction and is linearly polarized (Fig. 18.1). The electric and magnetic field have the form

$$\mathbf{E} = \begin{pmatrix} 0 \\ E_y(x,t) \\ 0 \end{pmatrix} \quad \mathbf{B} = \begin{pmatrix} 0 \\ 0 \\ B_z(x,t) \end{pmatrix}. \tag{18.2}$$

Maxwell's equations read in the absence of charges and currents

$$\operatorname{div}\mathbf{E} = \operatorname{div}\mathbf{B} = 0, \quad \operatorname{rot}\mathbf{E} = -\frac{\partial\mathbf{B}}{\partial t}, \quad \operatorname{rot}\mathbf{B} = \mu_0\varepsilon_0\frac{\partial\mathbf{E}}{\partial t}. \tag{18.3}$$

The fields (18.2) have zero divergence and satisfy the first two equations. Application of the third and fourth equation gives

$$\frac{\partial E_y}{\partial x} = -\frac{\partial B_z}{\partial t} \quad -\frac{\partial B_z}{\partial x} = \mu_0\varepsilon_0\frac{\partial E_y}{\partial t} \tag{18.4}$$

which can be combined to a one-dimensional wave equation

$$\frac{\partial^2 E_y}{\partial t^2} = c^2\frac{\partial^2 E_y}{\partial x^2} \tag{18.5}$$

with velocity $c = (\mu_0\varepsilon_0)^{-1/2}$.

Mechanical waves propagate through an elastic medium like air, water or an elastic solid. The material is subject to external forces deforming it and elastic forces which try to restore the deformation. As a result the atoms or molecules move around their equilibrium positions. As an example consider one-dimensional acoustic waves in an organ pipe (Fig. 18.2):

A mass element

$$dm = \varrho\, dV = \varrho A\, dx \tag{18.6}$$

at position x experiences an external force due to the air pressure which, according to Newton's law changes the velocity v of the element as described by Euler's equation[1]

---

[1] We consider only small deviations from the equilibrium values ϱ_0, p_0, $v_0 = 0$.

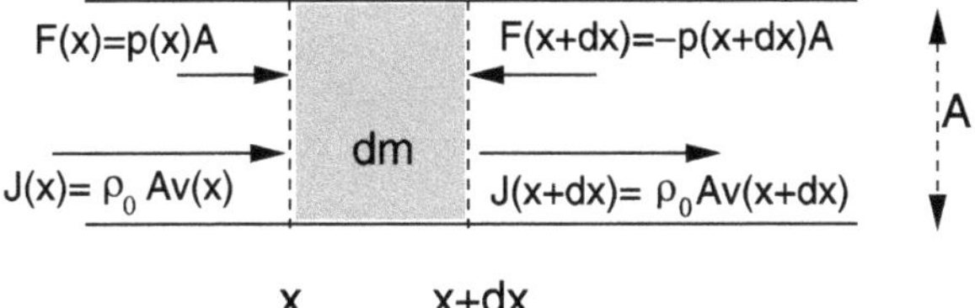

Fig. 18.2 (Acoustic waves in one dimension) A mass element $dm = \varrho A\, dx$ at position x experiences a total force $F = F(x) + F(x + dx) = -A\frac{\partial p}{\partial x}\, dx$. Due to the conservation of mass the change of the density $\frac{\partial \varrho}{\partial t}$ is given by the net flux $J = J(x) - J(x + dx) = -\varrho_0 A\frac{\partial v}{\partial x}\, dx$

$$\varrho_0 \frac{\partial}{\partial t} v = -\frac{\partial p}{\partial x}. \tag{18.7}$$

The pressure is a function of the density

$$\frac{p}{p_0} = \left(\frac{\varrho}{\varrho_0}\right)^n \quad \left(\frac{dp}{d\varrho}\right)_0 = n\frac{p_0}{\varrho_0} = c^2 \tag{18.8}$$

where $n = 1$ for an isothermal ideal gas and $n \approx 1.4$ for air under adiabatic conditions (no heat exchange), therefore

$$\varrho_0 \frac{\partial}{\partial t} v = -c^2 \frac{\partial \varrho}{\partial x}. \tag{18.9}$$

From the conservation of mass the continuity equation (11.10) follows

$$\frac{\partial}{\partial t}\varrho = -\varrho_0 \frac{\partial}{\partial x} v. \tag{18.10}$$

Combining the time derivative of (18.10) and the spatial derivative of (18.9) we obtain again the one-dimensional wave equation

$$\frac{\partial^2}{\partial t^2}\varrho = c^2 \frac{\partial^2}{\partial x^2}\varrho. \tag{18.11}$$

The wave equation can be factorized as

$$\left(\frac{\partial}{\partial t} + c\frac{\partial}{\partial x}\right)\left(\frac{\partial}{\partial t} - c\frac{\partial}{\partial x}\right)\varrho = \left(\frac{\partial}{\partial t} - c\frac{\partial}{\partial x}\right)\left(\frac{\partial}{\partial t} + c\frac{\partial}{\partial x}\right)\varrho = 0 \tag{18.12}$$

which shows that solutions of the advection equation

$$\left(\frac{\partial}{\partial t} \pm c\frac{\partial}{\partial x}\right)\varrho = 0 \tag{18.13}$$

are also solutions of the wave equation, which have the form

$$\varrho = f(x \pm ct). \tag{18.14}$$

In fact a general solution of the wave equation is given according to d'Alembert as the sum of two waves running to the left and right side with velocity c and a constant envelope (Fig. 18.3)

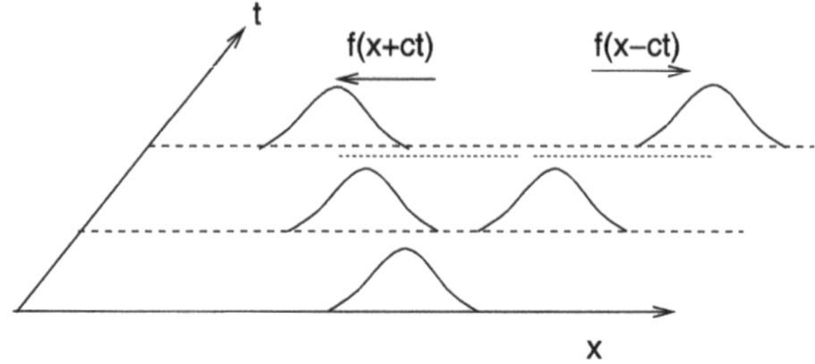

Fig. 18.3 d'Alembert solution to the wave equation

$$\varrho = f_1(x + ct) + f_2(x - ct). \tag{18.15}$$

A special solution of this kind is the plane wave solution

$$f(x,t) = e^{i\omega t \pm ikx}$$

with the dispersion relation

$$\omega = ck. \tag{18.16}$$

18.2 Spatial Discretization in One Dimension

We use the simplest finite difference expression for the spatial derivative (Sects. 3.4, 11.2)

$$\frac{\partial^2}{\partial x^2} f(x,t) = \frac{f(t, x + \Delta x) + f(t, x - \Delta x) - 2f(t,x)}{\Delta x^2} + O(\Delta x^2) \tag{18.17}$$

and a regular grid

$$x_m = m \Delta x \quad m = 1, 2 \ldots M \tag{18.18}$$

$$f_m = f(x_m). \tag{18.19}$$

This turns the wave equation into the system of ordinary differential equations (Sect. 11.2.3)

$$\frac{d^2}{dt^2} f_m = c^2 \frac{f_{m+1} + f_{m-1} - 2f_m}{\Delta x^2} \tag{18.20}$$

where f_0 and f_{M+1} have to be specified by suitable boundary conditions (Fig. 18.4). In matrix notation we have

$$\mathbf{f}(t) = \begin{pmatrix} f_1(t) \\ \vdots \\ f_M(t) \end{pmatrix} \tag{18.21}$$

$$\frac{d^2}{dt^2} \mathbf{f}(t) = \mathbf{A}\mathbf{f}(t) + \mathbf{S}(t) \tag{18.22}$$

where for

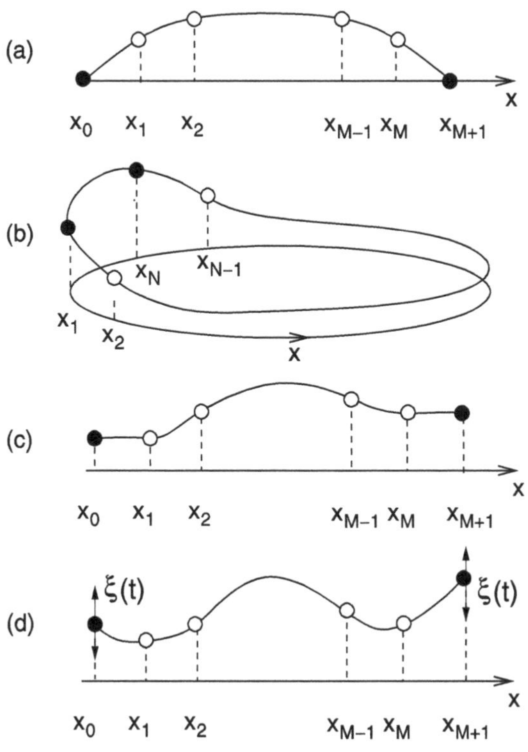

Fig. 18.4 (Boundary conditions for 1-dimensional waves) Additional boundary points x_0, x_{M+1} are used to realize the boundary conditions: (**a**) *fixed boundaries*: $f(x_0) = 0$, $\frac{\partial^2}{\partial x^2} f(x_1) = \frac{1}{\Delta x^2}(f(x_2) - 2f(x_1))$ or $f(x_{M+1}) = 0$, $\frac{\partial^2}{\partial x^2} f(x_M) = \frac{1}{\Delta x^2}(f(x_{M-1}) - 2f(x_M))$, (**b**) *periodic boundary conditions*: $x_0 \equiv x_M$, $\frac{\partial^2}{\partial x^2} f(x_1) = \frac{1}{\Delta x^2}(f(x_2) + f(x_M) - 2f(x_1))$, $x_{M+1} \equiv x_1$, $\frac{\partial^2}{\partial x^2} f(x_M) = \frac{1}{\Delta x^2}(f(x_{M-1}) + f(x_1) - 2f(x_{NM}))$, (**c**) *open boundaries*: $\frac{\partial}{\partial x} f(x_1) = \frac{f(x_2) - f(x_0)}{2\Delta x} = 0$, $\frac{\partial^2}{\partial x^2} f(x_1) = \frac{1}{\Delta x^2}(2f(x_2) - 2f(x_1))$ or $\frac{\partial}{\partial x} f(x_M) = \frac{f(x_{M+1}) - f(x_{M-1})}{2\Delta x} = 0$, $\frac{\partial^2}{\partial x^2} f(x_M) = \frac{1}{\Delta x^2}(2f(x_{M-1}) - 2f(x_M))$, (**d**) *moving boundaries*: $f(x_0, t) = \xi_0(t)$, $\frac{\partial^2}{\partial x^2} f(x_1) = \frac{1}{\Delta x^2}(f(x_2) - 2f(x_1) + \xi_0(t))$ or $f(x_{M+1}, t) = \xi_{M+1}(t)$, $\frac{\partial^2}{\partial x^2} f(x_M) = \frac{1}{\Delta x^2}(f(x_{M-1}) - 2f(x_M) + \xi_{N+1}(t))$

- fixed boundaries

$$
A = \begin{pmatrix}
-2 & 1 & & & & & \\
1 & -2 & 1 & & & & \\
& 1 & -2 & 1 & & & \\
& & \ddots & \ddots & \ddots & & \\
& & & 1 & -2 & 1 \\
& & & & 1 & -2
\end{pmatrix} \frac{c^2}{\Delta x^2} \quad \mathbf{S}(t) = 0 \qquad (18.23)
$$

- periodic boundaries[2]

$$A = \begin{pmatrix} -2 & 1 & & & & & 1 \\ 1 & -2 & 1 & & & & \\ & 1 & -2 & 1 & & & \\ & & \ddots & \ddots & \ddots & & \\ & & & 1 & -2 & 1 \\ 1 & & & & 1 & -2 \end{pmatrix} \frac{c^2}{\Delta x^2} \quad \mathbf{S}(t) = 0 \qquad (18.24)$$

- open boundaries

$$A = \begin{pmatrix} -2 & 2 & & & & \\ 1 & -2 & 1 & & & \\ & 1 & -2 & 1 & & \\ & & \ddots & \ddots & \ddots & \\ & & & 1 & -2 & 1 \\ & & & & 2 & -2 \end{pmatrix} \frac{c^2}{\Delta x^2} \quad \mathbf{S}(t) = 0 \qquad (18.25)$$

- moving boundaries

$$A = \begin{pmatrix} -2 & 1 & & & & \\ 1 & -2 & 1 & & & \\ & 1 & -2 & 1 & & \\ & & \ddots & \ddots & \ddots & \\ & & & 1 & -2 & 1 \\ & & & & 1 & -2 \end{pmatrix} \frac{c^2}{\Delta x^2} \quad \mathbf{S}(t) = \begin{pmatrix} \xi_0(t) \\ 0 \\ \vdots \\ \vdots \\ 0 \\ \xi_{N+1}(t) \end{pmatrix}.$$

$$(18.26)$$

A combination of different boundary conditions for both sides is possible.

Equation (18.20) corresponds to a series of mass points which are connected by harmonic springs (Fig. 18.5), a model, which is used in solid state physics to describe longitudinal acoustic waves [129].

18.3 Solution by an Eigenvector Expansion

For fixed boundaries (18.20) reads in matrix form

$$\frac{d^2}{dt^2} \mathbf{f}(t) = A \mathbf{f}(t) \qquad (18.27)$$

with the vector of function values:

---

[2] This corresponds to the boundary condition $f_0 = f_2$, $\frac{\partial}{\partial x} f(x_1) = 0$. Alternatively we could use $f_0 = f_1$, $\frac{\partial}{\partial x} f(x_{1/2}) = 0$ which replaces the 2 s in the first and last row by 1 s.

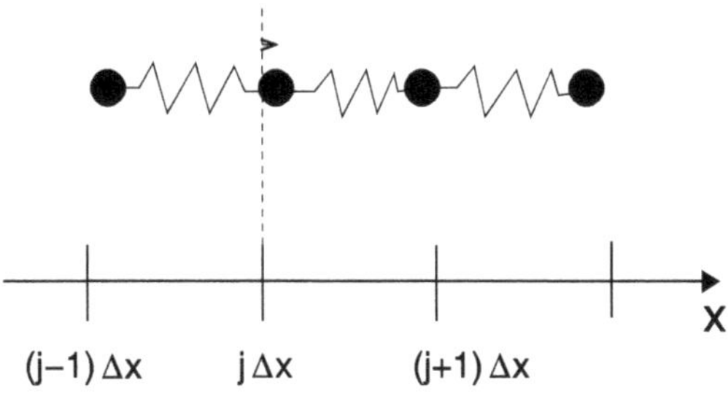

Fig. 18.5 (Atomistic model for longitudinal waves) A set of mass points m is connected by springs with stiffness K. The elongation of mass point number j from its equilibrium position $x_j = j \Delta x$ is ξ_j. The equations of motion $m\ddot{\xi}_j = -K(\xi_j - \xi_{j-1}) - K(\xi_j - \xi_{j+1})$ coincide with (18.20) with a velocity of $c = \Delta x \sqrt{\frac{k\Delta x}{m}}$

$$\mathbf{f}(t) = \begin{pmatrix} f_1(t) \\ \vdots \\ f_M(t) \end{pmatrix} \tag{18.28}$$

and the matrix

$$A = \begin{pmatrix} -2 & 1 & & & & \\ 1 & -2 & 1 & & & \\ & 1 & -2 & 1 & & \\ & & \ddots & \ddots & \ddots & \\ & & & 1 & -2 & 1 \\ & & & & 1 & -2 \end{pmatrix} \frac{c^2}{\Delta x^2} \tag{18.29}$$

which can be diagonalized exactly (Sect. 9.3). The two boundary points $f(0) = 0$ and $f((M+1)\Delta x) = 0$ can be added without any changes. The eigenvalues are

$$\lambda = 2\frac{c^2}{\Delta x^2}\left(\cos(k\Delta x) - 1\right) = -\frac{4c^2}{\Delta x^2}\sin^2\left(\frac{k\Delta x}{2}\right) = (i\omega_k)^2$$

$$k\Delta x = \frac{\pi l}{(M+1)}, \quad l = 1 \ldots M \tag{18.30}$$

with the frequencies

$$\omega_k = \frac{2c}{\Delta x}\sin\left(\frac{k\Delta x}{2}\right). \tag{18.31}$$

This result deviates from the dispersion relation of the continuous wave equation (18.11) $\omega_k = ck$ and approximates it only for $k\Delta x \ll 1$ (Fig. 18.6).

The general solution has the form (Sect. 11.2.4)

$$f_n(t) = \sum_{l=1}^{M}\left(C_{l+}e^{i\omega_l t} + C_{l-}e^{-i\omega_l t}\right)\sin\left(m\frac{\pi l}{(M+1)}\right). \tag{18.32}$$

Fig. 18.6 Dispersion of the discrete wave equation

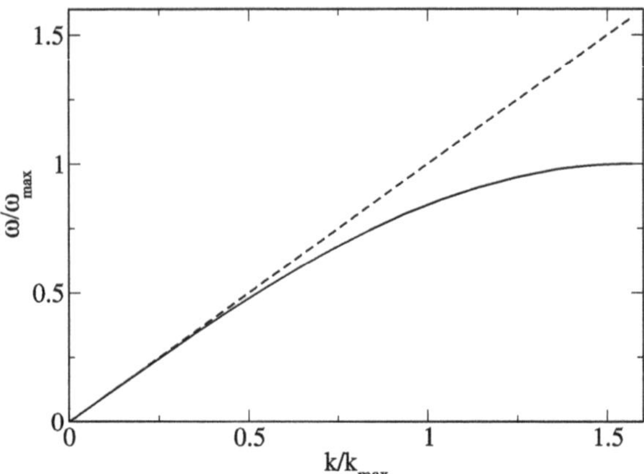

The initial amplitudes and velocities are

$$f_n(t=0) = \sum_{l=1}^{M} (C_{l+} + C_{l-}) \sin\left(m\frac{\pi l}{(M+1)}\right) = F_m$$

$$\frac{\mathrm{d}}{\mathrm{d}t} f_m(t=0, x_m) = \sum_{l=1}^{M} i\omega_l (C_{l+} - C_{l-}) \sin\left(m\frac{\pi l}{(M+1)}\right) = G_m \tag{18.33}$$

with F_m and G_m given. Different eigenfunctions of a tridiagonal matrix are mutually orthogonal

$$\sum_{m=1}^{M} \sin\left(m\frac{\pi l}{M+1}\right) \sin\left(m\frac{\pi l'}{M+1}\right) = \frac{M}{2}\delta_{l,l'} \tag{18.34}$$

and the coefficients $C_{l\pm}$ follow from a discrete Fourier transformation:

$$\begin{aligned}
\widetilde{F}_l &= \frac{1}{M} \sum_{m=1}^{M} \sin\left(m\frac{\pi l}{N+1}\right) F_m \\
&= \frac{1}{M} \sum_{m=1}^{M} \sum_{l'=1}^{M} (C_{l'+} + C_{l'-}) \sin\left(m\frac{\pi l'}{M+1}\right) \sin\left(m\frac{\pi l}{M+1}\right) \\
&= \frac{1}{2}(C_{l+} + C_{l-}) \tag{18.35}
\end{aligned}$$

$$\begin{aligned}
\widetilde{G}_l &= \frac{1}{M} \sum_{m=1}^{M} \sin\left(m\frac{\pi l}{N+1}\right) G_n \\
&= \frac{1}{M} \sum_{m=1}^{M} \sum_{l'=1}^{NM} i\omega_l (C_{l+} - C_{l-}) \sin\left(m\frac{\pi l'}{M+1}\right) \sin\left(m\frac{\pi l}{M+1}\right) \\
&= \frac{1}{2}i\omega_l (C_{l+} - C_{l-}) \tag{18.36}
\end{aligned}$$

$$C_{l+} = \widetilde{F}_l + \frac{1}{i\omega_l}\widetilde{G}_l$$
$$C_{l-} = \widetilde{F}_l - \frac{1}{i\omega_l}\widetilde{G}_l. \tag{18.37}$$

Finally the explicit solution of the wave equation is

$$f_m(t) = \sum_{l=1}^{M} 2\left(\widetilde{F}_l \cos(\omega_l t) + \frac{\widetilde{G}_l}{\omega_l}\sin(\omega_l t)\right)\sin\left(m\frac{\pi l}{M+1}\right). \tag{18.38}$$

Periodic or open boundaries can be treated similarly as the matrices can be diagonalized exactly (Sect. 9.3). For moving boundaries the expansion coefficients are time dependent (Sect. 11.2.4).

18.4 Discretization of Space and Time

Using the finite difference expression also for the second time derivative the fully discretized wave equation is

$$\frac{f(t+\Delta t, x) + f(t-\Delta t, x) - 2f(t, x)}{\Delta t^2}$$
$$= c^2 \frac{f(t, x+\Delta x) + f(t, x-\Delta x) - 2f(t, x)}{\Delta x^2} + O\left(\Delta x^2, \Delta t^2\right). \tag{18.39}$$

For a plane wave

$$f = e^{i(\omega t - kx)} \tag{18.40}$$

we find

$$e^{i\omega\Delta t} + e^{-i\omega\Delta t} - 2 = c^2 \frac{\Delta t^2}{\Delta x^2}\left(e^{ik\Delta x} + e^{-ik\Delta x} - 2\right) \tag{18.41}$$

which can be written as

$$\sin\frac{\omega\Delta t}{2} = \alpha \sin\frac{k\Delta x}{2} \tag{18.42}$$

with the so-called Courant-number [64]

$$\alpha = c\frac{\Delta t}{\Delta x}. \tag{18.43}$$

From (18.42) we see that the dispersion relation is linear only for $\alpha = 1$. For $\alpha \neq 1$ not all values of ω and k allowed (Fig. 18.7).

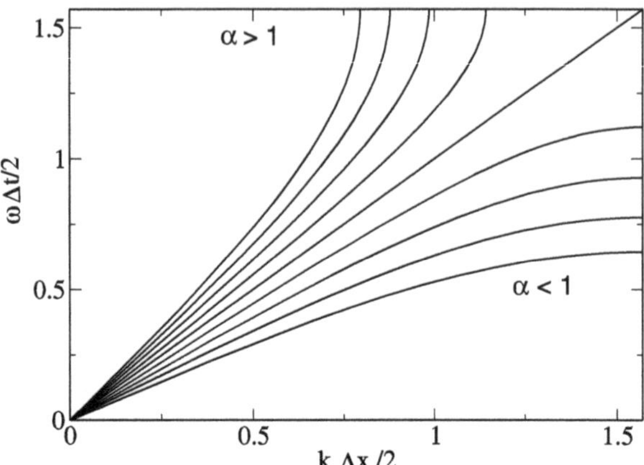

Fig. 18.7 (Dispersion of the discrete wave equation) Only for $\alpha = 1$ or for small values of $k\Delta x$ and $\omega\Delta t$ is the dispersion approximately linear. For $\alpha < 1$ only frequencies $\omega < \omega_{\max} = 2\arcsin(\alpha)/\Delta t$ are allowed whereas for $\alpha > 1$ the range of k-values is bounded by $k_{\max} = 2\arcsin(1/\alpha)/\Delta x$

18.5 Numerical Integration with a Two-Step Method

We solve the discrete wave equation (18.39) with fixed or open boundaries for

$$f(t + \Delta t, x) = 2f(t, x)\left(1 - \alpha^2\right) + \alpha^2\left(f(t, x + \Delta x) + f(t, x - \Delta x)\right)$$
$$- f(t - \Delta t, x) + O\left(\Delta t^2, \Delta x^2\right) \tag{18.44}$$

on the regular grids

$$x_m = m\Delta x \quad m = 1, 2 \ldots M \tag{18.45}$$

$$t_n = n\Delta t \quad n = 1, 2 \ldots N \tag{18.46}$$

$$\mathbf{f}_n = \begin{pmatrix} f_1^n \\ \vdots \\ f_M^n \end{pmatrix} = \begin{pmatrix} f(t_n, x_1) \\ \vdots \\ f(t_n, x_M) \end{pmatrix} \tag{18.47}$$

by applying the iteration

$$f_m^{n+1} = 2\left(1 - \alpha^2\right) f_m^n + \alpha^2 f_{m+1}^n + \alpha^2 f_{m-1}^n - f_m^{n-1}. \tag{18.48}$$

This is a two-step method which can be rewritten as a one-step method of double dimension

$$\begin{pmatrix} \mathbf{f}_{n+1} \\ \mathbf{f}_n \end{pmatrix} = T \begin{pmatrix} \mathbf{f}_n \\ \mathbf{f}_{n-1} \end{pmatrix} = \begin{pmatrix} 2 + \alpha^2 M & -1 \\ 1 & 0 \end{pmatrix} \begin{pmatrix} \mathbf{f}_n \\ \mathbf{f}_{n-1} \end{pmatrix} \tag{18.49}$$

with the tridiagonal matrix

$$M = \begin{pmatrix} -2 & a_1 & & & \\ 1 & -2 & 1 & & \\ & \ddots & \ddots & \ddots & \\ & & 1 & -2 & 1 \\ & & & a_N & -2 \end{pmatrix} \tag{18.50}$$

where a_1 and a_N have the values 1 for a fixed or 2 for an open end.

The matrix M has eigenvalues (Sect. 9.3)

$$\lambda = 2\cos(k\Delta x) - 2 = -4\sin^2\left(\frac{k\Delta x}{2}\right). \tag{18.51}$$

To simulate excitation of waves by a moving boundary we add one grid point with given elongation $\xi_0(t)$ and change the first equation into

$$f(t_{n+1}, x_1) = 2(1 - \alpha^2)f(t_n, x_1) + \alpha^2 f(t_n, x_2) + \alpha^2 \xi_0(t_n) - f(t_{n-1}, x_1). \tag{18.52}$$

Repeated iteration gives the series of function values

$$\begin{pmatrix} \mathbf{f}_1 \\ \mathbf{f}_0 \end{pmatrix}, \quad \begin{pmatrix} \mathbf{f}_2 \\ \mathbf{f}_1 \end{pmatrix} = T\begin{pmatrix} \mathbf{f}_1 \\ \mathbf{f}_0 \end{pmatrix}, \quad \begin{pmatrix} \mathbf{f}_3 \\ \mathbf{f}_2 \end{pmatrix} = T^2\begin{pmatrix} \mathbf{f}_1 \\ \mathbf{f}_0 \end{pmatrix}, \cdots \tag{18.53}$$

A necessary condition for stability is that all eigenvalues of T have absolute values smaller than one. Otherwise small perturbations would be amplified. The eigenvalue equation for T is

$$\begin{pmatrix} 2 + \alpha^2 M - \sigma & -1 \\ 1 & -\sigma \end{pmatrix}\begin{pmatrix} u \\ v \end{pmatrix} = \begin{pmatrix} 0 \\ 0 \end{pmatrix}. \tag{18.54}$$

We substitute the solution of the second equation

$$u = \sigma v \tag{18.55}$$

into the first equation and use the eigenvectors of M (Sect. 9.3) to obtain the eigenvalue equation

$$(2 + \alpha^2\lambda - \sigma)\sigma v - v = 0. \tag{18.56}$$

Hence σ is one of the two roots of

$$\sigma^2 - \sigma(\alpha^2\lambda + 2) + 1 = 0 \tag{18.57}$$

which are given by (Fig. 18.8)

$$\sigma = 1 + \frac{\alpha^2\lambda}{2} \pm \sqrt{\left(\frac{\alpha^2\lambda}{2} + 1\right)^2 - 1}. \tag{18.58}$$

From

$$\lambda = -4\sin^2\left(\frac{k\Delta x}{2}\right)$$

we find

$$-4 < \lambda < 0 \tag{18.59}$$

$$1 - 2\alpha^2 < \frac{\alpha^2\lambda}{2} + 1 < 1 \tag{18.60}$$

and the square root in (18.58) is imaginary if

$$-1 < \frac{\alpha^2\lambda}{2} + 1 < 1 \tag{18.61}$$

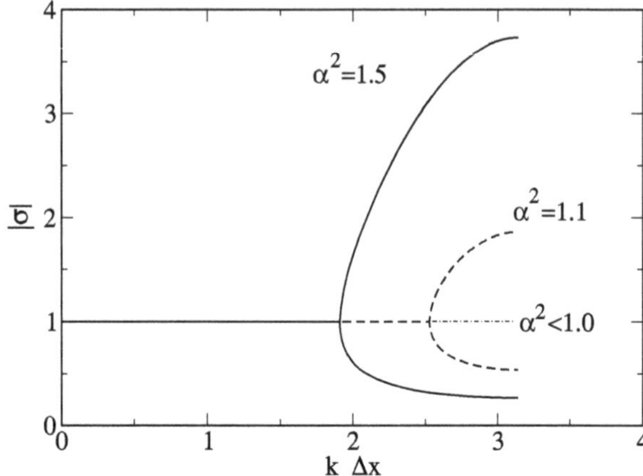

Fig. 18.8 (Stability regions of the two-step method) Instabilities appear for $|\alpha| > 1$. One of the two eigenvalues σ becomes unstable ($|\sigma| > 1$) for waves with large k-values

which is the case for

$$\sin^2\left(\frac{k\Delta x}{2}\right)\alpha^2 < 1. \tag{18.62}$$

This holds for all k only if

$$|\alpha| < 1. \tag{18.63}$$

But then

$$|\sigma|^2 = \left(1 + \frac{\alpha^2\lambda}{2}\right)^2 + \left(1 - \left(\frac{\alpha^2\lambda}{2} + 1\right)^2\right) = 1 \tag{18.64}$$

and the algorithm is (conditionally) stable. If on the other hand $|\alpha| > 1$ then for some k-values the square root is real. Here we have

$$1 + \frac{\alpha^2\lambda}{2} < -1 \tag{18.65}$$

and finally

$$1 + \frac{\alpha^2\lambda}{2} - \sqrt{\left(1 + \frac{\alpha^2\lambda}{2}\right)^2 - 1} < -1 \tag{18.66}$$

which shows that instabilities are possible in this case.

18.6 Reduction to a First Order Differential Equation

A general method to reduce the order of an ordinary differential equation (or a system of such) introduces the time derivatives as additional variables (Chap. 12). The spatially discretized one-dimensional wave equation (18.22) can be transformed into a system of double dimension

$$\frac{\mathrm{d}}{\mathrm{d}t}\mathbf{f}(t) = \mathbf{v}(t) \tag{18.67}$$

$$\frac{\mathrm{d}}{\mathrm{d}t}\mathbf{v}(t) = \frac{c^2}{\Delta x^2}M\mathbf{f}(t) + \mathbf{S}(t). \tag{18.68}$$

We use the improved Euler method (Sect. 12.5)

$$\mathbf{f}(t+\Delta t) = \mathbf{f}(t) + \mathbf{v}\left(t+\frac{\Delta t}{2}\right)\Delta t + O\left(\Delta t^3\right) \tag{18.69}$$

$$\mathbf{v}(t+\Delta t) = \mathbf{v}(t) + \left[\frac{c^2}{\Delta x^2}M\mathbf{f}\left(t+\frac{\Delta t}{2}\right) + \mathbf{S}\left(t+\frac{\Delta t}{2}\right)\right]\Delta t + O\left(\Delta t^3\right) \tag{18.70}$$

and two different time grids

$$\mathbf{f}_n = \mathbf{f}(t_n) \quad \mathbf{S}_n = \mathbf{S}(t_n) \quad n = 0, 1 \ldots \tag{18.71}$$

$$\mathbf{f}(t_{n+1}) = \mathbf{f}(t_n) + \mathbf{v}(t_{n+1/2})\Delta t \tag{18.72}$$

$$\mathbf{v}_n = \mathbf{v}(t_{n-1/2}) \quad n = 0, 1 \ldots \tag{18.73}$$

$$\mathbf{v}(t_{n+1/2}) = \mathbf{v}(t_{n-1/2}) + \left[\frac{c^2}{\Delta x^2}M\mathbf{f}(t_n) + \mathbf{S}(t_n)\right]\Delta t. \tag{18.74}$$

We obtain a leapfrog (Fig. 18.9) like algorithm (page 231)

$$\mathbf{v}_{n+1} = \mathbf{v}_n + \left[\frac{c^2}{\Delta x^2}M\mathbf{f}_n + \mathbf{S}_n\right]\Delta t \tag{18.75}$$

$$\mathbf{f}_{n+1} = \mathbf{f}_n + \mathbf{v}_{n+1}\Delta t \tag{18.76}$$

where the updated velocity (18.75) has to be inserted into (18.76). This can be combined into the iteration

$$\begin{pmatrix} \mathbf{f}_{n+1} \\ \mathbf{v}_{n+1} \end{pmatrix} = \begin{pmatrix} \mathbf{f}_n + \mathbf{v}_n\Delta t + [\frac{c^2}{\Delta x^2}M\mathbf{f}_n + \mathbf{S}_n]\Delta t^2 \\ \mathbf{v}_n + [\frac{c^2}{\Delta x^2}M\mathbf{f}_n + \mathbf{S}_n]\Delta t \end{pmatrix}$$

$$= \begin{pmatrix} 1 + \frac{c^2\Delta t^2}{\Delta x^2}M & \Delta t \\ \frac{c^2\Delta t}{\Delta x^2}M & 1 \end{pmatrix}\begin{pmatrix} \mathbf{f}_n \\ \mathbf{v}_n \end{pmatrix} + \begin{pmatrix} \mathbf{S}_n\Delta t^2 \\ \mathbf{S}_n\Delta t \end{pmatrix}. \tag{18.77}$$

Since the velocity appears explicitly we can easily add a velocity dependent damping like

$$-\gamma v(t_n, x_m) \tag{18.78}$$

which we approximate by

$$-\gamma v\left(t_n - \frac{\Delta t}{2}, x_m\right) \tag{18.79}$$

under the assumption of weak damping

$$\gamma\Delta t \ll 1. \tag{18.80}$$

Fig. 18.9 Leapfrog method

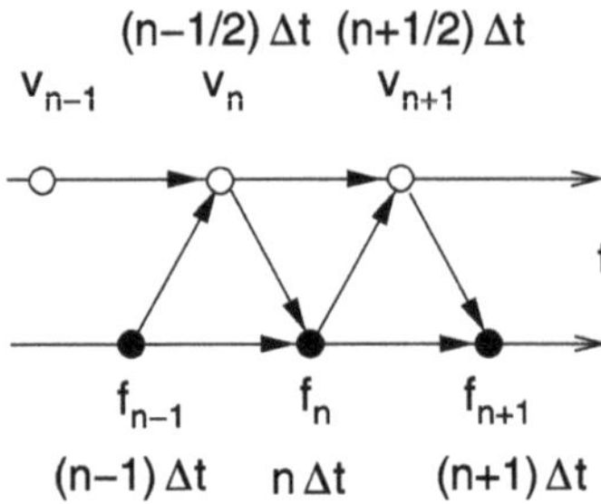

To study the stability of this algorithm we consider the homogeneous problem with fixed boundaries. With the Courant number $\alpha = \frac{c\Delta t}{\Delta x}$ (18.77) becomes

$$\begin{pmatrix} \mathbf{f}_{n+1} \\ \mathbf{v}_{n+1} \end{pmatrix} = \begin{pmatrix} 1 + \alpha^2 M & \Delta t(1 - \gamma \Delta t) \\ \frac{\alpha^2}{\Delta t} M & 1 - \gamma \Delta t \end{pmatrix} \begin{pmatrix} \mathbf{f}_n \\ \mathbf{v}_n \end{pmatrix}. \tag{18.81}$$

Using the eigenvectors and eigenvalues of M (Sect. 9.3)

$$\lambda = -4 \sin^2 \left(\frac{k \Delta x}{2} \right) \tag{18.82}$$

we find the following equation for the eigenvalues σ:

$$\left(1 + \alpha^2 \lambda - \sigma \right) u + \Delta t (1 - \gamma \Delta t) v = 0$$
$$\alpha^2 \lambda u + \Delta t (1 - \gamma \Delta t - \sigma) v = 0. \tag{18.83}$$

Solving the second equation for u and substituting into the first equation we have

$$\left[(1 + \alpha^2 \lambda - \sigma) \frac{\Delta t}{-\alpha^2 \lambda} (1 - \gamma \Delta t - \sigma) + \Delta t (1 - \gamma \Delta t) \right] = 0 \tag{18.84}$$

hence

$$(1 + \alpha^2 \lambda - \sigma)(1 - \gamma \Delta t - \sigma) - \alpha^2 \lambda (1 - \gamma \Delta t) = 0$$
$$\sigma^2 - \sigma (2 - \gamma \Delta t + \alpha^2 \lambda) + (1 - \gamma \Delta t) = 0 \tag{18.85}$$
$$\sigma = 1 - \frac{\gamma \Delta t}{2} + \frac{\alpha^2 \lambda}{2} \pm \sqrt{\left(1 - \frac{\gamma \Delta t}{2} + \frac{\alpha^2 \lambda}{2} \right)^2 - (1 - \gamma \Delta t)}.$$

Instabilities are possible if the square root is real and $\sigma < -1$ ($\sigma > 1$ is not possible). This is the case for

$$-1 + \frac{\gamma \Delta t}{2} \approx -\sqrt{1 - \gamma \Delta t} < 1 - \frac{\gamma \Delta t}{2} + \frac{\alpha^2 \lambda}{2} < \sqrt{1 - \gamma \Delta t} \approx 1 - \frac{\gamma \Delta t}{2} \tag{18.86}$$

$$-2 + \gamma \Delta t < \frac{\alpha^2 \lambda}{2} < 0. \tag{18.87}$$

The right inequality is satisfied, hence it remains

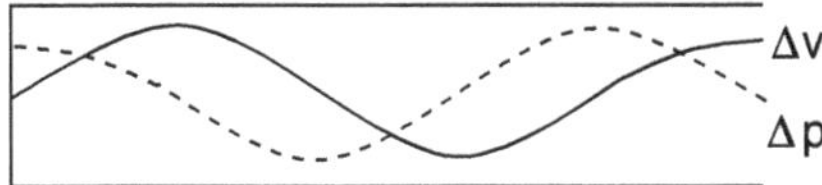

Fig. 18.10 (Standing waves in an organ pipe) At the closed (*left*) end the amplitude of the longitudinal velocity is zero whereas the amplitudes of pressure and density changes are extremal. This is reversed at the open (*right*) end

$$\alpha^2 \sin^2\left(\frac{k\Delta x}{2}\right) < 1 - \frac{\gamma \Delta t}{2}. \tag{18.88}$$

This holds for all k-values if it holds for the maximum of the sine-function

$$\alpha^2 < 1 - \frac{\gamma \Delta t}{2}. \tag{18.89}$$

This shows that inclusion of the damping term even favors instabilities.

18.7 Two-Variable Method

For the 1-dimensional wave equation (18.11) there exists another possibility to reduce the order of the time derivative by splitting it up into two first order equations similar to (18.9), (18.10)

$$\frac{\partial}{\partial t} f(t, x) = c \frac{\partial}{\partial x} g(t, x) \tag{18.90}$$

$$\frac{\partial}{\partial t} g(t, x) = c \frac{\partial}{\partial x} f(t, x). \tag{18.91}$$

Several algorithms can be applied to solve these equations [162]. We discuss only methods which are second order in space and time and are rather general methods to solve partial differential equations. The boundary conditions need some special care. For closed boundaries with $f(x_0) = 0$ obviously $\frac{\partial f}{\partial t}(x_0) = 0$ whereas $\frac{\partial f}{\partial x}(x_0)$ is finite. Hence a closed boundary for $f(t, x)$ is connected with an open boundary for $g(t, x)$ with $\frac{\partial g}{\partial x}(x_0) = 0$ and vice versa. This is well known from acoustics (Fig. 18.10).

18.7.1 Leapfrog Scheme

We use symmetric differences (Sect. 3.2) for the first derivatives

$$\frac{f(t + \frac{\Delta t}{2}, x) - f(t - \frac{\Delta t}{2}, x)}{\Delta t} = c \frac{g(t, x + \frac{\Delta x}{2}) - g(x - \frac{\Delta x}{2})}{\Delta x} + O(\Delta x^2, \Delta t^2)$$

$$\tag{18.92}$$

$$\frac{g(t + \frac{\Delta t}{2}, x) - g(t - \frac{\Delta t}{2}, x)}{\Delta t} = c \frac{f(t, x + \frac{\Delta x}{2}) - f(x - \frac{\Delta x}{2})}{\Delta x} + O(\Delta x^2, \Delta t^2)$$

$$\tag{18.93}$$

Fig. 18.11 (Simulation with the leapfrog method) A rectangular pulse is simulated with the two-variable leapfrog method. While for $\alpha = 1$ the pulse shape has not changed after 1000 steps, for smaller values the short wavelength components are lost due to dispersion

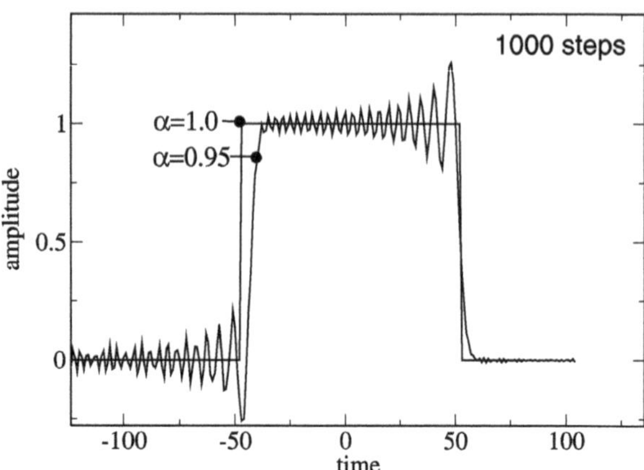

to obtain the following scheme

$$g(t_{n+1/2}, x_{m+1/2}) = g(t_{n-1/2}, x_{m+1/2}) + \alpha\big(f(t_n, x_{m+1})\big) - f(t_n, x_{m-1})$$

(18.94)

$$f(t_{n+1}, x_m) = f(t_n, x_m) + \alpha\big(g(t_{n+1/2}, x_{m+1/2}) - g(t_{n+1/2}, x_{m-1/2})\big).$$

(18.95)

Using different time grids for the two variables

$$\mathbf{f}_n = \begin{pmatrix} f_1^n \\ \vdots \\ f_M^n \end{pmatrix} = \begin{pmatrix} f(t_n, x_1) \\ \vdots \\ f(t_n, x_M) \end{pmatrix} \qquad \mathbf{g}_n = \begin{pmatrix} g_1^n \\ \vdots \\ g_M^n \end{pmatrix} = \begin{pmatrix} g(t_{n-1/2}, x_{1/2}) \\ \vdots \\ g(t_{n-1/2}, x_{M-1/2}) \end{pmatrix}$$

(18.96)

this translates into the algorithm (Fig. 18.11)

$$g_m^{n+1} = g_m^n + \alpha\big(f_m^n - f_{m-1}^n\big)$$

(18.97)

$$f_m^{n+1} = f_m^n + \alpha\big(g_{m+1}^{n+1} - g_m^{n+1}\big)$$

$$= f_m^n + \alpha\big(g_{m+1}^n - g_m^n\big) + \alpha^2\big(f_{m+1}^n - 2f_m^n + f_{m-1}^n\big).$$

(18.98)

To analyze the stability we insert

$$f_m^n = u e^{an} e^{ikm\Delta x} \qquad g_m^n = v e^{an} e^{ikm\Delta x}$$

(18.99)

and obtain the equations

$$Gv = v + \alpha u\big(1 - e^{-ik\Delta x}\big)$$

(18.100)

$$Gu = u + \alpha v\big(e^{ik\Delta x} - 1\big) + \alpha^2 u(2\cos k\Delta x - 2)$$

(18.101)

which in matrix form read

$$G\begin{pmatrix} u \\ v \end{pmatrix} = \begin{pmatrix} 1 + \alpha^2(2\cos k\Delta x - 2) & \alpha(e^{ik\Delta x} - 1) \\ \alpha(1 - e^{-ik\Delta x}) & 1 \end{pmatrix}\begin{pmatrix} u \\ v \end{pmatrix}.$$

(18.102)

The maximum amplification factor G is given by the largest eigenvalue, which is one of the roots of

$$\left(1 - 4\alpha^2 \sin^2\left(\frac{k\Delta x}{2}\right) - \sigma\right)(1 - \sigma) + 4\alpha^2 \sin^2\left(\frac{k\Delta x}{2}\right) = 0 \tag{18.103}$$

$$\left(1 - \sigma + \alpha^2\lambda^2\right)(1 - \sigma) - \alpha^2\lambda^2 = 0$$

$$\sigma = 1 - 2\alpha^2 \sin^2\left(\frac{k\Delta x}{2}\right) \pm \sqrt{\left(1 - 2\alpha^2 \sin^2\left(\frac{k\Delta x}{2}\right)\right)^2 - 1}. \tag{18.104}$$

The eigenvalues coincide with those of the two-step method (18.58).

18.7.2 Lax-Wendroff Scheme

The Lax-Wendroff scheme can be derived from the Taylor series expansion

$$f(t + \Delta t, x) = f(t, x) + \frac{\partial f(t, x)}{\partial t}\Delta t + \frac{1}{2}\Delta t^2 \frac{\partial^2 f(t, x)}{\partial t^2} + \cdots$$

$$= f(t, x) + c\Delta t \frac{\partial g(t, x)}{\partial x} + \frac{c^2\Delta t^2}{2}\frac{\partial^2 f(t, x)}{\partial t^2} + \cdots \tag{18.105}$$

$$g(t + \Delta t, x) = g(t, x) + \frac{\partial g(t, x)}{\partial t}\Delta t + \frac{1}{2}\Delta t^2 \frac{\partial^2 g(t, x)}{\partial t^2} + \cdots$$

$$= g(t, x) + c\Delta t \frac{\partial f(t, x)}{\partial x} + \frac{c^2\Delta t^2}{2}\frac{\partial^2 g(t, x)}{\partial t^2} + \cdots. \tag{18.106}$$

It uses symmetric differences on regular grids (18.45), (18.46) to obtain the iteration

$$f_m^{n+1} = f_m^n + c\Delta t \frac{g_{m+1}^n - g_{m-1}^n}{2\Delta x} + c^2\Delta t^2 \frac{f_{m+1}^n + f_{m-1}^n - 2f_m^n}{2\Delta x^2} \tag{18.107}$$

$$g_m^{n+1} = g_m^n + c\Delta t \frac{f_{m+1}^n - f_{m-1}^n}{2\Delta x} + c^2\Delta t^2 \frac{g_{m+1}^n + g_{m-1}^n - 2g_m^n}{2\Delta x^2} \tag{18.108}$$

$$\begin{pmatrix} \mathbf{f}^{n+1} \\ \mathbf{g}^{n+1} \end{pmatrix} = \begin{pmatrix} 1 + \frac{\alpha^2}{2}M & \frac{\alpha}{2}D \\ \frac{\alpha}{2}D & 1 + \frac{\alpha^2}{2}M \end{pmatrix}\begin{pmatrix} \mathbf{f}^n \\ \mathbf{g}^n \end{pmatrix} \tag{18.109}$$

with the tridiagonal matrix

$$D = \begin{pmatrix} 0 & 1 & & & \\ -1 & 0 & 1 & & \\ & \ddots & \ddots & \ddots & \\ & & -1 & 0 & 1 \\ & & & -1 & 0 \end{pmatrix}. \tag{18.110}$$

To analyze the stability we insert

$$f_m^n = u e^{\alpha n} e^{ikm\Delta x} \qquad g_m^n = v e^{\alpha n} e^{ikm\Delta x} \tag{18.111}$$

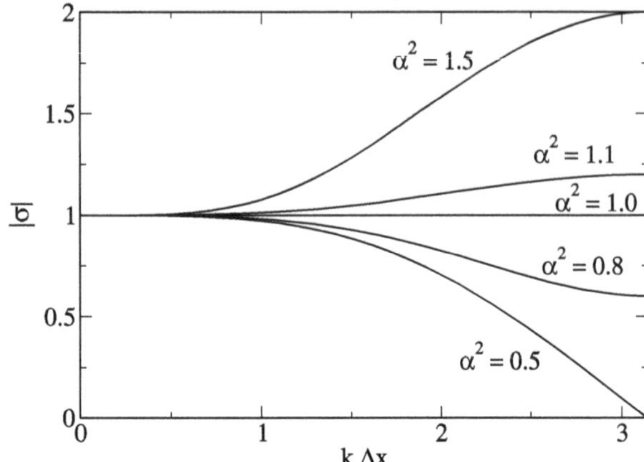

Fig. 18.12 (Stability region of the Lax-Wendroff method) Instabilities appear for $|\alpha| > 1$. In the opposite case short wavelength modes are damped

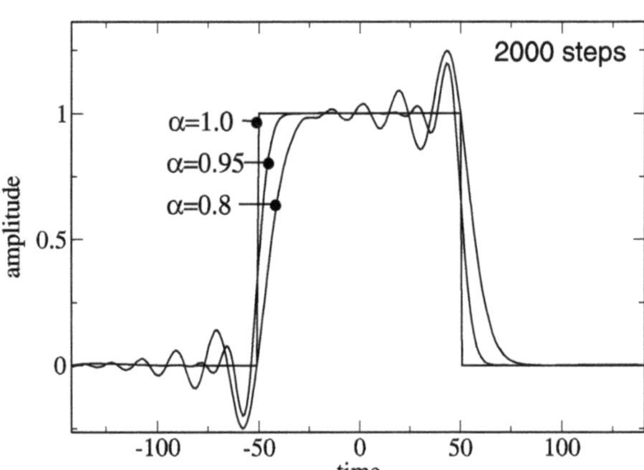

Fig. 18.13 (Simulation with the Lax-Wendroff method) A rectangular pulse is simulated with the two-variable Lax-Wendroff method. While for $\alpha = 1$ the pulse shape has not changed after 2000 steps, for smaller values the short wavelength components are lost due to dispersion and damping

and calculate the eigenvalues (compare with (18.102)) of

$$\begin{pmatrix} 1 + \alpha^2(\cos k\Delta x - 1) & i\alpha \sin k\Delta x \\ i\alpha \sin k\Delta x & 1 + \alpha^2(\cos k\Delta x - 1) \end{pmatrix} \tag{18.112}$$

which are given by

$$\sigma = 1 + \alpha^2(\cos k\Delta x - 1) \pm \sqrt{\alpha^2(\cos^2 k\Delta x - 1)}. \tag{18.113}$$

The root is always imaginary and

$$|\sigma|^2 = 1 + (\alpha^4 - \alpha^2)(\cos k\Delta x - 1)^2 \le 1 + 4(\alpha^4 - \alpha^2).$$

For $\alpha < 1$ we find $|\sigma| < 1$. The method is stable but there is wavelength dependent damping (Figs. 18.12, 18.13).

18.7.3 Crank-Nicolson Scheme

This method takes the average of the explicit and implicit Euler methods

$$f(t + \Delta t) = f(t) + \frac{c}{2}\left(\frac{\partial g}{\partial x}(t, x) + \frac{\partial g}{\partial x}(t + \Delta t, x)\right)\Delta t \qquad (18.114)$$

$$g(t + \Delta t) = g(t) + \frac{c}{2}\left(\frac{\partial f}{\partial x}(t, x) + \frac{\partial f}{\partial x}(t + \Delta t, x)\right)\Delta t \qquad (18.115)$$

and uses symmetric differences on the regular grids (18.45), (18.46) to obtain

$$f_m^{n+1} = f_m^n + \frac{\alpha}{4}\left(g_{m+1}^n - g_{m-1}^n + g_{m+1}^{n+1} - g_{m-1}^{n+1}\right) \qquad (18.116)$$

$$g_m^{n+1} = g_m^n + \frac{\alpha}{4}\left(f_{m+1}^n - f_{m-1}^n + f_{m+1}^{n+1} - f_{m-1}^{n+1}\right) \qquad (18.117)$$

which reads in matrix notation

$$\begin{pmatrix} \mathbf{f}_{n+1} \\ \mathbf{g}_{n+1} \end{pmatrix} = \begin{pmatrix} 1 & \frac{\alpha}{4}D \\ \frac{\alpha}{4}D & 1 \end{pmatrix} \begin{pmatrix} \mathbf{f}_n \\ \mathbf{g}_n \end{pmatrix} + \begin{pmatrix} & \frac{\alpha}{4}D \\ \frac{\alpha}{4}D & \end{pmatrix} \begin{pmatrix} \mathbf{f}_{n+1} \\ \mathbf{g}_{n+1} \end{pmatrix}. \qquad (18.118)$$

This equation can be solved formally by collecting terms at time t_{n+1}

$$\begin{pmatrix} 1 & -\frac{\alpha}{4}D \\ -\frac{\alpha}{4}D & 1 \end{pmatrix} \begin{pmatrix} \mathbf{f}_{n+1} \\ \mathbf{g}_{n+1} \end{pmatrix} = \begin{pmatrix} 1 & \frac{\alpha}{4}D \\ \frac{\alpha}{4}D & 1 \end{pmatrix} \begin{pmatrix} \mathbf{f}_n \\ \mathbf{g}_n \end{pmatrix} \qquad (18.119)$$

and multiplying with the inverse matrix from left

$$\begin{pmatrix} \mathbf{f}_{n+1} \\ \mathbf{g}_{n+1} \end{pmatrix} = \begin{pmatrix} 1 & -\frac{\alpha}{4}D \\ -\frac{\alpha}{4}D & 1 \end{pmatrix}^{-1} \begin{pmatrix} 1 & \frac{\alpha}{4}D \\ \frac{\alpha}{4}D & 1 \end{pmatrix} \begin{pmatrix} \mathbf{f}_n \\ \mathbf{g}_n \end{pmatrix}. \qquad (18.120)$$

Now, if $\mathbf{u}$ is an eigenvector of D with purely imaginary eigenvalue λ (Sect. 9.3)

$$\begin{pmatrix} 1 & \frac{\alpha}{4}D \\ \frac{\alpha}{4}D & 1 \end{pmatrix} \begin{pmatrix} \mathbf{u} \\ \pm\mathbf{u} \end{pmatrix} = \begin{pmatrix} (1 \pm \frac{\alpha}{4}\lambda)\mathbf{u} \\ (\frac{\alpha}{4}\lambda \pm 1)\mathbf{u} \end{pmatrix} = \left(1 \pm \frac{\alpha}{4}\lambda\right)\begin{pmatrix} \mathbf{u} \\ \pm\mathbf{u} \end{pmatrix} \qquad (18.121)$$

and furthermore

$$\begin{pmatrix} 1 & -\frac{\alpha}{4}D \\ -\frac{\alpha}{4}D & 1 \end{pmatrix} \begin{pmatrix} \mathbf{u} \\ \pm\mathbf{u} \end{pmatrix} = \begin{pmatrix} (1 \mp \frac{\alpha}{4}\lambda)\mathbf{u} \\ (-\frac{\alpha}{4}\lambda \pm 1)\mathbf{u} \end{pmatrix} = \left(1 \mp \frac{\alpha}{4}\lambda\right)\begin{pmatrix} \mathbf{u} \\ \pm\mathbf{u} \end{pmatrix} \qquad (18.122)$$

But, since the eigenvalue of the inverse matrix is the reciprocal of the eigenvalue, the eigenvalues of

$$T = \begin{pmatrix} 1 & -\frac{\alpha}{4}D \\ -\frac{\alpha}{4}D & 1 \end{pmatrix}^{-1} \begin{pmatrix} 1 & \frac{\alpha}{4}D \\ \frac{\alpha}{4}D & 1 \end{pmatrix} \qquad (18.123)$$

are given by

$$\sigma = \frac{1 \pm \frac{\alpha}{4}\lambda}{1 \mp \frac{\alpha}{4}\lambda}. \qquad (18.124)$$

Since λ is imaginary, we find $|\sigma| = 1$. The Crank-Nicolson method is stable and does not show damping like the Lax-Wendroff method. However, there is consid-

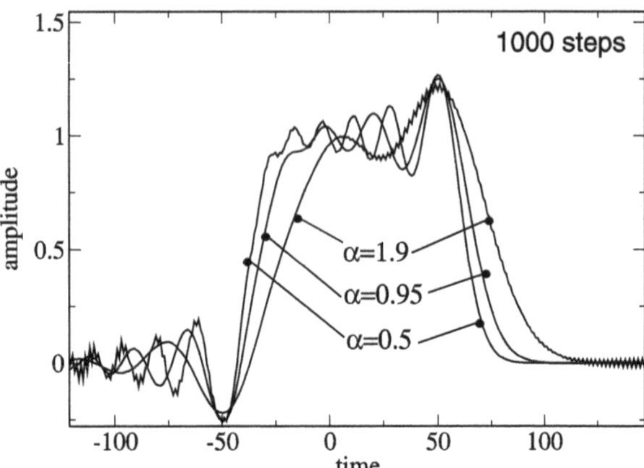

Fig. 18.14 (Simulation with the iterated Crank-Nicolson method) A rectangular pulse is simulated with the two-variable iterated Crank-Nicolson method. Only this method is stable for values $\alpha > 1$

erable dispersion. Solution of the linear system (18.119) is complicated and can be replaced by an iterative predictor-corrector method. Starting from the initial guess

$$\begin{pmatrix} {}^{(0)}\mathbf{f}_{n+1} \\ {}^{(0)}\mathbf{g}_{n+1} \end{pmatrix} = \begin{pmatrix} 1 & \frac{\alpha}{2}D \\ \frac{\alpha}{2}D & 1 \end{pmatrix} \begin{pmatrix} \mathbf{f}_n \\ \mathbf{g}_n \end{pmatrix} \tag{18.125}$$

we iterate

$$\begin{pmatrix} {}^{(0)}\mathbf{f}_{n+1/2} \\ {}^{(0)}\mathbf{g}_{n+1/2} \end{pmatrix} = \frac{1}{2}\begin{pmatrix} {}^{(0)}\mathbf{f}_{n+1} \\ {}^{(0)}\mathbf{g}_{n+1} \end{pmatrix} + \frac{1}{2}\begin{pmatrix} \mathbf{f}_n \\ \mathbf{g}_n \end{pmatrix} = \begin{pmatrix} 1 & \frac{\alpha}{4}D \\ \frac{\alpha}{4}D & 1 \end{pmatrix}\begin{pmatrix} \mathbf{f}_n \\ \mathbf{g}_n \end{pmatrix}$$

$$\begin{pmatrix} {}^{(1)}\mathbf{f}_{n+1} \\ {}^{(1)}\mathbf{g}_{n+1} \end{pmatrix} = \begin{pmatrix} \mathbf{f}_n \\ \mathbf{g}_n \end{pmatrix} + \begin{pmatrix} & \frac{\alpha}{2}D \\ \frac{\alpha}{2}D & \end{pmatrix}\begin{pmatrix} {}^{(0)}\mathbf{f}_{n+1/2} \\ {}^{(0)}\mathbf{g}_{n+1/2} \end{pmatrix} \tag{18.126}$$

$$= \begin{pmatrix} 1 & \frac{\alpha}{4}D \\ \frac{\alpha}{4}D & 1 \end{pmatrix}\begin{pmatrix} \mathbf{f}_n \\ \mathbf{g}_n \end{pmatrix} + \begin{pmatrix} & \frac{\alpha}{4}D \\ \frac{\alpha}{4}D & \end{pmatrix}\begin{pmatrix} {}^{(0)}\mathbf{f}_{n+1} \\ {}^{(0)}\mathbf{g}_{n+1} \end{pmatrix}$$

$$\begin{pmatrix} {}^{(1)}\mathbf{f}_{n+1/2} \\ {}^{(1)}\mathbf{g}_{n+1/2} \end{pmatrix} = \frac{1}{2}\begin{pmatrix} \mathbf{f}_n \\ \mathbf{g}_n \end{pmatrix} + \frac{1}{2}\begin{pmatrix} {}^{(1)}\mathbf{f}_{n+1} \\ {}^{(1)}\mathbf{g}_{n+1} \end{pmatrix}$$

$$= \begin{pmatrix} \mathbf{f}_n \\ \mathbf{g}_n \end{pmatrix} + \begin{pmatrix} & \frac{\alpha}{4}D \\ \frac{\alpha}{4}D & \end{pmatrix}\begin{pmatrix} {}^{(0)}\mathbf{f}_{n+1/2} \\ {}^{(0)}\mathbf{g}_{n+1/2} \end{pmatrix} \tag{18.127}$$

$$\begin{pmatrix} {}^{(2)}\mathbf{f}_{n+1} \\ {}^{(2)}\mathbf{g}_{n+1} \end{pmatrix} = \begin{pmatrix} \mathbf{f}_n \\ \mathbf{g}_n \end{pmatrix} + \begin{pmatrix} & \frac{\alpha}{2}D \\ \frac{\alpha}{2}D & \end{pmatrix}\begin{pmatrix} {}^{(1)}\mathbf{f}_{n+1/2} \\ {}^{(1)}\mathbf{g}_{n+1/2} \end{pmatrix}$$

$$= \begin{pmatrix} 1 & \frac{\alpha}{4}D \\ \frac{\alpha}{4}D & 1 \end{pmatrix}\begin{pmatrix} \mathbf{f}_n \\ \mathbf{g}_n \end{pmatrix} + \begin{pmatrix} & \frac{\alpha}{4}D \\ \frac{\alpha}{4}D & \end{pmatrix}\begin{pmatrix} {}^{(1)}\mathbf{f}_{n+1} \\ {}^{(1)}\mathbf{g}_{n+1} \end{pmatrix}. \tag{18.128}$$

In principle this iteration could be repeated more times, but as Teukolsky showed [250], two iterations are optimal for hyperbolic equations like the advection or wave equation. The region of stability is reduced (Figs. 18.14, 18.15) compared to the implicit Crank-Nicolson method. The eigenvalues are

$$^{(0)}\sigma = 1 \pm i\alpha \sin k\Delta x \quad |^{(0)}\sigma| > 1 \tag{18.129}$$

$$^{(1)}\sigma = 1 \pm i\alpha \sin k\Delta x - \frac{\alpha^2}{2}\sin^2 k\Delta x \quad |^{(1)}\sigma| > 1 \tag{18.130}$$

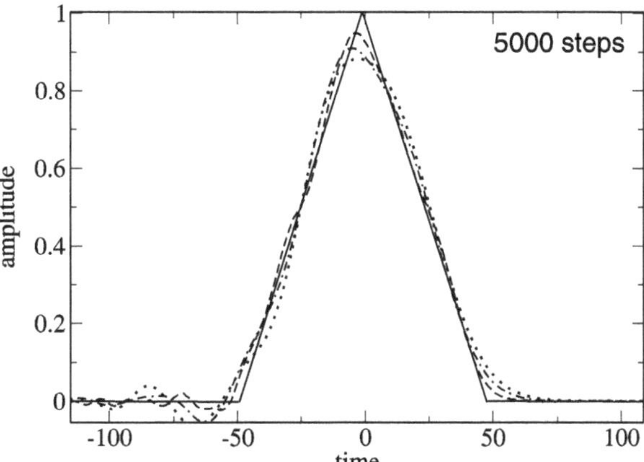

Fig. 18.15 (Simulation of a triangular pulse) A triangular pulse is simulated with different two-variable methods (*full curve*: initial conditions, *dashed curve*: leapfrog, *dash-dotted*: Lax-Wendroff, *dotted*: iterated Crank-Nicolson). This pulse contains less short wavelength components than the square pulse and shows much less deformation even after 5000 steps

$$^{(2)}\sigma = 1 - \frac{\alpha^2}{2}\sin^2 k\Delta x \pm \mathrm{i}\left(\alpha \sin k\Delta x - \frac{\alpha^3 \sin^3 k\Delta x}{4}\right)$$

$$\left|^{(2)}\sigma\right|^2 = 1 - \frac{\alpha^4 \sin^4 k\Delta x}{4} + \frac{\alpha^6 \sin^6 k\Delta x}{16} \le 1 \quad \text{for } |\alpha| \le 2.$$

(18.131)

18.8 Problems

Problem 18.1 (Waves on a damped string) In this computer experiment we simulate waves on a string with a moving boundary with the method from Sect. 18.6.

- Excite the left boundary with a continuous sine function and try to generate standing waves
- Increase the velocity until instabilities appear
- Compare reflection at open and fixed right boundary
- Observe the dispersion of pulses with different shape and duration
- The velocity can be changed by a factor n (refractive index) in the region $x > 0$. Observe reflection at the boundary $x = 0$

Problem 18.2 (Waves with the Fourier transform method) In this computer experiment we use the method from Sect. 18.3 to simulate waves on a string with fixed boundaries.

- Different initial excitations of the string can be selected.
- The dispersion can be switched off by using $\omega_k = ck$ instead of the proper eigenvalues (18.31).

Problem 18.3 (Two-variable methods) In this computer experiment we simulate waves with periodic boundary conditions. Different initial values (rectangular, triangular or Gaussian pulses of different widths) and methods (leapfrog, Lax-Wendroff, iterated Crank-Nicolson) can be compared.

Chapter 19
Diffusion

Diffusion is one of the simplest non-equilibrium processes. It describes the transport of heat [52, 95] and the time evolution of differences in substance concentrations [82]. In this chapter, the one-dimensional diffusion equation

$$\frac{\partial}{\partial t} f(t, x) = D \frac{\partial^2}{\partial x^2} f(t, x) + S(t, x) \tag{19.1}$$

is semi-discretized with finite differences. The time integration is performed with three different Euler methods. The explicit Euler method is conditionally stable only for small Courant number $\alpha = \frac{D \Delta t}{\Delta x^2} < 1/2$, which makes very small time steps necessary. The fully implicit method is unconditionally stable but its dispersion deviates largely from the exact expression. The Crank-Nicolson method is also unconditionally stable. However, it is more accurate and its dispersion relation is closer to the exact one. Extension to more than one dimension is easily possible, but the numerical effort increases drastically as there is no formulation involving simple tridiagonal matrices like in one dimension. The split operator approximation uses the one-dimensional method independently for each dimension. It is very efficient with almost no loss in accuracy. In a computer experiment the different schemes are compared for diffusion in two dimensions.

19.1 Particle Flux and Concentration Changes

Let $f(\mathbf{x}, t)$ denote the concentration of a particle species and $\mathbf{J}$ the corresponding flux of particles. Consider a small cube with volume h^3 (Fig. 19.1). The change of the number of particles within this volume is given by the integral form of the conservation law (11.10)

$$\frac{\partial}{\partial t} \int_V dV\, f(\mathbf{r}, t) + \oint_{\partial V} \mathbf{J}(\mathbf{r}, t)\, d\mathbf{A} = \int_V dV\, S(\mathbf{r}, t) \tag{19.2}$$

where the source term $S(\mathbf{r})$ accounts for creation or destruction of particles due to for instance chemical reactions. In Cartesian coordinates we have

P.O.J. Scherer, *Computational Physics*, Graduate Texts in Physics,
DOI 10.1007/978-3-319-00401-3_19,
© Springer International Publishing Switzerland 2013

Fig. 19.1 Flux through a volume element

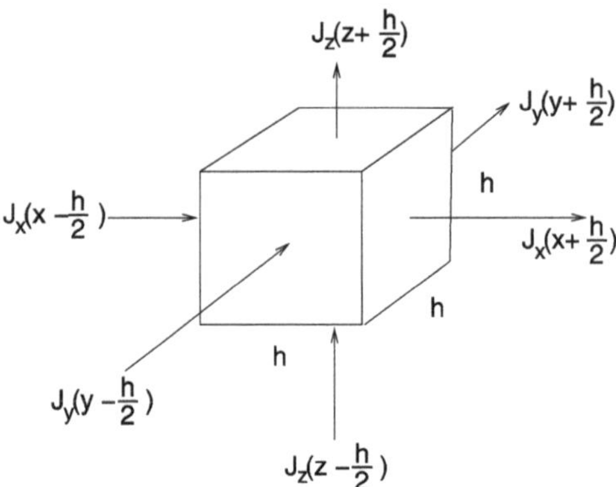

$$
\int_{x-h/2}^{x+h/2} dx' \int_{y-h/2}^{y+h/2} dy' \int_{z-h/2}^{z+h/2} dz' \left(\frac{\partial}{\partial t} f(x',y',z',t) - S(x',y',z',t) \right)
$$

$$
+ \int_{x-h/2}^{x+h/2} dx' \int_{y-h/2}^{y+h/2} dy' \left(J_z\left(x',y',z-\frac{h}{2}\right) - J_z\left(x',y',z+\frac{h}{2}\right) \right)
$$

$$
+ \int_{x-h/2}^{x+h/2} dx' \int_{z-h/2}^{z+h/2} dz' \left(J_y\left(x',y-\frac{h}{2},z'\right) - J_y\left(x',y+\frac{h}{2},z'\right) \right)
$$

$$
+ \int_{z-h/2}^{z+h/2} dz' \int_{y-h/2}^{y+h/2} dy' \left(J_z\left(x-\frac{h}{2},y',z'\right) - J_z\left(x+\frac{h}{2},y',z'\right) \right) = 0.
\tag{19.3}
$$

In the limit of small h this turns into the differential form of the conservation law

$$
h^3 \left(\frac{\partial}{\partial t} f(x,y,z,t) - S(x,y,z,t) \right) + h^2 \left(h\frac{\partial J_x}{\partial x} + h\frac{\partial J_y}{\partial y} + h\frac{\partial J_z}{\partial z} \right) = 0
\tag{19.4}
$$

or after division by h^3

$$
\frac{\partial}{\partial t} f(\mathbf{r},t) = -\operatorname{div} \mathbf{J}(\mathbf{r},t) + S(\mathbf{r},t).
\tag{19.5}
$$

Within the framework of linear response theory the flux is proportional to the gradient of f (Fig. 19.2),

$$
\mathbf{J} = -D \operatorname{grad} f.
\tag{19.6}
$$

Together we obtain the diffusion equation

$$
\frac{\partial f}{\partial t} = \operatorname{div}(D \operatorname{grad} f) + S
\tag{19.7}
$$

which in the special case of constant D simplifies to

$$
\frac{\partial f}{\partial t} = D\Delta f + S.
\tag{19.8}
$$

Fig. 19.2 Diffusion due to a concentration gradient

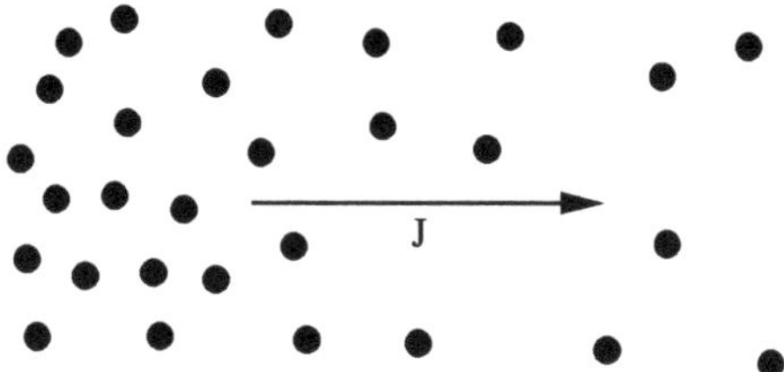

19.2 Diffusion in One Dimension

We will use the finite differences method which works well if the diffusion constant D is constant in time and space. We begin with diffusion in one dimension and use regular grids $t_n = n\Delta t$, $x_m = m\Delta x$, $f_m^n = f(t_n, x_m)$ and the discretized second derivative

$$\frac{\partial^2 f}{\partial x^2} = \frac{f(x+\Delta x) + f(x-\Delta x) - 2f(x)}{\Delta x^2} + O(\Delta x^2) \tag{19.9}$$

to obtain the semi-discrete diffusion equation

$$\dot{f}(t, x_m) = \frac{D}{\Delta x^2}\left(f(t, x_{m+1}) + f(t, x_{m-1}) - 2f(t, x_m)\right) + S(t, x_m) \tag{19.10}$$

or in matrix notation

$$\dot{\mathbf{f}}(t) = \frac{D}{\Delta x^2}M\mathbf{f}(t) + \mathbf{S}(t) \tag{19.11}$$

with the tridiagonal matrix

$$M = \begin{pmatrix} -2 & 1 & & & \\ 1 & -2 & 1 & & \\ & \ddots & \ddots & \ddots & \\ & & 1 & -2 & 1 \\ & & & 1 & -2 \end{pmatrix}. \tag{19.12}$$

Boundary conditions can be taken into account by introducing extra boundary points x_0, x_{M+1} (Fig. 19.3).

19.2.1 Explicit Euler (Forward Time Centered Space) Scheme

A simple Euler step (Sect. 12.3) makes the approximation

$$f_m^{n+1} - f_m^n = \dot{f}(t_n, x_m)\Delta t = D\frac{\Delta t}{\Delta x^2}\left(f_{m+1}^n + f_{m-1}^n - 2f_m^n\right) + S_m^n\Delta t. \tag{19.13}$$

For homogeneous boundary conditions $f = 0$ this becomes in matrix form

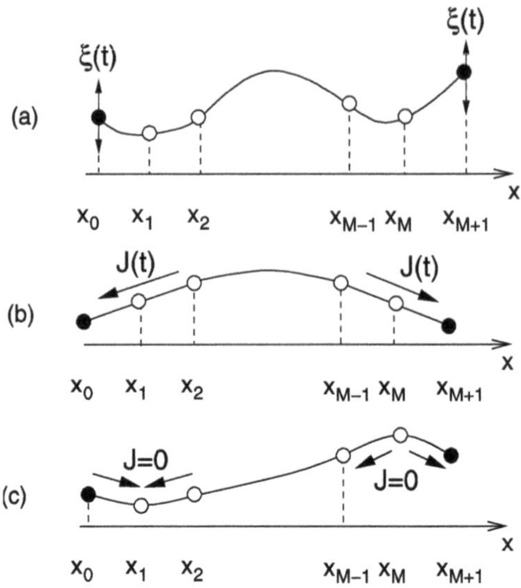

Fig. 19.3 (Boundary conditions for 1-dimensional diffusion) Additional boundary points x_0, x_{M+1} are used to realize the boundary conditions. (**a**) *Dirichlet boundary conditions*: the function values at the boundary are given $f(t, x_0) = \xi_0(t)$, $\frac{\partial^2}{\partial x^2} f(x_1) = \frac{1}{\Delta x^2}(f(x_2) - 2f(x_1) + \xi_0(t))$ or $f(t, x_{M+1}) = \xi_{M+1}(t)$, $\frac{\partial^2}{\partial x^2} f(x_M) = \frac{1}{\Delta x^2}(f(x_{M-1}) - 2f(x_M) + \xi_{M+1}(t))$. (**b**) *Neumann boundary conditions*: the flux through the boundary is given, hence the derivative $\frac{\partial f}{\partial x}$ at the boundary $f(t, x_0) = f(t, x_2) + 2\frac{\Delta x}{D} J_1(t)$, $\frac{\partial^2}{\partial x^2} f(x_1) = \frac{1}{\Delta x^2}(2f(x_2) - 2f(x_1) + 2\frac{\Delta x}{D} J_1(t))$ or $f(t, x_{M+1}) = f(t, x_{M-1}) - 2\frac{\Delta x}{D} J_M(t)$, $\frac{\partial^2}{\partial x^2} f(x_M) = \frac{1}{\Delta x^2}(2f(x_{M-1}) - 2f(x_M) - 2\frac{\Delta x}{D} J_M(t))$. (**c**) *No-flow boundary conditions*: there is no flux through the boundary, hence the derivative $\frac{\partial f}{\partial x} = 0$ at the boundary $f(t, x_0) = f(t, x_2)$, $\frac{\partial^2}{\partial x^2} f(x_1) = \frac{1}{\Delta x^2}(2f(x_2) - 2f(x_1))$ or $f(t, x_M) = f(t, x_{M-2})$, $\frac{\partial^2}{\partial x^2} f(x_M) = \frac{1}{\Delta x^2}(2f(x_{M-1}) - 2f(x_M))$

$$\begin{pmatrix} f_1^{n+1} \\ \vdots \\ f_M^{n+1} \end{pmatrix} = A \begin{pmatrix} f_1^n \\ \vdots \\ f_M^n \end{pmatrix} + \begin{pmatrix} S_1^n \Delta t \\ \vdots \\ S_M^n \Delta t \end{pmatrix} \tag{19.14}$$

with the tridiagonal matrix

$$A = \begin{pmatrix} 1 - 2D\frac{\Delta t}{\Delta x^2} & D\frac{\Delta t}{\Delta x^2} & & & \\ D\frac{\Delta t}{\Delta x^2} & 1 - 2D\frac{\Delta t}{\Delta x^2} & & & \\ & \ddots & \ddots & \ddots & \\ & & D\frac{\Delta t}{\Delta x^2} & 1 - 2D\frac{\Delta t}{\Delta x^2} & D\frac{\Delta t}{\Delta x^2} \\ & & & D\frac{\Delta t}{\Delta x^2} & 1 - 2D\frac{\Delta t}{\Delta x^2} \end{pmatrix} = 1 + \alpha M \tag{19.15}$$

where α is the Courant number for diffusion

$$\alpha = D\frac{\Delta t}{\Delta x^2}. \tag{19.16}$$

The eigenvalues of M are (compare (18.30))

$$\lambda = -4\sin^2\left(\frac{k\Delta x}{2}\right) \quad \text{with } k\Delta x = \frac{\pi}{M+1}, \frac{2\pi}{M+1}, \dots, \frac{M\pi}{M+1} \tag{19.17}$$

and hence the eigenvalues of A are given by

$$1 + \alpha\lambda = 1 - 4\alpha\sin^2\frac{k\Delta x}{2}. \tag{19.18}$$

The algorithm is stable if

$$|1 + \alpha\lambda| < 1 \quad \text{for all } \lambda \tag{19.19}$$

which holds if

$$-1 < 1 - 4\alpha\sin^2\frac{k\Delta x}{2} < 1. \tag{19.20}$$

The maximum of the sine function is $\sin(\frac{M\pi}{2(M+1)}) \approx 1$. Hence the right hand inequation is satisfied and from the left one we have

$$-1 < 1 - 4\alpha. \tag{19.21}$$

The algorithm is stable for

$$\alpha = D\frac{\Delta t}{\Delta x^2} < \frac{1}{2}. \tag{19.22}$$

The dispersion relation follows from inserting a plane wave ansatz

$$e^{i\omega\Delta t} = 1 - 4\alpha\sin^2\left(\frac{k\Delta x}{2}\right). \tag{19.23}$$

For $\alpha > 1/4$ the right hand side changes sign at

$$k_c\Delta x = 2\arcsin\sqrt{\frac{1}{4\alpha}}. \tag{19.24}$$

The imaginary part of ω has a singularity at k_c and the real part has a finite value of π for $k > k_c$ (Fig. 19.4 on page 356). Deviations from the exact dispersion

$$\omega = ik^2 \tag{19.25}$$

are large, except for very small k.

19.2.2 Implicit Euler (Backward Time Centered Space) Scheme

Next we use the backward difference

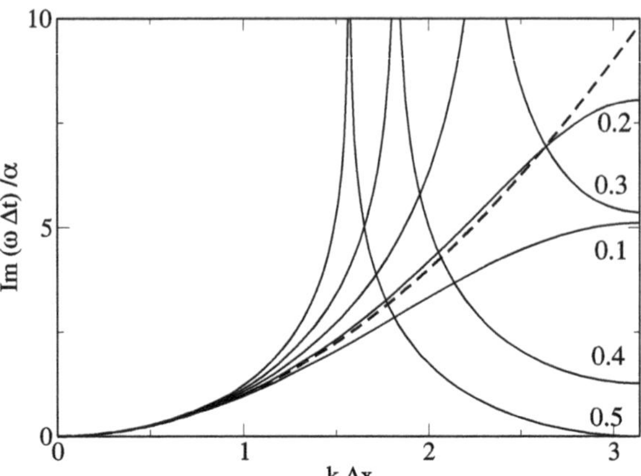

Fig. 19.4 (Dispersion of the explicit Euler method) The dispersion of the explicit method is shown for different values of the Courant number α and compared to the exact dispersion (*dashed curve*). The imaginary part of ω shows a singularity for $\alpha > 1/4$. Above the singularity ω is complex valued

$$f_m^{n+1} - f_m^n = \dot{f}(t_{n+1}, x_m)\Delta t$$
$$= D\frac{\partial^2 f}{\partial x^2}(t_{n+1}, x_m)\Delta t + S(t_{n+1}, x_m)\Delta t \tag{19.26}$$

to obtain the implicit method

$$f_m^{n+1} - \alpha\left(f_{m+1}^{n+1} + f_{m-1}^{n+1} - 2f_m^{n+1}\right) = f_m^n + S_m^{n+1}\Delta t \tag{19.27}$$

or in matrix notation

$$A\mathbf{f}_{n+1} = \mathbf{f}_n + \mathbf{S}_{n+1}\Delta t \quad \text{with } A = 1 - \alpha M \tag{19.28}$$

which can be solved formally by

$$\mathbf{f}_{n+1} = A^{-1}\mathbf{f}_n + A^{-1}\mathbf{S}_{n+1}\Delta t. \tag{19.29}$$

The eigenvalues of A are

$$\lambda(A) = 1 + 4\alpha \sin^2 \frac{k\Delta x}{2} > 1 \tag{19.30}$$

and the eigenvalues of A^{-1}

$$\lambda\left(A^{-1}\right) = \lambda(A)^{-1} = \frac{1}{1 + 4\alpha \sin^2 \frac{k\Delta x}{2}}. \tag{19.31}$$

The implicit method is unconditionally stable since

$$\left|\lambda\left(A^{-1}\right)\right| < 1. \tag{19.32}$$

The dispersion relation of the implicit scheme follows from

$$e^{i\omega\Delta t} = \frac{1}{1 + 4\alpha \sin^2\left(\frac{k\Delta x}{2}\right)}. \tag{19.33}$$

There is no singularity and ω is purely imaginary. Still, deviations from the exact expression are large (Fig. 19.5 on page 357).

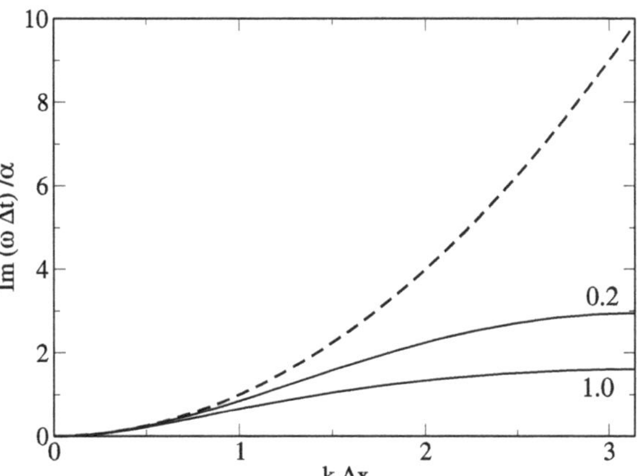

Fig. 19.5 (Dispersion of the implicit Euler method) The dispersion of the fully implicit method is shown for two different values of the Courant number α and compared to the exact dispersion (*dashed curve*)

Formally a matrix inversion is necessary. Numerically it is much more efficient to solve the tridiagonal system of equations (page 69).

$$(1 - \alpha M) f(t_{n+1}) = f(t_n) + S(t_{n+1}) \Delta t. \tag{19.34}$$

19.2.3 Crank-Nicolson Method

The Crank-Nicolson method [65] which is often used for diffusion problems, combines implicit and explicit methods. It uses the Heun method (Sect. 12.5) for the time integration

$$f_m^{n+1} - f_m^n = \frac{\Delta t}{2} \left(\frac{\partial f}{\partial t}(t_{n+1}, x_m) + \frac{\partial f}{\partial t}(t_n, x_m) \right) \tag{19.35}$$

$$= D \frac{\Delta t}{2} \left(\frac{\partial^2 f}{\partial x^2}(t_{n+1}, x_m) + \frac{\partial^2 f}{\partial x^2}(t_n, x_m) \right)$$

$$+ \left(S(t_n, x_m) + S(t_{n+1}, x_m) \right) \frac{\Delta t}{2} \tag{19.36}$$

$$= D \frac{\Delta t}{2} \left(\frac{f_{m+1}^n + f_{m-1}^n - 2 f_m^n}{\Delta x^2} + \frac{f_m^{n+1} + f_{m-1}^{n+1} - 2 f_m^{n+1}}{\Delta x^2} \right)$$

$$+ \frac{S_m^n + S_m^{n+1}}{2} \Delta t. \tag{19.37}$$

This approximation is second order both in time and space and becomes in matrix notation

$$\left(1 - \frac{\alpha}{2} M \right) \mathbf{f}_{n+1} = \left(1 + \frac{\alpha}{2} M \right) \mathbf{f}_n + \frac{\mathbf{S}_n + \mathbf{S}_{n+1}}{2} \Delta t \tag{19.38}$$

which can be solved by

$$\mathbf{f}_{n+1} = \left(1 - \frac{\alpha}{2}M\right)^{-1}\left(1 + \frac{\alpha}{2}M\right)\mathbf{f}_n + \left(1 - \frac{\alpha}{2}M\right)^{-1}\frac{\mathbf{S}_n + \mathbf{S}_{n+1}}{2}\Delta t.$$

$$(19.39)$$

Again it is numerically much more efficient to solve the tridiagonal system of equations (19.38) than to calculate the inverse matrix.

The eigenvalues of this method are

$$\lambda = \frac{1 + \frac{\alpha}{2}\mu}{1 - \frac{\alpha}{2}\mu} \quad \text{with } \mu = -4\sin^2\frac{k\Delta x}{2} \in [-4, 0]. \tag{19.40}$$

Since $\alpha\mu < 0$ it follows

$$1 + \frac{\alpha}{2}\mu < 1 - \frac{\alpha}{2}\mu \tag{19.41}$$

and hence

$$\lambda < 1. \tag{19.42}$$

On the other hand we have

$$1 > -1 \tag{19.43}$$
$$1 + \frac{\alpha}{2}\mu > -1 + \frac{\alpha}{2}\mu \tag{19.44}$$
$$\lambda > -1. \tag{19.45}$$

This shows that the Crank-Nicolson method is stable [251]. The dispersion follows from

$$\mathrm{e}^{i\omega\Delta t} = \frac{1 - 2\alpha\sin^2(\frac{k\Delta x}{2})}{1 + 2\alpha\sin^2(\frac{k\Delta x}{2})}. \tag{19.46}$$

For $\alpha > 1/2$ there is a sign change of the right hand side at

$$k_c\Delta x = 2\arcsin\sqrt{\frac{1}{2\alpha}}. \tag{19.47}$$

The imaginary part of ω has a singularity at k_c and ω is complex valued for $k > k_c$ (Fig. 19.6 on page 359).

19.2.4 Error Order Analysis

Taylor series gives for the exact solution

$$\begin{aligned}
\Delta f_{exact} &= \Delta t\,\dot{f}(t, x) + \frac{\Delta t^2}{2}\ddot{f}(t, x) + \frac{\Delta t^3}{6}\frac{\partial^3}{\partial t^3}f(t, x) + \cdots \\
&= \Delta t\left[Df''(t, x) + S(t, x)\right] \\
&\quad + \frac{\Delta t^2}{2}\left[D\dot{f}''(t, x) + \dot{S}(t, x)\right] + \cdots
\end{aligned} \tag{19.48}$$

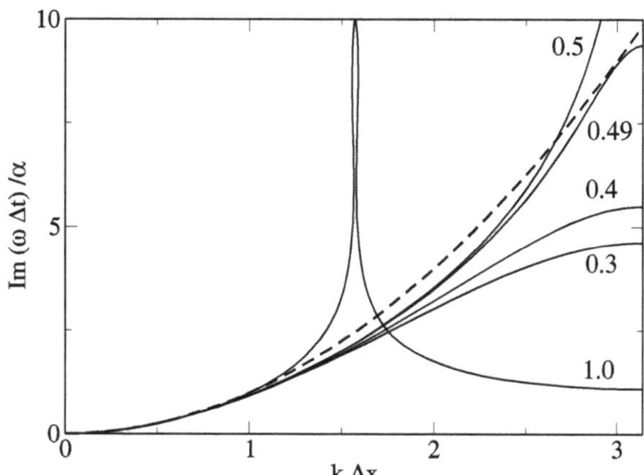

Fig. 19.6 (Dispersion of the Crank-Nicolson method) The dispersion of the Crank-Nicolson method is shown for different values of the Courant number α and compared to the exact dispersion (*dashed curve*). The imaginary part of ω shows a singularity for $\alpha > 1/2$. Above the singularity ω is complex valued. The exact dispersion is approached quite closely for $\alpha \approx 1/2$

whereas for the explicit method

$$\Delta f_{expl} = \alpha M f(t, x) + S(t, x)\Delta t$$

$$= D\frac{\Delta t}{\Delta x^2}\left(f(t, x + \Delta x) + f(t, x - \Delta x) - 2f(t, x)\right) + S(t, x)\Delta t$$

$$= D\frac{\Delta t}{\Delta x^2}\left(\Delta x^2 f''(t, x) + \frac{\Delta x^4}{12}f''''(t, x) + \cdots\right) + S(t, x)\Delta t$$

$$= \Delta f_{exact} + \frac{D\Delta t \Delta x^2}{12}f''''(t, x) - \frac{\Delta t^2}{2}\ddot{f}(t, x) + \cdots \tag{19.49}$$

and for the implicit method

$$\Delta f_{impl} = \alpha M f(t + \Delta t, x) + S(t + \Delta t, x)\Delta t$$

$$= D\frac{\Delta t}{\Delta x^2}\left(f(t + \Delta t, x + \Delta x) + f(t + \Delta t, x - \Delta x) - 2f(t + \Delta t, x)\right)$$
$$+ S(t + \Delta t, x)\Delta t$$

$$= D\frac{\Delta t}{\Delta x^2}\left(\Delta x^2 f''(t, x) + \frac{\Delta x^4}{12}f''''(t, x) + \cdots\right)$$

$$+ S(t, x)\Delta t + D\frac{\Delta t^2}{\Delta x^2}\left(\Delta x^2 \dot{f}''(t, x) + \frac{\Delta x^4}{12}\dot{f}''''(t, x) + \cdots\right)$$

$$+ \dot{S}(t, x)\Delta t^2$$

$$= \Delta f_{exact} + D\frac{\Delta t \Delta x^2}{12}f''''(t, x) + \frac{1}{2}\Delta t^2 \ddot{f}(t, x) + \cdots. \tag{19.50}$$

The Crank-Nicolson method has higher accuracy in Δt:

$$\Delta f_{CN} = \frac{\Delta f_{expl} + \Delta f_{impl}}{2} = \frac{D\Delta t \Delta x^2}{12}f''''(t, x) - \frac{\Delta t^3}{6}\frac{\partial^3 f}{\partial t^3} + \cdots. \tag{19.51}$$

19.2.5 Finite Element Discretization

In one dimension discretization with finite differences is very similar to discretization with finite elements, if Galerkin's method is applied on a regular grid (Chap. 11). The only difference is the non-diagonal form of the mass-matrix which has to be applied to the time derivative [88]. Implementation of the discretization scheme (11.170) is straightforward. The semi-discrete diffusion equation becomes

$$\frac{\partial}{\partial t}\left(\frac{1}{6}f(t, x_{m-1}) + \frac{2}{3}f(t, x_m) + \frac{1}{6}f(t, x_{m+1})\right)$$
$$= \frac{D}{\Delta x^2}\left(f(t, x_{m+1}) + f(t, x_{m-1}) - 2f(t, x_m)\right) + S(t, x_m) \qquad (19.52)$$

or in matrix form

$$\left(1 + \frac{1}{6}M\right)\dot{\mathbf{f}}(t) = \frac{D}{\Delta x^2}M\mathbf{f}(t) + \mathbf{S}(t). \qquad (19.53)$$

This can be combined with the Crank-Nicolson scheme to obtain

$$\left(1 + \frac{1}{6}M\right)(\mathbf{f}_{n+1} - \mathbf{f}_n) = \left(\frac{\alpha}{2}M\mathbf{f}_n + \frac{\alpha}{2}M\mathbf{f}_{n+1}\right) + \frac{\Delta t}{2}(\mathbf{S}_n + \mathbf{S}_{n+1})$$

$$(19.54)$$

or

$$\left[1 + \left(\frac{1}{6} - \frac{\alpha}{2}\right)M\right]\mathbf{f}_{n+1} = \left[1 + \left(\frac{1}{6} + \frac{\alpha}{2}\right)M\right]\mathbf{f}_n + \frac{\Delta t}{2}(\mathbf{S}_n + \mathbf{S}_{n+1}).$$

$$(19.55)$$

19.3 Split-Operator Method for Multidimensions

The simplest discretization of the Laplace operator in 3 dimensions is given by

$$\Delta f = \left(\frac{\partial^2}{\partial x^2} + \frac{\partial^2}{\partial y^2} + \frac{\partial^2}{\partial z^2}\right)f(t, x, y, z)$$
$$= \frac{1}{\Delta x^2}(M_x + M_y + M_z)f(t, x, y, z) \qquad (19.56)$$

where

$$\frac{1}{\Delta x^2}M_x f(t, x, y, z) = \frac{f(t, x + \Delta x, y, z) + f(t, x - \Delta x, y, z) - 2f(t, x, y, z)}{\Delta x^2}$$

$$(19.57)$$

etc. denote the discretized second derivatives. Generalization of the Crank-Nicolson method for the 3-dimensional problem gives

$$f(t_{n+1}) = \left(1 - \frac{\alpha}{2}M_x - \frac{\alpha}{2}M_y - \frac{\alpha}{2}M_z\right)^{-1}\left(1 + \frac{\alpha}{2}M_x + \frac{\alpha}{2}M_y + \frac{\alpha}{2}M_z\right)f(t).$$

$$(19.58)$$

But now the matrices representing the operators M_x, M_y, M_z are not tridiagonal. To keep the advantages of tridiagonal matrices we use the approximations

$$\left(1 + \frac{\alpha}{2}M_x + \frac{\alpha}{2}M_y + \frac{\alpha}{2}M_z\right) \approx \left(1 + \frac{\alpha}{2}M_x\right)\left(1 + \frac{\alpha}{2}M_y\right)\left(1 + \frac{\alpha}{2}M_z\right)$$

$$(19.59)$$

$$\left(1 - \frac{\alpha}{2}M_x - \frac{\alpha}{2}M_y - \frac{\alpha}{2}M_z\right) \approx \left(1 - \frac{\alpha}{2}M_x\right)\left(1 - \frac{\alpha}{2}M_y\right)\left(1 - \frac{\alpha}{2}M_z\right)$$

$$(19.60)$$

and rearrange the factors to obtain

$$f(t_{n+1}) = \left(1 - \frac{\alpha}{2}M_x\right)^{-1}\left(1 + \frac{\alpha}{2}M_x\right)\left(1 - \frac{\alpha}{2}M_y\right)^{-1}\left(1 + \frac{\alpha}{2}M_y\right)$$

$$\times \left(1 - \frac{\alpha}{2}M_z\right)^{-1}\left(1 + \frac{\alpha}{2}M_z\right)f(t_n) \qquad (19.61)$$

which represents successive application of the 1-dimensional scheme for the three directions separately. The last step was possible since the operators M_i and M_j for different directions $i \neq j$ commute. For instance

$$\begin{aligned}
M_x M_y f &= M_x\big(f(x, y + \Delta x) + f(x, y - \Delta x) - 2f(x, y)\big) \\
&= \big(f(x + \Delta x, y + \Delta y) + f(x - \Delta x, y + \Delta x) \\
&\quad - 2f(x, y + \Delta x) + f(x + \Delta x, y - \Delta x) \\
&\quad + f(x - \Delta x, y - \Delta x) - 2f(x, y - \Delta x) \\
&\quad - 2f(x + \Delta x, y) - 2f(x - \Delta x, y) + 4f(x, y)\big) \\
&= M_y M_x f.
\end{aligned} \qquad (19.62)$$

The Taylor series of (19.58) and (19.61) coincide up to second order with respect to αM_i:

$$\left(1 - \frac{\alpha}{2}M_x - \frac{\alpha}{2}M_y - \frac{\alpha}{2}M_z\right)^{-1}\left(1 + \frac{\alpha}{2}M_x + \frac{\alpha}{2}M_y + \frac{\alpha}{2}M_z\right)$$

$$= 1 + \alpha(M_x + M_y + M_z) + \frac{\alpha^2}{2}(M_x + M_y + M_z)^2 + O(\alpha^3) \qquad (19.63)$$

$$\left(1 - \frac{\alpha}{2}M_x\right)^{-1}\left(1 + \frac{\alpha}{2}M_x\right)\left(1 - \frac{\alpha}{2}M_y\right)^{-1}\left(1 + \frac{\alpha}{2}M_y\right)$$

$$\times \left(1 - \frac{\alpha}{2}M_z\right)^{-1}\left(1 + \frac{\alpha}{2}M_z\right)$$

$$= \left(1 + \alpha M_x + \frac{\alpha^2 M_x^2}{2}\right)\left(1 + \alpha M_y + \frac{\alpha^2 M_y^2}{2}\right)\left(1 + \alpha M_z + \frac{\alpha^2 M_z^2}{2}\right) + O\left(\alpha^3\right)$$

$$= 1 + \alpha(M_x + M_y + M_z) + \frac{\alpha^2}{2}(M_x + M_y + M_z)^2 + O\left(\alpha^3\right). \tag{19.64}$$

Hence we have

$$f_{n+1} = \left(1 + D\Delta t\left(\Delta + \frac{\Delta x^2}{12}\Delta^2 + \cdots\right) + \frac{D^2 \Delta t^2}{2}\left(\Delta^2 + \cdots\right)\right)f_n$$
$$+ \left(1 + \frac{D\Delta t}{2}\Delta + \cdots\right)\frac{S_{n+1} + S_n}{2}\Delta t$$
$$= f_n + \Delta t(D\Delta f_n + S_n) + \frac{\Delta t^2}{2}\left(D^2\Delta^2 + D\Delta S_n + \dot{S}_n\right)$$
$$+ O\left(\Delta t \Delta x^2, \Delta t^3\right) \tag{19.65}$$

and the error order is conserved by the split operator method.

19.4 Problems

Problem 19.1 (Diffusion in 2 dimensions) In this computer experiment we solve
the diffusion equation on a two dimensional grid for

- an initial distribution $f(t = 0, x, y) = \delta_{x,0}\delta_{y,0}$
- a constant source $f(t = 0) = 0$, $S(t, x, y) = \delta_{x,0}\delta_{y,0}$

Compare implicit, explicit and Crank-Nicolson method.

Chapter 20
Nonlinear Systems

Nonlinear problems [123, 145] are of interest to physicists, mathematicians and also engineers. Nonlinear equations are difficult to solve and give rise to interesting phenomena like indeterministic behavior, multistability or formation of patterns in time and space. In the following we discuss recurrence relations like an iterated function [135]

$$x_{n+1} = f(x_n) \tag{20.1}$$

systems of ordinary differential equations like population dynamics models [178, 214, 245]

$$\dot{x}(t) = f(x, y)$$
$$\dot{y}(t) = g(x, y) \tag{20.2}$$

or partial differential equations like the reaction-diffusion equation [83, 112, 178]

$$\frac{\partial}{\partial t} c(x, t) = D \frac{\partial^2}{\partial x^2} c(x, t) + f(c) \tag{20.3}$$

where f and g are nonlinear in the mathematical sense.[1] We discuss fixed points of the logistic mapping and analyze their stability. A bifurcation diagram visualizes the appearance of period doubling and chaotic behavior as a function of a control parameter. The Lyapunov exponent helps to distinguish stable fixed points and periods from chaotic regions. For continuous-time models, the iterated function is replaced by a system of differential equations. For stable equilibria all eigenvalues of the Jacobian matrix must have a negative real part. We discuss the Lotka-Volterra model, which is the simplest model of predator-prey interactions and the Holling-Tanner model, which incorporates functional response. Finally we allow for spatial inhomogeneity and include diffusive terms to obtain reaction-diffusion systems, which show the phenomena of traveling waves and pattern formation. Computer experiments study orbits and bifurcation diagram of the logistic map, periodic oscillations

---

[1]Linear functions are additive $f(x + y) = f(x) + f(y)$ and homogeneous $f(\alpha x) = \alpha f(x)$.

P.O.J. Scherer, *Computational Physics*, Graduate Texts in Physics,
DOI 10.1007/978-3-319-00401-3_20,
© Springer International Publishing Switzerland 2013

Fig. 20.1 (Orbit of an iterated function) The sequence of points $(x_i, x_{i+1}), (x_{i+1}, x_{i+1})$ is plotted together with the curves $y = f(x)$ (*dashed*) and $y = x$ (*dotted*)

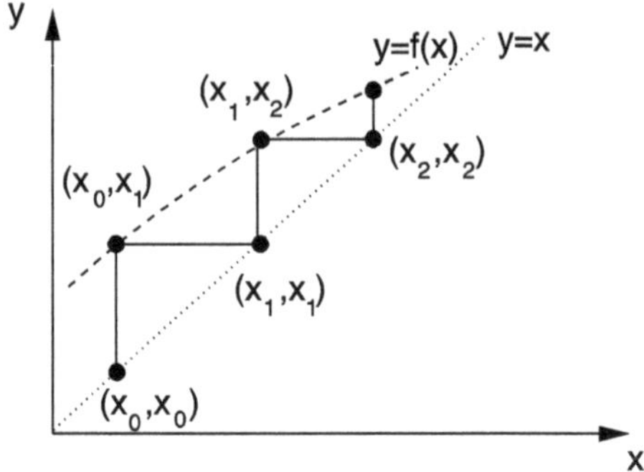

of the Lotka-Volterra model, oscillations and limit cycles of the Holling-Tanner model and finally pattern formation in the diffusive Lotka-Volterra model.

20.1 Iterated Functions

Starting from an initial value x_0 a function f is iterated repeatedly

$$
\begin{aligned}
x_1 &= f(x_0) \\
x_2 &= f(x_1) \\
&\;\vdots \\
x_{i+1} &= f(x_i).
\end{aligned}
\tag{20.4}
$$

The sequence of function values $x_0, x_1 \cdots$ is called the orbit of x_0. It can be visualized in a 2-dimensional plot by connecting the points

$$(x_0, x_1) \to (x_1, x_1) \to (x_1, x_2) \to (x_2, x_2) \to \cdots \to (x_i, x_{i+1}) \to (x_{i+1}, x_{i+1})$$

by straight lines (Fig. 20.1).

20.1.1 *Fixed Points and Stability*

If the equation

$$x^* = f\left(x^*\right) \tag{20.5}$$

has solutions x^*, then these are called fixed points. Consider a point in the vicinity of a fixed point

$$x = x^* + \varepsilon_0 \tag{20.6}$$

and make a Taylor series expansion

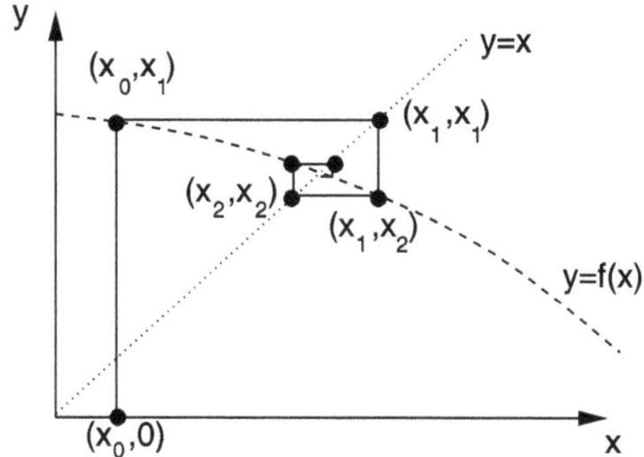

Fig. 20.2 (Attractive fixed point) The orbit of an attractive fixed point converges to the intersection of the curves $y = x$ and $y = f(x)$

$$f(x) = f\left(x^* + \varepsilon_0\right) = f\left(x^*\right) + \varepsilon_0 f'\left(x^*\right) + \cdots = x^* + \varepsilon_1 + \cdots \qquad (20.7)$$

with the notation

$$\varepsilon_1 = \varepsilon_0 f'\left(x^*\right). \qquad (20.8)$$

Repeated iteration gives[2]

$$f^{(2)}(x) = f\left(f(x)\right) = f\left(x^* + \varepsilon_1\right) + \cdots = x^* + \varepsilon_1 f'\left(x^*\right) = x^* + \varepsilon_2$$

$$\vdots \qquad (20.9)$$

$$f^{(n)}\left(x^*\right) = x^* + \varepsilon_n$$

with the sequence of deviations

$$\varepsilon_n = f'\left(x^*\right)\varepsilon_{n-1} = \cdots = \left(f'\left(x^*\right)\right)^n \varepsilon_0.$$

The orbit moves away from the fixed point for arbitrarily small ε_0 if $|f'(x^*)| > 1$ whereas the fixed point is attractive for $|f'(x^*)| < 1$ (Fig. 20.2).

Higher order fixed points are defined by iterating $f(x)$ several times. A fixed point of order n simultaneously solves

$$\begin{aligned} f\left(x^*\right) &\neq x^* \\ f^{(2)}\left(x^*\right) &\neq x^* \\ f^{(n-1)}\left(x^*\right) &\neq x^* \\ f^{(n)}\left(x^*\right) &= x^*. \end{aligned} \qquad (20.10)$$

The iterated function values cycle periodically (Fig. 20.3) through

$$x^* \to f\left(x^*\right) \to f^{(2)}\left(x^*\right) \cdots f^{(n-1)}\left(x^*\right).$$

This period is attractive if

$$\left| f'\left(x^*\right) f'\left(f\left(x^*\right)\right) f'\left(f^{(2)}\left(x^*\right)\right) \cdots f'\left(f^{(n-1)}\left(x^*\right)\right) \right| < 1.$$

---

[2] Here and in the following $f^{(n)}$ denotes an iterated function, not a derivative.

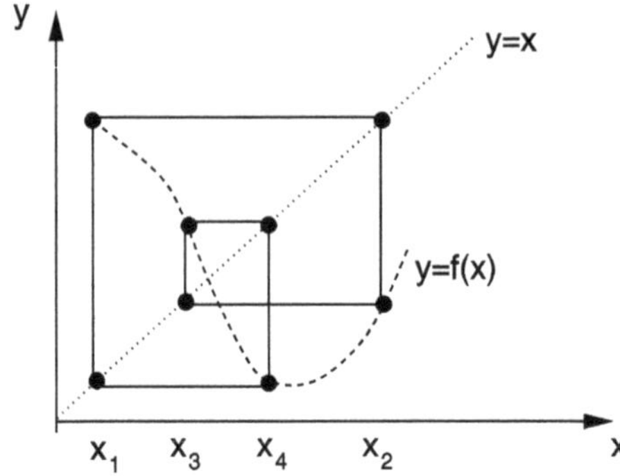

Fig. 20.3 (Periodic orbit) The orbit of an attractive fourth order fixed point cycles through the values $x_1 = f(x_4)$, $x_2 = f(x_1)$, $x_3 = f(x_2)$, $x_4 = f(x_3)$

20.1.2 The Lyapunov Exponent

Consider two neighboring orbits with initial values x_0 and $x_0 + \varepsilon_0$. After n iterations the distance is

$$\left| f\big(f(\cdots f(x_0))\big) - f\big(f(\cdots f(x_0 + \varepsilon_0))\big) \right| = |\varepsilon_0| e^{\lambda n} \tag{20.11}$$

with the so called Lyapunov exponent [161] λ which is useful to characterize the orbit. The Lyapunov exponent can be determined from

$$\lambda = \lim_{n \to \infty} \frac{1}{n} \ln \left(\frac{|f^{(n)}(x_0 + \varepsilon_0) - f^{(n)}(x_0)|}{|\varepsilon_0|} \right) \tag{20.12}$$

or numerically easier with the approximation

$$\begin{aligned}
\left| f(x_0 + \varepsilon_0) - f(x_0) \right| &= |\varepsilon_0| \left| f'(x_0) \right| \\
\left| f\big(f(x_0 + \varepsilon_0)\big) - f\big(f(x_0)\big) \right| &= \left| \big(f(x_0 + \varepsilon_0) - f(x_0)\big) \right| \left| f'(x_0 + \varepsilon_0) \right| \\
&= |\varepsilon_0| \left| f'(x_0) \right| \left| f'(x_0 + \varepsilon_0) \right| \tag{20.13} \\
\left| f^{(n)}(x_0 + \varepsilon_0) - f^{(n)}(x_0) \right| &= |\varepsilon_0| \left| f'(x_0) \right| \left| f'(x_1) \right| \cdots \left| f'(x_{n-1}) \right| \tag{20.14}
\end{aligned}$$

from

$$\lambda = \lim_{n \to \infty} \frac{1}{n} \sum_{i=0}^{n-1} \ln \left| f'(x_i) \right|. \tag{20.15}$$

For a stable fixed point

$$\lambda \to \ln \left| f'(x^*) \right| < 0 \tag{20.16}$$

and for an attractive period

$$\lambda \to \ln \left| f'(x^*) \, f'(f(x^*)) \ldots f'(f^{(n-1)}(x^*)) \right| < 0. \tag{20.17}$$

Orbits with $\lambda < 0$ are attractive fixed points or periods. If, on the other hand, $\lambda > 0$, the orbit is irregular and very sensitive to the initial conditions, hence is chaotic.

Fig. 20.4 (Reproduction rate of the logistic model) At low densities the growth rate has its maximum value r_0. At larger densities the growth rate declines and reaches $r = 0$ for $N = K$. The parameter K is called carrying capacity

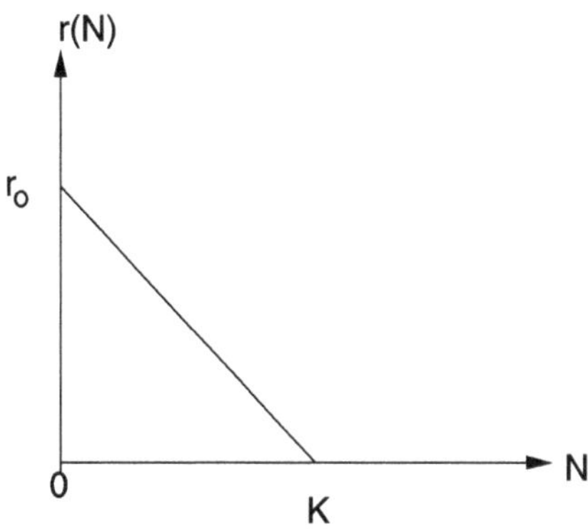

20.1.3 The Logistic Map

A population of animals is observed yearly. The evolution of the population density N is described in terms of the reproduction rate r by the recurrence relation

$$N_{n+1} = r\, N_n \tag{20.18}$$

where N_n is the population density in year number n. If r is constant, an exponential increase or decrease of N results.

The simplest model for the growth of a population which takes into account that the resources are limited is the logistic model by Verhulst [264]. He assumed that the reproduction rate r depends on the population density N in a simple way (Fig. 20.4)

$$r = r_0\left(1 - \frac{N}{K}\right). \tag{20.19}$$

The Verhulst model (20.19) leads to the iterated nonlinear function

$$N_{n+1} = r_0 N_n - \frac{r_0}{K} N_n^2 \tag{20.20}$$

with $r_0 > 0$, $K > 0$. We denote the quotient of population density and carrying capacity by the new variable

$$x_n = \frac{1}{K} N_n \tag{20.21}$$

and obtain an equation with only one parameter, the so called logistic mapping

$$x_{n+1} = \frac{1}{K} N_{n+1} = \frac{1}{K} r_0 N_n\left(1 - \frac{N_n}{K}\right) = r_0 \cdot x_n \cdot (1 - x_n). \tag{20.22}$$

20.1.4 Fixed Points of the Logistic Map

Consider an initial point in the interval

$$0 < x_0 < 1. \tag{20.23}$$

We want to find conditions on r to keep the orbit in this interval. The maximum value of x_{n+1} is found from

$$\frac{dx_{n+1}}{dx_n} = r(1 - 2x_n) = 0 \tag{20.24}$$

which gives $x_n = 1/2$ and $\max(x_{n+1}) = r/4$. If $r > 4$ then negative x_n appear after some iterations and the orbit is not bound by a finite interval since

$$\frac{|x_{n+1}|}{|x_n|} = |r|(1 + |x_n|) > 1. \tag{20.25}$$

The fixed point equation

$$x^* = rx^* - rx^{*2} \tag{20.26}$$

always has the trivial solution

$$x^* = 0 \tag{20.27}$$

and a further solution

$$x^* = 1 - \frac{1}{r} \tag{20.28}$$

which is only physically reasonable for $r > 1$, since x should be a positive quantity. For the logistic mapping the derivative is

$$f'(x) = r - 2rx \tag{20.29}$$

which for the first fixed point $x^* = 0$ gives $|f'(0)| = r$. This fixed point is attractive for $0 < r < 1$ and becomes unstable for $r > 1$. For the second fixed point we have $|f'(1 - \frac{1}{r})| = |2 - r|$, which is smaller than one in the interval $1 < r < 3$. For $r < 1$ no such fixed point exists. Finally, for $r_1 = 3$ the first bifurcation appears and higher order fixed points become stable.

Consider the fixed point of the double iteration

$$x^* = r\left(r(x^* - x^{*2}) - r^2(x^* - x^{*2})^2\right). \tag{20.30}$$

All roots of this fourth order equation can be found since we already know two of them. The remaining roots are

$$x_{1,2}^* = \frac{\frac{r+1}{2} \pm \sqrt{r^2 - 2r - 3}}{r}. \tag{20.31}$$

They are real valued if

$$(r - 1)^2 - 4 > 0 \rightarrow r > 3 \quad (\text{or } r < -1). \tag{20.32}$$

For $r > 3$ the orbit oscillates between x_1^* and x_2^* until the next period doubling appears for $r_2 = 1 + \sqrt{6}$. With increasing r more and more bifurcations appear and finally the orbits become chaotic (Fig. 20.5).

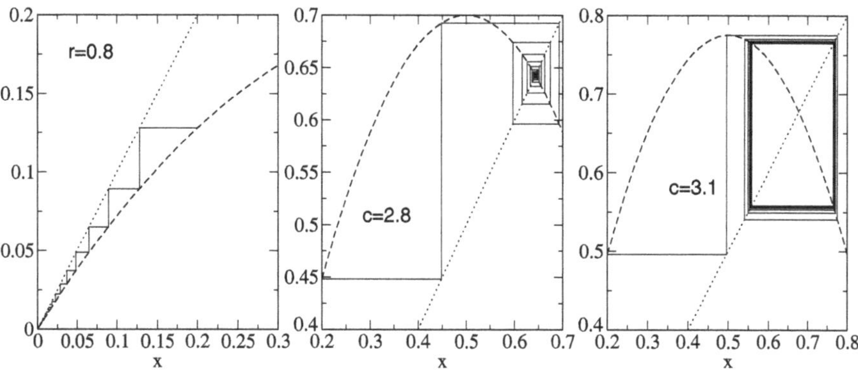

Fig. 20.5 (Orbits of the logistic map) *Left*: For $0 < r < 1$ the logistic map has the attractive fixed point $x^* = 0$. *Middle*: In the region $1 < r < 3$ this fixed point becomes unstable and another stable fixed point is at $x^* = 1 - 1/r$. *Right*: For $3 < r < 1 + \sqrt{6}$ the second order fixed point (20.31) is stable. For larger values of r more and more bifurcations appear

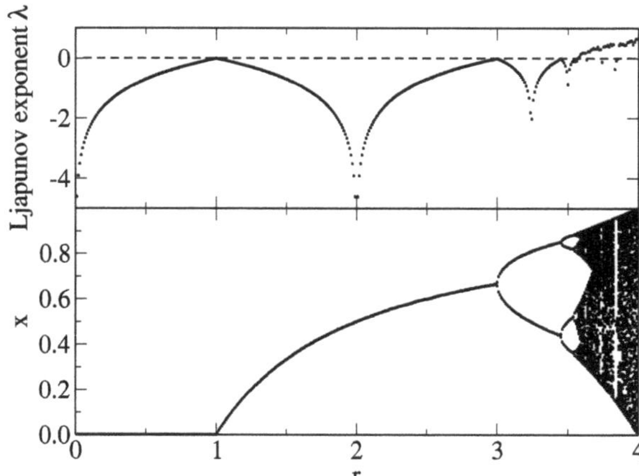

Fig. 20.6 (Bifurcation diagram of the logistic map) For different values of r the function is iterated 1100 times. The first 1000 iterations are dropped to allow the trajectory to approach stable fixed points or periods. The iterated function values $x_{1000} \cdots x_{1100}$ are plotted in an r–x diagram together with the estimate (20.15) of the Lyapunov exponent.
The first period doublings appear at $r = 3$ and $r = 1 + \sqrt{6}$. For larger values chaotic behavior is observed and the estimated Lyapunov exponent becomes positive. In some regions motion is regular again with negative Lyapunov exponent

20.1.5 Bifurcation Diagram

The bifurcation diagram visualizes the appearance of period doubling and chaotic behavior as a function of the control parameter r (Fig. 20.6).

20.2 Population Dynamics

If time is treated as a continuous variable, the iterated function has to be replaced
by a differential equation

$$\frac{dN}{dt} = f(N) \tag{20.33}$$

or, more generally by a system of equations

$$\frac{d}{dt}\begin{pmatrix} N_1 \\ N_2 \\ \vdots \\ N_n \end{pmatrix} = \begin{pmatrix} f_1(N_1 \cdots N_n) \\ f_2(N_1 \cdots N_n) \\ \\ f_n(N_1 \cdots N_n) \end{pmatrix}. \tag{20.34}$$

20.2.1 Equilibria and Stability

The role of the fixed points is now taken over by equilibria, which are solutions of

$$0 = \frac{dN}{dt} = f(N_{eq}) \tag{20.35}$$

which means roots of $f(N)$. Let us investigate small deviations from equilibrium
with the help of a Taylor series expansion. Inserting

$$N = N_{eq} + \xi \tag{20.36}$$

we obtain

$$\frac{d\xi}{dt} = f(N_{eq}) + f'(N_{eq})\xi + \cdots \tag{20.37}$$

but since $f(N_{eq}) = 0$, we have approximately

$$\frac{d\xi}{dt} = f'(N_{eq})\xi \tag{20.38}$$

with the solution

$$\xi(t) = \xi_0 \exp\{f'(N_{eq})t\}. \tag{20.39}$$

The equilibrium is only stable if $\mathfrak{Re}\, f'(N_{eq}) < 0$, since then small deviations
disappear exponentially. For $\mathfrak{Re}\, f'(N_{eq}) > 0$ deviations will increase, but the exponential behavior holds only for not too large deviations and saturation may appear.
If the derivative $f'(N_{eq})$ has a nonzero imaginary part then oscillations will be su-
perimposed. For a system of equations the equilibrium is defined by

$$\begin{pmatrix} f_1(N_1^{eq} \cdots N_n^{eq}) \\ f_2(N_1^{eq} \cdots N_n^{eq}) \\ \\ f_N(N_1^{eq} \cdots N_n^{eq}) \end{pmatrix} = \begin{pmatrix} 0 \\ 0 \\ \vdots \\ 0 \end{pmatrix} \tag{20.40}$$

Fig. 20.7 (Equilibria of the logistic model) The equilibrium $x_{eq} = 0$ is unstable since an infinitesimal deviation grows exponentially in time. The equilibrium $x_{eq} = 1$ is stable since initial deviations disappear exponentially

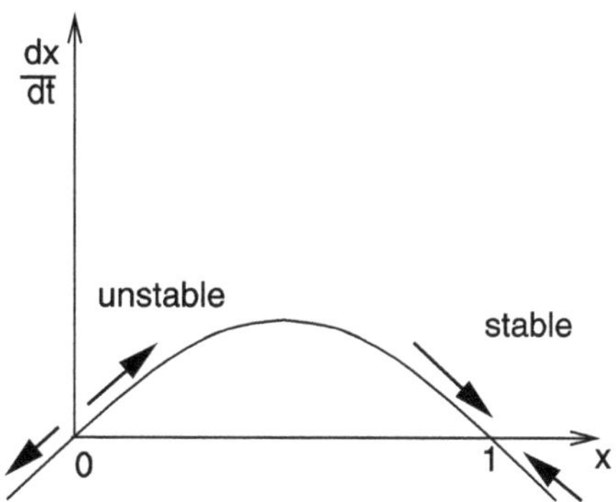

and if such an equilibrium exists, linearization gives

$$\begin{pmatrix} N_1 \\ N_2 \\ \vdots \\ N_n \end{pmatrix} = \begin{pmatrix} N_1^{eq} \\ N_2^{eq} \\ \vdots \\ N_n^{eq} \end{pmatrix} + \begin{pmatrix} \xi_1 \\ \xi_2 \\ \vdots \\ \xi_n \end{pmatrix} \tag{20.41}$$

$$\frac{d}{dt} \begin{pmatrix} \xi_1 \\ \xi_2 \\ \vdots \\ \xi_N \end{pmatrix} = \begin{pmatrix} \frac{\partial f_1}{\partial N_1} & \frac{\partial f_1}{\partial N_2} & \cdots & \frac{\partial f_1}{\partial N_n} \\ \frac{\partial f_2}{\partial N_1} & \frac{\partial f_2}{\partial N_2} & \cdots & \frac{\partial f_2}{\partial N_n} \\ \vdots & \vdots & \ddots & \vdots \\ \frac{\partial f_n}{\partial N_1} & \frac{\partial f_n}{\partial N_2} & \cdots & \frac{\partial f_n}{\partial N_n} \end{pmatrix} \begin{pmatrix} \xi_1 \\ \xi_2 \\ \vdots \\ \xi_n \end{pmatrix}. \tag{20.42}$$

The equilibrium is stable if all eigenvalues λ_i of the derivative matrix have a negative real part.

20.2.2 The Continuous Logistic Model

The continuous logistic model describes the evolution by the differential equation

$$\frac{dx}{dt} = r_0 x (1 - x). \tag{20.43}$$

To find possible equilibria we have to solve

$$x_{eq}(1 - x_{eq}) = 0 \tag{20.44}$$

which has the two roots $x_{eq} = 0$ and $x_{eq} = 1$ (Fig. 20.7).

The derivative f' is

$$f'(x) = \frac{d}{dx}\left(r_0 x(1 - x)\right) = r_0(1 - 2x). \tag{20.45}$$

Since $f'(0) = r_0 > 0$ and $f'(1) = -r_0 < 0$ only the second equilibrium is stable.

20.3 Lotka-Volterra Model

The model by Lotka [160] and Volterra [266] is the simplest model of predator-prey interactions. It has two variables, the density of prey (H) and the density of predators (P). The overall reproduction rate of each species is given by the difference of the birth rate r and the mortality rate m

$$\frac{dN}{dt} = (r - m)N$$

which both may depend on the population densities. The Lotka-Volterra model assumes that the prey mortality depends linearly on the predator density and the predator birth rate is proportional to the prey density

$$m_H = aP \qquad r_P = bH \tag{20.46}$$

where a is the predation rate coefficient and b is the reproduction rate of predators per 1 prey eaten. Together we end up with a system of two coupled nonlinear differential equations

$$\frac{dH}{dt} = f(H, P) = r_H H - aHP$$
$$\frac{dP}{dt} = g(H, P) = bHP - m_P P \tag{20.47}$$

where r_H is the intrinsic rate of prey population increase and m_P the predator mortality rate.

20.3.1 Stability Analysis

To find equilibria we have to solve the system of equations

$$f(H, P) = r_H H - aHP = 0$$
$$g(H, P) = bHP - m_P P = 0. \tag{20.48}$$

The first equation is solved by $H_{eq} = 0$ or by $P_{eq} = r_H/a$. The second equation is solved by $P_{eq} = 0$ or by $H_{eq} = m_P/b$. Hence there are two equilibria, the trivial one

$$P_{eq} = H_{eq} = 0 \tag{20.49}$$

and a nontrivial one

$$P_{eq} = \frac{r_H}{a} \qquad H_{eq} = \frac{m_P}{b}. \tag{20.50}$$

Linearization around the zero equilibrium gives

$$\frac{dH}{dt} = r_H H + \cdots \qquad \frac{dP}{dt} = -m_P P + \cdots. \tag{20.51}$$

This equilibrium is unstable since a small prey population will increase exponentially. Now expand around the nontrivial equilibrium:

$$P = P_{eq} + \xi, \qquad H = H_{eq} + \eta \tag{20.52}$$

$$\frac{d\eta}{dt} = \frac{\partial f}{\partial H}\eta + \frac{\partial f}{\partial P}\xi = (r_H - aP_{eq})\eta - aH_{eq}\xi = -\frac{am_P}{b}\xi \tag{20.53}$$

$$\frac{d\xi}{dt} = \frac{\partial g}{\partial H}\eta + \frac{\partial g}{\partial P}\xi = bP_{eq}\eta + (bH_{eq} - m_P)\xi = \frac{br_H}{a}\eta \tag{20.54}$$

or in matrix notation

$$\frac{d}{dt}\begin{pmatrix} \eta \\ \xi \end{pmatrix} = \begin{pmatrix} 0 & -\frac{am_P}{b} \\ \frac{br_H}{a} & 0 \end{pmatrix}\begin{pmatrix} \eta \\ \xi \end{pmatrix}. \tag{20.55}$$

The eigenvalues are purely imaginary

$$\lambda = \pm i\sqrt{m_H r_P} = \pm i\omega \tag{20.56}$$

and the corresponding eigenvectors are

$$\begin{pmatrix} i\sqrt{m_H r_P} \\ br_H/a \end{pmatrix}, \begin{pmatrix} am_P/b \\ i\sqrt{m_H r_P} \end{pmatrix}. \tag{20.57}$$

The solution of the linearized equations is then given by

$$\xi(t) = \xi_0 \cos\omega t + \frac{b}{a}\sqrt{\frac{r_P}{m_H}}\,\eta_0 \sin\omega t$$

$$\eta(t) = \eta_0 \cos\omega t - \frac{a}{b}\sqrt{\frac{m_H}{r_P}}\,\xi_0 \sin\omega t \tag{20.58}$$

which describes an ellipse in the $\xi-\eta$ plane (Fig. 20.8). The nonlinear equations (20.48) have a first integral

$$r_H \ln P(t) - aP(t) - bH(t) + m_P \ln H(t) = C \tag{20.59}$$

and therefore the motion in the $H-P$ plane is on a closed curve around the equilibrium which approaches an ellipse for small amplitudes ξ, η.

20.4 Functional Response

Holling [124, 125] studied predation of small mammals on pine sawflies. He suggested a very popular model of functional response. Holling assumed that the predator spends its time on two kinds of activities, searching for prey and prey handling (chasing, killing, eating, digesting). The total time equals the sum of time spent on searching and time spent on handling

$$T = T_{search} + T_{handling}. \tag{20.60}$$

Capturing prey is assumed to be a random process. A predator examines an area α per time and captures all prey found there. After spending the time T_{search} the

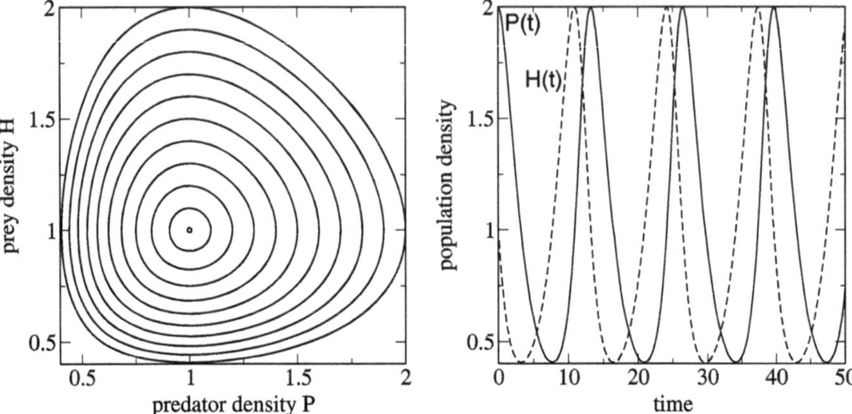

Fig. 20.8 (Lotka-Volterra model) The predator and prey population densities show periodic oscillations (*right*). In the H–P plane the system moves on a closed curve, which becomes an ellipse for small deviations from equilibrium (*left*)

Fig. 20.9 Functional response of Holling's model

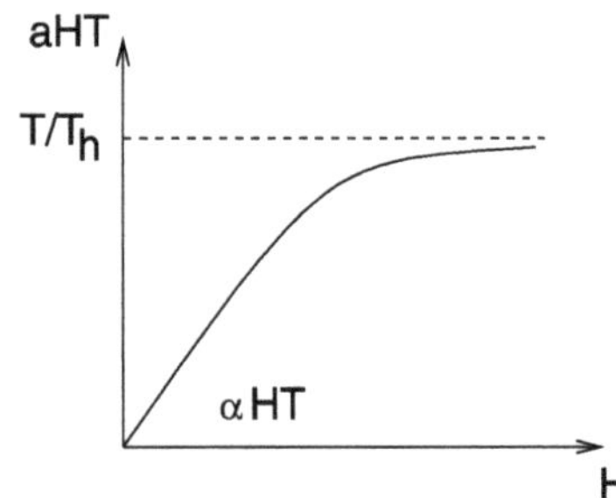

predator examined an area of αT_{search} and captured $H_T = H\alpha T_{search}$ prey. Hence the predation rate is

$$a = \frac{H_T}{HT} = \alpha\frac{T_{search}}{T} = \alpha\frac{1}{1 + T_{handling}/T_{search}}. \tag{20.61}$$

The handling time is assumed to be proportional to the number of prey captured

$$T_{handling} = T_h H\alpha T_{search} \tag{20.62}$$

where T_h is the handling time spent per one prey. The predation rate then is given by

$$a = \frac{\alpha}{1 + \alpha H T_h}. \tag{20.63}$$

At small densities handling time is unimportant and the predation rate is $a_0 = \alpha$ whereas at high prey density handling limits the number of prey captured and the predation rate approaches $a_\infty = \frac{1}{HT_h}$ (Fig. 20.9).

20.4.1 Holling-Tanner Model

We combine the logistic model with Holling's model for the predation rate [124, 125, 249]

$$\frac{dH}{dt} = r_H H\left(1 - \frac{H}{K_H}\right) - aHP$$
$$= r_H H\left(1 - \frac{H}{K_H}\right) - \frac{\alpha}{1+\alpha H T_h}HP = f(H,P) \qquad (20.64)$$

and assume that the carrying capacity of the predator is proportional to the density of prey

$$\frac{dP}{dt} = r_P P\left(1 - \frac{P}{K_P}\right) = r_P P\left(1 - \frac{P}{kH}\right) = g(H,P). \qquad (20.65)$$

Obviously there is a trivial equilibrium with $P_{eq} = H_{eq} = 0$. Linearization gives

$$\frac{dH}{dt} = r_H H + \cdots \qquad \frac{dP}{dt} = r_P P + \cdots \qquad (20.66)$$

which shows that this equilibrium is unstable. There is another trivial equilibrium with $P_{eq} = 0$, $H_{eq} = K_H$. Here we find:

$$\frac{d\eta}{dt} = r_H (K_H + \eta)\left(1 - \frac{K_H + \eta}{K_H}\right) - \frac{\alpha}{1+\alpha(K_H+\eta)T_h}(K_H+\eta)\xi$$
$$= r_H \eta - \frac{\alpha}{1+\alpha K_H T_h}K_H \xi + \cdots \qquad (20.67)$$

$$\frac{d\xi}{dt} = r_P \xi\left(1 - \frac{\xi}{k(K_H+\eta)}\right) = r_P \xi + \cdots \qquad (20.68)$$

$$\begin{pmatrix} \dot\eta \\ \dot\xi \end{pmatrix} = \begin{pmatrix} r_H & -\frac{\alpha}{1+\alpha K_H T_h}K_H \\ 0 & r_P \end{pmatrix}\begin{pmatrix} \eta \\ \xi \end{pmatrix} \qquad (20.69)$$

$$\lambda = r_H, r_P. \qquad (20.70)$$

Let us now look for nontrivial equilibria. The nullclines (Fig. 20.10) are the curves defined by $\frac{dH}{dt} = 0$ and $\frac{dP}{dt} = 0$, hence by

$$P = \frac{r_H}{\alpha}\left(1 - \frac{H}{K_H}\right)(1+\alpha H T_h) \qquad (20.71)$$
$$P = kH. \qquad (20.72)$$

The H-nullcline is a parabola at

$$H_m = \frac{\alpha T_h - K_H^{-1}}{2\alpha T_h K_H^{-1}} \qquad P_m = \frac{(\alpha T_h + K_H^{-1})^2}{4\alpha T_h K_H^{-1}} > 0. \qquad (20.73)$$

It intersects the H-axis at $H = K_H$ and $H = -1/\alpha T_h$ and the P-axis at $P = r_H/\alpha$. There is one intersection of the two nullclines at positive values of H and P which

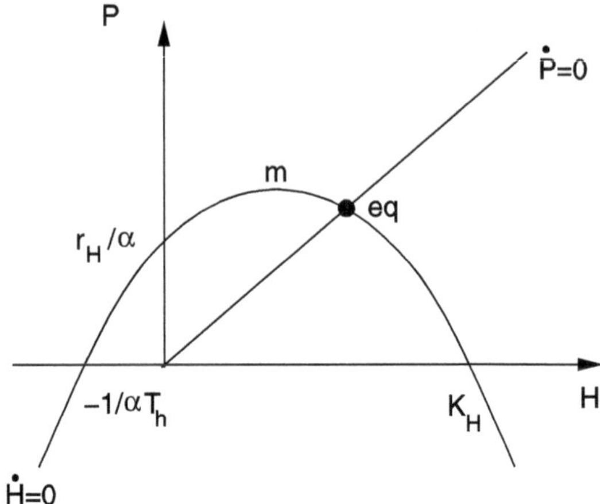

Fig. 20.10 Nullclines of the predator-prey model

corresponds to a nontrivial equilibrium. The equilibrium density H_{eq} is the positive root of

$$r_H \alpha T_h H_{eq}^2 + (r_H + \alpha k K_H - r_H K_H \alpha T_h) H_{eq} - r_H K_H = 0. \qquad (20.74)$$

It is explicitly given by

$$H_{eq} = -\frac{r_H + \alpha k K_H - r_H K_H \alpha T_h}{2 r_H \alpha T_h}$$
$$+ \frac{\sqrt{(r_H + \alpha k K_H - r_H K_H \alpha T_h)^2 + 4 r_H^2 \alpha T_h K_H}}{2 r_H \alpha T_h}. \qquad (20.75)$$

The prey density then follows from

$$P_{eq} = H_{eq} k. \qquad (20.76)$$

The matrix of derivatives has the elements

$$m_{HP} = \frac{\partial f}{\partial P} = -\frac{\alpha H_{eq}}{1 + \alpha T_h H_{eq}}$$

$$m_{HH} = \frac{\partial f}{\partial H} = r_H \left(1 - 2\frac{H_{eq}}{K_H}\right) - \frac{\alpha k H_{eq}}{1 + \alpha T_h H} + \frac{\alpha^2 H_{eq}^2 k T_h}{(1 + \alpha T_h H_{eq})^2} \qquad (20.77)$$

$$m_{PP} = \frac{\partial g}{\partial P} = -r_P$$

$$m_{PH} = \frac{\partial g}{\partial H} = r_P k$$

from which the eigenvalues are calculated as

$$\lambda = \frac{m_{HH} + m_{PP}}{2} \pm \sqrt{\frac{(m_{HH} + m_{PP})^2}{4} - (m_{HH} m_{PP} - m_{HP} m_{PH})}. \qquad (20.78)$$

Oscillations appear, if the squareroot is imaginary (Fig. 20.11).

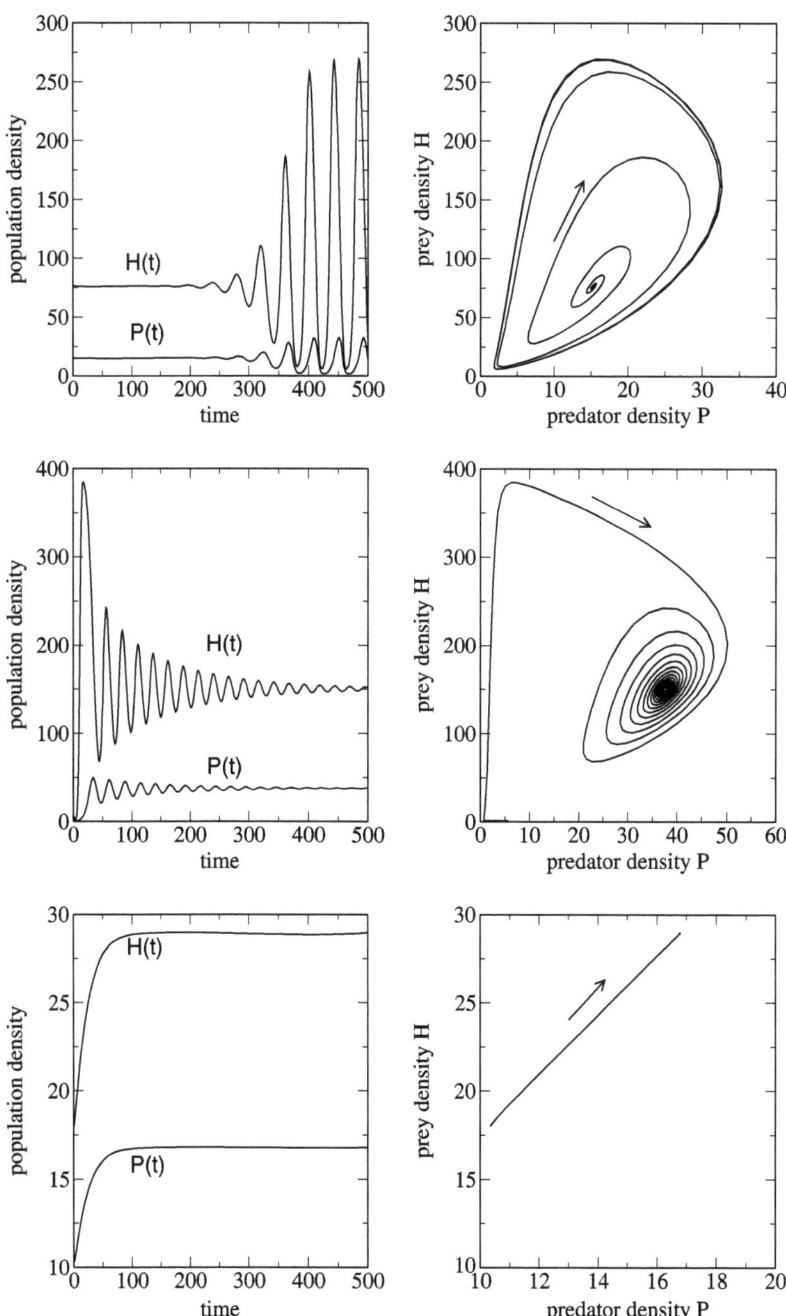

Fig. 20.11 (Holling-Tanner model) *Top*: evolution from an unstable equilibrium to a limit cycle, *middle*: a stable equilibrium is approached with oscillations, *bottom*: stable equilibrium without oscillations

20.5 Reaction-Diffusion Systems

So far we considered spatially homogeneous systems where the density of a population or the concentration of a chemical agent depend only on time. If we add spatial
inhomogeneity and diffusive motion, new and interesting phenomena like pattern
formation or traveling excitations can be observed.

20.5.1 General Properties of Reaction-Diffusion Systems

Reaction-diffusion systems are described by a diffusion equation[3] where the source
term depends nonlinearly on the concentrations

$$
\frac{\partial}{\partial t}
\begin{pmatrix} c_1 \\ \vdots \\ c_N \end{pmatrix}
=
\begin{pmatrix} D_1 & & \\ & \ddots & \\ & & D_N \end{pmatrix}
\Delta
\begin{pmatrix} c_1 \\ \vdots \\ c_N \end{pmatrix}
+
\begin{pmatrix} F_1(\{c\}) \\ \vdots \\ F_N(\{c\}) \end{pmatrix}.
\tag{20.79}
$$

20.5.2 Chemical Reactions

Consider a number of chemical reactions which are described by stoichiometric
equations

$$
\sum_i v_i A_i = 0.
\tag{20.80}
$$

The concentration of agent A_i is

$$
c_i = c_{i,0} + v_i x
\tag{20.81}
$$

with the reaction variable

$$
x = \frac{c_i - c_{i,0}}{v_i}
\tag{20.82}
$$

and the reaction rate

$$
r = \frac{\mathrm{d}x}{\mathrm{d}t} = \frac{1}{v_i}\frac{\mathrm{d}c_i}{\mathrm{d}t}
\tag{20.83}
$$

which, in general is a nonlinear function of all concentrations. The total concentration change due to diffusion and reactions is given by

$$
\frac{\partial}{\partial t} c_k = D_k \Delta c_k + \sum_j v_{kj} r_j = D_k \Delta c_k + F_k(\{c_i\}).
\tag{20.84}
$$

---

[3]We consider only the case, that different species diffuse independently and that the diffusion
constants do not depend on direction.

20.5.3 Diffusive Population Dynamics

Combination of population dynamics (20.2) and diffusive motion gives a similar set of coupled equations for the population densities

$$\frac{\partial}{\partial t} N_k = D_k \Delta N_k + f_k(N_1, N_2 \ldots N_n). \tag{20.85}$$

20.5.4 Stability Analysis

Since a solution of the nonlinear equations is not generally possible we discuss small deviations from an equilibrium solution N_k^{eq} [4] with

$$\frac{\partial}{\partial t} N_k = \Delta N_k = 0. \tag{20.86}$$

Obviously the equilibrium obeys

$$f_k(N_1 \cdots N_n) = 0 \quad k = 1, 2 \cdots n. \tag{20.87}$$

We linearize the equations by setting

$$N_k = N_k^{eq} + \xi_k \tag{20.88}$$

and expand around the equilibrium

$$\frac{\partial}{\partial t}
\begin{pmatrix} \xi_1 \\ \xi_2 \\ \vdots \\ \xi_n \end{pmatrix}
=
\begin{pmatrix} D_1 & & \\ & \ddots & \\ & & D_N \end{pmatrix}
\begin{pmatrix} \Delta\xi_1 \\ \Delta\xi_2 \\ \vdots \\ \Delta\xi_n \end{pmatrix}$$

$$+
\begin{pmatrix}
\frac{\partial f_1}{\partial N_1} & \frac{\partial f_1}{\partial N_2} & \cdots & \frac{\partial f_1}{\partial N_n} \\[4pt]
\frac{\partial f_2}{\partial N_1} & \frac{\partial f_2}{\partial N_2} & \cdots & \frac{\partial f_2}{\partial N_n} \\[4pt]
\vdots & \vdots & \ddots & \vdots \\[4pt]
\frac{\partial f_n}{\partial N_1} & \frac{\partial f_n}{\partial N_2} & \cdots & \frac{\partial f_n}{\partial N_n}
\end{pmatrix}
\begin{pmatrix} \xi_1 \\ \xi_2 \\ \vdots \\ \xi_n \end{pmatrix}
+ \cdots. \tag{20.89}$$

Plane waves are solutions of the linearized problem.[5] Using the ansatz

$$\xi_j = \xi_{j,0} e^{i(\omega t - \mathbf{k}\mathbf{x})} \tag{20.90}$$

we obtain

---

[4] We assume tacitly that such a solution exists.

[5] Strictly this is true only for an infinite or periodic system.

$$i\omega \begin{pmatrix} \xi_1 \\ \xi_2 \\ \vdots \\ \xi_n \end{pmatrix} = -k^2 D \begin{pmatrix} \xi_1 \\ \xi_2 \\ \vdots \\ \xi_n \end{pmatrix} + M_0 \begin{pmatrix} \xi_1 \\ \xi_2 \\ \vdots \\ \xi_n \end{pmatrix} \tag{20.91}$$

where M_0 denotes the matrix of derivatives and D the matrix of diffusion constants. For a stable plane wave solution $\lambda = i\omega$ is an eigenvalue of

$$M_k = M_0 - k^2 D \tag{20.92}$$

with

$$\Re(\lambda) \le 0. \tag{20.93}$$

If there are purely imaginary eigenvalues for some **k** they correspond to stable solutions which are spatially inhomogeneous and lead to formation of certain patterns. Interestingly, diffusion can lead to instabilities even for a system which is stable in the absence of diffusion [258].

20.5.5 Lotka-Volterra Model with Diffusion

As a simple example we consider again the Lotka-Volterra model. Adding diffusive terms we obtain the equations

$$\frac{\partial}{\partial t} \begin{pmatrix} H \\ P \end{pmatrix} = \begin{pmatrix} r_H H - a H P \\ b H P - m_P P \end{pmatrix} + \begin{pmatrix} D_H & \\ & D_P \end{pmatrix} \Delta \begin{pmatrix} H \\ P \end{pmatrix}. \tag{20.94}$$

There are two equilibria

$$H_{eq} = P_{eq} = 0 \tag{20.95}$$

and

$$P_{eq} = \frac{r_H}{a} \qquad H_{eq} = \frac{m_P}{b}. \tag{20.96}$$

The Jacobian matrix is

$$M_0 = \frac{\partial}{\partial C} F(C_0) = \begin{pmatrix} r_H - a P_{eq} & -a H_{eq} \\ b P_{eq} & b H_{eq} - m_P \end{pmatrix} \tag{20.97}$$

which gives for the trivial equilibrium

$$M_k = \begin{pmatrix} r_H - D_H k^2 & 0 \\ 0 & -m_P - D_P k^2 \end{pmatrix}. \tag{20.98}$$

One eigenvalue $\lambda_1 = -m_P - D_P k^2$ is negative whereas the second $\lambda_2 = r_H - D_H k^2$ is positive for $k^2 < r_H / D_H$. Hence this equilibrium is unstable against fluctuations with long wavelengths. For the second equilibrium we find:

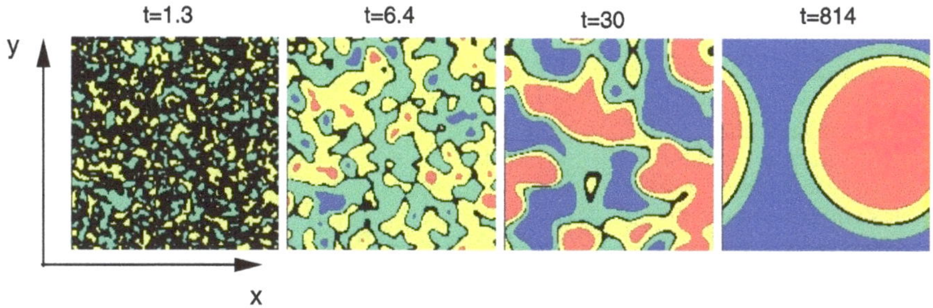

Fig. 20.12 (Lotka-Volterra model with diffusion) The time evolution is calculated for initial random fluctuations. Colors indicate the deviation of the predator concentration $P(x, y, t)$ from its average value (*blue*: $\Delta P < -0.1$, *green*: $-0.1 < \Delta P < -0.01$, *black*: $-0.01 < \Delta P < 0.01$, *yellow*: $0.01 < \Delta P < 0.1$, *red*: $\Delta P > 0.1$). Parameters as in Fig. 20.13

Fig. 20.13 (Dispersion of the diffusive Lotka-Volterra model) Real (*full curve*) and imaginary part (*broken line*) of the eigenvalue λ (20.100) are shown as a function of k. Parameters are $D_H = D_P = 1$, $m_P = r_H = a = b = 0.5$

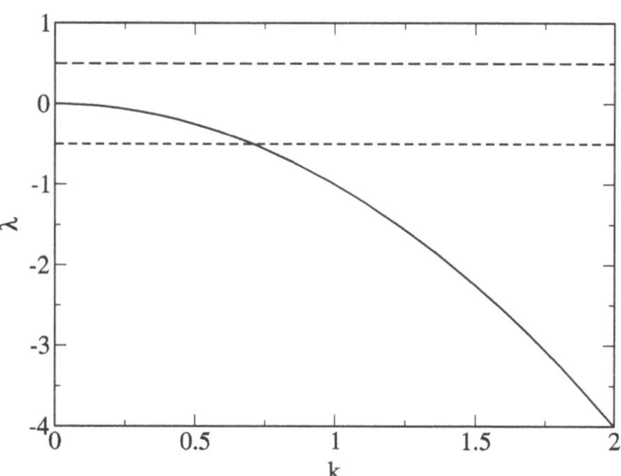

$$M_k = \begin{pmatrix} -D_H k^2 & -\frac{a m_P}{b} \\ \frac{b r_H}{a} & -D_P k^2 \end{pmatrix} \tag{20.99}$$

$$\mathrm{tr}(M_k) = -(D_H + D_P)k^2$$

$$\det(M_K) = m_P r_H + D_H D_P k^4 \tag{20.100}$$

$$\lambda = -\frac{D_H + D_P}{2}k^2 \pm \frac{1}{2}\sqrt{(D_H - D_P)^2 k^4 - 4 m_P r_H}.$$

For small k with $k^2 < 2\sqrt{m_P r_H}/|D_H - D_P|$ damped oscillations are expected whereas the system is stable against fluctuations with larger k (Figs. 20.12, 20.13, 20.14).

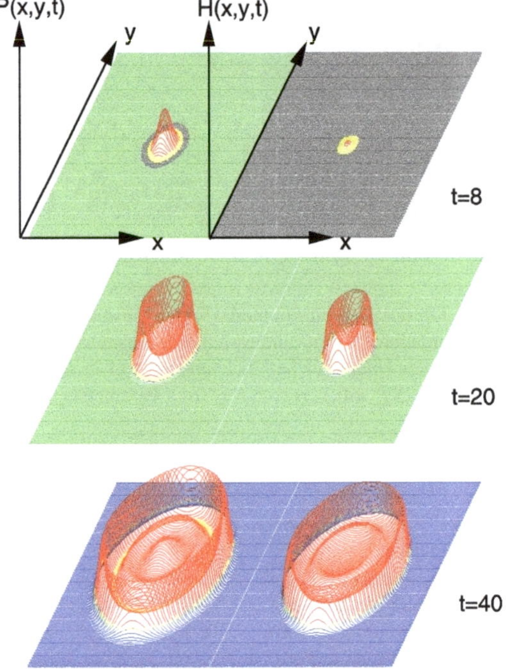

Fig. 20.14 (Traveling waves in the diffusive Lotka-Volterra model) Initially $P(x, y) = P_{eq}$ and $H(x, y)$ is peaked in the center. This leads to oscillations and a sharp wavefront moving away from the excitation. Color code and parameters as in Fig. 20.12

20.6 Problems

Problem 20.1 (Orbits of the iterated logistic map) This computer example draws orbits (Fig. 20.5) of the logistic map

$$x_{n+1} = r_0 \cdot x_n \cdot (1 - x_n). \tag{20.101}$$

You can select the initial value x_0 and the variable r.

Problem 20.2 (Bifurcation diagram of the logistic map) This computer example generates a bifurcation diagram of the logistic map (Fig. 20.6). You can select the range of r.

Problem 20.3 (Lotka-Volterra model) Equations (20.47) are solved with the improved Euler method (Fig. 20.8). The predictor step uses an explicit Euler step to calculate the values at $t + \Delta t/2$

$$H_{pr}\left(t + \frac{\Delta t}{2}\right) = H(t) + \left(r_H H(t) - a H(t) P(t)\right)\frac{\Delta t}{2} \tag{20.102}$$

$$P_{pr}\left(t + \frac{\Delta t}{2}\right) = P(t) + \left(b H(t) P(t) - m_p P(t)\right)\frac{\Delta t}{2} \tag{20.103}$$

and the corrector step advances time by Δt

$$H(t + \Delta t) = H(t) + \left(r_H H_{pr}\left(t + \frac{\Delta t}{2} \right) - a H_{pr}\left(t + \frac{\Delta t}{2} \right) P_{pr}\left(t + \frac{\Delta t}{2} \right) \right) \Delta t$$

$$\tag{20.104}$$

$$P(t + \Delta t) = P(t) + \left(b H_{pr}\left(t + \frac{\Delta t}{2} \right) P_{pr}\left(t + \frac{\Delta t}{2} \right) - m_p P_{pr}\left(t + \frac{\Delta t}{2} \right) \right) \Delta t.$$

$$\tag{20.105}$$

Problem 20.4 (Holling-Tanner model) The equations of the Holling-Tanner model (20.64), (20.65) are solved with the improved Euler method (see Fig. 20.11). The predictor step uses an explicit Euler step to calculate the values at $t + \Delta t/2$:

$$H_{pr}\left(t + \frac{\Delta t}{2} \right) = H(t) + f\big(H(t), P(t)\big) \frac{\Delta t}{2} \tag{20.106}$$

$$P_{pr}\left(t + \frac{\Delta t}{2} \right) = P(t) + g\big(H(t), P(t)\big) \frac{\Delta t}{2} \tag{20.107}$$

and the corrector step advances time by Δt:

$$H(t + \Delta t) = H(t) + f\left(H_{pr}\left(t + \frac{\Delta t}{2} \right), P_{pr}\left(t + \frac{\Delta t}{2} \right) \right) \Delta t \tag{20.108}$$

$$P(t + \Delta t) = P(t) + g\left(H_{pr}\left(t + \frac{\Delta t}{2} \right), P_{pr}\left(t + \frac{\Delta t}{2} \right) \right) \Delta t. \tag{20.109}$$

Problem 20.5 (Diffusive Lotka-Volterra model) The Lotka-Volterra model with diffusion (20.94) is solved in 2 dimensions with an implicit method (19.2.2) for the diffusive motion (Figs. 20.12, 20.14). The split operator approximation (19.3) is used to treat diffusion in x and y direction independently. The equations

$$\begin{pmatrix} H(t + \Delta t) \\ P(t + \Delta t) \end{pmatrix} = \begin{pmatrix} A^{-1} H(t) \\ A^{-1} P(t) \end{pmatrix} + \begin{pmatrix} A^{-1} f(H(t), P(t)) \Delta t \\ A^{-1} g(H(t), P(t)) \Delta t \end{pmatrix}$$

$$\approx \begin{pmatrix} A_x^{-1} A_y^{-1}[H(t) + f(H(t), P(t)) \Delta t] \\ A_x^{-1} A_y^{-1}[P(t) + g(H(t), P(t)) \Delta t] \end{pmatrix} \tag{20.110}$$

are equivalent to the following systems of linear equations with tridiagonal matrix (5.3):

$$A_y U = H(t) + f\big(H(t), P(t)\big) \Delta t \tag{20.111}$$

$$U = A_x H(t + \Delta t) \tag{20.112}$$

$$A_y V = P(t) + g\big(H(t), P(t)\big) \Delta t \tag{20.113}$$

$$V = A_x P(t + \Delta t). \tag{20.114}$$

Periodic boundary conditions are implemented with the method described in Sect. 5.4.

Chapter 21
Simple Quantum Systems

In this chapter we study simple quantum systems. A particle in a one-dimensional potential $V(x)$ is described by a wave packet which is a solution of the partial differential equation [232]

$$i\hbar \frac{\partial}{\partial t} \psi(x) = H\psi(x) = -\frac{\hbar^2}{2m} \frac{\partial^2}{\partial x^2} \psi(x) + V(x)\psi(x). \tag{21.1}$$

We discuss two approaches to discretize the second derivative. Finite differences are simple to use but their dispersion deviates largely from the exact relation, except high order differences are used. Pseudo-spectral methods evaluate the kinetic energy part in Fourier space and are much more accurate. The time evolution operator can be approximated by rational expressions like Cauchy's form which corresponds to the Crank-Nicholson method. These schemes are unitary but involve time consuming matrix inversions. Multistep differencing schemes have comparable accuracy but are explicit methods. Best known is second order differencing. Split operator methods approximate the time evolution operator by a product. In combination with finite differences for the kinetic energy this leads to the method of real-space product formula which can applied to wavefunctions with more than one component, for instance to study transitions between states. In a computer experiment we simulate a one-dimensional wave packet in a potential with one or two minima.

Few-state systems are described with a small set of basis states. Especially the quantum mechanical two-level system is often used as a simple model for the transition between an initial and a final state due to an external perturbation.[1] Its wavefunction has two components

$$|\psi\rangle = \begin{pmatrix} C_1 \\ C_2 \end{pmatrix} \tag{21.2}$$

which satisfy two coupled ordinary differential equations for the amplitudes $C_{1,2}$ of the two states

---

[1] For instance collisions or the electromagnetic radiation field.

P.O.J. Scherer, *Computational Physics*, Graduate Texts in Physics,
DOI 10.1007/978-3-319-00401-3_21,
© Springer International Publishing Switzerland 2013

$$i\hbar\frac{\mathrm{d}}{\mathrm{d}t}\begin{pmatrix} C_1 \\ C_2 \end{pmatrix} = \begin{pmatrix} H_{11} & H_{12} \\ H_{21} & H_{22} \end{pmatrix}\begin{pmatrix} C_1 \\ C_2 \end{pmatrix}. \tag{21.3}$$

In several computer experiments we study a two-state system in an oscillating field, a three-state system as a model for superexchange, the Landau-Zener model for curve-crossing and the ladder model for exponential decay. The density matrix formalism is used to describe a dissipative two-state system in analogy to the Bloch equations for nuclear magnetic resonance. In computer experiments we study the resonance line and the effects of saturation and power broadening. Finally we simulate the generation of a coherent superposition state or a spin flip by applying pulses of suitable duration. This is also discussed in connection with the manipulation of a qubit represented by a single spin.

21.1 Pure and Mixed Quantum States

Whereas pure states of a quantum system are described by a wavefunction, mixed states are described by a density matrix. Mixed states appear if the exact quantum state is unknown, for instance for a statistical ensemble of quantum states, a system with uncertain preparation history, or if the system is entangled with another system. A mixed state is different from a superposition state. For instance, the superposition

$$|\psi\rangle = C_0|\psi_0\rangle + C_1|\psi_1\rangle \tag{21.4}$$

of the two states $|\psi_0\rangle$ and $|\psi_1\rangle$ is a pure state, which can be described by the density operator

$$\begin{aligned} |\psi\rangle\langle\psi| = |C_0|^2|\psi_0\rangle\langle\psi_0| + |C_1|^2|\psi_1\rangle\langle\psi_1| \\ + C_0C_1^*|\psi_0\rangle\langle\psi_1| + C_0^*C_1|\psi_1\rangle\langle\psi_0| \end{aligned} \tag{21.5}$$

whereas the density operator

$$\rho = p_0|\psi_0\rangle\langle\psi_0| + p_1|\psi_1\rangle\langle\psi_1| \tag{21.6}$$

describes the mixed state of a system which is in the pure state $|\psi_0\rangle$ with probability p_0 and in the state $|\psi_1\rangle$ with probability $p_1 = 1 - p_0$. The expectation value of an operator A is in the first case

$$\begin{aligned} \langle A\rangle = \langle\psi|A|\psi\rangle = |C_0|^2\langle\psi_0|A|\psi_0\rangle + |C_1|^2\langle\psi_1 A|\psi_1\rangle \\ + C_0C_1^*\langle\psi_1|A|\psi_0\rangle + C_0^*C_1\langle\psi_0|A|\psi_1\rangle \end{aligned} \tag{21.7}$$

and in the second case

$$\langle A\rangle = p_0\langle\psi_0|A|\psi_0\rangle + p_1\langle\psi_1|A|\psi_1\rangle. \tag{21.8}$$

Both can be written in the form

$$\langle A\rangle = \mathrm{tr}(\rho A). \tag{21.9}$$

21.1.1 Wavefunctions

The time evolution of a quantum system is governed by the time dependent Schrödinger equation [230]

$$i\hbar \frac{\partial}{\partial t}|\psi\rangle = H|\psi\rangle \tag{21.10}$$

for the wavefunction ψ. The brackets indicate that $|\psi\rangle$ is a vector in an abstract Hilbert space [122]. Vectors can be added

$$|\psi\rangle = |\psi_1\rangle + |\psi_2\rangle = |\psi_1 + \psi_2\rangle \tag{21.11}$$

and can be multiplied with a complex number

$$|\psi\rangle = \lambda|\psi_1\rangle = |\lambda\psi_1\rangle. \tag{21.12}$$

Finally a complex valued scalar product of two vectors is defined[2]

$$C = \langle\psi_1|\psi_2\rangle \tag{21.13}$$

which has the properties

$$\langle\psi_1|\psi_2\rangle = \langle\psi_2|\psi_1\rangle^* \tag{21.14}$$

$$\langle\psi_1|\lambda\psi_2\rangle = \lambda\langle\psi_1|\psi_2\rangle = \left(\lambda^*\psi_1\big|\psi_2\right) \tag{21.15}$$

$$\langle\psi|\psi_1 + \psi_2\rangle = \langle\psi|\psi_1\rangle + \langle\psi|\psi_2\rangle \tag{21.16}$$

$$\langle\psi_1 + \psi_2|\psi\rangle = \langle\psi_1|\psi\rangle + \langle\psi_2|\psi\rangle. \tag{21.17}$$

21.1.2 Density Matrix for an Ensemble of Systems

Consider a thermal ensemble of systems. Their wave functions are expanded with respect to basis functions $|\psi_s\rangle$ as

$$|\psi\rangle = \sum_s C_s|\psi_s\rangle. \tag{21.18}$$

The ensemble average of an operator A is given by

$$\overline{\langle A\rangle} = \overline{\langle\psi A\psi\rangle} = \overline{\left(\sum_{s,s'} C_s^* \psi_s A C_{s'} \psi_{s'}\right)} \tag{21.19}$$

$$= \sum_{s,s'} \overline{C_s^* C_{s'}} A_{ss'} = \mathrm{tr}(\rho A) \tag{21.20}$$

---

[2]If, for instance the wavefunction depends on the coordinates of N particles, the scalar product is defined by $\langle\psi_n|\psi_{n'}\rangle = \int d^3r_1 \cdots d^3r_N \psi_n^*(r_1\cdots r_N)\psi_{n'}(r_1\cdots r_N)$.

with the density matrix

$$\rho_{s's} = \sum_{s,s'} \overline{C_s^* C_{s'}}. \tag{21.21}$$

The wave function of an N-state system is a linear combination

$$|\psi\rangle = C_1|\psi_1\rangle + C_2|\psi_2\rangle + \cdots + C_N|\psi_N\rangle. \tag{21.22}$$

The diagonal elements of the density matrix are the occupation probabilities

$$\rho_{11} = \overline{|C_1|^2} \qquad \rho_{22} = \overline{|C_2|^2} \cdots \qquad \rho_{NN} = \overline{|C_N|^2} \tag{21.23}$$

and the nondiagonal elements measure the correlation of two states[3]

$$\rho_{12} = \rho_{21}^* = \overline{C_2^* C_1}, \quad \dots. \tag{21.24}$$

21.1.3 Time Evolution of the Density Matrix

The expansion coefficients of

$$|\psi\rangle = \sum_s C_s|\psi_s\rangle \tag{21.25}$$

can be obtained from the scalar product

$$C_s = \langle\psi_s|\psi\rangle. \tag{21.26}$$

Hence we have

$$C_s^* C_{s'} = \langle\psi|\psi_s\rangle\langle\psi_{s'}|\psi\rangle = \langle\psi_{s'}|\psi\rangle\langle\psi|\psi_s\rangle \tag{21.27}$$

which can be considered to be the s', s matrix element of the projection operator $|\psi\rangle\langle\psi|$

$$C_s^* C_{s'} = \left(|\psi\rangle\langle\psi|\right)_{s's}. \tag{21.28}$$

The thermal average of $|\psi\rangle\langle\psi|$ is the statistical operator

$$\rho = \overline{|\psi\rangle\langle\psi|} \tag{21.29}$$

which is represented by the density matrix with respect to the basis functions $|\psi_s\rangle$

$$\rho_{s's} = \overline{|\psi\rangle\langle\psi|}_{s's} = \overline{C_s^* C_{s'}}. \tag{21.30}$$

From the Schrödinger equation

$$i\hbar|\dot\psi\rangle = H|\psi\rangle \tag{21.31}$$

we find

---

[3]They are often called the "coherence" of the two states.

Fig. 21.1 Potential well

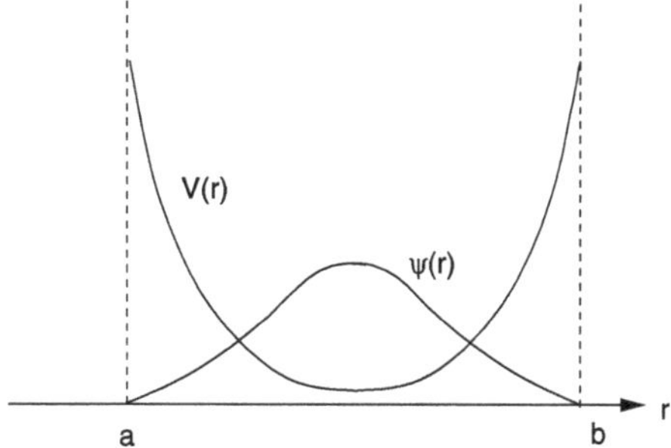

$$-i\hbar\langle\dot{\psi}| = \langle H\psi| = \langle\psi|H \tag{21.32}$$

and hence

$$i\hbar\dot{\rho} = i\hbar\left(\overline{|\dot{\psi}\rangle\langle\psi|} + \overline{|\psi\rangle\langle\dot{\psi}|}\right) = \overline{|H\psi\rangle\langle\psi|} - \overline{|\psi\rangle\langle H\psi|}. \tag{21.33}$$

Since the Hamiltonian H is identical for all members of the ensemble we end up with the Liouville-von Neumann equation

$$i\hbar\dot{\rho} = H\rho - \rho H = [H, \rho]. \tag{21.34}$$

With respect to a finite basis this becomes explicitly:

$$i\hbar\dot{\rho}_{ii} = \sum_{j} H_{ij}\rho_{ji} - \rho_{ij}H_{ji} = \sum_{j\neq i} H_{ij}\rho_{ji} - \rho_{ij}H_{ji} \tag{21.35}$$

$$i\hbar\dot{\rho}_{ik} = \sum_{j} H_{ij}\rho_{jk} - \rho_{ij}H_{jk}$$

$$= (H_{ii} - H_{kk})\rho_{ik} + H_{ik}(\rho_{kk} - \rho_{ii}) + \sum_{j\neq i,k}(H_{ij}\rho_{jk} - \rho_{ij}H_{jk}). \tag{21.36}$$

21.2 Wave Packet Motion in One Dimension

A quantum mechanical particle with mass m_p in a one-dimensional potential $V(x)$ is described by a complex valued wavefunction $\psi(x)$ (Fig. 21.1). We assume that the wavefunction is negligible outside an interval $[a, b]$. This is the case for a particle bound in a potential well i.e. a deep enough minimum of the potential or for a wave-packet with finite width far from the boundaries. Then the calculation can be restricted to the finite interval $[a, b]$ by applying the boundary condition

$$\psi(x) = 0 \quad \text{for } x \leq a \text{ or } x \geq b \tag{21.37}$$

or, if reflections at the boundary should be suppressed, transparent boundary conditions [9].

All observables (quantities which can be measured) of the particle are expectation values with respect to the wavefunction, for instance its average position is

$$\langle x \rangle = \langle \psi(x) x \psi(x) \rangle = \int_a^b dx \, \psi^*(x) x \psi(x). \tag{21.38}$$

The probability of finding the particle at the position x_0 is given by

$$P(x = x_0) = |\psi(x_0)|^2. \tag{21.39}$$

For time independent potential $V(x)$ the Schrödinger equation

$$i\hbar\dot{\psi} = H\psi = \left(-\frac{\hbar^2}{2m_p}\frac{\partial^2}{\partial x^2} + V(x)\right)\psi \tag{21.40}$$

can be formally solved by

$$\psi(t) = U(t, t_0)\psi(t_0) = \exp\left\{-\frac{i(t - t_0)}{\hbar}H\right\}\psi(t_0). \tag{21.41}$$

If the potential is time dependent, the more general formal solution is

$$\psi(t) = U(t, t_0)\psi(t_0) = \hat{T}_t \exp\left\{-\frac{i}{\hbar}\int_{t_0}^t H(\tau)\,d\tau\right\}\psi(t_0)$$

$$= \sum_{n=0}^{\infty} \frac{1}{n!}\left(\frac{-i}{\hbar}\right)^n \int_{t_0}^t dt_1 \int_{t_0}^t dt_2 \ldots \int_{t_0}^t dt_n \, \hat{T}_t\{H(t_1)H(t_2)\ldots H(t_n)\}$$

$$\tag{21.42}$$

where $\hat{T}_t$ denotes the time ordering operator. The simplest approach is to divide the time interval $0 \ldots t$ into a sequence of smaller steps

$$U(t, t_0) = U(t, t_{N-1}) \ldots U(t_2, t_1)U(t_1, t_0) \tag{21.43}$$

and to neglect the variation of the Hamiltonian during the small interval $\Delta t = t_{n+1} - t_n$ [158]

$$U(t_{n+1}, t_n) = \exp\left\{-\frac{i\Delta t}{\hbar}H(t_n)\right\}. \tag{21.44}$$

21.2.1 Discretization of the Kinetic Energy

The kinetic energy

$$T\psi(x) = -\frac{\hbar^2}{2m_p}\frac{\partial^2}{\partial x^2}\psi(x) \tag{21.45}$$

is a nonlocal operator in real space. It is most efficiently evaluated in Fourier space where it becomes diagonal

$$\mathcal{F}[T\psi](k) = \frac{\hbar^2 k^2}{2m_p}\mathcal{F}[\psi](k). \tag{21.46}$$

21.2.1.1 Pseudo-spectral Methods

The potential energy is diagonal in real space. Therefore, pseudo-spectral (Sect. 11.5.1) methods [93] use a Fast Fourier Transform algorithm (Sect. 7.3.2) to switch between real space and Fourier space. They calculate the action of the Hamiltonian on the wavefunction according to

$$H\psi(x) = V(x)\psi(x) + \mathcal{F}^{-1}\left[\frac{\hbar^2 k^2}{2m_p}\mathcal{F}[\psi](k)\right]. \tag{21.47}$$

21.2.1.2 Finite Difference Methods

In real space, the kinetic energy operator can be approximated by finite differences on a grid, like the simple 3-point expression (3.31)

$$-\frac{\hbar^2}{2m_p}\frac{\psi_{m+1}^n + \psi_{m-1}^n - 2\psi_m^n}{\Delta x^2} + O\left(\Delta x^2\right) \tag{21.48}$$

or higher order expressions (3.33)

$$-\frac{\hbar^2}{2m_p}\frac{-\psi_{m+2}^n + 16\psi_{m+1}^n - 30\psi_m^n + 16\psi_{m-1}^n - \psi_{m-2}^n}{12\Delta x^2} + O\left(\Delta x^4\right) \tag{21.49}$$

$$-\frac{\hbar^2}{2m_p}\frac{1}{\Delta x^2}\left(\frac{1}{90}\psi_{m+3}^n - \frac{3}{20}\psi_{m+2}^n + \frac{3}{2}\psi_{m+1}^n - \frac{49}{18}\psi_m^n\right.$$
$$\left. + \frac{3}{2}\psi_{m-1}^n - \frac{3}{20}\psi_{m-2}^n + \frac{1}{90}\psi_{m-3}^n\right) + O\left(\Delta x^6\right) \tag{21.50}$$

etc. [94]. However, finite differences inherently lead to deviations of the dispersion relation from (21.46). Inserting $\psi_m = e^{ikm\Delta x}$ we find

$$E(k) = \frac{\hbar^2}{2m_p}\frac{2(1 - \cos(k\Delta x))}{\Delta x^2} \tag{21.51}$$

for the 3-point expression (21.48),

$$E(k) = \frac{\hbar^2}{2m_p}\frac{15 - 16\cos(k\Delta x) + \cos(2k\Delta x)}{6\Delta x^2} \tag{21.52}$$

for the 5-point expression (21.49) and

$$\frac{\hbar^2}{2m_p}\frac{1}{\Delta x^2}\left(\frac{49}{18} - 3\cos(k\Delta x) + \frac{3}{10}\cos(2k\Delta x) - \frac{1}{45}\cos(3k\Delta x)\right) \tag{21.53}$$

for the 7-point expression (21.50). Even the 7-point expression shows large deviations for k-values approaching $k_{max} = \pi/\Delta x$ (Fig. 21.2). However, it has been shown that not very high orders are necessary to achieve the numerical accuracy of the pseudo-spectral Fourier method [113] and that finite difference methods may be even more efficient in certain applications [114].

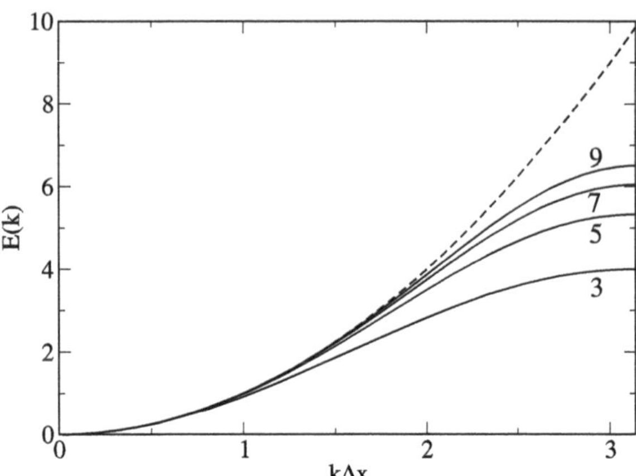

Fig. 21.2 (Dispersion of finite difference expressions) The dispersion relation of finite difference expressions of increasing order ((21.48), (21.49), (21.50) and the symmetric 9-point approximation [94]) are compared to the exact dispersion (21.46) of a free particle (*dashed curve*)

21.2.2 Time Evolution

A number of methods have been proposed [7, 51, 158, 278] to approximate the short time propagator (21.44). Unitarity is a desirable property since it guaranties stability and norm conservation even for large time steps. However, depending on the application, small deviations from unitarity may be acceptable in return for higher efficiency. The Crank-Nicolson (CN) method [104, 172, 173] is one of the first methods which have been applied to the time dependent Schrödinger equation. It is a unitary but implicit method and needs the inversion of a matrix which can become cumbersome in two or more dimensions or if high precision is required. Multistep methods [131, 132], especially second order [10] differencing (SOD) are explicit but only conditionally stable and put limits to the time interval Δt. Split operator methods (SPO) approximate the propagator by a unitary product of operators [14, 67, 68]. They are explicit and easy to implement. The real-space split-operator method has been applied to more complex problems like a molecule in a laser field [61]. Polynomial approximations, especially the Chebishev expansion [53, 248], have very high accuracy and allow for large time steps, if the Hamiltonian is time independent. However, they do not provide intermediate results and need many applications of the Hamiltonian. The short time iterative Lanczos (SIL) method [57, 153, 197] is very useful also for time dependent Hamiltonians. Even more sophisticated methods using finite elements and the discrete variable representation are presented for instance in [4, 226]. In the following we discuss three methods (CN, SOD, SPO) which are easy to implement and well suited to solve the time dependent Schrödinger equation for a mass point moving in a one-dimensional potential.

21.2.2.1 Rational Approximation

Taking the first terms of the Taylor expansion

$$U(t_{n+1}, t_n) = \exp\left\{-\frac{i\Delta t}{\hbar}H\right\} = 1 - \frac{i\Delta t}{\hbar}H + \cdots \qquad (21.54)$$

corresponds to a simple explicit Euler step

$$\psi(t_{n+1}) = \left(1 - \frac{i\Delta t}{\hbar}H\right)\psi(t_n). \qquad (21.55)$$

From the real eigenvalues E of the Hamiltonian we find the eigenvalues of the explicit method

$$\lambda = 1 - \frac{i\Delta t}{\hbar}E \qquad (21.56)$$

which all have absolute values

$$|\lambda| = \sqrt{1 + \frac{\Delta t^2 E^2}{\hbar^2}} > 1. \qquad (21.57)$$

Hence the explicit method is not stable.

Expansion of the inverse time evolution operator

$$U(t_n, t_{n+1}) = U(t_{n+1}, t_n)^{-1} = \exp\left\{+\frac{i\Delta t}{\hbar}H\right\} = 1 + \frac{i\Delta t}{\hbar}H + \cdots$$

leads to the implicit method

$$\psi(t_{n+1}) = \psi(t_n) - \frac{i\Delta t}{\hbar}H\psi(t_{n+1}) \qquad (21.58)$$

which can be rearranged as

$$\psi(t_{n+1}) = \left(1 + \frac{i\Delta t}{\hbar}H\right)^{-1}\psi(t_n). \qquad (21.59)$$

Now all eigenvalues have absolute values < 1. This method is stable but the norm of the wave function is not conserved. Combination of implicit and explicit method gives a method [172, 173] similar to the Crank-Nicolson method for the diffusion equation (Sect. 19.2.3)

$$\psi(t_{n+1}) - \psi(t_n) = -\frac{i\Delta t}{\hbar}H\left(\frac{\psi(t_{n+1})}{2} + \frac{\psi(t_n)}{2}\right). \qquad (21.60)$$

This equation can be solved for the new value of the wavefunction

$$\psi(t_{n+1}) = \left(1 + i\frac{\Delta t}{2\hbar}H\right)^{-1}\left(1 - i\frac{\Delta t}{2\hbar}H\right)\psi(t_n) \qquad (21.61)$$

which corresponds to a rational approximation[4] of the time evolution operator (Cayley's form)

---

[4]The Padé approximation (Sect. 2.4.1) of order [1, 1].

$$U(t_{n+1}, t_n) \approx \frac{1 - i\frac{\Delta t}{2\hbar} H}{1 + i\frac{\Delta t}{2\hbar} H}. \tag{21.62}$$

The eigenvalues of (21.62) all have an absolute value of

$$|\lambda| = \left| \left(1 + i\frac{E\Delta t}{2\hbar}\right)^{-1} \left(1 - i\frac{E\Delta t}{2\hbar}\right) \right| = \frac{\sqrt{1 + \frac{E^2 \Delta t^2}{4\hbar^2}}}{\sqrt{1 + \frac{E^2 \Delta t^2}{4\hbar^2}}} = 1. \tag{21.63}$$

It is obviously a unitary operator and conserves the norm of the wavefunction since

$$\left(\frac{1 - i\frac{\Delta t}{2\hbar} H}{1 + i\frac{\Delta t}{2\hbar} H}\right)^{\dagger} \left(\frac{1 - i\frac{\Delta t}{2\hbar} H}{1 + i\frac{\Delta t}{2\hbar} H}\right) = \left(\frac{1 + i\frac{\Delta t}{2\hbar} H}{1 - i\frac{\Delta t}{2\hbar} H}\right) \left(\frac{1 - i\frac{\Delta t}{2\hbar} H}{1 + i\frac{\Delta t}{2\hbar} H}\right) = 1 \tag{21.64}$$

as H is Hermitian $H^{\dagger} = H$ and $(1 + i\frac{\Delta t}{2\hbar} H)$ and $(1 - i\frac{\Delta t}{2\hbar} H)$ are commuting operators. From the Taylor series we find the error order

$$\left(1 + i\frac{\Delta t}{2\hbar} H\right)^{-1} \left(1 - i\frac{\Delta t}{2\hbar} H\right)$$

$$= \left(1 - i\frac{\Delta t}{2\hbar} H - \frac{\Delta t^2}{4\hbar^2} H^2 + \cdots\right) \left(1 - i\frac{\Delta t}{2\hbar} H\right)$$

$$= 1 - \frac{i\Delta t}{\hbar} H - \frac{\Delta t^2}{2\hbar^2} H^2 + \cdots = \exp\left(-\frac{i\Delta t}{\hbar} H\right) + O(\Delta t^3). \tag{21.65}$$

For practical application we rewrite [149]

$$\left(1 + i\frac{\Delta t}{2\hbar} H\right)^{-1} \left(1 - i\frac{\Delta t}{2\hbar} H\right) = \left(1 + i\frac{\Delta t}{2\hbar} H\right)^{-1} \left(-1 - i\frac{\Delta t}{2\hbar} H + 2\right)$$

$$= -1 + 2\left(1 + i\frac{\Delta t}{2\hbar} H\right)^{-1} \tag{21.66}$$

hence

$$\psi(t_{n+1}) = 2\left(1 + i\frac{\Delta t}{2\hbar} H\right)^{-1} \psi(t_n) - \psi(t_n) = 2\chi - \psi(t_n). \tag{21.67}$$

$\psi(t_{n+1})$ is obtained in two steps. First we have to solve

$$\left(1 + i\frac{\Delta t}{2\hbar} H\right)\chi = \psi(t_n). \tag{21.68}$$

Then $\psi(t_{n+1})$ is given by

$$\psi(t_{n+1}) = 2\chi - \psi(t_n). \tag{21.69}$$

We use the finite difference method (Sect. 11.2) on the grid

$$x_m = m\Delta x \quad m = 0 \cdots M \quad \psi_m^n = \psi(t_n, x_m) \tag{21.70}$$

and approximate the second derivative by

$$\frac{\partial^2}{\partial x^2}\psi(t_n, x_m) = \frac{\psi^n_{m+1} + \psi^n_{m-1} - 2\psi^n_m}{\Delta x^2} + O(\Delta x^2). \tag{21.71}$$

Equation (21.68) then becomes a system of linear equations

$$A\begin{bmatrix} \chi_0 \\ \chi_1 \\ \vdots \\ \chi_M \end{bmatrix} = \begin{bmatrix} \psi^n_0 \\ \psi^n_1 \\ \vdots \\ \psi^n_M \end{bmatrix} \tag{21.72}$$

with a tridiagonal matrix

$$A = 1 - i\frac{\hbar\Delta t}{4m_p\Delta x^2}\begin{pmatrix} 2 & -1 & & \\ -1 & 2 & \ddots & \\ & \ddots & \ddots & -1 \\ & & -1 & 2 \end{pmatrix} + i\frac{\Delta t}{2\hbar}\begin{pmatrix} V_0 & & & \\ & V_1 & & \\ & & \ddots & \\ & & & V_M \end{pmatrix}. \tag{21.73}$$

The second step (21.69) becomes

$$\begin{bmatrix} \psi^{n+1}_0 \\ \psi^{n+1}_1 \\ \vdots \\ \psi^{n+1}_M \end{bmatrix} = 2\begin{bmatrix} \chi_0 \\ \chi_1 \\ \vdots \\ \chi_M \end{bmatrix} - \begin{bmatrix} \psi^n_0 \\ \psi^n_1 \\ \vdots \\ \psi^n_M \end{bmatrix}. \tag{21.74}$$

Inserting a plane wave

$$\psi = e^{i(kx - \omega t)} \tag{21.75}$$

we obtain the dispersion relation (Fig. 21.3)

$$\frac{2}{\Delta t}\tan(\omega\Delta t/2) = \frac{\hbar}{2m_p}\left(\frac{2}{\Delta x}\sin\frac{k\Delta x}{2}\right)^2 \tag{21.76}$$

which we rewrite as

$$\omega\Delta t = 2\arctan\left[\frac{2\alpha}{\pi^2}\sin^2\frac{k\Delta x}{\pi}\frac{\pi}{2}\right] \tag{21.77}$$

with the dimensionless parameter

$$\alpha = \frac{\pi^2\hbar\Delta t}{2m_p\Delta x^2}. \tag{21.78}$$

For time independent potentials the accuracy of this method can be improved systematically [261] by using higher order finite differences for the spatial derivative (Sect. 21.2.1.2) and a higher order Padé approximation (Sect. 2.4.1) of order $[M, M]$ for the exponential function

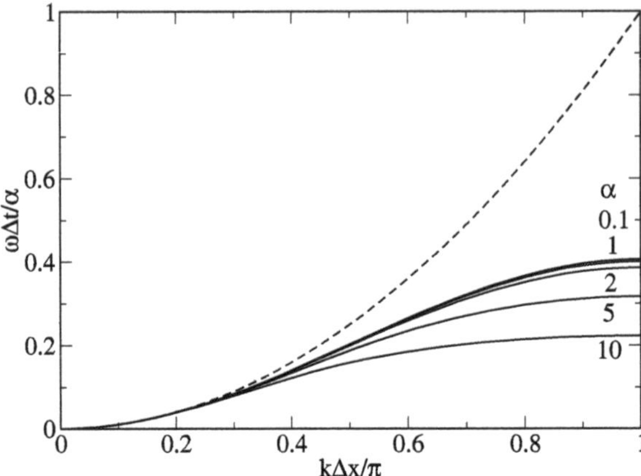

Fig. 21.3 (Dispersion of the Crank-Nicolson method) The dispersion relation of the Crank-Nicolson method (21.95) deviates largely from the exact dispersion (21.98), even for small values of the stability parameter α. The scaled frequency $\omega \Delta t / \alpha$ is shown as a function of $k \Delta x / \pi$ for $\alpha = 0.1, 1, 2, 5, 10$ (*solid curves*) and compared with the exact relation of a free particle $\omega \Delta t / \alpha = (k \Delta x / \pi)^2$ (*dashed curve*)

$$e^z = \prod_{s=1}^{M} \frac{1 - z/z_s^{(M)}}{1 + z/z_s^{(M)*}} + O\left(z^{2M+1}\right) \tag{21.79}$$

to approximate the time evolution operator

$$\exp\left(-\frac{i \Delta t}{\hbar} H\right) = \prod_{s=1}^{M} \frac{1 - (i \Delta t \, H/\hbar)/z_s^{(M)}}{1 + (i \Delta t \, H/\hbar)/z_s^{*(M)}} + O\left((\Delta t)^{2M+1}\right). \tag{21.80}$$

However, the matrix inversion can become very time consuming in two or more dimensions.

21.2.2.2 Second Order Differencing

Explicit methods avoid the matrix inversion. The method of second order differencing [10] takes the difference of forward and backward step

$$\psi(t_{n-1}) = U(t_{n-1}, t_n)\psi(t_n) \tag{21.81}$$

$$\psi(t_{n+1}) = U(t_{n+1}, t_n)\psi(t_n) \tag{21.82}$$

to obtain the explicit two-step algorithm

$$\psi(t_{n+1}) = \psi(t_{n-1}) + \left[U(t_{n+1}, t_n) - U^{-1}(t_n, t_{n-1})\right]\psi(t_n). \tag{21.83}$$

The first terms of the Taylor series give the approximation

$$\psi(t_{n+1}) = \psi(t_{n-1}) - 2\frac{i \Delta t}{\hbar} H \psi(t_n) + O\left((\Delta t)^3\right) \tag{21.84}$$

which can also be obtained from the second order approximation of the time derivative [150]

$$H\psi = i\hbar \frac{\partial}{\partial t}\psi = \frac{\psi(t + \Delta t) - \psi(t - \Delta t)}{2\Delta t}. \tag{21.85}$$

This two-step algorithm can be formulated as a discrete mapping

$$\begin{pmatrix} \psi(t_{n+1}) \\ \psi(t_n) \end{pmatrix} = \begin{pmatrix} -2\frac{i\Delta t}{\hbar}H & 1 \\ 1 & 0 \end{pmatrix} \begin{pmatrix} \psi(t_n) \\ \psi(t_{n-1}) \end{pmatrix} \tag{21.86}$$

with eigenvalues

$$\lambda = -\frac{iE_s\Delta t}{\hbar} \pm \sqrt{1 - \frac{E_s^2\Delta t^2}{\hbar^2}}. \tag{21.87}$$

For sufficiently small time step [158]

$$\Delta t < \frac{\hbar}{\max|E_s|} \tag{21.88}$$

the square root is real,

$$|\lambda|^2 = \frac{E_s^2\Delta t^2}{\hbar^2} + \left(1 - \frac{E_s^2\Delta t^2}{\hbar^2}\right) = 1 \tag{21.89}$$

and the method is conditionally stable and has the same error order as the Crank-Nicolson method (Sect. 21.2.2.1). Its big advantage is that it is an explicit method and does not involve matrix inversions. Generalization to higher order multistep differencing schemes is straightforward [131]. The method conserves [150] the quantities $\Re\langle\psi(t + \Delta t)|\psi(t)\rangle$ and $\Re\langle\psi(t + \Delta t)|H|\psi(t)\rangle$ but is not strictly unitary [10]. Consider a pair of wavefunctions at times t_0 and t_1 which obey the exact time evolution

$$\psi(t_1) = \exp\left\{-\frac{i\Delta t}{\hbar}H\right\}\psi(t_0) \tag{21.90}$$

and apply (21.84) to obtain

$$\psi(t_2) = \left[1 - 2\frac{i\Delta t}{\hbar}H\exp\left\{-\frac{i\Delta t}{\hbar}H\right\}\right]\psi(t_0) \tag{21.91}$$

which can be written as

$$\psi(t_2) = \mathcal{L}\,\psi(t_0) \tag{21.92}$$

where the time evolution operator L obeys

$$\mathcal{L}^\dagger\mathcal{L} = \left[1 + 2\frac{i\Delta t}{\hbar}H\exp\left\{+\frac{i\Delta t}{\hbar}H\right\}\right]\left[1 - 2\frac{i\Delta t}{\hbar}H\exp\left\{-\frac{i\Delta t}{\hbar}H\right\}\right]$$

$$= 1 - 4\frac{\Delta t}{\hbar}H\sin\left\{\frac{\Delta t}{\hbar}H\right\} + 4\left(\frac{\Delta t}{\hbar}H\right)^2.$$

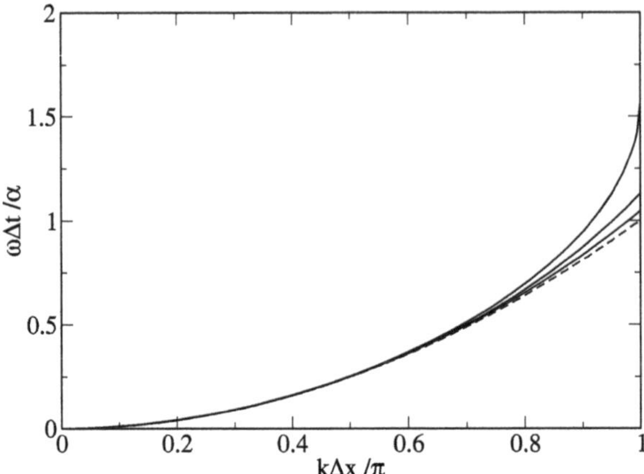

Fig. 21.4 (Dispersion of the Fourier method) The dispersion relation of the SOD-Fourier method (21.95) deviates from the exact dispersion (21.98) only for very high k-values and approaches it for small values of the stability parameter α. The scaled frequency $\omega \Delta t / \alpha$ is shown as a function of $k \Delta x / \pi$ for $\alpha = 0.5, 0.75, 1$ (*solid curves*) and compared with the exact relation of a free particle $\omega \Delta t / \alpha = (k \Delta x / \pi)^2$ (*dashed curve*)

Expanding the sine function we find the deviation from unitarity [10]

$$\mathcal{L}^\dagger \mathcal{L} - 1 = \frac{2}{3}\left(\frac{\Delta t}{\hbar} H\right)^4 + \cdots = O\big((\Delta t)^4\big) \tag{21.93}$$

which is of higher order than the error of the algorithm. Furthermore errors do not accumulate due to the stability of the algorithm (21.89). This also holds for deviations of the starting values from the condition (21.90).

The algorithm (21.84) can be combined with the finite differences method (Sect. 21.2.1.2)

$$\psi_m^{n+1} = \psi_m^{n-1} - 2\frac{i\Delta t}{\hbar}\left[V_m \psi_m^n - \frac{\hbar^2}{2m_p \Delta x^2}\big(\psi_{m+1}^n + \psi_{m-1}^n - 2\psi_m^n\big)\right] \tag{21.94}$$

or with the pseudo-spectral Fourier method [150]. This combination needs two Fourier transformations for each step but it avoids the distortion of the dispersion relation inherent to the finite difference method. Inserting the plane wave (21.75) into (21.84) we find the dispersion relation (Fig. 21.4) for a free particle ($V = 0$):

$$\omega = \frac{1}{\Delta t} \arcsin\left(\frac{\hbar \Delta t k^2}{2m_p}\right) = \frac{1}{\Delta t} \arcsin\left(\alpha \left(\frac{k \Delta x}{\pi}\right)^2\right). \tag{21.95}$$

For a maximum k-value

$$k_{\max} = \frac{\pi}{\Delta x} \tag{21.96}$$

the stability condition (21.88) becomes

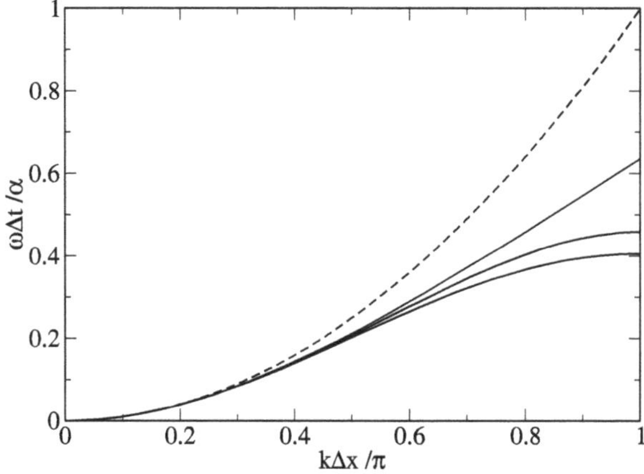

Fig. 21.5 (Dispersion of the finite difference method) The dispersion relation of the SOD-FD method (21.99) deviates largely from the exact dispersion (21.98), even for small values of the stability parameter α. The scaled frequency $\omega \Delta t / \alpha$ is shown as a function of $k \Delta x / \pi$ for $\alpha = \pi^2/4 \approx 2.467$, 1.85, 1.23, 0.2 (*solid curves*) and compared with the exact relation of a free particle $\omega \Delta t / \alpha = (k \Delta x / \pi)^2$ (*dashed curve*)

$$1 \geq \frac{\Delta t}{\hbar} \frac{\hbar^2 k_{\max}^2}{2m_p} = \alpha. \tag{21.97}$$

For small k the dispersion approximates the exact behavior

$$\omega = \frac{\hbar k^2}{2m_P}. \tag{21.98}$$

The finite difference method (21.94), on the other hand, has the dispersion relation (Fig. 21.5)

$$\omega = \frac{1}{\Delta t} \arcsin\left(\frac{4\alpha}{\pi^2} \sin^2\left(\frac{k\Delta x}{2} \right) \right) \tag{21.99}$$

and the stability limit

$$1 = \frac{\Delta t}{\hbar} E_{\max} = \frac{2\hbar \Delta t}{m_p \Delta x^2} = \frac{4\alpha}{\pi^2}. \tag{21.100}$$

The deviation from (21.98) is significant for $k \Delta x / \pi > 0.2$ even for small values of α [150].

21.2.2.3 Split-Operator Methods

The split-operator method approximates the exponential short time evolution operator as a product of exponential operators which are easier tractable. Starting from the Zassenhaus formula [164]

$$e^{\lambda(A+B)} = e^{\lambda A} e^{\lambda B} e^{\lambda^2 C_2} e^{\lambda^3 C_3} \ldots \tag{21.101}$$

$$C_2 = \frac{1}{2}[B, A] \qquad C_3 = \frac{1}{6}[C_2, A + 2B] \quad \ldots \tag{21.102}$$

approximants of increasing order can be systematically constructed [68, 246]

$$e^{\lambda(A+B)} = e^{\lambda A} e^{\lambda B} + O\left(\lambda^2\right) = e^{\lambda A} e^{\lambda B} e^{\lambda^2 C_2} + O\left(\lambda^3\right) \quad \ldots. \tag{21.103}$$

Since these approximants do not conserve time reversibility, often the symmetric expressions

$$e^{\lambda(A+B)} = e^{\lambda A/2} e^{\lambda B} e^{\lambda A/2} + O\left(\lambda^3\right)$$
$$= e^{\lambda A/2} e^{\lambda B/2} e^{\lambda^2 C_3/4} e^{\lambda B/2} e^{\lambda A/2} + O\left(\lambda^5\right) \quad \ldots \tag{21.104}$$

are preferred.

Split-Operator-Fourier Method Dividing the Hamiltonian into its kinetic and potential parts

$$H = T + V = -\frac{\hbar^2}{2m_P} \frac{\partial^2}{\partial x^2} + V(x) \tag{21.105}$$

the time evolution operator can be approximated by the time-symmetric expression

$$U(\Delta t) = e^{-\frac{i\Delta t}{2\hbar} T} e^{-\frac{i\Delta t}{\hbar} V} e^{-\frac{i\Delta t}{2\hbar} T} + O\left((\Delta t)^3\right) \tag{21.106}$$

where the exponential of the kinetic energy can be easily applied in Fourier space [80, 150]. Combining several steps (21.106) to integrate over a longer time interval, consecutive operators can be combined to simplify the algorithm

$$U(N\Delta t) = U^N(\Delta t) = e^{-\frac{i\Delta t}{2\hbar} T} \left(e^{-\frac{i\Delta t}{\hbar} V} e^{-\frac{i\Delta t}{\hbar} T}\right)^{N-1} e^{-\frac{i\Delta t}{\hbar} V} e^{-\frac{i\Delta t}{2\hbar} T}. \tag{21.107}$$

Real-Space Product Formulae Using the discretization (21.48) on a regular grid the time evolution operator becomes the exponential of a matrix

$$
\begin{aligned}
&U(\Delta t) \\
&= \exp\left\{ -i\Delta t \begin{pmatrix} \frac{V_0}{\hbar} + \frac{\hbar}{m_P \Delta x^2} & -\frac{\hbar}{2m_P \Delta x^2} & & \\ -\frac{\hbar}{2m_P \Delta x^2} & \frac{V_1}{\hbar} + \frac{\hbar}{m_P \Delta x^2} & -\frac{\hbar}{2m_P \Delta x^2} & \\ & & \ddots & \\ & & -\frac{\hbar}{2m_P \Delta x^2} & \frac{V_M}{\hbar} + \frac{\hbar}{m_P \Delta x^2} \end{pmatrix} \right\} \\
&= \exp\left\{ -i\Delta t \begin{pmatrix} \gamma_0 + 2\beta & -\beta & & \\ \beta & \gamma_1 + 2\beta & -\beta & \\ & & \ddots & \\ & & -\beta & \gamma_M + 2\beta \end{pmatrix} \right\}
\end{aligned} \tag{21.108}
$$

with the abbreviations

$$\gamma_m = \frac{1}{\hbar} V_m \quad \beta = \frac{\hbar}{2 m_P \Delta x^2}. \tag{21.109}$$

The matrix can be decomposed into the sum of two overlapping tridiagonal block matrices [61, 67][5]

$$H_o = \begin{pmatrix} \gamma_0 + 2\beta & -\beta & & \\ -\beta & \frac{1}{2}\gamma_1 + \beta & & \\ & & \frac{1}{2}\gamma_2 + \beta & -\beta \\ & & -\beta & \ddots \end{pmatrix} = \begin{pmatrix} A_1 & & & \\ & A_3 & & \\ & & \ddots & \\ & & & A_{M-1} \end{pmatrix} \tag{21.110}$$

$$H_e = \begin{pmatrix} 0 & 0 & & \\ 0 & \frac{1}{2}\gamma_1 + \beta & -\beta & \\ & -\beta & \frac{1}{2}\gamma_2 + \beta & 0 \\ & & 0 & \ddots \end{pmatrix} = \begin{pmatrix} 0 & & & \\ & A_2 & & \\ & & \ddots & \\ & & A_{M-2} & \\ & & & 0 \end{pmatrix} \tag{21.111}$$

The block structure simplifies the calculation of $e^{-i\Delta t H_o}$ and $e^{-i\Delta t H_e}$ tremendously since effectively only the exponential functions of 2×2 matrices

$$B_m(\tau) = e^{-i\tau A_m} \tag{21.112}$$

have to be calculated and the approximation to the time evolution operator

$$U(\Delta t) = e^{-i\Delta t H_o/2} e^{-i\Delta t H_e} e^{-i\Delta t H_o/2}$$

$$= \begin{pmatrix} B_1(\frac{\Delta t}{2}) & & & \\ & B_3(\frac{\Delta t}{2}) & & \\ & & \ddots & \end{pmatrix} \begin{pmatrix} 1 & & & \\ & B_2(\Delta t) & & \\ & & \ddots & \end{pmatrix}$$

$$\times \begin{pmatrix} B_1(\frac{\Delta t}{2}) & & & \\ & B_3(\frac{\Delta t}{2}) & & \\ & & \ddots & \end{pmatrix} \tag{21.113}$$

can be applied in real space without any Fourier transformation. To evaluate (21.112) the real symmetric matrix A_m is diagonalized by an orthogonal transformation (Sect. 9.2)

$$A = R^{-1} \tilde{A} R = R^{-1} \begin{pmatrix} \lambda_1 & 0 \\ 0 & \lambda_2 \end{pmatrix} R \tag{21.114}$$

and the exponential calculated from

$$e^{-i\tau A} = 1 - i\tau R^{-1} \tilde{A} R + \frac{(-i\tau)^2}{2} R^{-1} \tilde{A} R R^{-1} \tilde{A} R + \cdots$$

---

[5]For simplicity only the case of even M is considered.

$$= R^{-1}\left[1 - \mathrm{i}\tau\tilde{A} + \frac{(-\mathrm{i}\tau)^2}{2}\tilde{A}R + \cdots\right]R$$

$$= R^{-1}\mathrm{e}^{-\mathrm{i}\tau\tilde{A}}R = R^{-1}\begin{pmatrix} \mathrm{e}^{-\mathrm{i}\tau\lambda_1} & \\ & \mathrm{e}^{-\mathrm{i}\tau\lambda_2} \end{pmatrix}R. \tag{21.115}$$

21.2.3 Example: Free Wave Packet Motion

We simulate the free motion ($V = 0$) of a Gaussian wave packet along the x-axis (see Problem 21.1). To simplify the numerical calculation we set $\hbar = 1$ and $m_p = 1$ and solve the time dependent Schrödinger equation

$$\mathrm{i}\frac{\partial}{\partial t}\psi = -\frac{1}{2}\frac{\partial^2}{\partial x^2}\psi \tag{21.116}$$

for initial values given by a Gaussian wave packet with constant momentum

$$\psi_0(x) = \left(\frac{2}{a\pi}\right)^{1/4}\mathrm{e}^{\mathrm{i}k_0 x}\mathrm{e}^{-x^2/a}. \tag{21.117}$$

The exact solution can be easily found. Fourier transformation of (21.117) gives

$$\hat{\psi}_k(t = 0) = \frac{1}{\sqrt{2\pi}}\int_{-\infty}^{\infty} dx\,\mathrm{e}^{-\mathrm{i}kx}\psi_0(x)$$

$$= \left(\frac{a}{2\pi}\right)^{1/4}\exp\left\{-\frac{a}{4}(k - k_0)^2\right\}. \tag{21.118}$$

Time evolution in k-space is rather simple

$$\mathrm{i}\frac{\partial}{\partial t}\hat{\psi}_k = \frac{k^2}{2}\hat{\psi}_k \tag{21.119}$$

hence

$$\hat{\psi}_k(t) = \mathrm{e}^{-\mathrm{i}k^2 t/2}\hat{\psi}_k(t = 0) \tag{21.120}$$

and Fourier back transformation gives the solution of the time dependent Schrödinger equation in real space

$$\psi(t, x) = \frac{1}{\sqrt{2\pi}}\int_{-\infty}^{\infty} dx\,\mathrm{e}^{\mathrm{i}kx}\hat{\psi}_k(t)$$

$$= \left(\frac{2a}{\pi}\right)^{1/4}\frac{1}{\sqrt{a + 2\mathrm{i}t}}\exp\left\{-\frac{(x - \mathrm{i}\frac{ak_0}{2})^2 + \frac{ak_0^2}{4}(a + 2\mathrm{i}t)}{a + 2\mathrm{i}t}\right\}. \tag{21.121}$$

Finally, the probability density is given by a Gaussian

$$|\psi(t, x)|^2 = \sqrt{\frac{2a}{\pi}}\frac{1}{\sqrt{a^2 + 4t^2}}\exp\left\{-\frac{2a}{a^2 + 4t^2}(x - k_0 t)^2\right\} \tag{21.122}$$

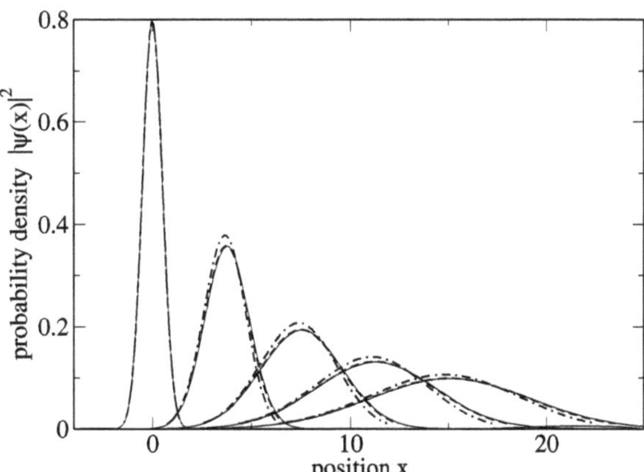

Fig. 21.6 (Conservation of norm and energy) The free motion of a Gaussian wave packet is simulated with the Crank-Nicolson method (CN) the second order differences method (SOD) with 3 point (21.48) 5 point (21.49) and 7-point (21.50) differences and with the real-space split-operator method (SPO). $\Delta t = 10^{-3}$, $\Delta x = 0.1$, $a = 1$, $k_0 = 3.77$

Fig. 21.7 (Free wave-packet motion) The free motion of a Gaussian wave packet is simulated. The probability density is shown for the initial Gaussian wave packet and at later times $t = 1, 2, 3, 4$. Results from the second order differences method with 3 point differences ((21.48), *dash-dotted*) and 5 point differences ((21.49), *dashed*) are compared with the exact solution ((21.122), *thin solid line*). $\Delta t = 10^{-3}$, $\Delta x = 0.1$, $a = 1$, $k_0 = 3.77$

which moves with constant velocity k_0 and kinetic energy

$$\int_{-\infty}^{\infty} dx\, \psi^*(x,t)\left(-\frac{\hbar^2}{2}\frac{\partial^2}{\partial x^2}\right)\psi(x,t) = \frac{1}{2}\left(k_0^2 + \frac{1}{a}\right). \tag{21.123}$$

Numerical results are shown in Figs. 21.6, 21.7 and Table 21.1.

21.3 Few-State Systems

In the following we discuss simple models which reduce the wavefunction to the superposition of a few important states, for instance an initial and a final state which are coupled by a resonant interaction. We approximate the solution of the time dependent Schrödinger equation as a linear combination

$$\left|\psi(t)\right\rangle \approx \sum_{j=1}^{M} C_j(t)\left|\phi_j\right\rangle \tag{21.124}$$

Table 21.1 (Accuracy of finite differences methods) The relative error of the kinetic energy (21.123) is shown as calculated with different finite difference methods

Method	E_{kin}	$\frac{E_{kin}-E_{kin}^{exact}}{E_{kin}^{exact}}$
Crank-Nicolson (CN) with 3 point differences	7.48608	-1.6×10^{-2}
Second Order Differences with 3 point differences (SOD3)	7.48646	-1.6×10^{-2}
Second Order Differences with 5 point differences (SOD5)	7.60296	-4.6×10^{-4}
Second Order Differences with 7 point differences (SOD7)	7.60638	-0.9×10^{-5}
Split-Operator method (SOP) with 3 point differences	7.48610	-1.6×10^{-2}
exact	7.60645	

of certain basis states $|\phi_1\rangle \cdots |\phi_M\rangle$[6] which are assumed to satisfy the necessary boundary conditions and to be orthonormalized

$$\langle \phi_i | \phi_j \rangle = \delta_{ij}. \qquad (21.125)$$

Applying the method of weighted residuals (Sect. 11.4) we minimize the residual

$$|R\rangle = i\hbar \sum_j \dot{C}_j(t)|\phi_j\rangle - \sum_j C_j(t)H|\phi_j\rangle \qquad (21.126)$$

by choosing the basis functions as weight functions (Sect. 11.4.4) and solving the system of ordinary differential equations

$$0 = R_j = \langle \phi_j | R \rangle = i\hbar \dot{C}_j - \sum_{j'} \langle \phi_j | H | \phi_{j'} \rangle C_{j'} \qquad (21.127)$$

which can be written

$$i\hbar \dot{C}_i = \sum_{j=1}^{M} H_{i,j} C_j(t) \qquad (21.128)$$

with the matrix elements of the Hamiltonian

$$H_{i,j} = \langle \phi_i | H | \phi_j \rangle. \qquad (21.129)$$

In matrix form (21.128) reads

$$i\hbar \begin{pmatrix} \dot{C}_1(t) \\ \vdots \\ \dot{C}_M(t) \end{pmatrix} = \begin{pmatrix} H_{1,1} & \cdots & H_{1,M} \\ \vdots & \ddots & \vdots \\ H_{M,1} & \cdots & H_{M,M} \end{pmatrix} \begin{pmatrix} C_1(t) \\ \vdots \\ C_M(t) \end{pmatrix} \qquad (21.130)$$

or more symbolically

$$i\hbar \dot{\mathbf{C}}(t) = \mathbf{H}\mathbf{C}(t). \qquad (21.131)$$

---

[6]This basis is usually incomplete.

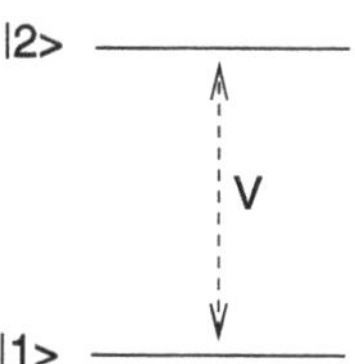

Fig. 21.8 Two-state system model

If the Hamilton operator does not depend explicitly on time ($H = $ const.) the formal solution of (21.131) is given by

$$\mathbf{C} = \exp\left\{\frac{t}{i\hbar}H\right\}\mathbf{C}(0). \tag{21.132}$$

From the solution of the eigenvalue problem

$$H\mathbf{C}_\lambda = \lambda\mathbf{C}_\lambda \tag{21.133}$$

(eigenvalues λ and corresponding eigenvectors $\mathbf{C}_\lambda$) we build the linear combination

$$\mathbf{C} = \sum_\lambda a_\lambda \mathbf{C}_\lambda e^{\frac{\lambda}{i\hbar}t}. \tag{21.134}$$

The amplitudes a_λ can be calculated from the set of linear equations

$$\mathbf{C}(0) = \sum_\lambda a_\lambda \mathbf{C}_\lambda. \tag{21.135}$$

In the following we use the 4th order Runge-Kutta method to solve (21.131) numerically whereas the explicit solution (21.132) will be used to obtain approximate analytical results for special limiting cases.

A time dependent Hamiltonian $H(t)$ appears in semiclassical models which treat some of the slow degrees of freedom as classical quantities, for instance an electron in the Coulomb field of (slowly) moving nuclei

$$H(t) = T_{el} + \sum_j \frac{-q_j e}{4\pi\varepsilon_0|\mathbf{r} - \mathbf{R}_j(t)|} + \sum_{j<j'} \frac{q_j q_{j'}}{4\pi\varepsilon_0|\mathbf{R_j}(t) - \mathbf{R}_{j'}(t)|} \tag{21.136}$$

or in a time dependent electromagnetic field

$$H(t) = T_{el} + V_{el} - e\mathbf{r}\mathbf{E}(t). \tag{21.137}$$

21.3.1 Two-State System

The two-state system (also known as two-level system or TLS; Fig. 21.8) is the simplest model of interacting states and is very often used in physics, for instance in the context of quantum optics, quantum information, spintronics and quantum dots.

Its interaction matrix is

$$H = \begin{pmatrix} E_1 & V \\ V & E_2 \end{pmatrix} \tag{21.138}$$

and the equations of motion are

$$i\hbar\dot{C}_1 = E_1 C_1 + V C_2$$
$$i\hbar\dot{C}_2 = E_2 C_2 + V C_1. \tag{21.139}$$

Equations (21.139) can be solved analytically but this involves some lengthy expressions. Let us therefore concentrate on two limiting cases:

(a) For $E_1 = E_2$ we add and subtract (21.139) to find

$$i\hbar\frac{d}{dt}(C_1 \pm C_2) = (E_1 \pm V)(C_1 \pm C_2) \tag{21.140}$$

with the solution

$$C_1 \pm C_2 = \big(C_1(0) \pm C_2(0)\big)e^{-it(E_1 \pm V)/\hbar}. \tag{21.141}$$

For initial conditions

$$C_1(0) = 1 \qquad C_2(0) = 0 \tag{21.142}$$

the explicit solution is given by

$$C_1 = e^{-itE_1/\hbar}\cos\frac{Vt}{\hbar} \qquad |C_1|^2 = \cos^2\frac{Vt}{\hbar} = \frac{1 + \cos\frac{2Vt}{\hbar}}{2} \tag{21.143}$$

$$C_2 = -ie^{-itE_1/\hbar}\sin\frac{Vt}{\hbar} \qquad |C_2|^2 = \sin^2\frac{Vt}{\hbar} = \frac{1 - \cos\frac{2Vt}{\hbar}}{2} \tag{21.144}$$

and the two-state system oscillates between the two states with the period

$$T = \frac{\pi\hbar}{V}. \tag{21.145}$$

(b) For $V \ll \Delta E = E_2 - E_1$[7] perturbation theory for the small quantity $V/\Delta E$ gives the following approximations:

$$\lambda_1 \approx E_1 - \frac{V^2}{\Delta E}$$
$$\lambda_2 \approx E_2 + \frac{V^2}{\Delta E} \tag{21.146}$$

$$\mathbf{C}_1 \approx \begin{pmatrix} 1 \\ \frac{V}{\Delta E} \end{pmatrix}$$
$$\mathbf{C}_2 \approx \begin{pmatrix} \frac{-V}{\Delta E} \\ 1 \end{pmatrix}. \tag{21.147}$$

For initial values $\mathbf{C}(0) = \begin{pmatrix} 1 \\ 0 \end{pmatrix}$ the amplitudes $a_{1,2}$ are calculated from

$$\begin{pmatrix} 1 \\ 0 \end{pmatrix} = \begin{pmatrix} a_1 - a_2\frac{V}{\Delta E} \\ a_1\frac{V}{\Delta E} + a_2 \end{pmatrix} \tag{21.148}$$

____

[7] We assume $E_2 > E_1$.

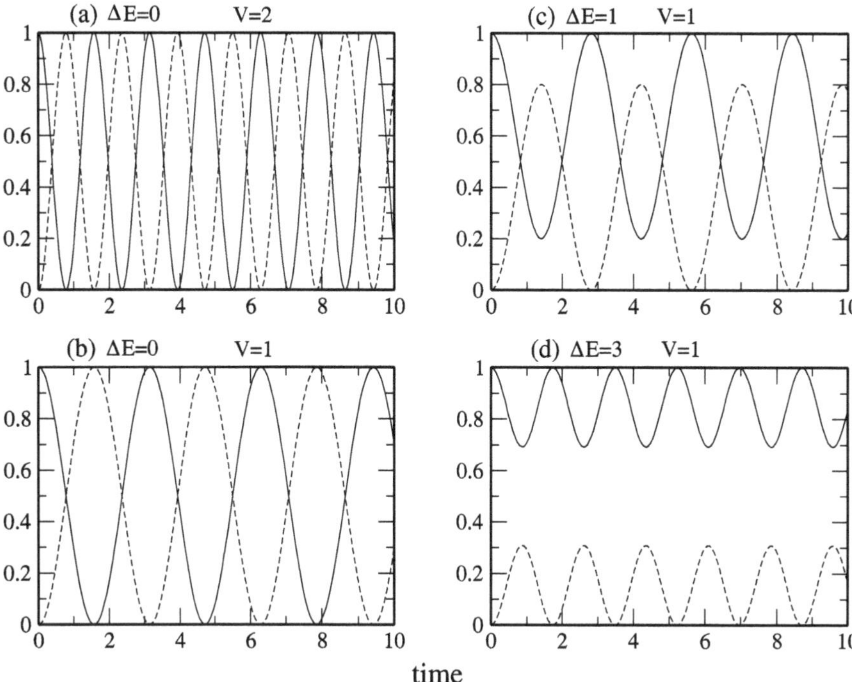

Fig. 21.9 (Numerical simulation of a two-state system) The equations of motion of the two-state system (21.139) are integrated with the 4th order Runge-Kutta method. For two resonant states the occupation probability of the initial state shows oscillations with the period (21.145) proportional to V^{-1}. With increasing energy gap $E_2 - E_1$ the amplitude of the oscillations decreases

which gives in lowest order

$$a_1 \approx 1 - \frac{V^2}{\Delta E^2}$$
$$a_2 \approx \frac{V^2}{\Delta E^2}. \tag{21.149}$$

The approximate solution is

$$\mathbf{C} = \begin{pmatrix} (1 - \frac{V^2}{\Delta E^2})e^{\frac{1}{i\hbar}(E_1 - \frac{V^2}{\Delta E^2})t} + \frac{V^2}{\Delta E^2}e^{\frac{1}{i\hbar}(E_2 + \frac{V^2}{\Delta E^2})t} \\ \frac{V}{\Delta E}e^{\frac{1}{i\hbar}(E_1 - \frac{V^2}{\Delta E^2})t} - \frac{V}{\Delta E}e^{\frac{1}{i\hbar}(E_2 + \frac{V^2}{\Delta E^2})t} \end{pmatrix} \tag{21.150}$$

and the occupation probability of the initial state is

$$|C_1|^2 \approx 1 - 2\frac{V^2}{\Delta E^2} + 2\frac{V^2}{\Delta E^2}\cos\left(\left(\Delta E + 2\frac{V^2}{\Delta E}\right)t\right). \tag{21.151}$$

Numerical results are shown in Fig. 21.9.

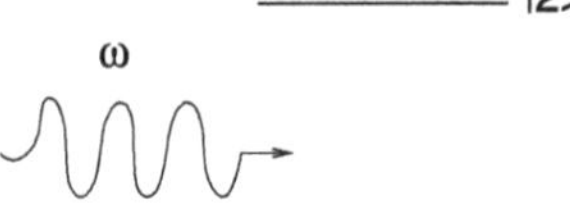

Fig. 21.10 Two-state system in an oscillating field

21.3.2 Two-State System with Time Dependent Perturbation

Consider now a 2-state system with an oscillating perturbation (for instance an atom or molecule in a laser field; Fig. 21.10)

$$H = \begin{pmatrix} E_1 & V(t) \\ V(t) & E_2 \end{pmatrix} \qquad V(t) = V_0 \cos \omega t. \tag{21.152}$$

The equations of motion are

$$i\hbar \dot{C}_1 = E_1 C_1 + V(t) C_2$$
$$i\hbar \dot{C}_2 = V(t) C_1 + E_2 C_2. \tag{21.153}$$

After the substitutions

$$C_1 = e^{\frac{E_1}{i\hbar} t} u_1$$
$$C_2 = e^{\frac{E_2}{i\hbar} t} u_2 \tag{21.154}$$
$$\omega_{21} = \frac{E_2 - E_1}{\hbar} \tag{21.155}$$

they become

$$i\hbar \dot{u}_1 = V(t) e^{\frac{E_2 - E_1}{i\hbar} t} u_2 = \frac{V_0}{2} \left(e^{-i(\omega_{21}-\omega)t} + e^{-i(\omega_{21}+\omega)t} \right) u_2$$
$$i\hbar \dot{u}_2 = V(t) e^{\frac{E_1 - E_2}{i\hbar} t} u_1 = \frac{V_0}{2} \left(e^{i(\omega_{21}-\omega)t} + e^{i(\omega_{21}+\omega)t} \right) u_1. \tag{21.156}$$

At larger times the system oscillates between the two states.[8] Applying the rotating wave approximation for $\omega \approx \omega_{21}$ we neglect the fast oscillating perturbation

$$i\hbar \dot{u}_1 = \frac{V_0}{2} e^{-i(\omega_{21}-\omega)t} u_2 \tag{21.157}$$

$$i\hbar \dot{u}_2 = \frac{V_0}{2} e^{i(\omega_{21}-\omega)t} u_1 \tag{21.158}$$

and substitute

$$u_1 = a_1 e^{-i(\omega_{21}-\omega)t} \tag{21.159}$$

to have

---

[8] So called Rabi oscillations.

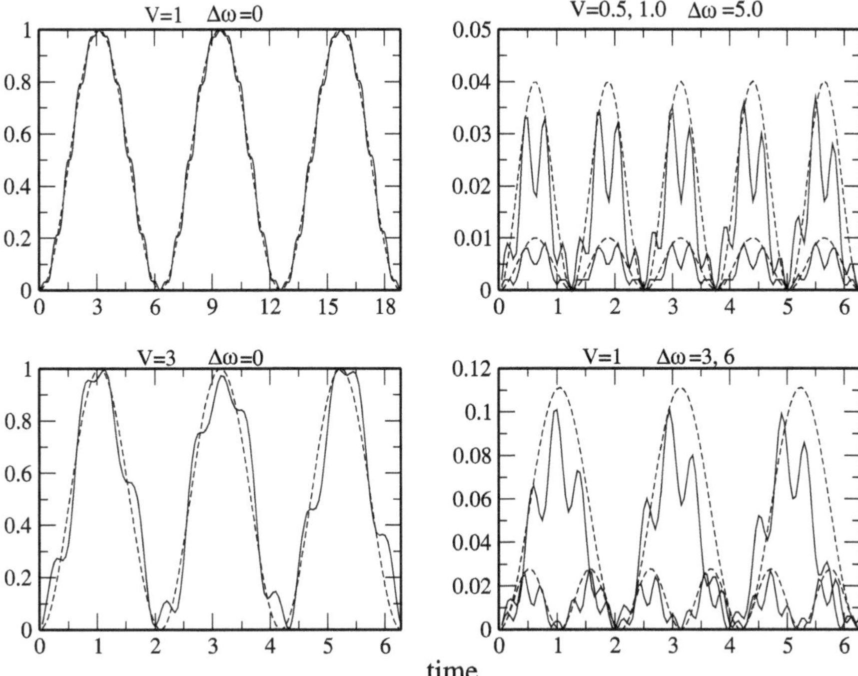

Fig. 21.11 (Simulation of a two-state system in an oscillating field) The equations of motion (21.153) are integrated with the 4th order Runge-Kutta method. At resonance the system oscillates between the two states with the frequency $V/\hbar$. The dashed curves show the corresponding solution of a two-state system with constant coupling (Sect. 21.3.1)

$$i\hbar\big(\dot{a}_1 - a_1 i(\omega_{21} - \omega)\big)e^{-i(\omega_{21}-\omega)t} = \frac{V_0}{2}e^{-i(\omega_{21}-\omega)t}u_2 \qquad (21.160)$$

$$i\hbar\dot{u}_2 = \frac{V_0}{2}e^{i(\omega_{21}-\omega)t}e^{-i(\omega_{21}-\omega)t}a_1 \qquad (21.161)$$

or

$$i\hbar\dot{a}_1 = \hbar(\omega - \omega_{21})a_1 + \frac{V_0}{2}u_2 \qquad (21.162)$$

$$i\hbar\dot{u}_2 = \frac{V_0}{2}a_1 \qquad (21.163)$$

which shows that the system behaves approximately like a two-state system with a constant interaction $V_0/2$ and an energy gap $\hbar(\omega_{21} - \omega) = E_2 - E_1 - \hbar\omega$ (a comparison with a full numerical calculation is shown in Fig. 21.11).

Fig. 21.12 Superexchange
model

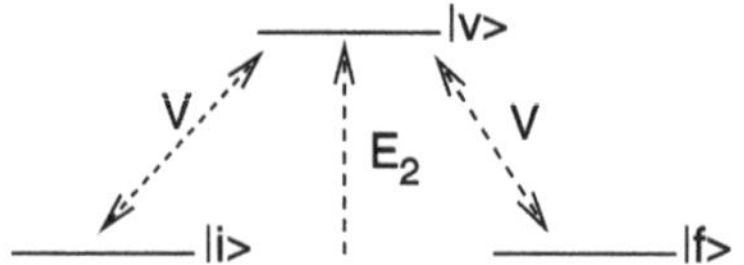

21.3.3 Superexchange Model

The concept of superexchange was originally formulated for magnetic interactions
[8] and later introduced to electron transfer theory [117]. It describes an indirect
interaction through high energy intermediates (Fig. 21.12). In the simplest case, we
have to consider two isoenergetic states i and f which do not interact directly but
via coupling to an intermediate state v.

The interaction matrix is

$$H = \begin{pmatrix} 0 & V_1 & 0 \\ V_1 & E_2 & V_2 \\ 0 & V_2 & 0 \end{pmatrix}. \tag{21.164}$$

For simplification we choose $V_1 = V_2$.

Let us first consider the special case of a resonant intermediate state $E_2 = 0$:

$$H = \begin{pmatrix} 0 & V & 0 \\ V & 0 & V \\ 0 & V & 0 \end{pmatrix}. \tag{21.165}$$

Obviously one eigenvalue is $\lambda_1 = 0$ and the corresponding eigenvector is

$$\mathbf{C}_1 = \begin{pmatrix} 1 \\ 0 \\ -1 \end{pmatrix}. \tag{21.166}$$

The two remaining eigenvalues are solutions of

$$0 = \det \begin{vmatrix} -\lambda & V & 0 \\ V & -\lambda & V \\ 0 & V & -\lambda \end{vmatrix} = \lambda(-\lambda^2 + 2V^2) \tag{21.167}$$

which gives

$$\lambda_{2,3} = \pm\sqrt{2}V. \tag{21.168}$$

The eigenvectors are

$$\mathbf{C}_{2,3} = \begin{pmatrix} 1 \\ \pm\sqrt{2} \\ 1 \end{pmatrix}. \tag{21.169}$$

From the initial values

$$\mathbf{C}(0) = \begin{pmatrix} a_1 + a_2 + a_3 \\ \sqrt{2}a_2 - \sqrt{2}a_3 \\ -a_1 + a_2 + a_3 \end{pmatrix} = \begin{pmatrix} 1 \\ 0 \\ 0 \end{pmatrix} \tag{21.170}$$

the amplitudes are calculated as

$$a_1 = \frac{1}{2}\, a_2 = a_3 = \frac{1}{4} \tag{21.171}$$

and finally the solution is

$$\mathbf{C} = \frac{1}{2}\begin{pmatrix} 1 \\ 0 \\ -1 \end{pmatrix} + \frac{1}{4}\begin{pmatrix} 1 \\ \sqrt{2} \\ 1 \end{pmatrix} e^{\frac{1}{i\hbar}\sqrt{2}Vt} + \frac{1}{4}\begin{pmatrix} 1 \\ -\sqrt{2} \\ 1 \end{pmatrix} e^{-\frac{1}{i\hbar}\sqrt{2}Vt}$$

$$= \begin{pmatrix} \frac{1}{2} + \frac{1}{2}\cos\frac{\sqrt{2}V}{\hbar}t \\ \frac{\sqrt{2}}{2}i\,\sin\frac{\sqrt{2}V}{\hbar}t \\ -\frac{1}{2} + \frac{1}{2}\cos\frac{\sqrt{2}V}{\hbar}t \end{pmatrix}. \tag{21.172}$$

Let us now consider the case of a distant intermediate state $V \ll |E_2|$. $\lambda_1 = 0$ and the corresponding eigenvector still provide one solution. The two other eigenvalues are approximately given by

$$\lambda_{2,3} = \pm\sqrt{\frac{E_2^2}{4} + 2V^2} + \frac{E_2}{2} \approx \frac{E_2}{2} \pm \frac{E_2}{2}\left(1 + \frac{4V^2}{E_2^2}\right) \tag{21.173}$$

$$\lambda_2 \approx E_2 + \frac{2V^2}{E_2} \qquad \lambda_3 \approx -\frac{2V^2}{E_2} \tag{21.174}$$

and the eigenvectors by

$$\mathbf{C}_2 \approx \begin{pmatrix} 1 \\ \frac{E_2}{V} + \frac{2V}{E_2} \\ 1 \end{pmatrix} \qquad \mathbf{C}_3 \approx \begin{pmatrix} 1 \\ -\frac{2V}{E_2} \\ 1 \end{pmatrix}. \tag{21.175}$$

From the initial values

$$\mathbf{C}(0) = \begin{pmatrix} 1 \\ 0 \\ 0 \end{pmatrix} = \begin{pmatrix} a_1 + a_2 + a_3 \\ a_2\lambda_2 + a_3\lambda_3 \\ -a_1 + a_2 + a_3 \end{pmatrix} \tag{21.176}$$

we calculate the amplitudes

$$a_1 = \frac{1}{2} \qquad a_2 \approx \frac{V^2}{E_2^2} \qquad a_3 \approx \frac{1}{2}\left(1 - \frac{2V^2}{E_2^2}\right) \tag{21.177}$$

and finally the solution

$$\mathbf{C} \approx \begin{pmatrix} \frac{1}{2}\left(1 + e^{-\frac{1}{i\hbar}\frac{2V^2}{E_2}t}\right) \\ \frac{V}{E_2}e^{\frac{1}{i\hbar}E_2 t} - \frac{2V}{E_2}e^{-\frac{1}{i\hbar}\frac{2V^2}{E_2}t} \\ \frac{1}{2}\left(-1 + e^{-\frac{1}{i\hbar}\frac{2V^2}{E_2}t}\right) \end{pmatrix}. \tag{21.178}$$

The occupation probability of the initial state is

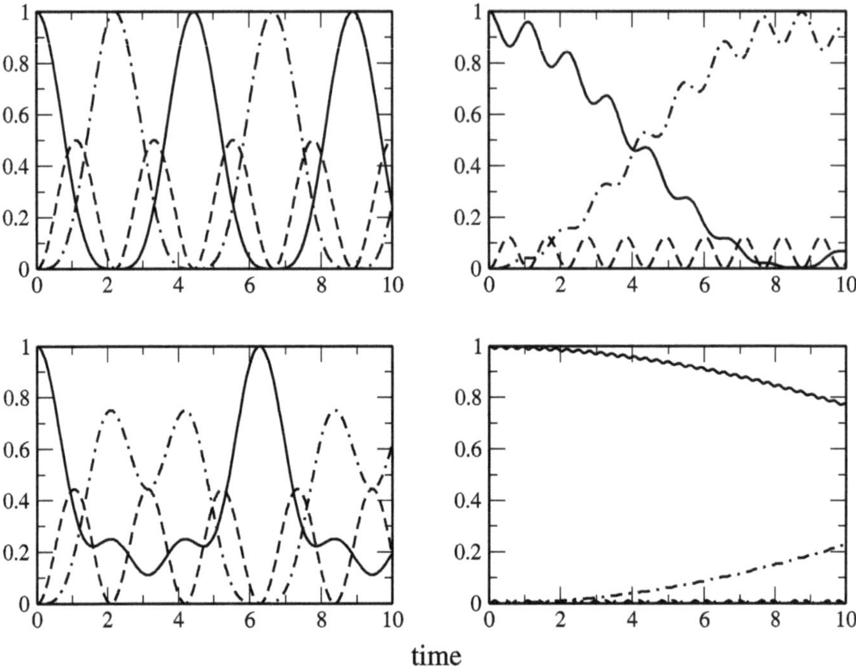

time

Fig. 21.13 (Numerical simulation of the superexchange model) The equations of motion for the model (21.164) are solved numerically with the 4th order Runge-Kutta method. The energy gap is varied to study the transition from the simple oscillation with $\omega = \sqrt{2}V/\hbar$ (21.172) to the effective two-state system with $\omega = V_{eff}/\hbar$ (21.179). Parameters are $V_1 = V_2 = 1$, $E_1 = E_3 = 0$, $E_2 = 0, 1, 5, 20$. The occupation probability of the initial (*solid curves*), virtual intermediate (*dashed curves*) and final (*dash-dotted curves*) state are shown

$$|C_1|^2 = \frac{1}{4}\left|1 + e^{-\frac{1}{i\hbar}\frac{2V^2}{E_2}t}\right|^2 = \cos^2\left(\frac{V^2}{\hbar E_2}t\right) \tag{21.179}$$

which shows that the system behaves like a 2-state system with an effective interaction of

$$V_{eff} = \frac{V^2}{E_2}. \tag{21.180}$$

Numerical results are shown in Fig. 21.13.

21.3.4 Ladder Model for Exponential Decay

For time independent Hamiltonian the solution (21.132) of the Schrödinger equation is a sum of oscillating terms and the quantum recurrence theorem [29] states that the system returns to the initial state arbitrarily closely after a certain time T_r. However, if the initial state is coupled to a larger number of final states, the recurrence time can become very long and an exponential decay observed over a large period. The ladder

Fig. 21.14 Ladder model

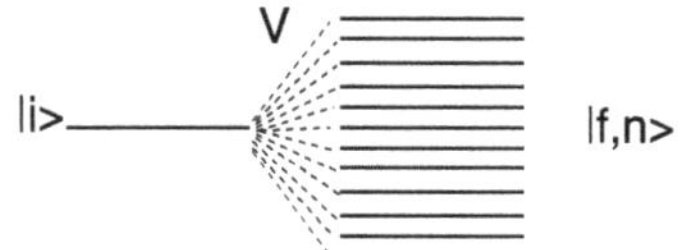

model [27, 242] considers an initial state $|0\rangle$ interacting with a manifold of states $|1\rangle \cdots |n\rangle$, which do not interact with each other and are equally spaced (Fig. 21.14)

$$H = \begin{pmatrix} 0 & V & \cdots & V \\ V & E_1 & & \\ \vdots & & \ddots & \\ V & & & E_n \end{pmatrix} \qquad E_j = E_1 + (j-1)\Delta E. \tag{21.181}$$

The equations of motion are

$$i\hbar \dot{C}_0 = V \sum_{j=1}^{n} C_j \tag{21.182}$$

$$i\hbar \dot{C}_j = E_j C_j + V C_0.$$

For the special case $\Delta E = 0$ we simply have

$$\ddot{C}_0 = -\frac{V^2}{\hbar^2} n C_0 \tag{21.183}$$

with an oscillating solution

$$C_0 \sim \cos\left(\frac{V\sqrt{n}}{\hbar} t\right). \tag{21.184}$$

Here the n states act like one state with an effective coupling of $V\sqrt{n}$.

For the general case $\Delta E \neq 0$ we substitute

$$C_j = u_j e^{\frac{E_j}{i\hbar} t} \tag{21.185}$$

and have

$$i\hbar \dot{u}_j e^{\frac{E_j}{i\hbar} t} = V C_0. \tag{21.186}$$

Integration gives

$$u_j = \frac{V}{i\hbar} \int_{t_0}^{t} e^{-\frac{E_j}{i\hbar} t'} C_0(t') \, dt' \tag{21.187}$$

and therefore

$$C_j = \frac{V}{i\hbar} \int_{t_0}^{t} e^{i\frac{E_j}{\hbar}(t'-t)} C_0(t') \, dt'. \tag{21.188}$$

With the definition

$$E_j = j * \hbar \Delta\omega \tag{21.189}$$

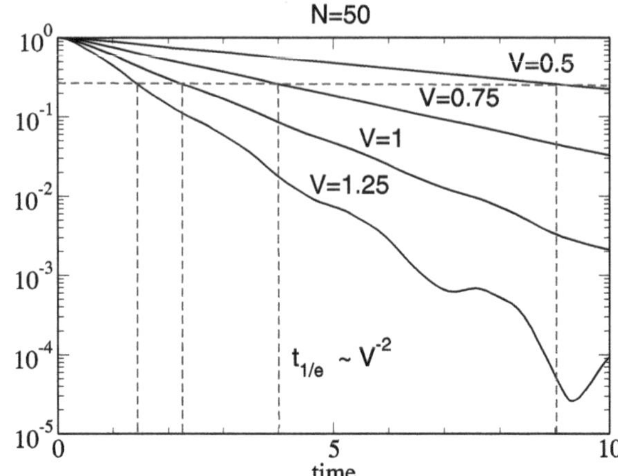

Fig. 21.15 (Numerical solution of the ladder model) The time evolution of the ladder model (21.182) is calculated with the 4th order Runge-Kutta method for $N = 50$ states and different values of the coupling V

we have

$$\dot{C}_0 = \frac{V}{\mathrm{i}\hbar}\sum_{j=1}^{n}C_j = -\frac{V^2}{\hbar^2}\sum_j \int_{t_0}^{t} e^{\mathrm{i}j\Delta\omega(t'-t)}C_0(t')\,dt'. \tag{21.190}$$

We replace the sum by an integral over the continuous variable

$$\omega = j\Delta\omega \tag{21.191}$$

and extend the integration range to $-\infty\cdots\infty$. Then the sum becomes approximately a delta function

$$\sum_{j=-\infty}^{\infty} e^{\mathrm{i}j\Delta\omega(t'-t)}\Delta j \;\rightarrow\; \int_{-\infty}^{\infty} e^{\mathrm{i}\omega(t'-t)}\frac{d\omega}{\Delta\omega} = \frac{2\pi}{\Delta\omega}\delta(t-t') \tag{21.192}$$

and the final result is an exponential decay law

$$\dot{C}_0 = -\frac{2\pi V^2}{\hbar^2\Delta\omega}C_0 = -\frac{2\pi V^2}{\hbar}\rho(E)C_0 \tag{21.193}$$

with the density of final states

$$\rho(E) = \frac{1}{\hbar\Delta\omega} = \frac{1}{\Delta E}. \tag{21.194}$$

Numerical results are shown in Fig. 21.15.

21.3.5 Landau-Zener Model

This model describes crossing of two states (Fig. 21.16), for instance for colliding atoms or molecules [154, 282]. It is assumed that the interaction V is constant near the crossing point and that the nuclei move classically with constant velocity

Fig. 21.16 Slow atomic collision

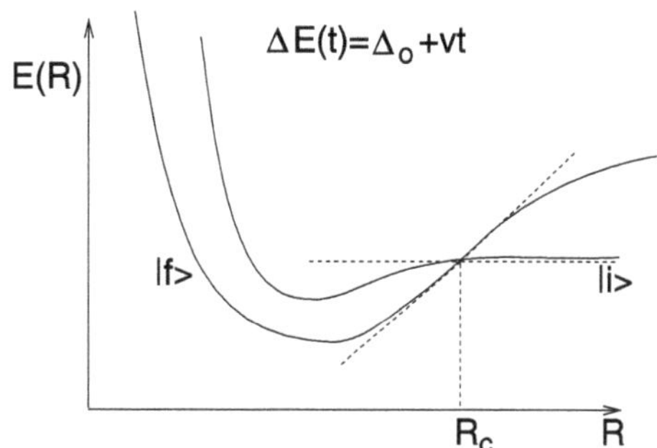

Fig. 21.17 Multiple passage of the interaction region

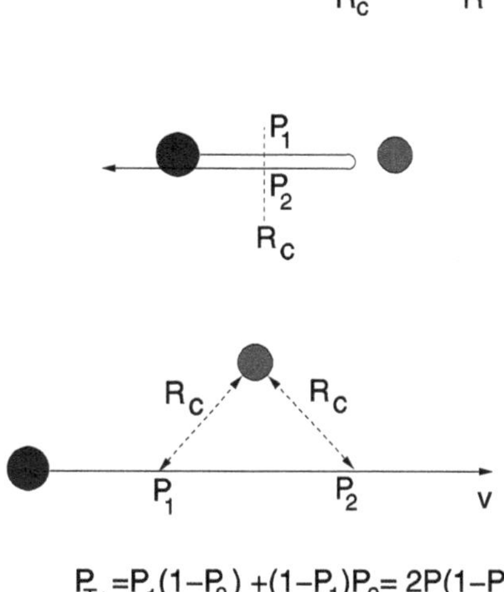

$$H = \begin{pmatrix} 0 & V \\ V & \Delta E(t) \end{pmatrix} \qquad \Delta E(t) = \Delta E_0 + vt. \tag{21.195}$$

For small interaction V or large velocity $\frac{\partial}{\partial t}\Delta E = \dot{Q}\frac{\partial}{\partial Q}\Delta E$ the transition probability can be calculated with perturbation theory to give

$$P = \frac{2\pi V^2}{\hbar \frac{\partial}{\partial t}\Delta E}. \tag{21.196}$$

This expression becomes invalid for small velocities. Here the system stays on the adiabatic potential surface, i.e. $P \to 1$. Landau and Zener found the following expression which is valid in both limits:

$$P_{LZ} = 1 - \exp\left(-\frac{2\pi V^2}{\hbar \frac{\partial}{\partial t}\Delta E}\right). \tag{21.197}$$

In case of collisions multiple crossing of the interaction region has to be taken into account (Fig. 21.17). Numerical results are shown in Fig. 21.18.

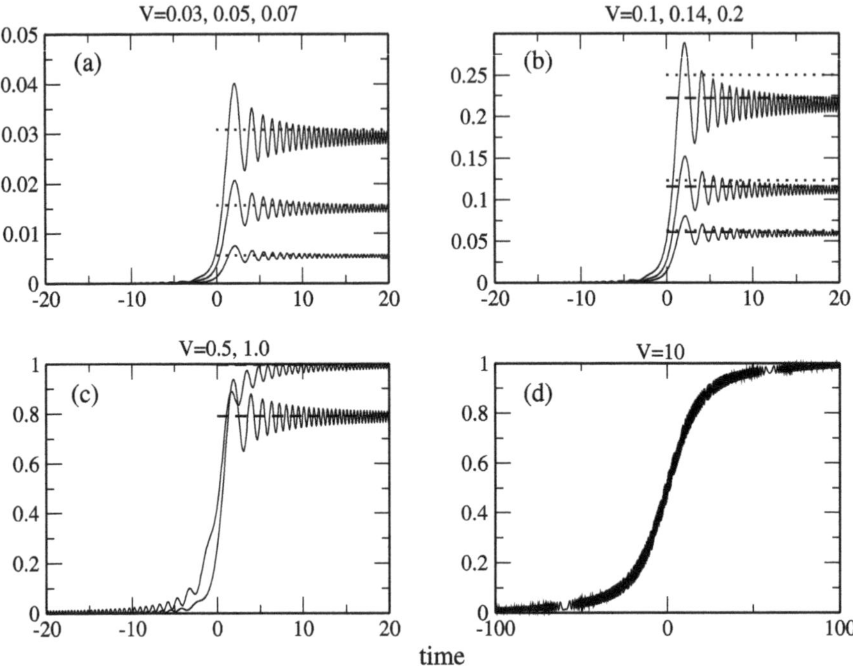

Fig. 21.18 (Numerical solution of the Landau-Zener model) Numerical calculations (*solid curves*) are compared with the Landau-Zener probability ((21.197), *dashed lines*) and the approximation ((21.196), *dotted lines*). The velocity is $d\Delta E/dt = 1$

21.4 The Dissipative Two-State System

A two-state quantum system coupled to a thermal bath serves as a model for magnetic resonance phenomena, coherent optical excitations [263, 280] and, quite recently, for a qubit, the basic element of a future quantum computer [73, 183]. Its quantum state can not be described by a single wavefunction. Instead mixed quantum states have to be considered which can be conveniently described within the density matrix formalism [232].

21.4.1 Equations of Motion for a Two-State System

The equations of motion for a two-state system are

$$i\hbar\dot{\rho}_{11} = H_{12}\rho_{21} - \rho_{12}H_{21} \tag{21.198}$$

$$i\hbar\dot{\rho}_{22} = H_{21}\rho_{12} - \rho_{21}H_{12} \tag{21.199}$$

$$i\hbar\dot{\rho}_{12} = (H_{11} - H_{22})\rho_{12} + H_{12}(\rho_{22} - \rho_{11}) \tag{21.200}$$

$$-i\hbar\dot{\rho}_{21} = (H_{11} - H_{22})\rho_{21} + H_{21}(\rho_{22} - \rho_{11}) \tag{21.201}$$

which can be arranged as a system of linear equations[9]

$$i\hbar \begin{pmatrix} \dot{\rho}_{11} \\ \dot{\rho}_{22} \\ \dot{\rho}_{12} \\ \dot{\rho}_{21} \end{pmatrix} = \begin{pmatrix} 0 & 0 & -H_{21} & H_{12} \\ 0 & 0 & H_{21} & -H_{12} \\ -H_{12} & H_{12} & H_{11} - H_{22} & 0 \\ H_{21} & -H_{21} & 0 & H_{22} - H_{11} \end{pmatrix} \begin{pmatrix} \rho_{11} \\ \rho_{22} \\ \rho_{12} \\ \rho_{21} \end{pmatrix}.$$

$$(21.202)$$

21.4.2 The Vector Model

The density matrix is Hermitian

$$\rho_{ij} = \rho_{ji}^{*} \tag{21.203}$$

its diagonal elements are real valued and due to conservation of probability

$$\rho_{11} + \rho_{22} = \text{const.} \tag{21.204}$$

Therefore the four elements of the density matrix can be specified by three real parameters, which are usually chosen as

$$x = 2\Re\rho_{21} \tag{21.205}$$

$$y = 2\Im\rho_{21} \tag{21.206}$$

$$z = \rho_{11} - \rho_{22} \tag{21.207}$$

and satisfy the equations

$$\frac{d}{dt}2\Re(\rho_{21}) = -\frac{1}{\hbar}\big((H_{11} - H_{22})2\Im(\rho_{21}) + 2\Im(H_{12})(\rho_{11} - \rho_{22})\big) \tag{21.208}$$

$$\frac{d}{dt}2\Im(\rho_{21}) = \frac{1}{\hbar}\big((H_{11} - H_{22})2\Re(\rho_{21}) - 2\Re(H_{12})(\rho_{11} - \rho_{22})\big) \tag{21.209}$$

$$\frac{d}{dt}(\rho_{11} - \rho_{22}) = \frac{2}{\hbar}\big(\Im(H_{12})2\Re(\rho_{21}) + \Re H_{12}2\Im(\rho_{21})\big). \tag{21.210}$$

Together they form the Bloch vector

$$\mathbf{r} = \begin{pmatrix} x \\ y \\ z \end{pmatrix} \tag{21.211}$$

which is often used to visualize the time evolution of a two-state system [81]. In terms of the Bloch vector the density matrix is given by

$$\begin{pmatrix} \rho_{11} & \rho_{12} \\ \rho_{21} & \rho_{22} \end{pmatrix} = \begin{pmatrix} \frac{1+z}{2} & \frac{x-iy}{2} \\ \frac{x+iy}{2} & \frac{1-z}{2} \end{pmatrix} = \frac{1}{2}(1 + \mathbf{r}\boldsymbol{\sigma}) \tag{21.212}$$

---

[9]The matrix of this system corresponds to the Liouville operator.

with the Pauli matrices

$$\sigma_x = \begin{pmatrix} 0 & 1 \\ 1 & 0 \end{pmatrix}, \qquad \sigma_y = \begin{pmatrix} 0 & -i \\ i & 0 \end{pmatrix}, \qquad \sigma_z = \begin{pmatrix} 1 & \\ & -1 \end{pmatrix}. \qquad (21.213)$$

From (21.208), (21.209), (21.210) we obtain the equation of motion

$$\frac{d}{dt} \begin{pmatrix} x \\ y \\ z \end{pmatrix} = \begin{pmatrix} -y\,\frac{H_{11}-H_{22}}{\hbar} - z\,\frac{2\Im(H_{12})}{\hbar} \\ x\,\frac{H_{11}-H_{22}}{\hbar} - z\,\frac{2\Re(H_{12})}{\hbar} \\ x\,\frac{2\Im(H_{12})}{\hbar} + y\,\frac{2\Re(H_{12})}{\hbar} \end{pmatrix} \qquad (21.214)$$

which can be written as a cross product

$$\frac{d}{dt}\mathbf{r} = \boldsymbol{\omega} \times \mathbf{r} \qquad (21.215)$$

with

$$\boldsymbol{\omega} = \begin{pmatrix} \frac{2}{\hbar}\Re H_{12} \\ -\frac{2}{\hbar}\Im H_{12} \\ \frac{1}{\hbar}(H_{11} - H_{22}) \end{pmatrix}. \qquad (21.216)$$

Any normalized pure quantum state of the two-state system can be written as [75]

$$|\psi\rangle = \begin{pmatrix} C_1 \\ C_2 \end{pmatrix} = \cos\frac{\theta}{2} \begin{pmatrix} 1 \\ 0 \end{pmatrix} + e^{i\phi}\sin\frac{\theta}{2} \begin{pmatrix} 0 \\ 1 \end{pmatrix} \qquad (21.217)$$

corresponding to the density matrix

$$\rho = \begin{pmatrix} \cos^2\frac{\theta}{2} & e^{-i\phi}\sin\frac{\theta}{2}\cos\frac{\theta}{2} \\ e^{i\phi}\sin\frac{\theta}{2}\cos\frac{\theta}{2} & \sin^2\frac{\theta}{2} \end{pmatrix}. \qquad (21.218)$$

The Bloch vector

$$\mathbf{r} = \begin{pmatrix} \cos\phi\,\sin\theta \\ \sin\phi\,\sin\theta \\ \cos\theta \end{pmatrix} \qquad (21.219)$$

represents a point on the unit sphere (the Bloch sphere, Fig. 21.19). Mixed states correspond to the interior of the Bloch sphere with the fully mixed state $\rho = \begin{pmatrix} 1/2 & 0 \\ 0 & 1/2 \end{pmatrix}$ represented by the center of the sphere (Fig. 21.19).

21.4.3 The Spin-$\frac{1}{2}$ System

An important example of a two-state system is a particle with spin $\frac{1}{2}$. Its quantum state can be described by a two-component vector

$$\begin{pmatrix} C_1 \\ C_2 \end{pmatrix} = C_1 \begin{pmatrix} 1 \\ 0 \end{pmatrix} + C_2 \begin{pmatrix} 0 \\ 1 \end{pmatrix} \qquad (21.220)$$

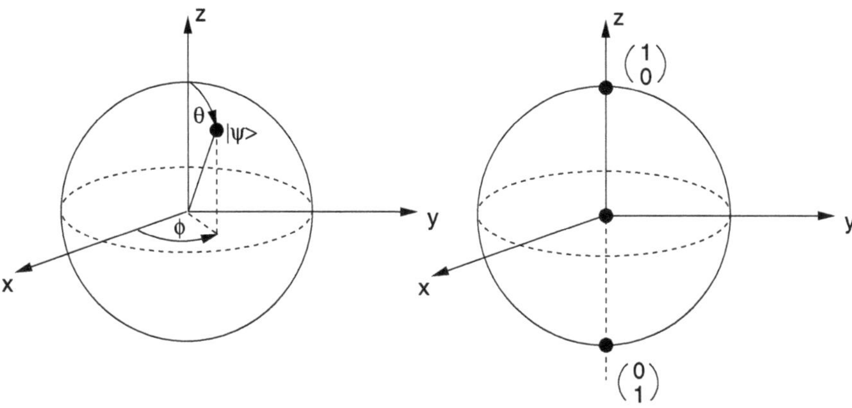

Fig. 21.19 (Bloch sphere) *Left*: Any pure quantum state of a two-state system can be represented by a point on the Bloch sphere. *Right*: The poles represent the basis states. Mixed quantum states correspond to the interior of the sphere, the center represents the fully mixed state

where the two unit vectors are eigenvectors of the spin component in z-direction corresponding to the eigenvalues $s_z = \pm\frac{\hbar}{2}$. The components of the spin operator are given by the Pauli matrices

$$S_i = \frac{\hbar}{2}\sigma_i \tag{21.221}$$

and have expectation values

$$\langle \mathbf{S}\rangle = \frac{\hbar}{2}\begin{pmatrix} C_1^* & C_2^* \end{pmatrix}\begin{pmatrix} \sigma_x \\ \sigma_y \\ \sigma_z \end{pmatrix}\begin{pmatrix} C_1 \\ C_2 \end{pmatrix} = \hbar\begin{pmatrix} \frac{C_1^*C_2+C_2^*C_1}{2} \\ \frac{C_1^*C_2-C_2^*C_1}{2i} \\ \frac{|C_1|^2-|C_2|^2}{2} \end{pmatrix}. \tag{21.222}$$

The ensemble average for a system of spin-$\frac{1}{2}$ particles is given by the Bloch vector

$$\overline{\langle \mathbf{S}\rangle} = \hbar\begin{pmatrix} \frac{\rho_{21}+\rho_{12}}{2} \\ \frac{\rho_{21}-\rho_{12}}{2i} \\ \frac{\rho_{11}-\rho_{22}}{2} \end{pmatrix} = \frac{\hbar}{2}\mathbf{r}. \tag{21.223}$$

The Hamiltonian of a spin-$\frac{1}{2}$ particle in a magnetic field $\mathbf{B}$ is

$$H = -\gamma\frac{\hbar}{2}\boldsymbol{\sigma}\mathbf{B} = -\gamma\frac{\hbar}{2}\begin{pmatrix} B_z & B_x - iB_y \\ B_x + iB_y & -B_z \end{pmatrix} \tag{21.224}$$

from which the following relations are obtained

$$\gamma B_x = -\frac{2}{\hbar}\Re H_{12} \tag{21.225}$$

$$\gamma B_y = \frac{2}{\hbar}\Im H_{12} \tag{21.226}$$

$$\gamma B_z = -\frac{H_{11} - H_{22}}{\hbar} \tag{21.227}$$

$$\omega = -\gamma \mathbf{B}. \tag{21.228}$$

The average magnetization

$$\mathbf{m} = \gamma \overline{\langle \mathbf{S} \rangle} = \gamma \frac{\hbar}{2} \mathbf{r} \tag{21.229}$$

obeys the equation of motion

$$\frac{\mathrm{d}}{\mathrm{d}t} \mathbf{m} = -\gamma \mathbf{B} \times \mathbf{m}. \tag{21.230}$$

21.4.4 Relaxation Processes—The Bloch Equations

Relaxation of the nuclear magnetization due to interaction with the environment was first described phenomenologically by Bloch in 1946 [28]. A more rigorous description was given later [201, 269] and also applied to optical transitions [179]. Recently electron spin relaxation has attracted much interest in the new field of spintronics [284] and the dissipative two-state system has been used to describe the decoherence of a qubit [46].

21.4.4.1 Phenomenological Description

In thermal equilibrium the density matrix is given by a canonical distribution

$$\rho^{eq} = \frac{e^{-\beta H}}{\mathrm{tr}(e^{-\beta H})} \tag{21.231}$$

which for a two-state system without perturbation

$$H_0 = \begin{pmatrix} \frac{\Delta}{2} & \\ & -\frac{\Delta}{2} \end{pmatrix} \tag{21.232}$$

becomes

$$\rho^{eq} = \begin{pmatrix} \frac{e^{-\beta\Delta/2}}{e^{\beta\Delta/2}+e^{-\beta\Delta/2}} & \\ & \frac{e^{\beta\Delta/2}}{e^{\beta\Delta/2}+e^{-\beta\Delta/2}} \end{pmatrix} \tag{21.233}$$

where, as usually $\beta = 1/k_B T$. If the energy gap is very large $\Delta \gg k_B T$ like for an optical excitation, the equilibrium state is the state with lower energy[10]

$$\rho^{eq} = \begin{pmatrix} 0 & 0 \\ 0 & 1 \end{pmatrix}. \tag{21.234}$$

---

[10]We assume $\Delta \geq 0$, such that the equilibrium value of $z = \rho_{11} - \rho_{22}$ is negative. Eventually, the two states have to be exchanged.

The phenomenological model assumes that deviations of the occupation difference from its equilibrium value

$$\rho_{11}^{eq} - \rho_{22}^{eq} = -\tanh\left(\frac{\Delta}{2k_B T}\right) \tag{21.235}$$

decay exponentially with a time constant T_1 (for NMR this is the spin-lattice relaxation time)

$$\frac{d}{dt}\Big|_{Rel}(\rho_{11} - \rho_{22}) = -\frac{1}{T_1}\left[(\rho_{11} - \rho_{22}) - \left(\rho_{11}^{eq} - \rho_{22}^{eq}\right)\right]. \tag{21.236}$$

The coherence of the two states decays exponentially with a time constant T_2 which is closely related to T_1 in certain cases[11] but can be much smaller than T_1 if there are additional dephasing mechanisms. The equation

$$\frac{d}{dt}\Big|_{Rel}\rho_{12} = -\frac{1}{T_2}\rho_{12} \tag{21.237}$$

describes the decay of the transversal polarization due to spatial and temporal differences of different spins (spin-spin relaxation), whereas for an optical excitation or a qubit it describes the loss of coherence of a single two-state system due to interaction with its environment.

The combination of (21.230) and the relaxation terms (21.236), (21.237) gives the Bloch equations [28] which were originally formulated to describe the time evolution of the macroscopic polarization

$$\frac{d\mathbf{m}}{dt} = -\gamma \mathbf{B} \times \mathbf{m} - R(\mathbf{m} - \mathbf{m}_{eq}) \qquad R = \begin{pmatrix} \frac{1}{T_2} & 0 & 0 \\ 0 & \frac{1}{T_2} & 0 \\ 0 & 0 & \frac{1}{T_1} \end{pmatrix}. \tag{21.238}$$

For the components of the Bloch vector they read explicitly

$$\frac{d}{dt}\begin{pmatrix} x \\ y \\ z \end{pmatrix} = \begin{pmatrix} -1/T_2 & -\frac{1}{\hbar}(H_{11} - H_{22}) & -\frac{2}{\hbar}\Im H_{12} \\ \frac{1}{\hbar}(H_{11} - H_{22}) & -1/T_2 & -\frac{2}{\hbar}\Re H_{12} \\ \frac{2}{\hbar}\Im H_{12} & \frac{2}{\hbar}\Re H_{12} & -1/T_1 \end{pmatrix}\begin{pmatrix} x \\ y \\ z \end{pmatrix}$$

$$+ \begin{pmatrix} 0 \\ 0 \\ z_{eq}/T_1 \end{pmatrix}. \tag{21.239}$$

21.4.5 The Driven Two-State System

The Hamiltonian of a two-state system (for instance an atom or molecule) in an oscillating electric field $Ee^{-i\omega_f t}$ with energy splitting Δ and transition dipole moment μ is

---

[11] For instance $T_2 = 2T_1$ for pure radiative damping.

$$H = \begin{pmatrix} \frac{\Delta}{2} & -\mu E e^{-i\omega_f t} \\ -\mu E e^{i\omega_f t} & -\frac{\Delta}{2} \end{pmatrix}. \tag{21.240}$$

The corresponding magnetic field

$$B_x = \frac{2}{\gamma\hbar}\mu E \cos\omega_f t \tag{21.241}$$

$$B_y = \frac{2}{\gamma\hbar}\mu E \sin\omega_f t \tag{21.242}$$

$$B_z = -\frac{\Delta}{\gamma\hbar} \tag{21.243}$$

is that of a typical NMR experiment with a constant component along the z-axis and a rotating component in the xy-plane.

21.4.5.1 Free Precession

Consider the special case $B_z = \text{const.}$, $B_x = B_y = 0$. The corresponding Hamiltonian matrix is diagonal

$$H = \begin{pmatrix} \frac{\hbar\Omega_0}{2} & 0 \\ 0 & -\frac{\hbar\Omega_0}{2} \end{pmatrix} \tag{21.244}$$

with the Larmor-frequency

$$\Omega_0 = \frac{\Delta}{\hbar} = -\gamma B_0. \tag{21.245}$$

The equations of motion for the density matrix are

$$\frac{\partial}{\partial t}(\rho_{11} - \rho_{22}) = -\frac{(\rho_{11} - \rho_{22}) - (\rho_{11}^{eq} - \rho_{22}^{eq})}{T_1} \tag{21.246}$$

$$i\hbar\frac{\partial}{\partial t}\rho_{12} = \hbar\Omega_0\rho_{12} - i\hbar\frac{1}{T_2}\rho_{12} \tag{21.247}$$

with the solution

$$(\rho_{11} - \rho_{22}) = \left(\rho_{11}^{eq} - \rho_{22}^{eq}\right) + \left[\left(\rho_{11}(0) - \rho_{22}(0)\right) - \left(\rho_{11}^{eq} - \rho_{22}^{eq}\right)\right]e^{-t/T_1} \tag{21.248}$$

$$\rho_{12} = \rho_{12}(0)e^{-i\Omega_0 t - t/T_2}. \tag{21.249}$$

The Bloch vector

$$\mathbf{r} = \begin{pmatrix} (x_0\cos\Omega_0 t - y_0\sin\Omega_0 t)e^{-t/T_2} \\ (y_0\cos\Omega_0 t + x_0\sin\Omega_0 t)e^{-t/T_2} \\ z^{eq} + (z_0 - z^{eq})e^{-t/T_1} \end{pmatrix} \tag{21.250}$$

is subject to damped precession around the z-axis with the Larmor frequency (Fig. 21.20).

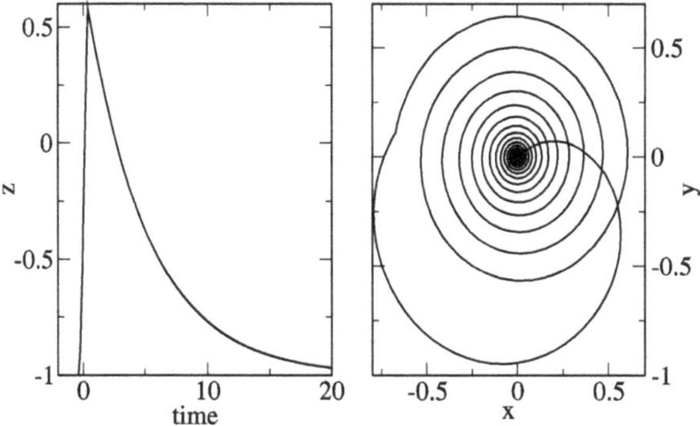

Fig. 21.20 (Free precession) The Bloch equations (21.239) are numerically solved with the 4th order Runge-Kutta method. After excitation with a short resonant pulse the free precession is observed. *Left*: The occupation difference $z = \rho_{11} - \rho_{22}$ decays exponentially to its equilibrium value. *Right*: In the xy-plane the Bloch vector moves on a spiral towards the equilibrium position ($x = 0$, $y = 0$)

21.4.5.2 Stationary Solution for Monochromatic Excitation

For the two-state system (21.240) with

$$H_{11} - H_{22} = \Delta = \hbar\Omega_0 \tag{21.251}$$

$$H_{12} = V_0(\cos\omega_f t - i\sin\omega_f t) \tag{21.252}$$

the solution of the Bloch equations (21.238)

$$\frac{d}{dt}\begin{pmatrix} x \\ y \\ z \end{pmatrix} = \begin{pmatrix} -1/T_2 & -\Omega_0 & \frac{2V_0}{\hbar}\sin\omega_f t \\ \Omega_0 & -1/T_2 & -\frac{2V_0}{\hbar}\cos\omega_f t \\ -\frac{2V_0}{\hbar}\sin\omega_f t & \frac{2V_0}{\hbar}\cos\omega_f t & -1/T_1 \end{pmatrix}\begin{pmatrix} x \\ y \\ z \end{pmatrix} + \begin{pmatrix} 0 \\ 0 \\ z_{eq}/T_1 \end{pmatrix} \tag{21.253}$$

can be found explicitly [280]. We transform to a coordinate system which rotates around the z-axis ((13.3) on page 243) with angular velocity ω_f

$$\begin{pmatrix} x' \\ y' \\ z' \end{pmatrix} = \begin{pmatrix} \cos(\omega_f t) & \sin(\omega_f t) & 0 \\ -\sin(\omega_f t) & \cos(\omega_f t) & 0 \\ 0 & 0 & 1 \end{pmatrix}\begin{pmatrix} x \\ y \\ z \end{pmatrix} = A(t)\begin{pmatrix} x \\ y \\ z \end{pmatrix}. \tag{21.254}$$

Then

$$\frac{d}{dt}\begin{pmatrix} x' \\ y' \\ z' \end{pmatrix} = \dot{A}\begin{pmatrix} x \\ y \\ z \end{pmatrix} + A\frac{d}{dt}\begin{pmatrix} x \\ y \\ z \end{pmatrix} = (\dot{A}A^{-1} + AKA^{-1})\begin{pmatrix} x' \\ y' \\ z' \end{pmatrix} + A\begin{pmatrix} 0 \\ 0 \\ \frac{z_{eq}}{T_1} \end{pmatrix} \tag{21.255}$$

with

$$K = \begin{pmatrix} -1/T_2 & -\Omega_0 & \frac{2V_0}{\hbar}\sin\omega_f t \\ \Omega_0 & -1/T_2 & -\frac{2V_0}{\hbar}\cos\omega_f t \\ -\frac{2V_0}{\hbar}\sin\omega_f t & \frac{2V_0}{\hbar}\cos\omega_f t & -1/T_1 \end{pmatrix}. \tag{21.256}$$

The matrix products are

$$\dot{A}A^{-1} = W = \begin{pmatrix} 0 & \omega_f & 0 \\ -\omega_f & 0 & 0 \\ 0 & 0 & 0 \end{pmatrix} \qquad AKA^{-1} = \begin{pmatrix} -1/T_2 & -\Omega_0 & 0 \\ \Omega_0 & -1/T_2 & -\frac{2V_0}{\hbar} \\ 0 & \frac{2V_0}{\hbar} & -1/T_1 \end{pmatrix}$$

$$\tag{21.257}$$

and the equation of motion simplifies to

$$\begin{pmatrix} \dot{x}' \\ \dot{y}' \\ \dot{z}' \end{pmatrix} = \begin{pmatrix} -\frac{1}{T_2} & \omega_f - \Omega_0 & 0 \\ \Omega_0 - \omega_f & -\frac{1}{T_2} & -\frac{2V_0}{\hbar} \\ 0 & \frac{2V_0}{\hbar} & -\frac{1}{T_1} \end{pmatrix} \begin{pmatrix} x' \\ y' \\ z' \end{pmatrix} + \begin{pmatrix} 0 \\ 0 \\ \frac{z^{eq}}{T_1} \end{pmatrix}. \tag{21.258}$$

For times short compared to the relaxation times the solution is approximately given by harmonic oscillations. The generalized Rabi frequency Ω_R follows from [96]

$$i\Omega_R x' = (\omega_f - \Omega_0)y' \tag{21.259}$$

$$i\Omega_R y' = (\Omega_0 - \omega_f)x' - \frac{2V_0}{\hbar}z' \tag{21.260}$$

$$i\Omega_R z' = \frac{2V_0}{\hbar}y' \tag{21.261}$$

as

$$\Omega_R = \sqrt{(\Omega_0 - \omega_f)^2 + \left(\frac{2V_0}{\hbar}\right)^2}. \tag{21.262}$$

At larger times these oscillations are damped and the stationary solution is approached (Fig. 21.21) which is given by

$$\frac{z^{eq}}{1 + 4\frac{V_0^2}{\hbar^2}T_1 T_2 + T_2^2(\omega_f - \Omega_0)^2} \begin{pmatrix} 2T_2^2 \frac{V_0}{\hbar}(\Omega_0 - \omega_f) \\ -2T_2\frac{V_0}{\hbar} \\ 1 + T_2^2(\omega_f - \Omega_0)^2 \end{pmatrix}. \tag{21.263}$$

The occupation difference

$$z = \rho_{11} - \rho_{22} = z^{eq}\left(1 - \frac{4\frac{V_0^2}{\hbar^2}T_1 T_2}{1 + 4\frac{V_0^2}{\hbar^2}T_1 T_2 + T_2^2(\omega_f - \Omega_0)^2}\right) \tag{21.264}$$

has the form of a Lorentzian. The line width increases for higher intensities (power broadening)

$$\Delta\omega = \frac{1}{T_2}\sqrt{1 + 4\frac{V_0^2}{\hbar^2}T_1 T_2} \tag{21.265}$$

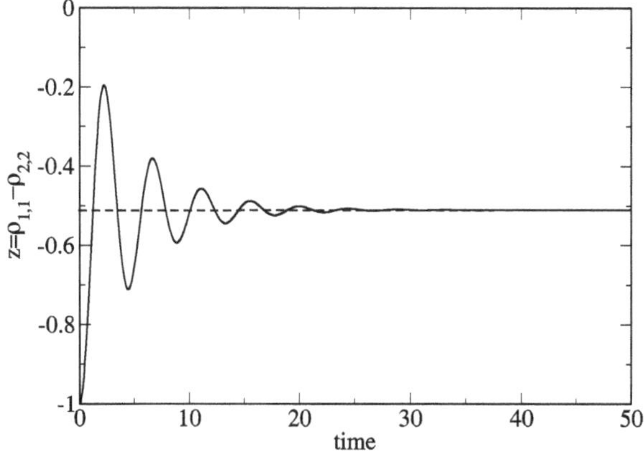

Fig. 21.21 (Monochromatic Excitation) The Bloch equations are solved numerically with the 4th order Runge-Kutta method for a monochromatic perturbation with $\omega = 4$, $V_0 = 0.5$. Parameters of the two-state system are $\omega_0 = 5$, $z_{eq} = -1.0$ and $T_1 = T_2 = 5.0$. The occupation difference $z = \rho_{11} - \rho_{22}$ initially shows Rabi oscillations which disappear at larger times where the stationary value $z = -0.51$ is reached

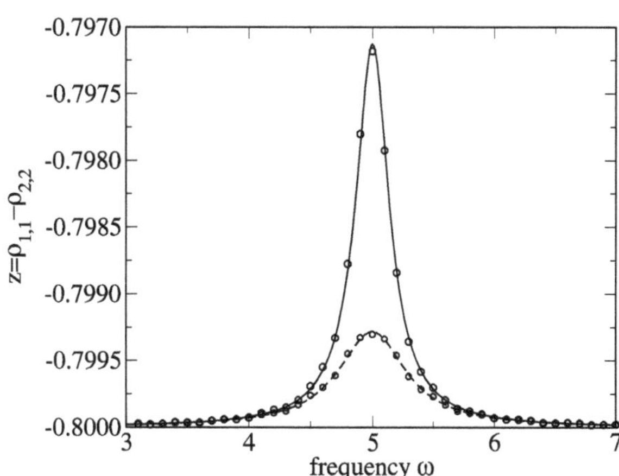

Fig. 21.22 (Resonance line) The equations of motion of the two-state system including relaxation terms are integrated with the 4th order Runge-Kutta method until a steady state is reached. Parameters are $\omega_0 = 5$, $z_{eq} = -0.8$, $V = 0.01$ and $T_1 = T_2 = 3.0, 6.9$. The change of the occupation difference is shown as a function of frequency (*circles*) and compared with the steady state solution (21.263)

and the maximum

$$\frac{z(\Omega_0)}{z^{eq}} = \frac{1}{1 + 4\frac{V_0^2}{\hbar^2} T_1 T_2} \tag{21.266}$$

approaches zero (saturation), Figs. 21.22, 21.23.

21.4.5.3 Excitation by a Resonant Pulse

For a resonant pulse with real valued envelope $V_0(t)$ and initial phase angle Φ_0

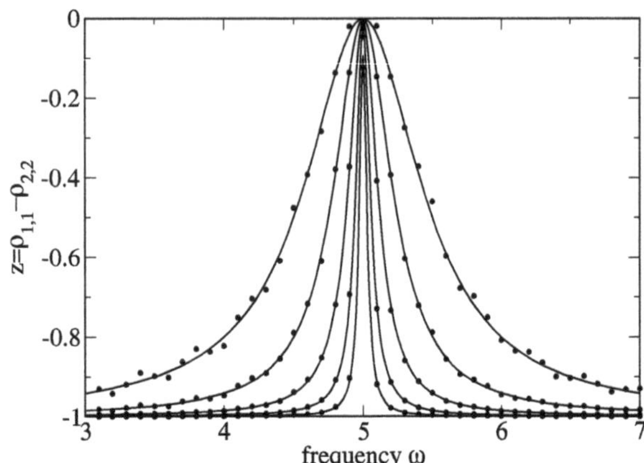

Fig. 21.23 (Power saturation and broadening) The resonance line is investigated as a function of the coupling strength V and compared with the stationary solution (21.263) to observe the broadening of the line width (21.265). Parameters are $\omega_0 = 5$, $z_{eq} = -1.0$, $T_1 = T_2 = 100$ and $V = 0.5$, 0.25, 0.125, 0.0625, 0.03125

$$H_{12} = V_0(t)e^{-i(\Omega_0 t + \Phi_0)}$$

the equation of motion in the rotating system is

$$\begin{pmatrix} \dot{x'} \\ \dot{y'} \\ \dot{z'} \end{pmatrix} = \begin{pmatrix} -\frac{1}{T_2} & 0 & -\frac{2V_0(t)}{\hbar}\sin\Phi_0 \\ 0 & -\frac{1}{T_2} & -\frac{2V_0(t)}{\hbar}\cos\Phi_0 \\ \frac{2V_0(t)}{\hbar}\sin\Phi_0 & \frac{2V_0(t)}{\hbar}\cos\Phi_0 & -\frac{1}{T_1} \end{pmatrix} \begin{pmatrix} x' \\ y' \\ z' \end{pmatrix} + \begin{pmatrix} 0 \\ 0 \\ \frac{z^{eq}}{T_1} \end{pmatrix}. \tag{21.267}$$

If the relaxation times are large compared to the pulse duration this describes approximately a rotation around an axis in the xy-plane (compare with (13.24))

$$\frac{d}{dt}\mathbf{r'} \approx W(t)\mathbf{r'} = \frac{2V_0(t)}{\hbar}W_0\mathbf{r'} \tag{21.268}$$

$$W_0 = \begin{pmatrix} 0 & 0 & -\sin\Phi_0 \\ 0 & 0 & -\cos\Phi_0 \\ \sin\Phi_0 & \cos\Phi_0 & 0 \end{pmatrix}. \tag{21.269}$$

Since the axis is time independent, a formal solution is given by

$$\mathbf{r'}(t) = e^{W\int_{t_0}^{t}\frac{2V_0(t')}{\hbar}dt'}\mathbf{r'}(0) = e^{W_0\Phi(t)}\mathbf{r'}(0) \tag{21.270}$$

with the phase angle

$$\Phi(t) = \int_{t_0}^{t}\frac{2V_0(t')}{\hbar}dt'. \tag{21.271}$$

Now, since

$$W_0^2 = \begin{pmatrix} -\sin^2\Phi_0 & -\sin\Phi_0\cos\Phi_0 & 0 \\ -\sin\Phi_0\cos\Phi_0 & -\cos^2\Phi_0 & 0 \\ 0 & 0 & -1 \end{pmatrix} \tag{21.272}$$

$$W_0^3 = -W_0 \tag{21.273}$$

$$W_0^4 = -W_0^2 \tag{21.274}$$

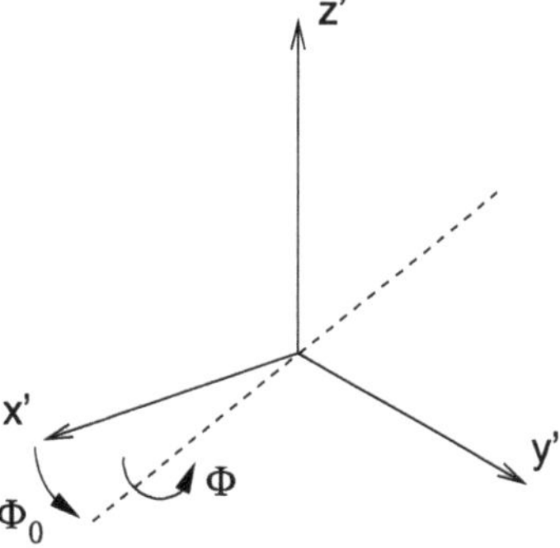

Fig. 21.24 (Rotation of the Bloch vector by a resonant pulse) A resonant pulse rotates the Bloch vector by the angle Φ around an axis in the $x'y'$-plane

the Taylor series of the exponential function in (21.270) can be summed up

$$e^{W_0\Phi}$$

$$= 1 + \Phi W_0 + \frac{1}{2}\Phi^2 W_0^2 + \frac{1}{3!}\Phi^3 W_0^3 + \cdots$$

$$= 1 + W_0^2\left(\frac{\Phi^2}{2} - \frac{\Phi^4}{4!} + \cdots\right) + W_0\left(\Phi - \frac{\Phi^3}{3!} + \cdots\right)$$

$$= 1 + W_0^2(1 - \cos\Phi) + W_0\sin\Phi$$

$$= \begin{pmatrix} 1 - \sin^2\Phi_0(1 - \cos\Phi) & -\sin\Phi_0\cos\Phi_0(1 - \cos\Phi) & -\sin\Phi_0\sin\Phi \\ -\sin\Phi_0\cos\Phi_0(1 - \cos\Phi) & 1 - \cos^2\Phi_0(1 - \cos\Phi) & -\cos\Phi_0\sin\Phi \\ \sin\Phi_0\sin\Phi & \cos\Phi_0\sin\Phi & \cos\Phi \end{pmatrix}$$

$$= \begin{pmatrix} \cos\Phi_0 & \sin\Phi_0 & 0 \\ -\sin\Phi_0 & \cos\Phi_0 & 0 \\ 0 & 0 & 1 \end{pmatrix} \begin{pmatrix} 1 & 0 & 0 \\ 0 & \cos\Phi & -\sin\Phi \\ 0 & \sin\Phi & \cos\Phi \end{pmatrix}$$

$$\times \begin{pmatrix} \cos\Phi_0 & -\sin\Phi_0 & 0 \\ \sin\Phi_0 & \cos\Phi_0 & 0 \\ 0 & 0 & 1 \end{pmatrix}. \tag{21.275}$$

The result is a rotation about the angle Φ around an axis in the xy-plane determined by Φ_0 (Fig. 21.24), especially around the x-axis for $\Phi_0 = 0$ and around the y-axis for $\Phi_0 = \frac{\pi}{2}$.

After a π-pulse ($\Phi = \pi$) the z-component changes its sign

$$\mathbf{r}' = \begin{pmatrix} \cos(2\Phi_0) & -\sin(2\Phi_0) & 0 \\ -\sin(2\Phi_0) & -\cos(2\Phi_0) & 0 \\ 0 & 0 & -1 \end{pmatrix} \mathbf{r}(0). \tag{21.276}$$

The transition between the two basis states $z = -1$ and $z = 1$ corresponds to a spin flip (Fig. 21.25). On the other hand, a $\pi/2$-pulse transforms the basis states into a coherent mixture

$$\mathbf{r}' = \begin{pmatrix} 1 - \sin^2\Phi_0 & -\sin\Phi_0\cos\Phi_0 & -\sin\Phi_0 \\ -\sin\Phi_0\cos\Phi_0 & 1 - \cos^2\Phi_0 & -\cos\Phi_0 \\ \sin\Phi_0 & \cos\Phi_0 & 0 \end{pmatrix} \mathbf{r}(0). \tag{21.277}$$

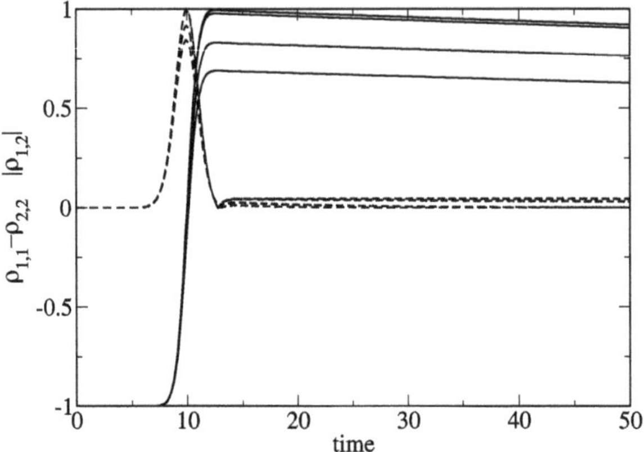

Fig. 21.25 (Spin flip by a π-pulse) The equations of motion of the Bloch vector (21.253) are solved with the 4th order Runge-Kutta method for an interaction pulse with a Gaussian shape. The pulse is adjusted to obtain a spin flip. The influence of dephasing processes is studied. $T_1 = 1000$, $t_p = 1.8$, $V_0 = 0.25$. The occupation difference $\rho_{11} - \rho_{22} = z$ (*solid curves*) and the coherence $|\rho_{12}| = \frac{1}{2}\sqrt{x^2 + y^2}$ (*broken curves*) are shown for several values of the dephasing time $T_2 = 5, 10, 100, 1000$

21.4.6 Elementary Qubit Manipulation

Whereas a classical bit can be only in one of two states

$$\text{either} \begin{pmatrix} 1 \\ 0 \end{pmatrix} \quad \text{or} \begin{pmatrix} 0 \\ 1 \end{pmatrix} \tag{21.278}$$

the state of a qubit is a quantum mechanical superposition

$$|\psi\rangle = C_0 \begin{pmatrix} 1 \\ 0 \end{pmatrix} + C_1 \begin{pmatrix} 0 \\ 1 \end{pmatrix}. \tag{21.279}$$

The time evolution of the qubit is described by a unitary transformation

$$|\psi\rangle \rightarrow U|\psi\rangle \tag{21.280}$$

which is represented by a complex 2×2 unitary matrix that has the general form (see also Sect. 13.14)

$$U = \begin{pmatrix} \alpha & \beta \\ -\mathrm{e}^{\mathrm{i}\varphi}\beta^* & \mathrm{e}^{\mathrm{i}\varphi}\alpha^* \end{pmatrix} \quad |\alpha|^2 + |\beta|^2 = 1, \quad \det U = \mathrm{e}^{\mathrm{i}\varphi}. \tag{21.281}$$

The Bloch vector is transformed with an orthogonal matrix A, which can be found from (21.212) and the transformed density matrix $U\rho U^{-1}$

$$\mathbf{r} \rightarrow A\mathbf{r} \quad A = \begin{pmatrix} \Re((\alpha^2 - \beta^2)\mathrm{e}^{-\mathrm{i}\varphi}) & \Im((\alpha^2 + \beta^2)\mathrm{e}^{-\mathrm{i}\varphi}) & -2\Re(\alpha\beta\mathrm{e}^{-\mathrm{i}\varphi}) \\ \Im((\beta^2 - \alpha^2)\mathrm{e}^{-\mathrm{i}\varphi}) & \Re((\alpha^2 + \beta^2)\mathrm{e}^{-\mathrm{i}\varphi}) & 2\Im(\alpha\beta\mathrm{e}^{-\mathrm{i}\varphi}) \\ 2\Re(\alpha^*\beta) & 2\Im(\alpha^*\beta) & (|\alpha|^2 - |\beta|^2) \end{pmatrix}. \tag{21.282}$$

Any single qubit transformation can be realized as a sequence of rotations around just two axes [126, 183, 263]. In the following we consider some simple transformations, so called quantum gates [274].

21.4.6.1 Pauli-Gates

Of special interest are the gates represented by the Pauli matrices $U = \sigma_i$ since any complex 2×2 matrix can be obtained as a linear combination of the Pauli matrices and the unit matrix (Sect. 13.14). For all three of them $\det U = -1$ and $\varphi = \pi$.

The X-gate

$$U_X = \sigma_x = \begin{pmatrix} 0 & 1 \\ 1 & 0 \end{pmatrix} \tag{21.283}$$

corresponds to rotation by π radians around the x-axis ((21.276) with $\Phi_0 = 0$)

$$A_X = \begin{pmatrix} 1 & 0 & 0 \\ 0 & -1 & 0 \\ 0 & 0 & -1 \end{pmatrix}. \tag{21.284}$$

It is also known as NOT-gate since it exchanges the two basis states. Similarly, the Y-gate

$$U_Y = \sigma_y = \begin{pmatrix} 0 & -i \\ i & 0 \end{pmatrix}$$

rotates the Bloch vector by π radians around the y-axis (21.276 with $\Phi_0 = \pi/2$)

$$A_Y = \begin{pmatrix} -1 & 0 & 0 \\ 0 & 1 & 0 \\ 0 & 0 & -1 \end{pmatrix} \tag{21.285}$$

and the Z-gate

$$U_Z = \sigma_z = \begin{pmatrix} 1 & 0 \\ 0 & -1 \end{pmatrix} \tag{21.286}$$

by π radians around the z-axis

$$A_Z = \begin{pmatrix} -1 & 0 & 0 \\ 0 & -1 & 0 \\ 0 & 0 & 1 \end{pmatrix}. \tag{21.287}$$

This rotation can be replaced by two successive rotations in the xy-plane

$$A_Z = A_X A_Y. \tag{21.288}$$

The corresponding transformation of the wavefunction produces an overall phase shift of $\pi/2$ since the product of the Pauli matrices is $\sigma_x \sigma_y = i\sigma_z$, which is not relevant for observable quantities.

21.4.6.2 Hadamard Gate

The Hadamard gate is a very important ingredient for quantum computation. It transforms the basis states into coherent superpositions and vice versa. It is described by the matrix

$$U_H = \begin{pmatrix} \frac{1}{\sqrt{2}} & \frac{1}{\sqrt{2}} \\ \frac{1}{\sqrt{2}} & -\frac{1}{\sqrt{2}} \end{pmatrix} \tag{21.289}$$

with $\det U_H = -1$ and

$$A_H = \begin{pmatrix} 0 & 0 & 1 \\ 0 & -1 & 0 \\ 1 & 0 & 0 \end{pmatrix} \tag{21.290}$$

which can be obtained as the product

$$A_H = \begin{pmatrix} 0 & 0 & -1 \\ 0 & 1 & 0 \\ 1 & 0 & 0 \end{pmatrix} \begin{pmatrix} 1 & 0 & 0 \\ 0 & -1 & 0 \\ 0 & 0 & -1 \end{pmatrix} \tag{21.291}$$

of a rotation by π radians around the x-axis and a second rotation by $\pi/2$ radians around the y-axis. The first rotation corresponds to the X-gate and the second to (21.277) with $\Phi_0 = \pi/2$

$$U = \begin{pmatrix} \frac{1}{\sqrt{2}} & \frac{1}{\sqrt{2}} \\ -\frac{1}{\sqrt{2}} & \frac{1}{\sqrt{2}} \end{pmatrix}. \tag{21.292}$$

21.5 Problems

Problem 21.1 (Wave packet motion) In this computer experiment we solve the Schrödinger equation for a particle in the potential $V(x)$ for an initially localized Gaussian wave packet $\psi(t = 0, x) \sim \exp(-a(x - x_0)^2)$. The potential is a box, a harmonic parabola or a fourth order double well. Initial width and position of the wave packet can be varied.

- Try to generate the time independent ground state wave function for the harmonic oscillator.
- Observe the dispersion of the wave packet for different conditions and try to generate a moving wave packet with little dispersion.
- Try to observe tunneling in the double well potential.

Problem 21.2 (Two-state system) In this computer experiment a two-state system is simulated. Amplitude and frequency of an external field can be varied as well as the energy gap between the two states (see Fig. 21.9).

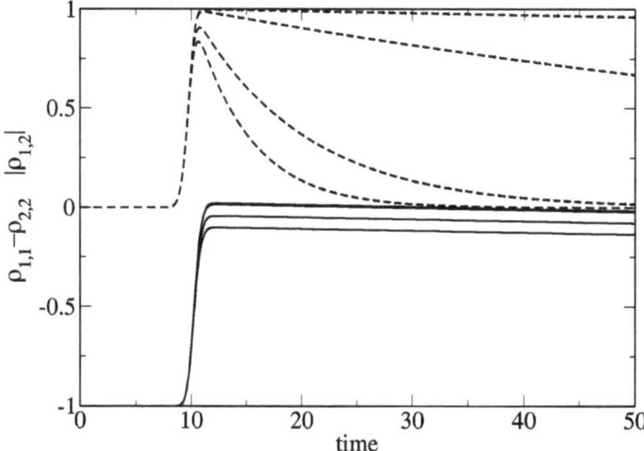

Fig. 21.26 (Generation of a coherent mixture by a $\pi/2$-pulse) The equations of motion of the Bloch vector (21.253) are solved with the 4th order Runge-Kutta method for an interaction pulse with a Gaussian shape. The pulse is adjusted to obtain a coherent mixture. The influence of dephasing processes is studied. $T_1 = 1000$, $t_p = 0.9$, $V_0 = 0.25$. The occupation difference $\rho_{11} - \rho_{22} = z$ (*solid curves*) and the coherence $|\rho_{12}| = \frac{1}{2}\sqrt{x^2 + y^2}$ (*broken curves*) are shown for several values of the dephasing time $T_2 = 5, 10, 100, 1000$

- Compare the time evolution at resonance and away from it.

Problem 21.3 (Three-state system) In this computer experiment a three-state system is simulated.

- Verify that the system behaves like an effective two-state system if the intermediate state is higher in energy than initial and final states (see Fig. 21.13).

Problem 21.4 (Ladder model) In this computer experiment the ladder model is simulated. The coupling strength and the spacing of the final states can be varied.

- Check the validity of the exponential decay approximation (see Fig. 21.15).

Problem 21.5 (Landau-Zener model) This computer experiment simulates the Landau Zener model. The coupling strength and the nuclear velocity can be varied (see Fig. 21.18).

- Try to find parameters for an efficient crossing of the states.

Problem 21.6 (Resonance line) In this computer experiment a two-state system with damping is simulated. The resonance curve is calculated from the steady state occupation probabilities (see Figs. 21.22, 21.23).

- Study the dependence of the line width on the intensity (power broadening).

Problem 21.7 (Spin flip) The damped two-state system is now subject to an external pulsed field (see Figs. 21.25, 21.26, 21.27).

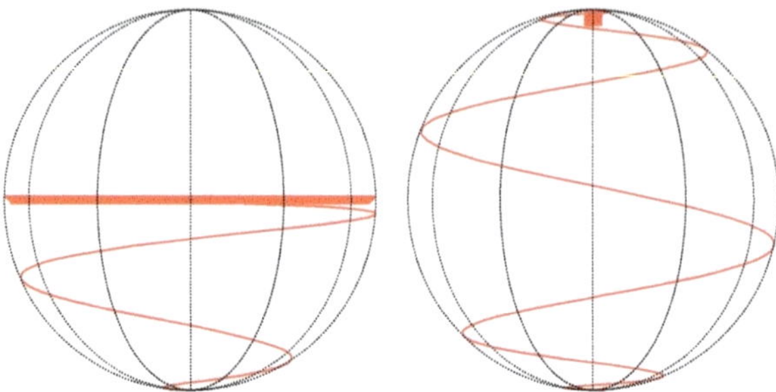

Fig. 21.27 (Motion of the Bloch vector during $\frac{\pi}{2}$ and π pulses) The trace of the Bloch vector is shown in the laboratory system. *Left*: $\frac{\pi}{2}$-pulse as in Fig. 21.26 with $T_2 = 1000$. *Right*: π-pulse as in Fig. 21.25 with $T_2 = 1000$

- Try to produce a coherent superposition state ($\pi/2$ pulse) or a spin flip (π pulse).
- Investigate the influence of decoherence.

Appendix I
Performing the Computer Experiments

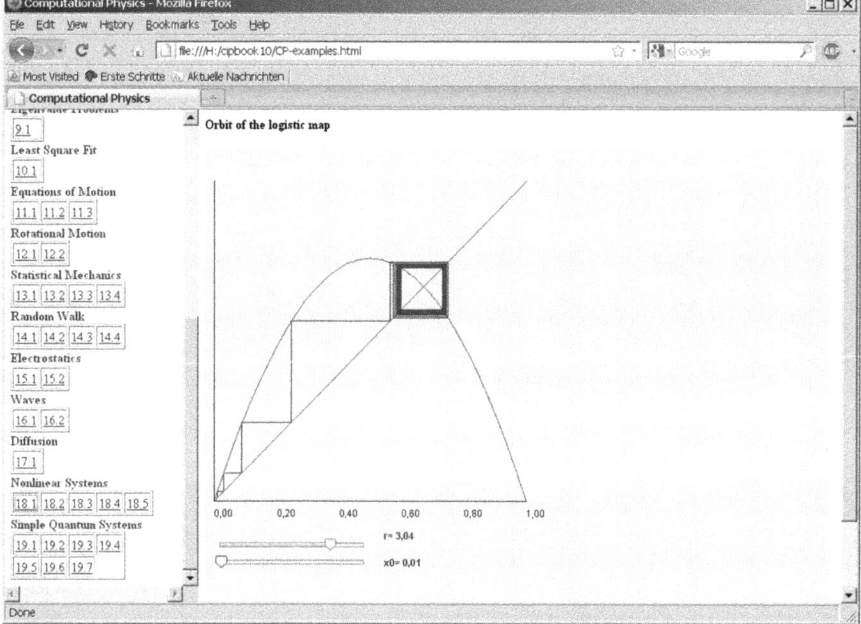

The computer experiments are realized as Java-applets which can be run in any browser that has the Java plug-in installed without installing anything else. They are written in a C-like fashion which improves the readability for readers who are not so familiar with object oriented programming. The source code can be studied most conveniently with the netbeans environment which is open source and allows quick generation of graphical user interfaces.

After downloading and unzipping the zipped file from extras.springer.com you have two options:

P.O.J. Scherer, *Computational Physics*, Graduate Texts in Physics,
DOI 10.1007/978-3-319-00401-3,
© Springer International Publishing Switzerland 2013

Run a program in your Browser

Open the file CP-examples.html in your browser. If the Java plug-in is installed properly you can start any one of the programs by simply clicking its number in the left hand frame.

Open a program with the netbeans environment

If you have the netbeans environment installed, you can import any of the programs as a separate project by opening the corresponding folder in the directory HTML/code/. You may have a look at the source code and compile and run it

```
103
104     public void precision() {
105         double one, x, y, p;
106         double powers[];
107         int n, i;
108         df = new DecimalFormat("##0.00000000000000000");
109         powers = new double[100];
110         one = 1.0;
111         x = 1.0;
112         n = 0;
113         powers[0] = one;
114
115         // add 1+ 2^(-n)   until the result equals 1
116         // we have to keep the compiler from optimizing too much !
117
118         jTextArea1.setTabSize(10);
119         jTextArea1.append("      n \t  x     \t    1+x  \t  (1+x)-1\n");
120
121         do {
122
123
124             y = x + one;
125             jTextArea1.append("2^-" + n + "   " + df.format(x) + "    " + d
126
127             x = x / 2.0;
128             n = n + 1;
129             powers[n] = x;
130         } while (y > one);
131
132         jTextArea1.append(" machine precision reached\n");
133
134
135
136
137     }
138 }
```

Appendix II
Methods and Algorithms

Purpose	Method	Comments	Pages
Interpolation	Lagrange polynomial	explicit form, easy to evaluate	17
	Barycentric Lagrange polynomial	for evaluation at many points	17
	Newton's divided differences	new points added easily	18
	Neville method	for evaluation at one point	20
	Spline interpolation	smoother, less oscillatory	22
	Rational interpolation	smoother, less oscillatory, often less coefficients necessary	25, 28
	Padé approximation	often better than Taylor series	25
	Barycentric rational interpolation	easy to evaluate	27
	Rational interpolation without poles	alternative to splines, analytical	31
	Multivariate interpolation	multidimensional	32
	Trigonometric interpolation	periodic functions	116
Differentiation	One-sided difference quotient	low error order	37
	Central difference quotient	higher error order	38
	Extrapolation	high accuracy	221
	Higher derivatives	finite difference methods	41
	Partial derivatives	finite difference methods	42

P.O.J. Scherer, *Computational Physics*, Graduate Texts in Physics,
DOI 10.1007/978-3-319-00401-3,
© Springer International Publishing Switzerland 2013

Purpose	Method	Comments	Pages
Integration	Newton-Cotes formulae	equally spaced points	46
	Trapezoidal rule	simple, closed interval	46
	Midpoint rule	simple, open interval	48
	Simpson's rule	more accurate	46
	Composite Newton-Cotes rules	for larger intervals	48
	Extrapolation (Romberg)	high accuracy	49
	Clenshaw-Curtis expressions	suitable for adaptive and multidimensional quadrature	50
	Gaussian integration	high accuracy if polynomial approximation possible	52
	Monte Carlo integration	high dimensional integrals	139
Linear equations	Gaussian elimination (LU reduction)	standard method for linear equations and matrix inversion	60
	QR decomposition	numerically more stable	64
	Iterative solution	large sparse systems	73
	Jacobi relaxation	converges for diagonally dominant matrices, parallel computation possible	73
	Gauss-Seidel relaxation	converges for symmetric positive definite or diagonal dominant matrices, no extra storage	74
	Chessboard (black-red)	two independent subgrids, especially for Poisson equation	307
	Damping and Successive over-relaxation	speeds up convergence for proper relaxation parameter	75
	Multigrid	fast convergence but more complicated	307
	Conjugate gradients	for symmetric positive definite matrices, preconditioning often necessary	76
	Special LU decomposition	Tridiagonal linear equations	69
	Sherman-Morrison formula	Cyclic tridiagonal systems	71
Root finding	Bisection	reliable but slow continuous functions	84
	Regula falsi (false position)	speed and robustness between bisection and interpolation	85
	Newton-Raphson	continuous derivative necessary, converges fast if starting point is close to a root	85
	Interpolation (secant)	no derivative necessary, but slower than Newton	87
	Inverse interpolation	mainly used by combined methods	88
	Dekker's combined method	Combination of bisection and secant method	91
	Brent's combined method	Combination of bisection, secant, and quadratic inverse interpolation methods, very popular	92
	Chandrupatla's combined method	Uses quadratic interpolation whenever possible, faster than Brent's method, especially for higher order roots	95

Purpose	Method	Comments	Pages
Multidimensional root finding	Newton-Raphson	Needs full Hessian	97
	Quasi-Newton (Broyden)	Hessian not needed, no matrix inversion	98
Function minimization	Ternary search	no gradient needed, very simple, for unimodal functions	99
	Golden section search (Brent)	faster than ternary search but more complicated	101
Multidimensional minimization	Steepest descent	simple but slow	106
	Conjugate gradients	faster than steepest descent	107
	Newton-Raphson	fast, if starting point close to minimum, needs full Hessian	107
	Quasi-Newton (BFGS,DFP)	Hessian not needed, very popular	108
Fourier transformation	Görtzel's algorithm	efficient if only some Fourier components are needed	120
	Fast Fourier transform	much faster than direct discrete Fourier transform	121
Random numbers	Linear congruent mapping	simple pseudo-random number generator	135
	Marsaglia-Zamann	higher quality random numbers but more complicated	135
	RN with given distribution	inverse of cumulative distribution function needed	136
	Random points on unit sphere	random directions	137
	Gaussian RN (Box-Muller)	Gaussian random numbers	138
Thermodynamic average	Simple sampling	inefficient	141
	Importance sampling	samples preferentially important configurations	142
	Metropolis algorithm	generates configurations according to a canonical distribution	142
Eigenvalue problems	Direct solution	only for very small dimension	148
	Tridiagonal matrices	explicit solutions for some special tridiagonal matrices	150
	Jacobi	simple but not very efficient	148
	QL	efficient method for not too large matrices, especially in combination with tridiagonalization by Householder transformations	156
	Lanczos	iterative method for very large matrices or if only a few eigenvalues are needed	159
	Singular value decomposition (SVD)	Generalization for arbitrary matrices	167

Purpose	Method	Comments	Pages
Data fitting	Least square fit	fit a model function to a set of data	162
	Linear least square fit with normal equations	simple but less accurate	163
	Linear fit with orthogonalization	better numerical stability	165
	Linear fit with SVD	expensive but more reliable, also for rank deficient matrices	172
	Low rank matrix approximation	data compression, total linear least squares	170
Discretization	Method of lines Eigenvector expansion	continuous time, discretized space	183
	Finite differences	simplest discretization, uniform grids	180
	Finite volumes	partial differential equations with a divergence term (conservation laws), flux conservative, allows unstructured meshes and discontinuous material parameters	185
	Finite elements	very flexible and general discretization method but also more complicated	196
	Spectral methods	expansion with global basis functions, mostly polynomials and Fourier sums, less expensive than finite elements but not as accurate for discontinuous material parameters and complicated geometries	193
	Dual grid	for finite volumes	185, 314
	Weighted residuals	general method to determine the expansion coefficients	190
	Point collocation	simplest criterion, often used for nonlinear problems and spectral methods	191
	Sub-domains	more general than finite volumes	191
	Least square	popular for computational fluid dynamics and electrodynamics	192
	Galerkin	most widely used criterion, leads often to symmetric matrices	192
	Fourier pseudo-spectral method	very useful whenever a Laplacian is involved, reduces dispersion	193
	Boundary elements	if the Green's function is available	204
Time evolution	Explicit forward Euler	low error order and unstable, mainly used as predictor step	210
	Implicit backward Euler	low error order but stable, used for stiff problems and as corrector step	212
	Improved Euler (Heun, predictor-corrector)	higher error order	213
	Nordsieck predictor-corrector	implicit method, has been used for molecular dynamics	215
	Gear predictor-corrector	optimized for molecular dynamics	217
	Explicit Runge Kutta (2nd, 3rd, 4th)	general and robust methods, easy step size and quality control	217

Purpose	Method	Comments	Pages
	Extrapolation (Gragg-Bulirsch-Stör)	very accurate and very slow	221
	Explicit Adams-Bashforth	high error order but not self-starting, for smooth functions, can be used as predictor	222
	Implicit Adams-Moulton	better stability than explicit method, can be used as corrector	223
	Backward differentiation (Gear)	implicit, especially for stiff problems	223
	Linear multistep predictor-corrector	General class, includes Adams-Bashforth-Moulton and Gear methods	224
	Verlet integration	symplectic, time reversible, for molecular dynamics	225
	Position Verlet	less popular	227
	Velocity Verlet	often used	227
	Störmer-Verlet	if velocities are not needed	228
	Beeman's method	velocities more accurate than for Störmer-Verlet	230
	Leapfrog	simple but two different grids	231, 231, 343
	Crank-Nicolson	implicit, stable, diffusion and Schrödinger equation	357, 347
	Lax-Wendroff	hyperbolic differential equations	345
	Two-step	differential equation with second order time derivative	338
	Reduction to a first order equation	Derivatives treated as additional variables	340
	Two-variable	transforms wave equation into a system of two first order equations	343
	Split operator	approximates an operator by a product	360, 226, 399
Unitary time evolution	Rational approximation	implicit, unitary	392
	Second order differencing	explicit, not exactly unitary	396
	Split operator Fourier	low dispersion, needs fast Fourier transformation	399
	Real space product formula	fast but less accurate, useful for wavepackets in coupled states	399
Rotation	Reorthogonalization	restore orthogonality of rotation matrix	250
	Quaternions	optimum parametrization of the rotation matrix	256
	Euler angles	numerical singularities	255
	Explicit method	low accuracy, reorthogonalization needed	250
	Implicit method	higher accuracy, orthogonal transformation	251
Molecular dynamics	Force field gradients	needed for molecular dynamics	270
	Normal mode analysis	small amplitude motion around an equilibrium	274
	Behrendsen thermostat	simple method to control temperature	281
	Langevin dynamics	Brownian motion	301

References

1. D.J. Acheson, *Elementary Fluid Dynamics* (Oxford University Press, London, 1990)
2. N. Ahmed, T. Natarajan, K.R. Rao, IEEE Trans. Comput. **23**, 90 (1974)
3. M.P. Allen, D.J. Tildesley, *Computer Simulation of Liquids* (Oxford University Press, London, 1989). ISBN 0-19-855645-4
4. I. Allonso-Mallo, N. Reguera, J. Comp. Physiol. **220**, 409 (2006)
5. V.S. Allured, C.M. Kelly, C.R. Landis, J. Am. Chem. Soc. **113**, 1 (1991)
6. W. Ames, *Numerical Methods for Partial Differential Equations* (Academic Press, San Diego, 1992)
7. Z.A. Anastassi, T.E. Simos, Phys. Rep. **482–483**, 1 (2009)
8. P.W. Anderson, Phys. Rev. **79**, 350 (1950)
9. X. Antoine, A. Arnold, C. Besse, M. Erhardt, A. Schädle, Commun. Comput. Phys. **4**, 729 (2008)
10. A. Askar, A.S. Cakmak, J. Chem. Phys. **68**, 2794 (1978)
11. J.S. Bader et al., J. Chem. Phys. **106**, 2372 (1997)
12. C.T.H. Baker, Numer. Math. **15**, 315 (1970)
13. G.A. Baker Jr., P. Graves-Morris, *Padé Approximants* (Cambridge University Press, New York, 1996)
14. A.D. Bandrauk, H. Shen, Chem. Phys. Lett. **176**, 428 (1991)
15. E.J. Bautista, J.M. Seminario, Int. J. Quant. Chem. **108**, 180 (2008)
16. B. Beckermann, Numer. Math. **85**, 553 (2000)
17. D. Beeman, J. Comp. Physiol. **20**, 130 (1976)
18. G. Beer, I. Smith, C. Duenser, *The Boundary Element Method with Programming: For Engineers and Scientists* (Springer, Berlin, 2008)
19. H. Bekker, H.J.C. Berendsen, W.F. van Gunsteren, J. Comput. Chem. **16**, 527 (1995)
20. A. Ben-Israel, T.N.E. Greville, *Generalized Inverses* (Springer, Berlin, 2003). ISBN 0-387-00293-6
21. H.J.C. Berendsen, J.P.M. Postma, W.F. van Gunsteren, A. DiNola, J.R. Haak, J. Chem. Phys. **81**, 3684 (1984)
22. A. Bermann, S. Gueron, Math. Gaz. **86**, 274 (2002)
23. J.P. Berrut, Comput. Math. Appl. **14**, 1 (1988)
24. J.-P. Berrut, L.N. Trefethen, Barycentric Lagrange interpolation. SIAM Rev. **46**(3), 501–517 (2004)
25. J.P. Berrut, R. Baltensperger, H.D. Mittelmann, Int. Ser. Numer. Math. **151**, 27 (2005)
26. K. Binder, Ising model, in *Encyclopedia of Mathematics*, Suppl. vol. 2, ed. by R. Hoksbergen (Kluwer Academic, Dordrecht, 2000), pp. 279–281
27. M. Bixon, J. Jortner, J. Chem. Phys. **48**, 715 (1986)
28. F. Bloch, Nuclear induction. Phys. Rev. **70**, 460 (1946)

P.O.J. Scherer, *Computational Physics*, Graduate Texts in Physics,
DOI 10.1007/978-3-319-00401-3,

29. P. Bocchieri, A. Loinger, Phys. Rev. **107**, 337 (1957)
30. P. Bochev, M. Gunzburger, *Least Squares Finite Element Methods* (Springer, Berlin, 2009)
31. E. Bonomi, J.-L. Lutton, SIAM Rev. **26**, 551 (1984)
32. A.H. Boschitsch, M.O. Fenley, H.X. Zhou, J. Phys. Chem. B **106**, 2741 (2002)
33. K.J. Bowers, E. Chow, H. Xu, R.U. Dror, M.P. Eastwood, B.A. Gregersen, J.L. Klepeis, I. Kolossvary, M.A. Moraes, F.D. Sacerdoti, J.K. Salmon, Y. Shan, D.E. Shaw, Scalable algorithms for molecular dynamics simulations on commodity clusters, in *SC 2006 Conference*. Proceedings of the ACM/IEEE (2006), p. 43
34. G.E.P. Box, M.E. Muller, Ann. Math. Stat. **29**, 610 (1958)
35. J.P. Boyd, *Chebyshev and Fourier Spectral Methods* (Dover, New York, 2001)
36. R.P. Brent, Comput. J. **14**, 422 (1971)
37. R.P. Brent, *Algorithms for Minimization without Derivatives* (Prentice Hall, Englewood Cliffs, 1973), Chap. 4
38. I.N. Bronshtein, K.A. Semendyayev, *Handbook of Mathematics*, 3rd edn. (Springer, New York, 1997), p. 892
39. B.R. Brooks, R.E. Bruccoleri, B.D. Olafson, D.J. States, S. Swaminathan, M. Karplus, J. Comput. Chem. **4**, 187 (1983)
40. B.R. Brooks, C.L. Brooks III, A.D. Mackerell Jr., L. Nilsson, R.J. Petrella, B. Roux, Y. Won, G. Archontis, C. Bartels, S. Boresch, A. Caflisch, L. Caves, Q. Cui, A.R. Dinner, M. Feig, S. Fischer, J. Gao, M. Hodoscek, W. Im, K. Kuzcera, T. Lazaridis, J. Ma, V. Ovchinnikov, E. Paci, R.W. Pastor, C.B. Post, J.Z. Pu, M. Schaefer, B. Tidor, R.M. Venable, H.L. Woodcock, X. Wu, W. Yang, D.M. York, M. Karplus, J. Comput. Chem. **30**, 1545 (2009)
41. R. Brown, Philos. Mag. **4**, 161 (1828)
42. C.G. Broyden, Math. Comput. **19**, 577 (1965)
43. C.G. Broyden, J. Inst. Math. Appl. **6**, 76 (1970)
44. R.E. Bruccoleri, J. Novotny, M.E. Davis, K.A. Sharp, J. Comput. Chem. **18**, 268 (1997)
45. R.L. Burden, J.D. Faires, *Numerical Analysis* (Brooks Cole, Boston, 2010), Chap. 2
46. G. Burkard, R.H. Koch, D.P. DiVincenzo, Phys. Rev. B **69**, 64503 (2004)
47. J.C.P. Bus, T.J. Dekker, ACM Trans. Math. Softw. **1**, 330 (1975)
48. J.C. Butcher, *The Numerical Analysis of Ordinary Differential Equations: Runge-Kutta and General Linear Methods* (Wiley, New York, 1987)
49. R.E. Caflisch, Monte Carlo and quasi-Monte Carlo methods. Acta Numer. **7**, 1–49 (1998)
50. R. Car, M. Parrinello, Phys. Rev. Lett. **55**, 2471 (1985)
51. A. Castro, M.A.L. Marques, A. Rubio, J. Chem. Phys. **121**, 3425 (2004)
52. Y.A. Cengel, *Heat transfer—A Practical Approach*, 2nd edn. (McGraw-Hill, New York, 2003), p. 26. ISBN 0072458933, 9780072458930
53. C. Cerjan, K.C. Kulander, Comput. Phys. Commun. **63**, 529 (1991)
54. T.R. Chandrupatla, Adv. Eng. Softw. **28**, 145 (1997)
55. H.W. Chang, S.E. Liu, R. Burridge, Linear Algebra Appl. **430**, 999 (2009)
56. D.L. Chapman, Philos. Mag. **25**, 475 (1913)
57. R. Chen, H. Guo, J. Chem. Phys. **111**, 9944 (1999)
58. M. Christen, P.H. Hünenberger, D. Bakowies, R. Baron, R. Bürgi, D.P. Geerke, T.N. Heinz, M.A. Kastenholz, V. Kräutler, C. Oostenbrink, C. Peter, D. Trzesniak, W.F. van Gunsteren, J. Comput. Chem. **26**, 1719 (2005)
59. S. Chynoweth, U.C. Klomp, L.E. Scales, Comput. Phys. Commun. **62**, 297 (1991)
60. C.W. Clenshaw, A.R. Curtis, Numer. Math. **2**, 197 (1960)
61. L.A. Collins, J.D. Kress, R.B. Walker, Comput. Phys. Commun. **114**, 15 (1998)
62. J.W. Cooley, J.W. Tukey, Math. Comput. **19**, 297 (1965)
63. W.D. Cornell, P. Cieplak, C.I. Bayly, I.R. Gould, K.M. Merz Jr., D.M. Ferguson, D.C. Spellmeyer, T. Fox, J.W. Caldwell, P.A. Kollman, J. Am. Chem. Soc. **117**, 5179 (1995)
64. R. Courant, K. Friedrichs, H. Lewy, Math. Ann. **100**, 32 (1928)
65. J. Crank, P. Nicolson, Proc. Camb. Philol. Soc. **43**, 50 (1947)
66. J.W. Daniel, W.B. Gragg, L. Kaufmann, G.W. Stewart, Math. Comput. **30**, 772 (1976)
67. H. De Raedt, Comput. Phys. Rep. **7**, 1 (1987)

68. H. De Raedt, B. De Raedt, Phys. Rev. A **28**, 3575 (1983)
69. P. Debye, E. Hückel, Phys. Z. **24**, 185 (1923)
70. T.J. Dekker, Finding a zero by means of successive linear interpolation, in *Constructive Aspects of the Fundamental Theorem of Algebra*, ed. by B. Dejon, P. Henrici (Wiley-Interscience, London, 1969)
71. D.P. Derrarr et al., in *A Practical Approach to Microarray Data Analysis* (Kluwer Academic, Norwell, 2003), p. 91
72. DGESVD routine from the freely available software package LAPACK, http://www.netlib.org/lapack
73. L. Diosi, *A Short Course in Quantum Information Theory* (Springer, Berlin, 2007)
74. P. Duhamel, M. Vetterli, Signal Process. **19**, 259 (1990)
75. S.E. Economou, T.L. Reinecke, in *Optical Generation and Control of Quantum Coherence in Semiconductor Nanostructures*, ed. by G. Slavcheva, P. Roussignol (Springer, Berlin, 2010), p. 62
76. A. Einstein, Ann. Phys. **17**, 549 (1905)
77. A. Einstein, *Investigations on the Theory of Brownian Movement* (Dover, New York, 1956)
78. J.R. Errington, P.G. Debenedetti, S. Torquato, J. Chem. Phys. **118**, 2256 (2003)
79. R. Eymard, T. Galloue, R. Herbin, Finite volume methods, in *Handbook of Numerical Analysis*, vol. 7, ed. by P.G. Ciarlet, J.L. Lions (Elsevier, Amsterdam, 2000), pp. 713–1020
80. M.D. Feit, J.A. Fleck Jr., A. Steiger, J. Comp. Physiol. **47**, 412 (1982)
81. R.P. Feynman, F.L. Vernon, R.W. Hellwarth, J. Appl. Phys. **28**, 49 (1957)
82. A. Fick, Philos. Mag. **10**, 30 (1855)
83. P.C. Fife, *Mathematical Aspects of Reacting and Diffusing Systems* (Springer, Berlin, 1979)
84. J. Fish, T. Belytschko, *A First Course in Finite Elements* (Wiley, New York, 2007)
85. G.S. Fishman, *Monte Carlo: Concepts, Algorithms, and Applications* (Springer, New York, 1995). ISBN 038794527X
86. R. Fletcher, Comput. J. **13**, 317 (1970)
87. C.A.J. Fletcher, *Computational Galerkin Methods* (Springer, Berlin, 1984)
88. C.A.J. Fletcher, *Computational Techniques for Fluid Dynamics*, vol. I, 2nd edn. (Springer, Berlin, 1991)
89. R. Fletcher, C. Reeves, Comput. J. **7**, 149 (1964)
90. M.S. Floater, K. Hormann, Numer. Math. **107**, 315 (2007)
91. F. Fogolari, A. Brigo, H. Molinari, J. Mol. Recognit. **15**, 377 (2002)
92. J.A. Ford, *Improved Algorithms of Illinois-Type for the Numerical Solution of Nonlinear Equations* (University of Essex Press, Essex, 1995), Technical Report CSM-257
93. B. Fornberg, Geophysics **52**, 4 (1987)
94. B. Fornberg, Math. Comput. **51**, 699 (1988)
95. J. Fourier, *The Analytical Theory of Heat* (Cambridge University Press, Cambridge, 1878), reissued by Cambridge University Press, 2009. ISBN 978-1-108-00178-6
96. M. Fox, *Quantum Optics: An Introduction* (Oxford University Press, London, 2006)
97. D. Frenkel, B. Smit, *Understanding Molecular Simulation: From Algorithms to Applications* (Academic Press, San Diego, 2002). ISBN 0-12-267351-4
98. B.G. Galerkin, On electrical circuits for the approximate solution of the Laplace equation. Vestnik Inzh. **19**, 897–908 (1915)
99. A.E. Garcia, Phys. Rev. Lett. **86**, 2696 (1992)
100. C.W. Gear, Math. Comput. **21**, 146 (1967)
101. C.W. Gear, *Numerical Initial Value Problems in Ordinary Differential Equations* (Prentice Hall, Englewood Cliffs, 1971)
102. C.W. Gear, IEEE Trans. Circuit Theory **18**, 89–95 (1971)
103. G. Goertzel, Am. Math. Mon. **65**, 34 (1958)
104. A. Goldberg, H.M. Schey, J.L. Schwartz, Am. J. Phys. **35**, 177 (1967)
105. D. Goldfarb, Math. Comput. **24**, 23 (1970)
106. H. Goldstein, *Klassische Mechanik* (Akademische Verlagsgesellschaft, Frankfurt am Main, 1974)

107. G. Golub, W. Kahan, J. Soc. Ind. Appl. Math., Ser. B Numer. Anal. **2**, 205 (1965)
108. G.H. Golub, C.F. Van Loan, *Matrix Computations*, 3rd edn. (Johns Hopkins University Press, Baltimore, 1976). ISBN 978-0-8018-5414-9
109. G.H. Golub, J.H. Welsch, Math. Comput. **23**, 221–230 (1969)
110. G.L. Gouy, J. Phys. **9**, 457 (1910)
111. W.B. Gragg, SIAM J. Numer. Anal. **2**, 384 (1965)
112. P. Grindrod, *Patterns and Waves: The Theory and Applications of Reaction-Diffusion Equations* (Clarendon Press, Oxford, 1991)
113. R. Guantes, S.C. Farantos, J. Chem. Phys. **111**, 10827 (1999)
114. R. Guantes, A. Nezis, S.C. Farantos, J. Chem. Phys. **111**, 10836 (1999)
115. J.M. Haile, *Molecular Dynamics Simulation: Elementary Methods* (Wiley, New York, 2001). ISBN 0-471-18439-X
116. E. Hairer, C. Lubich, G. Wanner, Acta Numer. **12**, 399 (2003)
117. J. Halpern, L.E. Orgel, Discuss. Faraday Soc. **29**, 32 (1960)
118. J.-P. Hansen, L. Verlet, Phys. Rev. **184**, 151 (1969)
119. F.j. Harris, Proc. IEEE **66**, 51 (1978)
120. M.T. Heath, Multigrid methods, in *Scientific Computing: An Introductory Survey* (McGraw-Hill, New York, 2002), p. 478 ff. §11.5.7, Higher Education
121. M.R. Hestenes, E. Stiefel, J. Res. Natl. Bur. Stand. **49**, 435 (1952)
122. D. Hilbert, L. Nordheim, J. von Neumann, Math. Ann. **98**, 1 (1927)
123. D. Hinrichsen, A.J. Pritchard, *Mathematical Systems Theory I—Modelling, State Space Analysis, Stability and Robustness* (Springer, Berlin, 2005). ISBN 0-978-3-540-441250
124. C.S. Holling, Can. Entomol. **91**, 293 (1959)
125. C.S. Holling, Can. Entomol. **91**, 385 (1959)
126. F.-Y. Hong, S.-J. Xiong, Chin. J. Phys. **46**, 379 (2008)
127. B.K.P. Horn, H.M. Hilden, S. Negahdaripour, J. Opt. Soc. Am. A **5**, 1127 (1988)
128. P.H. Huenenberger, Adv. Polym. Sci. **173**, 105 (2005)
129. H. Ibach, H. Lüth, *Solid-State Physics: An Introduction to Principles of Materials Science*. Advanced Texts in Physics (Springer, Berlin, 2003)
130. IEEE 754-2008, Standard for Floating-Point Arithmetic, IEEE Standards Association, 2008
131. T. Iitaka, Phys. Rev. E **49**, 4684 (1994)
132. T. Iitaka, N. Carjan, T. Strottman, Comput. Phys. Commun. **90**, 251 (1995)
133. R.Z. Iqbal, Master thesis, School of Mathematics, University of Birmingham, 2008
134. E. Ising, Beitrag zur Theorie des Ferromagnetismus. Z. Phys. **31**, 253–258 (1925). doi:10.1007/BF02980577
135. V.I. Istratescu, *Fixed Point Theory: An Introduction* (Reidel, Dordrecht 1981). ISBN 90-277-1224-7
136. IUPAC-IUB Commission on Biochemical Nomenclature, Biochemistry **9**, 3471 (1970)
137. H. Jeffreys, B.S. Jeffreys, Lagrange's interpolation formula, in *Methods of Mathematical Physics*, 3rd edn. (Cambridge University Press, Cambridge, 1988), p. 260, § 9.011
138. H. Jeffreys, B.S. Jeffreys, Divided differences, in *Methods of Mathematical Physics*, 3rd edn. (Cambridge University Press, Cambridge, 1988), pp. 260–264, § 9.012
139. H. Jeffreys, B.S. Jeffreys, *Methods of Mathematical Physics*, 3rd edn. (Cambridge University Press, Cambridge, 1988), pp. 305–306
140. B.-n. Jiang, *The Least-Squares Finite Element Method* (Springer, Berlin, 1998)
141. F. John, Duke Math. J. **4**, 300 (1938)
142. J.K. Johnson, J.A. Zollweg, K.E. Gubbins, Mol. Phys. **78**, 591 (1993)
143. A.H. Juffer et al., J. Phys. Chem. B **101**, 7664 (1997)
144. E.I. Jury, *Theory and Application of the Z-Transform Method* (Krieger, Melbourne, 1973). ISBN 0-88275-122-0
145. K. Khalil Hassan, *Nonlinear Systems* (Prentice Hall, Englewood Cliffs, 2001). ISBN 0-13-067389-7
146. S.A. Khrapak, M. Chaudhuri, G.E. Morfill, Phys. Rev. B **82**, 052101 (2010)
147. J. Kiefer, Proc. Am. Math. Soc. **4**, 502–506 (1953)

148. J.G. Kirkwood, J. Chem. Phys. **2**, 351 (1934)
149. S.E. Koonin, C.M. Dawn, *Computational Physics* (Perseus Books, New York, 1990). ISBN 978-0201127799
150. D. Kosloff, R. Kosloff, J. Comp. Physiol. **52**, 35 (1983)
151. S. Kouachi, Electron. J. Linear Algebra **15**, 115 (2006)
152. T. Kühne, M. Krack, F. Mohamed, M. Parrinello, Phys. Rev. Lett. **98**, 066401 (2007)
153. C. Lanczos, J. Res. Natl. Bur. Stand. **45**, 255 (1951)
154. L. Landau, Phys. Sov. Union **2**, 46–51 (1932)
155. H.P. Langtangen, *Computational Partial Differential Equations: Numerical Methods and Diffpack Programming* (Springer, Berlin, 2003)
156. C. Lavor, Physica D **227**, 135 (2007)
157. A. Leach, *Molecular Modelling: Principles and Applications*, 2nd edn. (Prentice Hall, Harlow, 2001). ISBN 978-0582382107
158. C. Leforestier, R.H. Bisseling, C. Cerjan, M.D. Feit, R. Friesner, A. Guldberg, A. Hammerich, G. Jolicard, W. Karrlein, H.-D. Meyer, N. Lipkin, O. Roncero, R. Kosloff, J. Comp. Physiol. **94**, 59 (1991)
159. D. Levesque, L. Verlet, Phys. Rev. A **2**, 2514 (1970)
160. A.J. Lotka, *Elements of Physical Biology* (Williams and Wilkins, Baltimore, 1925)
161. A.M. Lyapunov, *Stability of Motion* (Academic Press, New York, 1966)
162. L. Lynch, Numerical integration of linear and nonlinear wave equations. Bachelor thesis, Florida Atlantic University, Digital Commons@University of Nebraska-Lincoln, 2004
163. A.D. MacKerell Jr., B. Brooks, C.L. Brooks III, L. Nilsson, B. Roux, Y. Won, M. Karplus, CHARMM: The energy function and its parameterization with an overview of the program, in *The Encyclopedia of Computational Chemistry*, vol. 1, ed. by P.v.R. Schleyer et al. (Wiley, Chichester, 1998), pp. 271– 277
164. W. Magnus, Commun. Appl. Math. **7**, 649 (1954)
165. J. Mandel, Am. Stat. **36**, 15 (1982)
166. A.A. Markov, Theory of Algorithms [Translated by Jacques J. Schorr-Kon and PST staff] Imprint Moscow, Academy of Sciences of the USSR, 1954 [Jerusalem, Israel Program for Scientific Translations, 1961; available from Office of Technical Services, United States Department of Commerce] Added t.p. in Russian Translation of Works of the Mathematical Institute, Academy of Sciences of the USSR, v. 42. Original title: Teoriya algorifmov. [QA248.M2943 Dartmouth College library. U.S. Dept. of Commerce, Office of Technical Services, number OTS 60-51085] (1954)
167. A.A. Markov, Extension of the limit theorems of probability theory to a sum of variables connected in a chain, in *Dynamic Probabilistic Systems, vol. 1: Markov Chains* (Wiley, New York, 1971), reprinted in Appendix B of Howard, R.
168. G. Marsaglia, A. Zaman, Ann. Appl. Probab. **1**, 462 (1991)
169. G.J. Martyna, M.E. Tuckerman, J. Chem. Phys. **102**, 8071 (1995)
170. W.L. Mattice, U.W. Suter, *Conformational Theory of Large Molecules* (Wiley-Interscience, New York, 1994). ISBN 0-471-84338-5
171. B.M. McCoy, T.T. Wu, *The Two-Dimensional Ising Model* (Harvard University Press, Cambridge, 1973). ISBN 0674914406
172. E.A. McCullough Jr., R.E. Wyatt, J. Chem. Phys. **51**, 1253 (1969)
173. E.A. McCullough Jr., R.E. Wyatt, J. Chem. Phys. **54**, 3592 (1971)
174. N. Metropolis, S. Ulam, J. Am. Stat. Assoc. **44**, 335 (1949)
175. N. Metropolis, A. Rosenbluth, M. Rosenbluth, A. Teller, E. Teller, J. Chem. Phys. **21**, 1087 (1953)
176. C. Muguruma, Y. Okamoto, M. Mikami, Croat. Chem. Acta **80**, 203 (2007)
177. D.E. Muller, Math. Tables Other Aids Comput. **10**, 208 (1956)
178. J.D. Murray, *Mathematical Biology: I. An Introduction*, vol. 2, 3rd edn. (Springer, Berlin, 2002). ISBN 0-387-95223-3
179. S.J. Nettel, A. Kempicki, Am. J. Phys. **47**, 987 (1979)
180. E.H. Neville, J. Indian Math. Soc. **20**, 87 (1933)

181. A. Nicholls, B. Honig, J. Comput. Chem. **12**, 435 (1990)
182. J.J. Nicolas, K.E. Gubbins, W.B. Streett, D.J. Tildesley, Mol. Phys. **37**, 1429 (1979)
183. M. Nielsen, I. Chuang, *Quantum Computation and Quantum Information* (Cambridge University Press, Cambridge, 2000)
184. A. Nordsieck, Math. Comput. **16**, 22 (1962)
185. M. Novelinkova, in: *WDS'11, Proceedings of Contributed Papers, Part I* (2011), p. 67
186. G. Nürnberger, *Approximation by Spline Functions* (Springer, Berlin, 1989). ISBN 3-540-51618-2
187. H.J. Nussbaumer, *Fast Fourier Transform and Convolution Algorithms* (Springer, Berlin, 1990)
188. M. Oevermann, R. Klein, J. Comp. Physiol. **219**, 749 (2006)
189. M. Oevermann, C. Scharfenberg, R. Klein, J. Comput. Phys. **228**(14), 5184–5206 (2009). doi:10.1016/j.jcp.2009.04.018
190. I.P. Omelyan, Phys. Rev. **58**, 1169 (1998)
191. I.P. Omelyan, Comput. Phys. Commun. **109**, 171 (1998)
192. I.P. Omelyan, Comput. Phys. **12**, 97 (1998)
193. I.P. Omelyan, I.M. Mryglod, R. Folk, Comput. Phys. Commun. **151**, 272 (2003)
194. L. Onsager, Phys. Rev. **65**, 117 (1944)
195. E.L. Ortiz, SIAM J. Numer. Anal. **6**, 480 (1969)
196. J.M. Papakonstantinou, A historical development of the $(n + 1)$-point secant method. M.A. thesis, Rice University Electronic Theses and Dissertations, 2007
197. T.J. Park, J.C. Light, J. Chem. Phys. **85**, 5870 (1986)
198. B.N. Parlett, *The Symmetric Eigenvalue Problem* (Society for Industrial and Applied Mathematics, Philadelphia, 1998)
199. K. Pearson, The problem of the random walk. Nature **72**, 294 (1905)
200. J. Peiro, S. Sherwin, in *Handbook of Materials Modeling*, vol. 1, ed. by S. Yip (Springer, Berlin, 2005), pp. 1–32
201. D. Pines, C.P. Slichter, Phys. Rev. **100**, 1014 (1955)
202. H. Pollard, Q. J. Pure Appl. Math. **49**, 1 (1920)
203. W.H. Press, S.A. Teukolsky, W.T. Vetterling, B.P. Flannery, LU decomposition and its applications, in *Numerical Recipes: The Art of Scientific Computing*, 3rd edn. (Cambridge University Press, Cambridge, 2007), pp. 48–55
204. W.H. Press, S.A. Teukolsky, W.T. Vetterling, B.P. Flannery, Cholesky decomposition, in *Numerical Recipes: The Art of Scientific Computing*, 3rd edn. (Cambridge University Press, Cambridge, 2007), pp. 100–101
205. W.H. Press, S.A. Teukolsky, W.T. Vetterling, B.P. Flannery, Cyclic tridiagonal systems, in *Numerical Recipes: The Art of Scientific Computing*, 3rd edn. (Cambridge University Press, Cambridge, 2007), p. 79
206. W.H. Press, S.A. Teukolsky, W.T. Vetterling, B.P. Flannery, Relaxation mehods for boundary value problems, in *Numerical Recipes: The Art of Scientific Computing*, 3rd edn. (Cambridge University Press, Cambridge, 2007), pp. 1059–1065
207. W.H. Press, S.A. Teukolsky, W.T. Vetterling, B.P. Flannery, Golden section search in one dimension, in *Numerical Recipes: The Art of Scientific Computing*, 3rd edn. (Cambridge University Press, New York, 2007). ISBN 978-0-521-88068-8, Sect. 10.2
208. W.H. Press, S.A. Teukolsky, W.T. Vetterling, B.P. Flannery, Second-order conservative equations, in *Numerical Recipes: The Art of Scientific Computing*, 3rd edn. (Cambridge University Press, Cambridge, 2007), Sect. 17.4
209. A. Quarteroni, R. Sacco, F. Saleri, *Numerical Mathematics*, 2nd edn. (Springer, Berlin, 2007)
210. S.S. Rao, *The Finite Element Method in Engineering* (Elsevier, Amsterdam, 2011)
211. K.R. Rao, P. Yip, *Discrete Cosine Transform: Algorithms, Advantages, Applications* (Academic Press, Boston, 1990)
212. D.C. Rapaport, *The Art of Molecular Dynamics Simulation* (Cambridge University Press, Cambridge, 2004). ISBN 0-521-44561-2

213. A.K. Rappe, C.J. Casewit, K.S. Colwell, W.A. Goddar, W.M. Skiff, J. Am. Chem. Soc. **115**, 10024 (1992)
214. E. Renshaw, *Modelling Biological Populations in Space and Time* (Cambridge University Press, Cambridge, 1991). ISBN 0-521-44855-7
215. J. Rice, *Mathematical Statistics and Data Analysis*, 2nd edn. (Duxbury Press, N. Scituate, 1995). ISBN 0-534-20934-3
216. J.A. Richards, *Remote Sensing Digital Image Analysis* (Springer, Berlin, 1993)
217. L.F. Richardson, Philos. Trans. R. Soc. Lond. Ser. A **210**, 307–357 (1911)
218. R.D. Richtmyer, *Principles of Modern Mathematical Physics I* (Springer, New York, 1978)
219. H. Risken, *The Fokker-Planck Equation* (Springer, Berlin, 1989)
220. C.P. Robert, G. Casella, *Monte Carlo Statistical Methods*, 2nd edn. (Springer, New York, 2004). ISBN 0387212396
221. T.D. Romo et al., Proteins **22**, 311 (1995)
222. I.R. Savage, E. Lukacs, in *Contributions to the Solution of Systems of Linear Equations and the Determination of Eigenvalues*, ed. by O. Taussky. National Bureau of Standards, Applied Mathematics Series, vol. 39 (1954)
223. G. Schaftenaar, J.H. Noordik, J. Comput.-Aided Mol. Des. **14**, 123 (2000)
224. T. Schlick, J. Comput. Chem. **10**, 951 (1989)
225. T. Schlick, *Molecular Modeling and Simulation* (Springer, Berlin, 2002). ISBN 0-387-95404-X
226. B.I. Schneider, J.-A. Collins, J. Non-Cryst. Solids **351**, 1551 (2005)
227. C. Schneider, W. Werner, Math. Comput. **47**, 285 (1986)
228. I.J. Schoenberg, Q. Appl. Math. **4**, 45–99 and 112–141 (1946)
229. P. Schofield, Comput. Phys. Commun. **5**, 17 (1973)
230. E. Schrödinger, Phys. Rev. **28**, 1049 (1926)
231. F. Schwabl, *Statistical Mechanics* (Springer, Berlin, 2003)
232. F. Schwabl, *Quantum Mechanics*, 4th edn. (Springer, Berlin, 2007)
233. L.F. Shampine, IMA J. Numer. Anal. **3**, 383 (1983)
234. L.F. Shampine, L.S. Baca, Numer. Math. **41**, 165 (1983)
235. D.F. Shanno, Math. Comput. **24**, 647 (1970)
236. J. Sherman, W.J. Morrison, Ann. Math. Stat. **20**, 621 (1949)
237. T. Simonson, Rep. Prog. Phys. **66**, 737 (2003)
238. G. Skollermo, Math. Comput. **29**, 697 (1975)
239. B. Smit, J. Chem. Phys. **96**, 8639 (1992)
240. A. Sommariva, Comput. Math. Appl. **65**(4), 682–693 (2013). doi:10.1016/j.camwa.2012.12.004 [MATLAB CODES (zip file)]
241. R. Sonnenschein, A. Laaksonen, E. Clementi, J. Comput. Chem. **7**, 645 (1986)
242. B.I. Stepanov, V.P. Grobkovskii, *Theory of Luminescence* (Butterworth, London, 1986)
243. G.W. Stewart, SIAM Rev. **35**, 551 (1993)
244. J. Stör, R. Bulirsch, *Introduction to Numerical Analysis*, 3rd revised edn. (Springer, New York, 2010). ISBN 978-1441930064
245. S.H. Strogatz, *Nonlinear Dynamics and Chaos: Applications to Physics, Biology, Chemistry, and Engineering* (Perseus Books, Reading, 2001). ISBN 0-7382-0453-6
246. M. Suzuki, Commun. Math. Phys. **51**, 183 (1976)
247. P.N. Swarztrauber, R.A. Sweet, in *Handbook of Fluid Dynamics and Fluid Machinery*, ed. by J.A. Schetz, A.E. Fuhs (Wiley, New York, 1996)
248. H. Tal-Ezer, R. Kosloff, J. Chem. Phys. **81**, 3967 (1984)
249. J.T. Tanner, Ecology **56**, 855 (1975)
250. S.A. Teukolsky, Phys. Rev. D **61**(8), 087501 (2000). doi:10.1103/PhysRevD.61.087501
251. J.W. Thomas, *Numerical Partial Differential Equations: Finite Difference Methods*. Texts in Applied Mathematics, vol. 22 (Springer, Berlin, 1995)
252. J.Y. Tjalling, Historical development of the Newton-Raphson method. SIAM Rev. **37**, 531 (1995)
253. L.N. Trefethen, SIAM Rev. **50**, 67–87 (2008)

254. L.N. Trefethen, D. Bau III, *Numerical Linear Algebra* (Society for Industrial and Applied Mathematics, Philadelphia, 1997), p. 1
255. S.-H. Tsai et al., Braz. J. Phys. **34**, 384 (2004)
256. T. Tsang, H. Tang, Phys. Rev. A **15**, 1696 (1977)
257. M. Tuckerman, B.J. Berne, J. Chem. Phys. **97**, 1990 (1992)
258. A.M. Turing, Philos. Trans. R. Soc. Lond. B **237**, 37 (1952)
259. R.E. Tuzun, D.W. Noid, B.G. Sumpter, Macromol. Theory Simul. **5**, 771 (1996)
260. R.E. Tuzun, D.W. Noid, B.G. Sumpter, J. Comput. Chem. **18**, 1804 (1997)
261. W. van Dijk, F.M. Toyama, Phys. Rev. E **75**, 036707 (2007)
262. W.F. van Gunsteren, S.R. Billeter, A.A. Eising, P.H. Hünenberger, P. Krüger, A.E. Mark, W.R.P. Scott, I.G. Tironi, *Biomolecular Simulation: The GROMOS96 Manual and User Guide* (vdf Hochschulverlag AG an der ETH Zürich and BIOMOS b.v., Zürich, Groningen, 1996)
263. L.M.K. Vandersypen, I.L. Chuang, Rev. Mod. Phys. **76**, 1037 (2004)
264. P.F. Verhulst, Mem. Acad. R. Sci. Belles Lettres Bruxelles **18**, 1–42 (1845)
265. L. Verlet, Phys. Rev. **159**, 98 (1967)
266. V. Volterra, Mem. R. Accad. Naz. Lincei **2**, 31 (1926)
267. J. Waldvogel, BIT Numer. Math. **46**, 195 (2006)
268. Z. Wang, B.R. Hunt, Appl. Math. Comput. **16**, 19 (1985)
269. R.K. Wangsness, F. Bloch, Phys. Rev. **89**, 728 (1953)
270. H. Watanabe, N. Ito, C.K. Hu, J. Chem. Phys. **136**, 204102 (2012)
271. W. Werner, Math. Comput. **43**, 205 (1984)
272. H. Wilf, *Mathematics for the Physical Sciences* (Wiley, New York, 1962)
273. J.H. Wilkinson, J. ACM **8**, 281 (1961)
274. C.P. Williams, *Explorations in Quantum Computing* (Springer, Berlin, 2011)
275. J. Wolberg, *Data Analysis Using the Method of Least Squares: Extracting the Most Information from Experiments* (Springer, Berlin, 2005)
276. L.C. Wrobel, M.H. Aliabadi, *The Boundary Element Method* (Wiley, New York, 2002)
277. G. Wunsch, *Feldtheorie* (VEB Technik, Berlin, 1973)
278. K. Yabana, G.F. Bertsch, Phys. Rev. B **54**, 4484 (1996)
279. S. Yang, M.K. Gobbert, Appl. Math. Lett. **22**, 325 (2009)
280. A. Yariv, *Quantum Electronics* (Wiley, New York, 1975)
281. W.C. Yueh, Appl. Math. E-Notes **5**, 66 (2005)
282. C. Zener, Proc. R. Soc. Lond. A **137**(6), 696–702 (1932)
283. Y. Zhu, A.C. Cangellaris, *Multigrid Finite Element Methods for Electromagnetic Field Modeling* (Wiley, New York, 2006), p. 132 ff. ISBN 0471741108
284. I. Zutic, J. Fabian, S. Das Sarma, Rev. Mod. Phys. **76**, 323 (2004)

Index

A

Adams-Bashforth, 222, 234
Adams-Moulton, 223
Amplification factor, 345
Angular momentum, 246–248, 254
Angular velocity, 242–244
Attractive fixed point, 365
Auto-correlation, 302
Average extension, 301
Average of measurements, 134

B

Backward difference, 37
Backward differentiation, 223
Backward substitution, 61
Ballistic motion, 285
Beeman, 230
BFGS, 110
Bicubic spline interpolation, 35
Bifurcation, 368
Bifurcation diagram, 369
Bilinear interpolation, 32, 35
Binomial distribution, 133
Bio-molecules, 315
Biopolymer, 296
Birth rate, 372
Bisection, 84
Bloch equations, 420, 421
Bloch vector, 417
Bond angle, 265
Bond length, 264
Boundary conditions, 178
Boundary element, 318, 324, 327
Boundary element method, 204
Boundary potential, 321
Boundary value problems, 178
Box Muller, 138

Brent, 92
Brownian motion, 285, 293, 301, 303
Broyden, 99

C

Calculation of π, 138
Carrying capacity, 367, 375
Cartesian coordinates, 264
Cavity, 318, 322, 325, 326
Cayley-Klein, 256, 257
Central difference quotient, 38
Central limit theorem, 133, 144, 294, 297
Chain, 296
Chandrupatla, 95
Chaotic behavior, 369
Characteristic polynomial, 151
Charged sphere, 309, 314, 317
Chebyshev, 51
Chemical reactions, 378
Chessboard method, 308
Circular orbit, 211, 232
Clenshaw-Curtis, 50
Coin, 133
Collisions, 285, 302, 415
Composite midpoint rule, 48
Composite Newton-Cotes formulas, 48
Composite Simpson's rule, 48
Composite trapezoidal rule, 48
Computer experiments, 433
Concentration, 351
Condition number, 77
Configuration integral, 141
Conjugate gradients, 76, 107
Conservation laws, 179
Continuous logistic model, 371
Control parameter, 369
Control volumes, 186

P.O.J. Scherer, *Computational Physics*, Graduate Texts in Physics,
DOI 10.1007/978-3-319-00401-3,
© Springer International Publishing Switzerland 2013